MODERN APPROACHES TO THE TAXONOMY OF RED AND BROWN ALGAE

THE SYSTEMATICS ASSOCIATION
SPECIAL VOLUME NO. 10

MODERN APPROACHES TO THE TAXONOMY OF RED AND BROWN ALGAE

Proceedings of an International Symposium held at the Polytechnic of North London

Edited by
D. E. G. IRVINE
Department of Biology and Geology, Polytechnic of North London
and
J. H. PRICE
Department of Botany, British Museum (Natural History)

1978

Published for the
SYSTEMATICS ASSOCIATION
by
ACADEMIC PRESS · LONDON · NEW YORK · SAN FRANCISCO

ACADEMIC PRESS INC. (LONDON) LTD.
24-28 Oval Road
London NW1 7DX

U.S. Edition published by
ACADEMIC PRESS INC.
111 Fifth Avenue
New York, New York 10003

Library of Congress Catalog Card Number: 77-92824
ISBN: 0 12 374050 9

PRINTED IN GREAT BRITAIN BY
T. & A. CONSTABLE LTD., EDINBURGH

Contributors and Participants

Contributors are indicated by an asterisk.

ALLEN, Dr M. A., *Royal Masonic School for Girls, Rickmansworth Park, Herts, England.*

ALVIN, Dr K. L., *Botany Department, Imperial College of Science and Technology, London SW7 2AZ, England.*

ÅSEN, P. A., *Department of Botany, Kristiansand Museum, Box 479, N-4601 Kristiansand S., Norway.*

*BLUNDEN, Dr G., *School of Pharmacy, Portsmouth Polytechnic, King Henry I Street, Portsmouth, Hants PO1 2DZ, England.*

BOALCH, Dr G. T., *The Laboratory, Citadel Hill, Plymouth, Devon PL1 2PB, England.*

*BONEY, Professor A. D., *Department of Botany, The University, Glasgow G12 8QQ, Scotland.*

BURROWS, Dr E. M., *Box Cottage, 4 Main Street, Broadmayne, nr Dorchester, Dorset DT2 8EB, England.*

BUTLER, Mrs Y. M. (see CHAMBERLAIN).

*CALLOW, Dr J. A., *Department of Plant Sciences, Baines Wing, The University, Leeds LS2 9JT, Yorkshire, England.*

*CALLOW, Dr M. E., *Department of Plant Sciences, Baines Wing, The University, Leeds LS2 9JT, Yorkshire, England.*

*CHAMBERLAIN, Y. M. (Mrs Y. M. Butler), *4 Shelley Close, Itchen Abbas, Winchester, Hampshire, England.*

*CHAPMAN, Dr A. R. O., *Biology Department, Dalhousie University, Halifax, Nova Scotia, Canada.*

COOKE, P., *Department of Botany, University College, Galway, Ireland.*

DAVEY, Dr K. W., *British Museum (Natural History), Cromwell Road, London SW7 5BD, England.*

DE VALÉRA, Professor M., *Department of Botany, University College, Galway, Ireland.*

*DUCKETT, Dr J. G., *School of Plant Biology, University College of N. Wales, Bangor, Gwynedd, LL57 2UW, Wales.*

ETHERINGTON, D., *Department of Biology and Geology, Polytechnic of North London, Holloway Road, London N7 8DB, England.*

ETHERINGTON, Mrs O. E. J., *Department of Biology and Geology, Polytechnic of North London, Holloway Road, London N7 8DB, England.*

*EVANS, Dr L. V., *Department of Plant Sciences, Baines Wing, The University, Leeds LS2 9JT, Yorkshire, England.*

★Farnham, W. F., *The Marine Laboratory, Portsmouth Polytechnic, Hayling Island, Hants, PO11 0DG, England.*

★Fletcher, Dr R. L., *The Marine Laboratory, Portsmouth Polytechnic, Hayling Island, Hants PO11 0DG, England.*

★Garbary, D., *Department of Botany, University of Liverpool, P.O. Box 147, Liverpool L69 3BX, England.*

★Guiry, Dr M. D., *The Marine Laboratory, Portsmouth Polytechnic, Hayling Island, Hants PO11 0DG, England.*

Hiscock, Mrs S., *Oil Pollution Research Unit, Orielton Field Centre, Pembroke, Dyfed, Wales.*

Huxley, R., *Department of Biology and Geology, Polytechnic of North London, Holloway Road, London N7 8DB, England.*

Irvine, Dr D. E. G., *Department of Biology and Geology, Polytechnic of North London, Holloway Road, London N7 8DB, England.*

Irvine, Mrs L. M., *Department of Botany, British Museum (Natural History), Cromwell Road, London SW7 5BD, England.*

John, Dr D. M., *Department of Botany, University of Ghana, P.O. Box 55, Legon, Ghana, W. Africa.*

Knight, S. J. T., *Scottish Marine Biological Association, Dunstaffnage Marine Research Laboratory, P.O. Box 3, Oban, Argyll PA34 4AD, Scotland.*

Lambe, Dr E., *Department of Botany, University College, Galway, Ireland.*

Lex, Dr M., *Natural Environment Research Council, Alhambra House, 27-33 Charing Cross Road, London WC2H 0AX, England.*

★McCandless, Professor E. L., *Department of Biology, McMaster University, 1280 Main Street W., Hamilton, Ontario L85 4KI, Canada.*

★Magne, Professor F., *Université de Paris VI, Laboratoire de Cryptogamie, 9 Quai St. Bernard 75005, Paris, France.*

★Pankhurst, R. J., *Department of Botany, British Museum (Natural History), Cromwell Road, London SW7 5BD, England.*

★Peel, Dr M. C., *Department of Biology, Queen Elizabeth College, Campden Hill Road, London W8 7AH, England.*

★Percival, Dr E., *Department of Chemistry, Royal Holloway College, University of London, Egham Hill, Egham, Surrey TW20 0EX, England.*

Picken, M. J., *Scottish Marine Biological Association, Dunstaffnage Marine Research Laboratory, P.O. Box 3, Oban, Argyll PA34 4AD, Scotland.*

★Price, J. H., *Department of Botany, British Museum (Natural History), Cromwell Road, London SW7 5BD, England.*

★Prud'homme van Reine, Drs W. F., *Rijksherbarium, Schelpenkade 6, Leiden, Netherlands.*

★Roberts, Dr M., *Department of Botany, University of Durham, Durham DH1 3LE, England.*

★Rogers, Dr D. J., *School of Pharmacy, Portsmouth Polytechnic, King Henry I Street, Portsmouth, Hants PO1 2DZ, England.*

★Rueness, Dr J., *Universitet et i Oslo, Institutt for Marinbiologi og Limnologi, Avd. Marin Botanikk, Postboks 1069, Blindern, Oslo 3, Norway.*

★Russell, Dr G., *Department of Botany, University of Liverpool, P.O. Box 147 Liverpool L69 3BX, England.*

TAYLOR, Professor A. R. A., *Department of Biology, University of New Brunswick, Fredericton, New Brunswick, Canada.*

*TITTLEY, I., *Department of Botany, British Museum (Natural History), Cromwell Road, London SW7 5BD, England.*

VREELAND, Dr V., *Department of Botany, University of California, Berkeley, California 94720, U.S.A.*

YOUEL, A. J., *Information Retrieval Ltd., 1 Falconberg Court, London W1V 5FG, England.*

Preface

The benthic marine algae, which include most of the red algae (Rhodophyta) and brown algae (Phaeophyta) offer challenging problems to the taxonomist. They are often difficult to study in the field, growing below the low tide mark, in inaccessible crevices, or exposed to dangerously powerful wave action. Many are morphologically variable (we avoid the word "plastic": cf. Chapman's paper!), producing widely differing growth forms under diverse habitat conditions; by contrast, different taxa may produce closely similar plants when growing under similar conditions. Fertile stages may be rare or transient, and different life-history phases of the same species may not bear the least resemblance to each other. Their habitat requirements may be difficult, indeed virtually impossible, to reproduce under laboratory conditions, and preservation of specimens under conditions allowing their customary structure to be re-examined later may also pose serious problems.

Given these difficulties, it is not surprising that marine algal taxonomy got off to a slow and confused start; that the full life-histories of many common and important species even yet are imperfectly known; that species new to science can still be found even in such apparently well-worked areas as the English Channel; and that many of the fundamental criteria used by earlier taxonomists are now being seriously questioned.

The development of increasingly sophisticated subaqua apparatus and techniques has revolutionized field study of the subtidal benthic algae, whilst improved microscopic and culture facilities have provided fresh insights into the ultrastructure and life-histories of many species. Nevertheless, the potential for morphological variation and the inevitable artificiality of the imposed environment (emphasized by many contributors) indicate that conclusions derived from culture work alone must be treated with caution; there is considerable need to identify taxonomic criteria less readily modified by environmental factors, as well as for critical assessment of the variability potential itself.

The concept of the present Symposium derived from a belief that personal interchange of information is more immediately effective and stimulatory than correspondence or the printed word and that most Symposia including the benthic algae are now too large to allow full rein to informal discussion. The Symposium was therefore organized to bring together workers on widely differing aspects of the red and brown algae; to evaluate the taxonomic potential of new data emerging from their researches; and to encourage interdisciplinary approaches. We believe that these aims were realized and that the papers read at this Symposium provide a valuable contribution to algal taxonomy in themselves, while also indicating promising lines for future taxonomic research. It is worth mentioning that no attempt has been made to standardize the taxon level used by authors for the major divisions of algae. Some have consistently used class endings, others—division endings. Many used either class or division as it suited them. As there is hardly room for misunderstanding we have left things as they were.

Few Societies nowadays have either the finance or the available space to permit authors to be as detailed as stimulatory exchange often requires. We therefore wish to thank Academic Press, and the Systematics Association in whose series the volume appears, for placing few obstacles in the way of editors and authors, thus achieving what we hope will be regarded as a fruitful permanent record of the sessions. The value of the Symposium was due in large measure to the co-operation and support given by all concerned, so that even the concurrent student sit-in at the Polytechnic of North London, was not allowed to disrupt proceedings. We are grateful to our many helpers, but special mention must be made of the assistance given by various members of the Department of Biology and Geology, notably the Head of the Department (Mr D. Etherington), Mrs O. E. J. Etherington, Mr Richard Havely, and Mrs J. Bucklew. The Head of the Department of Geography (Dr P. F. Brandon) kindly allowed the Symposium to be held in his Department. The Director of the Polytechnic (Mr T. G. Miller) welcomed the participants and ably opened the proceedings of the Symposium. We are also grateful to the Sponsors of the Symposium for the financial backing which made it possible, i.e. to the Royal Society, the Systematics Association, the Natural Environment Research Council and the British Phycological Society.

D. E. G. IRVINE
J. H. PRICE

August 1978

Contents

1 | Taxonomy of Red and Brown Algae

A. D. BONEY

Department of Botany, University of Glasgow, Scotland

Abstract: A review is given of the historical background to present day concepts of taxonomy in red and brown algae. The several opinions on the stabilities of the systems of classification are discussed and the new approaches of recent years reviewed.

Taxonomy is the science of classification. To quote Lam (1959): "Taxonomy is born out of classification, and classification is born out of necessity". Classification is a necessary basic feature of our life because of the great variety of natural phenomena. We apply classification procedures at all times in order to simplify the wide diversity of our experiences in relation to our environment. According to Dr Joel Hedgpeth, man's oldest profession was that of a taxonomist, since his first task in the Garden was not to work for a living—that necessity came after the Fall—but to name every beast of the field and fowl of the air. As botanists we would add ". . . and all plants of the field and water" (especially those bearing fruit!). In similar vein, Turrill (1964) pointed out that whilst animal nomenclature started with Adam (Genesis 2, 19–20), perhaps Eve dealt with plants, since plant nomenclature is said to be in better general shape than animal nomenclature. Heywood (1976), in tracing the development of systems of classification, has stated that each tribe or community would have developed a "folk taxonomy" for those plants and animals whose characteristics and properties (nutritive, toxic, medical) were well known to the tribe members. A "folk taxonomy" based on the doctrine of signatures would have been a more specialized development of limited working value. Two examples of "folk taxonomies" linked with marine algae can be cited; first, the naming of marine algae of economic value in tropical South and East Asia (Zaneveld, 1959) and, secondly, the several forms of edible seaweed ("limu") used in

Systematics Association Special Volume No. 10, "Modern Approaches to the Taxonomy of Red and Brown Algae," edited by D. E. G. Irvine and J. H. Price, 1978, pp. 1–19, Academic Press, London and New York.

Hawaii (Abbott and Williamson, 1974). The limitations of a "folk taxonomy" are clearly very closely linked with the numbers of organisms to be classified. An estimate of the numbers of described species of plants is given in Table I. Bold (1970) gives the approximate total of living species of all plants as 350 000, of which 19 000 are algae. The very approximate number of species of red algae is 5524 according to Dixon (1973). Papenfuss (1951) states that the number of species of brown algae is about 1500, whilst Cronquist (1960) gives 2000 species.

TABLE I. Estimated numbers of described species (Grant 1963)

Flowering plants	286 000	Fungi and slime moulds	40 400
Gymnosperms	640	Protista	30 000
Ferns and fern allies	10 000	Blue–green algae	1 400
Bryophytes	23 000	Bacteria	1 630
Algae (green, brown and red)	8 675	Viruses	200

With such numbers the guidelines of our systems of classification need to be well defined.

A closer look at the terminology involved is instructive. "Classification" is derived from the Latin "classis"—a group or multitude—in fact, a term originally applied to each of the six ancient divisions of the Roman people. "Taxonomy" (a version of de Candolle's "taxonomie" of 1813) is derived from the Greek "taxis"—order or arrangement—and means "the rendering of order" (Lam 1959). By strictly applying the rules of taxonomy we are practising a "law of order" (Turrill 1964)—a structure of the system of living things, or . . . "the science of the structure of the multiformity of living nature" (Danser 1950). Such a system of living things brings us to a further term. "Systematics" is derived from the Greek "systema"—that which is "put together" (e.g. Linnaeus's "Systema *naturae*"). Systematics is a process of putting together with the purpose of obtaining an "organized or connected" group of objects. Thus—to quote Lam (1959)—"classification is an art rather than a science, taxonomy is the science behind the art, and a system is the result of both". Heywood (1976) has described the interrelationships of the three terms in more detail. Systematics is the "scientific study of the diversity and differentiation of organisms and the various relationships which exist between them". Taxonomy is "that part of systematics which deals with the study of classification, including its bases, principles, procedures and rules". The two terms are often regarded as interchangeable. Classification is the . . . "process of ordering plants in groups which are arranged hierarchically". Savory (1962), whilst pointing out the significance of systematics in arranging both the plant and animal kingdoms into a convenient and logical order, describes the term "taxonomy" as . . . "an invention, a piece of biological jargon, a synonym with almost the same meaning". Wherever we look for precise definitions, the proximity of the two terms is evident. Thus, the opening words of Turrill's (1964) first sentence are "Plant taxonomy, or systematic botany . . ."

What are the declared goals of modern taxonomy? Davis and Heywood (1963) have put forward the following aims as those most widely accepted:

1. To provide a convenient method of identification and communication.
2. To provide a classification which as far as possible expresses the natura relationships of organisms.
3. To detect evolution at work, discovering its processes and interpreting its results.

To achieve such aims, these two authors propose a four-phased approach:

1. *The pioneer or exploratory phase:* concerned primarily with the study of herbarium material; identification, morphology, some information on occurrence, distribution and (for algae) possibly spore production.

2. *Consolidation phase:* increased knowledge with the subsequent elimination of arbitrary judgments of the pioneer phase, and frequent reduction of many names to synonymy.

3. *Biosystematic phase:* after the acquisition of cytological and biosystematic information, with the emphasis changing to micro-evolution.

4. *Encyclopedic phase:* the phase of co-ordination of 1–3.

Turrill's (1938) concept of the development of taxonomic knowledge envisages a gradient from "alpha" to "omega". The alpha taxonomy corresponds to Davis and Heywood's pioneer phase and the omega to the encyclopedic stage—a natural classification based on all possible characters and relevant data. In two recent reviews with reference to marine algal taxonomy, it is very clear that opinions have placed the status of such taxonomic studies as little beyond the "pioneer phase" of Davis and Heywood, or the "alpha" state of Turrill (Dixon, 1970b; Russell, 1973). Dixon (1970b) further points out that this lack of a taxonomic understanding of the algae is in part due to an unfortunate accident—the development of a pronounced aversion to taxonomic studies in botany at about the beginning of the present century. This would appear an odd circumstance when considered in the light of Prescott's (1951) description of 1800–1875 as the "golden age of plant taxonomy". The aversion to such studies manifested at the end of the nineteenth century and the beginning of the present one may well have been a reaction against the over-specialization on taxonomic studies of earlier years. Bower's (1938) description of his own undergraduate experience of the official teaching of botany at Cambridge in 1875 ("... moribund in the summer and actually dead in winter . . .") is illuminating. "Official" botany at Cambridge at that time was concerned principally with ". . . splitting analytically the varieties of *Rubus*". This was in marked contrast to the exciting developments in experimental botany then making great strides in the continental laboratories of botanists such as Sachs and de Bary. Aspiring young British botanists (such as F. O. Bower and S. H. Vines) sought their scientific inspirations abroad, and their experiences, even for short periods of time in University vacations, profoundly influenced their subsequent careers. These were, in many cases, careers of great importance in their influences on subsequent botanical teaching and research in the British Isles. Nor can we ignore the influence both at that time and in later years of the "Hofmeisterian approach" to the sequences of spore-producing and gamete-producing generations in cryptogams. This pursuit of the "new botany" (sometimes called "The Cause" by its enthusiastic disciples) was one factor leading to the eclipse of the traditional systematic botany in Britain. If, however, there was a "golden age of plant taxonomy" in the last century, to what extent was this reflected in the taxonomy of algae, and, in particular, the taxonomy of red and brown algae?

Linnaeus's "Species Plantarum" (1753), in terms of its contribution to algal taxonomy has been dismissed as being of relatively little significance, mainly because many of his "algae" were in fact liverworts (Prescott, 1951; Smith, 1951). However, his genera *Conferva* (filamentous algae), *Ulva* (membranaceous algae) and *Fucus* (algae with cartilaginous thalli) remained in this circumscription for some 50 years. The contributions of Lamouroux in the first two decades of the nineteenth century are also important. We see in Lamouroux's work the early recognition of chloroplast pigment differences in the major groups of algae. We also see early recognition of a taxonomic problem which is ever with us regarding red and brown algae—that of polymorphism. Stackhouse's (1797) *Chondrus crispus* was renamed *Fucus polymorphus* by Lamouroux (1805) and this name was in turn accepted by Stackhouse (1809) with the creation of a superfluous genus *Polymorpha*. Later Lamouroux (1813) accepted Stackhouse's genus *Chondrus*, and gathered all Stackhouse's species into the one *Chondrus polymorphus*. This polymorphism of *Chondrus* so impressed Lamouroux that he regarded the plant as a veritable "proteus" of seaweeds. More recently Thomas (1938) described 21 forms of *Chondrus*, eight of which were of his own naming.

Much importance is attached to the contributions of C. A. Agardh at the University of Lund in Sweden, and those of his son J. G. Agardh. Their recognition of the systematic value of the "cystocarp" in the taxonomy of red algae is acknowledged as a significant contribution. Attention has been drawn, however, to the lack of field studies by these and others (e.g. Kützing), with the result that each morphological variant came to be regarded as a distinct taxon, (Dixon, 1970b). This lack of field experience cannot be said of W. H. Harvey, however. His recognition of three main "series" of algae—Chlorospermeae, Melanospermeae and Rhodospermeae—based on the visual recognition of pigment arrays—represents the first biochemical classification of major plant groups (Harvey, 1836). The Diatomaceae constituted the fourth series. J. G. Agardh's (1842) delimitation of the groups Zoospermeae (blue–green and green algae), Fucoideae and Florideae follows similar lines. In his short life of 55 years, Harvey collected extensively in South Africa, America and Australia as well as in the United Kingdom and the monuments to his abilities as a collector, artist and describer of marine algae stand in "Phycologia Britannica" (1846–1851) and "Phycologia Australica" (1858–1863). His acute powers as a field botanist in the recognition of variations in growth at different seasons and in different environments is a lesson for us all today. Asa Gray's comment in his obituary of Harvey is sufficient tribute—"He was a keen observer and a capital describer".

Major taxonomic groupings of algae appeared to have followed many paths in the middle years of the century. Something almost of despair can be sensed in the sections on Algae in Sach's "Lehrbuch", where he commented that the classification of the algae appeared to be in the utmost confusion. In the 4th edition (cf. Vines's edited translation, 1882) he followed the proposals of Cohn (1872):

Gruppe Thallophyten:

- Klasse Protophyten (blue–green algae (Cyanophyceen), bacteria, yeasts);
- Klasse Zygosporeen (isogamous—green algae, diatoms, myxomycetes);
- Klasse Oosporeen (oogamous—including Oedogonieen, Fucaceen);
- Klasse Carposporeen (with a complex "fruiting" body—including Coleochaeteen, Florideen, Characeen, Ascomyceten, Aecidiomyceten, Basidiomyceten).

In an English translation of Sach's "Lehrbuch" by Bennett and Thiselton Dyer (1875) no attempt is made to give a systematic account of the algae. Whilst paying tribute to the research of earlier years, it was stated that

> ". . . these researches have not yet been carried sufficiently far to construct a new and complete classification of the Algae corresponding to the present requirements of science. The discovery of alternation of generations and polymorphism in some sections justifies the supposition that certain forms not as accurately known may be

merely conditions of development of unknown cycles of forms, although hitherto considered distinct species and genera".

This comment is in what became one of the foremost botanical textbooks of its day in the English language.

Eichler's system of 1880 was used in a number of later German "Handbücher" and "Lehrbücher" on Botany. Eichler recognized two major plant "groups", the Cryptogamen and Phanerogamen. In the Cryptogamen three further "groups" were recognized:

Thallophyten
  Algen Cyanophyceen Chlorophyceen
    Phaeophyceen Rhodophyceen
  Fungi
Bryophyten
Pteridophyten

Eichler's recognition of the four principal classes of algae was based on colour, vegetative habits of the organisms and their reproduction. The class Rhodophyceae had, in fact, been earlier recognized by Ruprecht (1850).

By the time that the appropriate part of Engler and Prantl's "Die natürlichen Pflanzenfamilien" appeared in 1896, the numerous researches of the preceding 20–30 years were beginning to result in some semblance of taxonomic regularity. Thus Schmitz, whose pioneering and critical work on "cystocarp" structure (1883, 1889), laid some reliable foundations to the taxonomy of red algae, contributed the section on the class Rhodophyceae as joint author with Hauptfleisch. Two "subclasses" were described, the Bangiales and the Florideae, with the latter name being derived from the Floridées of Lamouroux (1813). In further subdividing these subclasses, four "families" were described for the Bangiales (Bangiaceae, Rhodochaetaceae, Compsopogonaceae, Thoreaceae) of which the last three were designated as being "of uncertain position". Within the Florideae, four orders were described, Nemalionales, Gigartinales, Rhodymeniales and Cryptonemiales. These four orders were based on the fertilization and post-fertilization changes on which the earlier critical investigations had been largely the work of Schmitz (1883).

J. G. Agardh's use of red algal fruiting bodies in taxonomy was largely restricted to descriptions of external form. Whilst Bornet and Thuret (1867) described accurately the process of gamete fusion in red algae, Schmitz (1883) was the first to show clearly the relationship between the fertilized carpogonium and the subsequent carposporophyte phase, in which, for a small number of the Florideae, the carposporophyte develops directly from the fertilized carpogonium, whilst in the majority the fertilized carpogonium forms cells or filaments which fuse with another cell prior to carposporophyte development. This cell involved in the second fusion was termed the "auxiliary cell" by Schmitz, although he interpreted this erroneously as a "second fertilization process" (Dixon and Irvine, 1977). Schmitz's researches formed the basis for all subsequent interpretations of ordinal status in the red algae attributed to his subclass Florideae. The four orders described above were increased to five by the establishment of the order Ceramiales by Oltmanns (1904) by removing genera of three of the families in Schmitz's Rhodymeniales. The order Gelidiales was later established by Kylin (1923), with the result that six orders of Florideae became recognized (Drew, 1951; Kylin, 1956). Schmitz and Hauptfleisch's (1896) difficulties with their "subclass" Bangiales have been mentioned. De Toni (1897) renamed this group Bangioideae, and Skuja (1939) proposed four orders: Porphyridiales, Goniotrichales, Bangiales and Compsopogonales with the suggestion that a further order, the Rhodochaetales, might be justified (Dixon and Irvine, 1977). It is interesting to note some correlation here with Schmitz and Hauptfleisch's views, although they had expressed some doubts regarding the groupings they had proposed.

Kjellman (1897) reviewed the brown algae in "Die natürlichen Pflanzenfamilien". Numerous contributions to the taxonomy of brown algae had been made in the preceding years, notably Thuret and Bornet's elegant studies in their "Études Phycologiques" (1878). Kjellman divided the class Phaeophyceae into two "series", the Phaeosporeae and the Cyclosporeae, the latter containing one family, the Fucaceae. The Phaeosporeae he further divided into "subgroups" Zoogonicae (motile reproductive cells) and Acinetae (reproductive cells sometimes non-motile). The former "subgroup" contained 17 families, and the latter subgroup one—the Tilopteridaceae. Oltmanns (1904), in his monographs on the algae, used Kjellman's system without change. Kylin (1933) divided the brown algae into three classes based on life cycles:

Class Isogeneratae (similar gametophyte and sporophyte phases)
- Orders Ectocarpales
  - Sphacelariales
  - Cutleriales
  - Tilopteridales
  - Dictyotales

Class Heterogeneratae (dissimilar gametophyte and sporophyte phases)
- sub-class Haplostichineae
  - Orders Chordariales
    - Sporochnales
    - Desmarestiales

sub-class Polystichinae

Orders Dictyosiphonales

Punctariales

Laminariales

Class Cyclosporeae (gametophyte phase practically eliminated)

Order Fucales

Vegetative organization of the thallus was the basis for separating the two subclasses in the Heterogeneratae.

Fritsch (1945) rejected separation into subclasses and described nine orders in the one class Phaeophyceae:

Ectocarpales	Sporochnales	Sphacelariales
Tilopteridales	Desmarestiales	Dictyotales
Cutleriales	Laminariales	Fucales

The Punctariales, Dictyosiphonales and Chordariales of Kylin he merged with the Ectocarpales. Fritsch considered that heteromorphic life histories were derivatives of isomorphic life histories by divergent development of the two phases. Similarly a wide variety of thallus forms were derivable from simple filamentous types. But Fritsch also pointed out that this classification of the Ectocarpales was "essentially one of convenience".

Kylin's (1933) system of classification of brown algae was subsequently used by Papenfuss (1951, 1955), Fott (1959), Chadefaud (1960) and Chapman (1962). All later proposals regarding classification at ordinal level seem to have revolved around certain problematical groups, e.g. the establishment of the order Scytosiphonales (Feldmann, 1949; Christensen, 1962; Parke and Dixon, 1964), for a family previously included in the Dictyosiphonales; rejection of the order Chordariales, with incorporation of the family in the Ectocarpales (Christensen, 1962; Parke and Dixon, 1964); rejection of the order Tilopteridales and its incorporation in the Dictyosiphonales (Christensen, 1962). Scagel (1966) returned to the idea of two subclasses, based this time more on the distinctive and atypical flagellation that exists in the Fucales. He suggested two subclasses, the Cyclosporidae to receive the Fucales with all other orders being placed in the Phaeophycidae. This separation is thus based on a fundamental aspect, distinct from a life history sequence.

With few exceptions, the names of orders in both the red and brown algae appear to have been a constant feature for some years now. The Nemalionales has become the Nemaliales (Dixon, 1970a) and the Gelidiales merged with the Nemaliales (Dixon, 1961; Parke and Dixon, 1964). Are we then to assume their stability, or is this a compromise situation of the best arrangement possible in the present state of knowledge?

For the red algae, the six orders defined by Kylin (1956) have been strictly adhered to by Papenfuss (1964, 1966), with the comment "... the genera that have been studied are referable to fairly clear-cut Orders" (Papenfuss, 1966). The presence of a few seemingly intermediate types "was to be expected with a phylogenetically related group of organisms". Dixon (1964), however, emphasized that the apparent stability of this ordinal system is "largely an illusion", and that one only, the Ceramiales, is at all well defined. The need for a complete revision, based on new concepts and criteria, was called for. Papenfuss (1966) did point out that detailed knowledge of the structure and reproduction of red algae was then available for only about a quarter of the genera and a very small percentage of the species. Any additional knowledge would in turn necessitate revisions at all taxonomic levels, ordinal, generic and specific. The point made by Dixon (1970a) is a valid one however. Many conditions and sublittoral populations remain virtually unexplored from a marine algal viewpoint in S. America, Africa, Asia etc. There will be need for a concerted effort over the years to remedy this deficiency—for which piecemeal additions will be inadequate. The question remains whether these revisions will depend on the same basic taxonomic criteria, and whether any entirely new approaches are possible. One revision has been proposed in recent years, not through the discovery of a new genus, or by application of entirely new criteria, but as a result of a more critical examination of a long "established" species—the red alga *Palmaria palmata* for *Rhodymenia palmata* (Guiry and Irvine, 1974; Guiry, 1974), with the subsequent establishment of a new order, the Palmariales (Guiry and Irvine, in the press; Parke and Dixon, 1976; Dixon and Irvine, 1977).*

Whilst much discussion has centred around taxa at ordinal and family level, reference should also be made to the status of the major groupings. Such taxa must be named divisions, since "phyla" are not used in plant classification. Recognition of the distinctiveness of the red algae has long been acknowledged. The division Rhodophyta (Wettstein, 1901) has been similarly long accepted, with either the single class Rhodophyceae and two subclasses, the Bangioideae or Bangiophycidae and the Florideae or Florideophycidae; or two classes, the Bangiophyceae and the Florideophyceae (Cronquist, 1960). The conferring of divisional status on the brown algae has been more difficult. Opinions have varied between recognition of the division Phaeophyta (e.g. Papenfuss, 1951) or placing the brown algae as a class (Phaeophyceae) within a larger division, for example Chadefaud's Chromophycophyta (1960) and Christensen's Chromophyta (1962), along with other classes of algae distinguishable from green algae of the major division Chlorophycophyta (Chadefaud, 1960) or the Chlorophyta (Christensen, 1962). Whilst Scagel (1966) considered divisional

---

* But see p. 138 of this book (Ed.).

status for the brown algae unnecessary, Russell (1973) has retained it and the same principle is used in the most recent check-list of the British marine algae (Parke and Dixon, 1976), in which divisional status is conferred on a number of algal groups previously described at class level.

Papenfuss (1946) named seven "phyla" of algae, based principally on Pascher's (1931) eight "phyla" but excluding the blue–green algae which were placed in the division Schizophyta with the bacteria. The introduction of the connotation "phykos" by Papenfuss was put forward as having ". . . the added merit of being explicit as to the kinds of plants to which they apply". Thus the red and brown algae were named Rhodophycophyta and Phaeophycophyta. Such a modification has been declared unnecessary (Smith, 1951) and the terminology has not received extensive usage since. The proposal that brown, red and green algae should be placed in a single division, the Euphycophyta (Chapman, 1962) has similarly not been generally accepted. Whilst Scagel (1966) has considered divisional status for brown algae as being unnecessary, it has become a recognized taxon (Russell, 1973; Parke and Dixon, 1976).

Major groupings of organisms at the levels of divisions or phyla can be simply defined (quantitatively) as the maximum number of organisms having the smallest possible number of common features. One may quote the example of the Protozoa, in which the common feature is the non-cellular habit, or organisms in whose "bodies" no one nucleus has control solely over a specialized part of the cytoplasm. Cronquist's (1960) delimitation of both the Rhodophyta and Phaeophyta places emphasis on pigmentation, plant body construction, reproductive cells and reserve substances, all workable criteria for those with sufficient experience. Nevertheless, Christensen's (1962) groupings of Chlorophyta and Chromophyta has the attraction of being based on the presence or absence of chlorophyll *b*.

The distinctiveness of the red and brown algae in terms of their relationship to one another needs little emphasis. The recent suggestion (Simon-Bichard-Bréaud, 1971) that the spermatangia of *Bonnemaisonia hamifera* contained an "appareil cinétique" in a "cryptflagellaire"—an observation of profound potential phylogenetic importance—has not been verified in subsequent ultrastructural studies, the most recent being by Young (1977) on the spermatangia of the related *B. nootkana*. The recorded presence of fucoxanthin in the red alga *Callithamnion pikeanum* is undoubtedly due to epiphytic "contaminants" (as stated by Strain, 1951).

The taxonomic isolation of the red algae has always raised questions regarding their phylogenetic interrelationships. Whilst pigmentation and lack of flagellated cells and other biochemical features places them close to the blue–green algae, their eucaryote nature is a strong argument against such a relationship. Regarding relationships with other divisions of algae, either the membrane-bound organelles and chlorophyll *a* have evolved independently and there is no close relationship, or the eucaryotic condition has evolved only once from a blue–green ancestral stock. In this case, there was an early divergence of the red algae. In contrast there are a number of common features indicating inter-relationships between the brown and other algae, notably certain algae in the division Chrysophyta (cf. Smith, 1951). Dodge (1973) places the class Phaeophyceae in the Chrysophyta. The recent description in the division Prochlorophyta of procaryote organisms with chlorophylls *a* and *b* (Lewin, 1976) is a pertinent reminder of change even at these taxonomic levels.

It is an interesting point of comparison that whilst the taxonomy of bacteria tends to be morphological with the major groupings, that of the lower taxa tends to be biochemical. The reverse generally holds for the algae. In the red and brown algae morphology and life history sequences are important at both genus and species level. Many of the data of recent years which have begun to unravel the links between reproduction, morphological phases and environmental factors (temperature, photoperiod and light quality), have brought significant taxonomic contributions as additional bonuses. Recent reviews on life histories have dealt critically with these aspects for both the red algae (Dixon, 1973) and the brown algae (Wynne and Loiseaux, 1976). To be credible, culture-based life history investigations require painstaking approaches to the experiments, e.g. reliable identification of the alga and life-history phase involved, with retention of voucher specimens if possible; ensurance that there are no contaminants, particularly of organisms of the same class; and verification of the source of the inoculum (spore types, propagulae, thallus fragments). The question can always be raised regarding the possibility of culture artifacts giving rise to a spurious "culture taxonomy". Culture methods of recent years have become more sophisticated with the recognition that careful definition of media, proper use of suitable artificial light sources and reliable control of environmental conditions are essential. As pointed out by Wynne and Loiseaux (1976) it is possible nowadays to compare and evaluate results from different laboratories in many different parts of the world. From a taxonomic point of view, culture methods have both given answers and raised new questions. Notably most problems of interpretation have arisen with taxa in the most primitive orders of red and brown algae (Nemaliales and Ectocarpales); similarly the life histories of members of the Bangiophyceae remain largely problematical (Dixon, 1973) although some attempt has been made to bring order to the *Porphyra* complex (Conway *et al.*, 1975).

Russell (1973) has pointed out the need for a re-examination of the orders Ectocarpales and Scytosiphonales, since culture studies have linked life histories of taxa in the Scytosiphonales with encrusting forms previously regarded as distinct entities in the Ectocarpales. A number of examples are known also, in the red algae, in which culture studies have linked morphologically distinct entities as components of a common life history. This will be a continuing process of great taxonomic significance.

Reference to a "culture-based taxonomy", made earlier, needs amplification. It is a fact that plant forms can sometimes arise in culture which bear no resemblance to existing taxa. The significance of such forms should be regarded with great caution. Form variation is a particularly well demonstrated feature observed in culture studies with certain representatives of the genus *Audouinella*, notably with certain endophytic and endozoic species (White and Boney, 1969, 1971; Boney, 1972a). External form is still a major taxonomic criterion in red and brown algae. We still lack information for many common algae on the seasonal and environmental variations in form. The earlier confusions between seasonal growth forms and distinct taxonomic entities have been well illustrated with the genera *Grateloupia*, *Pterocladia* and *Gelidium* (Dixon, 1966). Lack of data on this fundamental aspect of algal taxonomy will allow little opportunity to move on from Davis and Heywood's "pioneer phase" or Turrill's "alpha" stage. For higher taxonomic levels the call has been made for a phylogenetic approach, with the use of new criteria (e.g. Dixon, 1973).

This need for novel criteria in delimiting algal taxa has been mentioned by several authorities. One such attempt by Stoloff (1962) distinguished between red algae producing agars, carrageenans and gelans (Stoloff and Silva, 1957)—the nature of the κ-carrageenan fractions being important in distinguishing between carrageenan and gelan. Stoloff's (1962) proposals are summarized in Table II. Whilst the order Gelidiales has been incorporated in the Nemaliales (Dixon, 1961; Parke and Dixon, 1964), Stoloff's proposals included the establishment of a new order, the Hypneales, to include *Hypnea*, *Furcellaria* and *Agardhiella*. Yaphe and Duckworth (1971) have stated that these early attempts at obtaining data of taxonomic significance were of limited value (e.g. Yaphe, 1957, 1959), since the major criteria evaluated were gel strengths and the presence of characteristic oligosaccharides.

Elegant techniques are now available to determine the "fine structures" of agar and carrageenan families of polysaccharides and these compounds may yet provide a criterion for taxonomic studies. For the brown algae the chemistry of cell wall compounds and intercellular matrices has been studied in terms of the ratio of mannuronic to guluronic acid in the alginates, (Fischer and Dorfel,

TABLE II. Extractives from various genera of red algae (Data from Stoloff, 1962)

Order	Genus	Extractive		
		Agar	Carrageenan	Gelan
GELIDIALES	*Gelidium* (18 species)	+	−	−
	Pterocladia (3 species)	+	−	−
	Suhria (1 species)	+	−	−
	Acanthopeltis (1 species)	+	−	−
	Yatabella (1 species)	−	+	−
CRYPTONEMIALES	*Endocladia* (1 species)	−	+	−
	Gloiopeltis (3 species)	−	+	−
	Aeodes (1 species)	−	+	−
GIGARTINALES	*Gracilaria* (8 species)	+	−	−
	Corallopsis (1 species)	+	−	−
	Furcellaria (1 species)	−	−	+
	Agardhiella (1 species)	−	−	+
	Eucheuma (3 species)	−	+	−
	Hypnea (2 species)	−	−	+
	Ahnfeltia (1 species)	+	−	−
	Phyllophora (2 species)	+	−	−
	Gymnogongrus (2 species)	−	+	−
	Chondrus (2 species)	−	+	−
	Gigartina (13 species)	−	+	−
	Iridophycus (3 species)	−	+	−
	Rhodoglossum (1 species)	−	+	−

1955). These results have indicated that the lower ratios are obtained with members of the Ectocarpales and the highest with the Fucales. Corrected values allowing for the higher rate of destruction of guluronic acid makes such a conclusion less certain (Lewin, 1974) and Nisizawa *et al.* (1969), using specific enzyme hydrolysis methods, have shown that these ratios vary not only from species to species but also from one part of the thallus to another in the same organism. Attempts at linking the type of phycobilin pigment (by spectrophotometric analysis) with various orders in the red algae have not proved successful (Hirose *et al.*, 1969). Polyacrylamide gel electrophoresis of soluble cell proteins of red algae is an interesting development of some taxonomic potential (Mallery and Richardson, 1971). The banding patterns of the protein moieties of the phycobilins show striking correlations with taxonomic positions. The simplest patterns are seen with representatives of the Bangiophyceae, and the most complex with members of the Ceramiales. More species need to be examined but the method holds promise. Compounds of low molecular weight have been identified in both red and brown algae. These include the mono- and di-iodotyrosines in certain families of brown algae and chlorinated and brominated compounds in the family Rhodomelaceae (Ceramiales) (Lewin, 1974). The extent of their taxonomic significance needs to be examined. A similar statement can be made regarding certain fluorescent substances of unknown chemical composition obtained as extracellular products from five *Audouinella* species (Boney, 1972b). There is need for a continuing search for suitable biochemical criteria, notably on a "presence or absence basis", since quantitative data on the components of certain compounds do not appear to be successful when applied as taxonomic criteria.

The major contributions to algal taxonomy made by ultrastructural studies over the last 25 years need little emphasis here. Dodge's (1973) preliminary assessment of algal fine structure and phylogeny, and the tentative conclusions drawn, are principally concerned with relationships at class level. It is at the lower taxonomic levels that the critical information will be necessary. Ultrastructural studies have been little used in the red algae. Nor can we ignore the rôle that cytological studies with light microscopy can still play. Until relatively recent times, genetic studies on algae were largely confined to unicellular green representatives. Now that successful cross-fertilizations have been effected using the non-motile gametes of certain red algae, a new field of potential taxonomic value has been opened, applicable also to the brown algae. The application of methods of numerical taxonomy has been used in the study of speciation in *Alaria* (Widdowson, 1971) and more recently with the British representatives of the Phaeophyta (Russell and Fletcher, 1975). This last study has again drawn attention to the need for re-thinking on the various orders, and returns again to a conservative viewpoint of the order Ectocarpales (cf. Fritsch, 1945).

As stated by Heywood (1976), a unique feature of taxonomy is that none of its sources need become obsolete. Morphology and anatomy cannot be dispensed with, although biochemical, cytological and other chemical data may be available. And whilst we may agree that all taxonomic decisions are subjective, such decisions must be based on as much data as possible and should be subject to regular review (Russell and Fletcher, 1975). With such vast areas of the coastal waters in parts of the world unexplored and with the attendant gaps in our knowledge of marine floras, there is no short-term answer to this problem of the "pioneer stage" or "alpha phase" of taxonomy for marine algae. Both the lack of trained personnel and the shortage of money will further delay any resolution of these logistic problems. Perhaps an answer may be sought in concerted studies on the marine ecology of key areas, similar to the Aldabra project of recent years. Meanwhile there is much that can be done in our continued search for novel criteria which will still be applicable no matter what the extent may be of future floristic studies. Mendeleev's great contribution to the classification of the elements was not only in putting every element in its right place in the order of things but also knowing where the gaps had to be left to be filled in the light of later information. This is an attitude of mind to be recommended to algal taxonomists. In conclusion, the following quotation is a description by a taxonomist of a taxonomist at work: W. N. P. Barbellion (B. T. Cummings), then working at the B.M. (Nat. Hist.), in his "Journal of a Disappointed Man" (1919), made the following entry for 15 July 1914:

> "... Spent the day measuring the legs and antennae of lice to two places of decimals. To the lay mind how fantastic this must seem. Indeed I hope it is fantastic. I do not mind being thought odd. It seems almost fitting that an incurable dilettante like myself should earn his livelihood by measuring the legs of lice."

In this we see both the painstaking accumulation of information which is the accepted lot of the taxonomist and an example of a source of denigration he or she often suffers at the hands of their non-taxonomic colleagues who accept the second part of the paragraph as true, despite the fact that Barbellion was writing this with his tongue in his cheek.

A dilettante approach can no longer be of much assistance in present day studies on the taxonomy of red and brown algae.

REFERENCES

Abbott, I. A. and Williamson, E. H. (1974). "Limu: an Ethnobotanical Study of Some Edible Hawaiian Seaweeds" (2nd edition). Pacific Tropical Botanical Garden, Kauai.

AGARDH, J. G. (1842). "Algae maris Mediterranei et Adriaticae." Fortin, Masson et Cie, Parisiis.

BARBELLION, W. N. P. (1919). "The Journal of a Disappointed Man." Chatto and Windus, London.

BOLD, H. C. (1970). "The Plant Kingdom" (3rd edition). Prentice-Hall, New Jersey.

BONEY, A. D. (1972a). *In vitro* growth of the endophyte *Acrochaetium bonnemaisoniae* (Batt.) J. et G. Feldmann. *Nova Hedwigia* **23**, 173–186.

BONEY, A. D. (1972b). Fluorescent substances from *Acrochaetium* species (Nemaliales, sub-class Florideophycidae, Rhodophyceae). *J. nat. Hist.* **6**, 47–53.

BORNET, E. and THURET, G. (1867). Recherches sur la fécondation des Floridées. *Annls Sci. nat.* (*Bot.*) Sér. 5, **7**, 137–166.

BOWER, F. O. (1938). "Sixty Years of Botany in Britain 1875–1935." Macmillan, London.

CHADEFAUD, M. (1960). "Traité de Botanique systematique", Vol. 1. Masson et Cie, Paris.

CHAPMAN, V. J. (1962). "The Algae." Macmillan and Co. Ltd., London.

CHRISTENSEN, T. (1962). Systematisk Botanik Nr. 2 Alger. *In* "Botanik", Bd II (T. W. Böcher, M. Lange and T. Sørensen, eds). Munksgaard, København.

COHN, F. [J.] (1872). Conspectus familiarum cryptogamarum secundum methodum naturalem dispositarum. *Hedwigia* **11**, 17–20.

CONWAY, E., MUMFORD, T. F. and SCAGEL, R. F. (1975). The genus *Porphyra* in British Columbia and Washington. *Syesis* **8**, 185–244.

CRONQUIST, A. (1960). The divisions and classes of plants. *Bot. Rev.* **26**, 425–482.

DANSER, B. H. (1950). A theory of systematics. *Bibl. Biotheor.* **4**, 116–180.

DAVIS, P. H. and HEYWOOD, V. H. (1963). "Principles of Angiosperm Taxonomy." Oliver and Boyd, Edinburgh and London.

DE TONI, G. B. (1897). "Sylloge Algarum." Sumptibus auctoris, Patavii.

DIXON, P. S. (1961). On the classification of the Florideae with particular reference to the position of the Gelidiaceae. *Botanica mar.* **3**, 1–16.

DIXON, P. S. (1964). Classification of Rhodophyta. *Abstr. Int. bot. Congr.* **10**, 105.

DIXON, P. S. (1966). On the form of the thallus in the Florideophyceae. *In* "Trends in Plant Morphogenesis" (E. G. Cutter, ed.), pp. 45–63. Longmans, Green and Co. Ltd., London.

DIXON, P. S. (1970a). The Rhodophyta: some aspects of their biology II. *Oceanogr. mar. Biol. ann. Rev.* **8**, 307–352.

DIXON, P. S. (1970b). A critique of the taxonomy of marine algae. *Ann. N.Y. Acad. Sci.* **175**, 617–622.

DIXON, P. S. (1973). "Biology of the Rhodophyta." Oliver and Boyd, Edinburgh and London.

DIXON, P. S. and IRVINE, L. M. (1977). "Seaweeds of the British Isles. Volume 1. Rhodophyta." British Museum (Natural History), London.

DODGE, J. D. (1973). "The Fine Structure of Algal Cells." Academic Press, London and New York.

DREW, K. M. (1951). Rhodophyta. *In* "Manual of Phycology . . ." (G. M. Smith, ed.), pp. 167–191. Chronica Botanica Company, Waltham (Mass.).

EICHLER, A. W. (1880). "Syllabus der Vorlesungen über specielle und medicinisch-pharmaceutische Botanik." Gebrüder Borntraeger, Berlin.

Engler, A. and Prantl, K. (1896). "Die naturlichen Pflanzenfamilien" Teil 1, Abt. 2. Wilhelm Engelmann, Leipzig.

Feldmann, J. (1949). L'ordre des scytosiphonales. *Mém. Soc. Hist. nat. Afrique Nord, hors-sér.* **2**, 103–115.

Fischer, F. G. and Dorfel, H. (1955). Die polyuronsäuren der Braunalgen. *Hoppe. Seyler's Z. Physiol. Chem.* **302**, 186–203.

Fott, B. (1959). "Algenkunde." Gustav Fischer, Jena.

Fritsch, F. E. (1945). "Structure and Reproduction of the Algae", Vol. 2. Cambridge University Press, Cambridge.

Grant, V. (1963). "The Origin of Adaptations." Columbia University Press, New York and London.

Guiry, M. D. (1974). A preliminary consideration of the taxonomic position of *Palmaria palmata* (Linnaeus) Stackhouse—*Rhodymenia palmata* (Linnaeus) Greville. *J. mar. biol. Ass. U.K.* **54**, 509–528.

Guiry, M. D. and Irvine, D. E. G. (1974). A critical assessment of sub-ordinal classification in the Rhodymeniales. *Abstr. int. Seaweed Symp.* **8A**, 34.

Guiry, M. D. and Irvine, D. E. G. (1978). A critical assessment of infraordinal classification in the Rhodymeniales. *Proc. int. Seaweed Symp.* **8** (in the press).

Harvey, W. H. (1836). Algae. *In* "Flora Hibernica" (by J. T. Mackay), Vol. 2. William Curry Junior and Co., Dublin.

Harvey, W. H. (1846–51). "Phycologia britannica: . . ." Reeve and Benham, London.

Harvey, W. H. (1858–63). "Phycologia australica; . . ." Lovell Reeve and Co., London.

Heywood, V. H. (1976). "Plant Taxonomy" (2nd edition). Arnold, London.

Hirose, H., Kumano, S. and Madono, K. (1969). Spectroscopic studies on phycoerythrins from cyanophycean and rhodophycean algae with special reference to their phylogenetical relations. *Bot. Mag. Tokyo* **82**, 197–203.

Kjellman, F. R. (1897). Phaeophyceae. *In* "Die natürlichen Pflanzenfamilien" by A. Engler and K. Prantl. Wilhelm Engelmann, Leipzig.

Kylin, H. (1923). Studien über die Entwicklungsgeschichte der Florideen. *K. svenska VetenskAkad. Handl.* **63**(11), 1–139.

Kylin, H. (1933). Uber die Entwicklungsgeschichte einiger Phaeophyceen. *Lunds. Univ. Årsskr. N.F. Avd.* **2**, **29**(7), 1–102.

Kylin, H. (1956). "Die Gattungen der Rhodophyceen." C.W.K. Gleerups, Lund.

Lam, H. J. (1959). Taxonomy: general principles and angiosperms. *In* "Vistas in Botany" (W. B. Turrill, ed.), pp. 3–75. Pergamon Press, London.

Lamouroux, [J. V. F.] (1805). "Dissertations sur plusieurs espèces de Fucus, . . ." Raymond Noubel, Agen.

Lamouroux, [J. V. F.] (1813). Essai sur les genres de la famille de Thalassiophytes non articulées. *Ann. Mus. natn. Hist. nat., Paris* **20**, 21–47, 115–139, 267–293.

Lewin, R. A. (1974). Biochemical taxonomy. *In* "Algal Physiology and Biochemistry" (W. D. P. Stewart, ed.), pp. 1–39. Blackwell Scientific Publications, Oxford, London, Edinburgh, Melbourne.

Lewin, R. A. (1976). Prochlorophyta as a proposed new division of algae. *Nature, Lond.* **281**, 697.

Linnaeus, C. (1753). "Species plantarum, . . .", Tom. 1, 2. Impensis Laurentii Salvii, Holmii.

MALLERY, C. H. and RICHARDSON, [W.] N. (1971). Disc gel electrophoresis of proteins of Rhodophyta. *Proc. int. Seaweed Symp.* **7**, 292–300.

NISIZAWA, K., FUJIBAYASHI, S. and HABE, H. (1969). On the fine structure of alginic acid. *Proc. int. Seaweed Symp.* **6**, 553–565.

OLTMANNS, F. (1904). "Morphologie und Biologie der Algen", Bd. 1–2. Gustav Fischer, Jena.

PAPENFUSS, G. F. (1946). Proposed names for the phyla of algae. *Bull. Torrey bot. Club* **73**, 217–218.

PAPENFUSS, G. F. (1951). Phaeophyta. *In* "Manual of Phycology . . ." (G. M. Smith, ed.), pp. 119–158. Chronica Botanica Company, Waltham (Mass.).

PAPENFUSS, G. F. (1955). Classification of the algae. *In* "A Century of Progress in the Natural Sciences, 1853–1953", pp. 115–224. Calif. Acad. Sci., San Francisco.

PAPENFUSS, G. F. (1964). An evaluation of the present system of classification of the Florideophycidae into orders. *Abstr. int. bot. Congr.* **10**, 105.

PAPENFUSS, G. F. (1966). A review of the present system of classification of the Florideophycidae. *Phycologia* **5**, 247–255.

PARKE, M. and DIXON, P. S. (1964). A revised check-list of British marine algae. *J. mar. biol. Ass. U.K.* **44**, 499–542.

PARKE, M. and DIXON, P. S. (1976). Check-list of British marine algae—third revision. *J. mar. biol. Ass. U.K.* **56**, 527–594.

PASCHER, A. (1931). Systematische Ubersicht über die mit Flagellaten in Zusammenhang stehenden Algenreihen und Versuch einer Einreihung dieser Algenstämme in die Stämme des Pflanzenreiches. *Beih. Bot. Centralbl.* **48**, 317–332.

PRESCOTT, G. (1951). History of phycology. *In* "Manual of Phycology . . ." (G. M. Smith, ed.), pp. 1–11. Chronica Botanica Company, Waltham (Mass.).

RUPRECHT, F. J. (1850). Algae Ochotenses. . . . Tange des Ochotskischen Meeres. *In* "Sibirischer Reise" (A. T. von Middendorff, ed.), Botanik, 1(2), pp. 1(193)–243(435). Kaiserlichen Akademie der Wissenschaften, St Petersburg.

RUSSELL, G. (1973). The Phaeophyta: a synopsis of some recent developments. *Oceanogr. Mar. Biol. Ann. Rev.*, **11**, 45–88.

RUSSELL, G. and FLETCHER, R. L. (1975). A numerical taxonomic study of the British Phaeophyta. *J. mar. biol. Ass. U.K.* **55**, 763–784.

SACHS, J. VON (1875). "Text Book of Botany, Morphological and Physiological." Translated and annotated by A. W. Bennett and W. T. T. Dyer. Clarendon Press Oxford.

SACHS, J. VON (1882). "Textbook of Botany, Morphological and Physiological" (2nd edition) (S. H. Vines, ed.). Clarendon Press, Oxford.

SAVORY, T. H. (1962). "Naming the Living World—an Introduction to the Principles of Biological Nomenclature." English Universities Press, London.

SCAGEL, R. F. (1966). The Phaeophyceae in perspective. *Oceanogr. Mar. Biol. Ann. Rev.* **4**, 123–194.

SCHMITZ, F. (1883). Untersuchungen über dei Befruchtung der Florideen. *Sber. Acad. Wiss.* **1883**, 215–258.

SCHMITZ, F. (1889). Systematische Übersicht der bischer bekannten Gattungen der Florideen. *Flora Jena* **72**, 435–456.

SCHMITZ, F. and HAUPTFLEISCH, P. (1896). Rhodophyceae. *In* "Die natürlichen Pflanzenfamilien . . ." (by A. Engler and K. Prantl). Vol. 1. (2), pp. 298–306. Wilhelm, Engelmann, Leipzig.

SIMON-BICHARD-BRÉAUD, J. (1971). Un appareil cinétique dans le gamétocystes mâles de *Bonnemaisonia hamifera*, Hariot (Rhodophycée). *C.r. hebd. Séanc. Acad. Sci., Paris* **274**, 1485–1488.

SKUJA, H. (1939). Versuch einer systematischen Einteilung der Bangioideenoder Protoflorideen. *Acta Horti bot. Univ. latv.* **11**/**12**, 23–40.

SMITH, G. M. (1951). The classification of algae. *In* "Manual of Phycology . . ." (G. M. Smith, ed.), pp. 13–19. Chronica Botanica Company, Waltham (Mass.).

STACKHOUSE, J. (1797). "Nereis Britannica." Stackhouse, Bath.

STACKHOUSE, J. (1809). Tentamen marino-cryptogamicum. *Mém. Soc. Nat. Moscou* **2**, 50–97.

STOLOFF, L. (1962). Algal classification—an aid to improved industrial utilization. *Econ. Bot.* **16**, 86–94.

STOLOFF, L. and SILVA, P. (1957). An attempt to determine possible taxonomic significance of the properties of water-extractable polysaccharides in red algae. *Econ. Bot.* **11**, 327–330.

STRAIN, H. H. (1951). The pigments of algae. *In* "Manual of Phycology . . ." (G. M. Smith, ed.), pp. 243–262. Chronica Botanica Company, Waltham (Mass.).

THOMAS, M. (1938). Der Formenkreis von *Chondrus crispus* und seine ökologische Bedingtheit. *Hedwigia* **77**, 137–210.

THURET, G. and BORNET, E. (1878). "Études phycologiques." Libraire de l'Académie de Médecine, Paris.

TURRILL, W. B. (1938). The expansion of taxonomy. *Biol. Revs* **13**, 342–373.

TURRILL, W. B. (1964). Preface. "Vistas in Botany", Vol. IV, vii–xi. Pergamon Press, London.

WETTSTEIN, R. VON (1901). "Handbuch der systematischen Botanik." I. Band. Frenz Deuticke, Leipzig and Wien.

WHITE, E. B. and BONEY, A. D. (1969). Experiments with some endophytic and endozoic *Acrochaetium* species. *J. exp. mar. Biol. Ecol.* **3**, 246–274.

WHITE, E. B. and BONEY, A. D. (1971) ["1970"]. *In situ* and *in vitro* studies on some endophytic and endozoic *Acrochaetium* species. *Nova Hedwigia* **19**, 841–881.

WIDDOWSON, I. B. (1971). A statistical analysis of variation in the brown alga *Alaria*. *Syesis* **4**, 125–143.

WYNNE, M. J. and LOISEAUX, S. (1976). Recent advances in life history studies of the Phaeophyta. *Phycologia* **15**, 435–452.

YAPHE, W. (1957). The use of agarase from *Pseudomonas atlantica* in the identification of agar in marine algae (Rhodophyceae). *Can. J. Microbiol.* **3**, 987.

YAPHE, W. (1959). The determination of κ-carrageenan as a factor in the classification of the Rhodophyceae. *Can. J. Bot.* **37**, 751–757.

YAPHE, W. and DUCKWORTH, M. (1971). The relationship between structures and biological properties of agars. *Proc. int. Seaweed Symp.* **7**, 15–22.

YOUNG, D. N. (1977). A note on the absence of flagellar structures in spermatia of *Bonnemaisonia*. *Phycologia* **16**, 219–222.

ZANEVELD, J. S. (1959). The utilization of marine algae in tropical South and East Asia. *Econ. Bot.* **13**, 89–131.

2 | Haemagglutinins in British Marine Algae and their Possible Taxonomic Value

G. BLUNDEN[1], D. J. ROGERS[1] and W. F. FARNHAM[2]

School of Pharmacy[1] and Department of Biological Sciences[2], Portsmouth Polytechnic, England

Abstract: Extracts prepared from over 100 species of freshly collected British marine algae have been examined for the presence of haemagglutinins. Tests were conducted in 0·9% sodium chloride solution both with and without the addition of bovine albumin and also in 0·9% sodium chloride solution using papain-treated erythrocytes. Compared to the results obtained in sodium chloride solution alone, the addition of bovine albumin increased the number of species giving positive haemagglutination reactions. The use of papain-treated cells also increased the number of extracts giving positive results. The extracts of certain species reacted in the albumin medium only, the most pronounced example of this being *Solieria chordalis*. The strongest haemagglutination reactions were produced by the extracts of *Codium fragile*, *Cystoseira tamariscifolia*, *Fucus spiralis*, *Halidrys siliquosa*, *Pelvetia canaliculata*, *Spongonema tomentosum*, *Stictyosiphon tortilis*, *Carpomitra costata*, *Ptilota plumosa* and *Solieria chordalis*. Certain orders, for example Fucales, are characterized by most species containing haemagglutinins, whereas others, for example Laminariales, are characterized by most species lacking haemagglutinins.

INTRODUCTION

Gold and Balding (1975) reviewed the occurrence of haemagglutinins in plants and animals. Many of these investigations had been carried out with the intention of detecting substances possessing blood-group specificity rather than for any taxonomic purpose. It has been known for some time that haemagglutinins could be extracted from angiosperms, particularly from their seeds. The Leguminosae have proved to be rich sources of haemagglutinins (Bird, 1959), and Toms and Weston (1971) suggested that phytohaemagglutinins could be useful in chemotaxonomy.

Systematics Association Special Volume No. 10, "Modern Approaches to the Taxonomy of Red and Brown Algae", edited by D. E. G. Irvine and J. H. Price, 1978, pp. 21–45, Academic Press, London and New York.

Barrett and Howe (1968) screened 22 species of lichens, mainly belonging to the Lecanorales, and found that all possessed some haemagglutinic activity. The possibility of both interspecific and intraspecific differences was indicated. In the basidiomycete fungi, some clustering of haemagglutinic activity has been indicated at the family or genus level (Gold and Balding, 1975) and a similar clustering seems to occur in the mollusc family Helicidae, which is characterized by anti-A specificity (Gold and Balding, 1975).

The presence of haemagglutinins in marine algae was first reported by Boyd *et al.* (1966), who examined 24 tropical species. They found that most species were inactive and those that were active reacted only weakly. The cell sap from *Codium isthmocladum* Vickers, *Sargassum natans* (L.) J. Meyen and *Turbinaria turbinata* (L.) O. Kuntze agglutinated A, B and O erythrocytes. The three species of *Dictyota* examined, *Dictyopteris delicatula* Lamour., *Sargassum rigidulum* Kütz. and *Padina vickersiae* Hoyt, reacted preferentially with human A and O erythrocytes and were considered, therefore, to show A+H specificity. Detailed specificity studies have not been reported on any of the species listed. We have previously tested over 100 species, subspecies and varieties of British marine algae for haemagglutinic activity, but obtained positive results with only 19 (Blunden *et al.*, 1975). The activity of all the extracts giving positive results was low, the highest titre being obtained with *Ptilota plumosa*. This species was also the only one which showed human blood-group specificity, reacting preferentially with erythrocytes carrying the B antigen.

Rogers (1974) demonstrated that haemagglutinins from marine algae may be detected more readily in a bovine albumin medium or by pre-treatment of the erythrocytes with proteolytic enzymes. As a result, most of the British species examined earlier, as well as some which were not investigated, have been tested for haemagglutinic activity using these more sensitive methods. Moreover, instead of using mainly dry material as earlier, fresh samples were used.

The few haemagglutinins that have been examined biochemically were found to be composed principally of protein; as "protein taxonomy", including serology, has been applied with varying degrees of success to the angiosperms (Gibbs, 1974, pp. 384–386) it was considered that the distribution of haemagglutinins in marine algae may have taxonomic significance.

MATERIALS AND METHODS

Materials

The collection areas and dates of collection of the algal samples are listed in Table I. All species, after collection, were placed in insulated flasks containing ice and the plant material was examined immediately on arrival in the laboratory or was frozen rapidly at −20°C and stored at this temperature until tested. Each collection was carefully sorted to reduce contamination by extraneous material, such as epiphytes. The nomenclature of the algae follows that of Parke and Dixon (1976).

Extraction

Collected material was first washed in 0·9% sodium chloride solution to remove seawater and then 2 g was ground to a paste in a mortar containing 1 g of acid-washed sand and 6 ml 0·9% sodium chloride solution. The extracts were centrifuged to remove gross particulate matter and the supernatant fluid was collected.

TESTS FOR HAEMAGGLUTINIC ACTIVITY

(*a*) Four series of doubling dilutions from 1 : 2 to 1 : 2048 were prepared in 0·9% sodium chloride solution from the supernatant fluid of the extract, each dilution having a final volume of 0·1 ml. An equal volume of a 2% washed suspension of erythrocytes of one blood-group was added to each tube of one series and left for 2 h at 18°C before examination for haemagglutination. The erythrocytes used were from individuals with blood-groups A_1 rr, B rr, O rr and O R_1R_2. The greatest dilution showing agglutinated erythrocytes was recorded as the titre of the extract against the blood-group of erythrocytes contained in that tube. The dilutions showing no apparent agglutination were examined microscopically.

(*b*) *Bovine albumin tests.* Doubling dilutions of extracts were prepared as described above and erythrocytes were added to them. The tubes were then allowed to stand at 18°C for 1 h. These tubes were then inclined carefully without disturbing the erythrocytes which had settled to the bottom and 0·1 ml 30% bovine albumin was allowed to run down the inside of each tube. The tubes were allowed to stand at 18°C for a further 30 min before examination for agglutination. The dilutions showing no apparent agglutination were examined microscopically.

(*c*) *Enzyme treatment of erythrocytes.* Enzyme treatment of erythrocytes was carried out with papain according to the method described by Mollison (1972). After this treatment, 0·1 ml of a 4% suspension of appropriate erythrocytes in 0·9% sodium chloride solution was added to a series of doubling dilutions of extract. The tubes were allowed to stand at 18°C for 2 h before examination for agglutination was carried out. The dilutions showing no agglutination were examined microscopically.

TABLE 1. Haemagglutinic activities of British seaweeds

Species	Collection area and date	Titres of extract with blood group								
		0·9% sodium chloride solution			albumin addition			papain pre-treatment of erythrocytes		
		A	B	O	A	B	O	A	B	O
CHLOROPHYCEAE										
ULVALES										
Ulvaceae										
Enteromorpha compressa (L.) Grev.	Isle of Shuna, Argyll; July, 1975	–	–	–	–	–	–	–	–	–
E. intestinalis (L.) Link	Kimmeridge, Dorset; April, 1976	–	–	–	–	–	–	–	–	1 : 2
E. compressa+*E. intestinalis*	Bull Bay, Anglesey; Oct., 1975	–	–	–	–	1 : 2	1 : 8	1 : 2	1 : 2	1 : 8
	Kimmeridge, Dorset; April, 1976	–	–	–	–	–	–	–	–	–
Ulva lactuca L.	Southsea, Hants; June, 1975	–	–	–	1 : 2	1 : 8	1 : 4	1 : 64	1 : 64	1 : 128
	Lepe, Hants; July, 1975	–	–	–	1 : 2	–	1 : 2	1 : 16	1 : 32	1 : 16
CLADOPHORALES										
Cladophoraceae										
Chaetomorpha capillaris (Kütz.) Børg.	Isle of Shuna, Argyll; July, 1975	1 : 2	–	1 : 2	1 : 2	–	1 : 2	1 : 8	1 : 4	1 : 4

– no haemagglutination observed

C. linum (O.F. Müll.) Kütz.	Kimmeridge, Dorset; April, 1976	—	—	—	—	—	—	—	—	—
Cladophora pellucida (Huds.) Kütz.	Kimmeridge, Dorset; April, 1976	—	—	—	—	—	—	—	—	—
C. rupestris (L.) Kütz.	Isle of Shuna, Argyll; July, 1975	—	—	—	1 : 16	1 : 8	1 : 32	—	—	—
CODIALES Codiaceae *Codium fragile* (Sur.) Hariot subsp. *atlanticum* (Cotton) Silva	Isle of Shuna, Argyll; July, 1975	1 : 4	1 : 4	1 : 4	1 : 256	1 : 16	1 : 32	1 : 256	1 : 256	1 : 128
C. fragile (Sur.) Hariot subsp. *tomentosoides* (Goor) Silva	Lulworth, Dorset; Sept., 1975	1 : 16	1 : 32	1 : 64	1 : 64	1 : 32	1 : 32	1 : 512	1 : 512	1 : 512
CHAROPHYCEAE CHARALES Characeae *Chara vulgaris* L., contaminated with a small amount of *Tolypothrix*? *lanata* (Desv.) Wartm.	Freshwater quarry pit, Hayling Island, Hants; Aug., 1975	—	—	—	1 : 8	1 : 2	1 : 4	1 : 32	—	—
Lamprothamnium papulosum (Wallr.) J. Gr.	Lymington, Hants; Sept., 1976	—	—	—	—	—	—	—	—	—

TABLE I—*continued*

Species	Collection area and date	Titres of extract with blood group								
		0·9% sodium chloride solution			albumin addition			papain pre-treatment of erythrocytes		
		A	B	O	A	B	O	A	B	O
PHAEOPHYCEAE										
ECTOCARPALES										
Ectocarpaceae										
Pilayella littoralis (L.) Kjellm.	Cowes, I.O.W.; March, 1975	–	–	–	–	–	–	–	–	–
Spongonema tomentosum (Huds.) Kütz.	Isle of Shuna, Argyll; July, 1975	–	–	–	–	–	–	1 : 256	1 : 512	1 : 512
Elachistaceae										
Elachista fucicola (Vell.) Aresch.	Aberffraw, Anglesey; Oct., 1975	–	–	–	–	–	–	–	–	–
Corynophlaeaceae										
Leathesia difformis (L.) Aresch.	Isle of Shuna, Argyll July, 1975	–	–	–	–	–	–	–	–	–
Chordariaceae										
Chordaria flagelliformis (O. F. Müll.) C. Ag.	Lulworth, Dorset; Sept., 1975	–	–	–	–	–	–	–	–	–
Spermatochnaceae										
Stilophora rhizodes (Turn.) J. Ag.	Loch Ine, Co. Cork; Sept., 1975	–	–	–	–	–	–	–	–	–

Striariaceae *Stictyosiphon tortilis* (Rupr.) Reinke	Kimmeridge, Dorset; April, 1976	1 : 2	1 : 2	1 : 2	1 : 2	1 : 2	1 : 2	1 : 512	1 : 256	1 : 512
Punctariaceae *Asperococcus fistulosus* (Huds.) Hook.	Isle of Shuna, Argyll; July, 1975	—	—	—	—	—	—	—	—	—
Dictyosiphonaceae *Dictyosiphon foeniculaceus* (Huds.) Grev.	Isle of Shuna, Argyll; July, 1975	—	—	—	—	—	—	1 : 1	—	1 : 1
Scytosiphonaceae *Colpomenia peregrina* Sauv.	Bembridge, I.O.W., June, 1975	—	—	—	—	—	—	—	—	—
Scytosiphon lomentaria (Lyngb.) Link	Lulworth, Dorset; Sept., 1975	—	—	—	—	—	—	—	—	—
DESMARESTIALES Desmarestiaceae *Desmarestia aculeata* (L.) Lamour.	Southsea, Hants.; June, 1975	—	—	—	1 : 2	1 : 8	1 : 4	1 : 8⁺	1 : 8⁺	1 : 4⁺
D. ligulata (Lightf.) Lamour.	Solent, Hants.; June, 1976	—	—	—	—	—	—	—	—	—
Sporochnaceae *Carpomitra costata* (Stackh.) Batt.	Plymouth, Devon; Dec., 1975	—	—	—	—	—	1 : 2	1 : 128	1 : 512	1 : 256

+ haemolysis observed.

TABLE I—*continued*

Species	Collection area and date	Titres of extract with blood group								
		0·9% sodium chloride solution			albumin addition			papain pre-treatment of erythrocytes		
		A	B	O	A	B	O	A	B	O
LAMINARIALES										
Chordaceae										
Chorda filum (L.) Stackh.	Isle of Shuna, Argyll; July, 1975	—	—	—	—	—	—	—	—	—
Laminariaceae										
Laminaria digitata (Huds.) Lamour.	Southsea, Hants.; June, 1975	—	—	—	—	—	—	—	—	—
	Bembridge, I.O.W., Sept., 1975	—	—	—	—	—	—	—	—	—
L. hyperborea (Gunn.) Fosl.	Kimmeridge, Dorset; April, 1976	—	—	—	—	—	—	—	—	—
L. ochroleuca Pyl.	Scilly Isles, Cornwall; July, 1976	—	—	—	—	—	—	—	—	—
L. saccharina (L.) Lamour.	Littlehampton, Sussex; May, 1975	—	—	—	—	—	—	—	—	—
	Southsea, Hants.; June, 1975	—	—	—	—	—	—	—	—	—
	Shoreham, Sussex; Dec., 1976	—	—	—	—	1 : 2	—	1 : 16	1 : 8	1 : 8

Saccorhiza polyschides (Lightf.) Batt.	Kimmeridge, Dorset; April, 1975	—	—	—	—	—	—	—	—	—
SPHACELARIALES Cladostephaceae *Cladostephus spongiosus* (Huds.) C. Ag. f. *spongiosus*	Bembridge, I.O.W.; Oct., 1976	—	—	—	—	—	—	—	—	—
DICTYOTALES Dictyotaceae *Dictyopteris membranacea* (Stackh.) Batt.	Kimmeridge, Dorset; Aug., 1976	—	—	—	—	—	—	1 : 2	1 : 4	1 : 2
Dictyota dichotoma (Huds.) Lamour.	Isle of Shuna, Argyll; July, 1975	—	—	—	—	—	—	—	—	—
Padina pavonia (L.) Lamour.	Bembridge, I.O.W.; Oct., 1976	—	—	—	—	—	—	—	—	—
FUCALES Fucaceae *Ascophyllum nodosum* (L.) Le Jol.	Bull Bay, Anglesey; Oct., 1975	1 : 32	1 : 16	1 : 16	1 : 32	1 : 16	1 : 32	1 : 32	1 : 32	1 : 32
Fucus ceranoides L.	Isle of Shuna, Argyll; July, 1975	—	—	—	—	—	—	1 : 32	1 : 16	1 : 32
F. serratus L.	Southsea, Hants.; June, 1975	1 : 2	1 : 2	1 : 2	1 : 4	1 : 4	1 : 4	1 : 4	1 : 8	1 : 8
F. spiralis L.	Southsea, Hants.; June, 1975	1 : 2	1 : 4	1 : 4	1 : 2	1 : 4	1 : 4	1 : 256	1 : 128	1 : 128

TABLE I—*continued*

Species	Collection area and date	Titres of extract with blood group								
		0·9% sodium chloride solution			albumin addition			papain pre-treatment of erythrocytes		
		A	B	O	A	B	O	A	B	O
F. vesiculosus L.	Emsworth, Hants.; Sept., 1975	1 : 8	1 : 16	1 :8	–	–	1 : 2	1 : 8	1 : 4	1 : 32
Pelvetia canaliculata (L.) Dcne et Thur.	Isle of Shuna, Argyll; July, 1975	–	1 : 2	–	–	–	–	1 : 64	1 : 128	1 : 256
Himanthaliaceae *Himanthalia elongata* (L.) S. F. Gray	Isle of Shuna, Argyll; July, 1975	–	–	–	1 : 32	1 : 16	1 : 16	1 : 64	1 : 64	1 : 64
Cystoseiraceae *Bifurcaria bifurcata* Ross	Plymouth, Devon; Nov., 1976	–$^{+}$	–$^{+}$	–$^{+}$	–$^{+}$	–	–$^{+}$	1 : 128^{+}	1 : 256^{+}	1 : 128^{+}
Cystoseira baccata (Gmel.) S. Silva	Kimmeridge, Dorset; June, 1976	1 : 1	1 : 1	1 : 1	1 : 64	1 : 64	1 : 128	1 : 32	1 : 128	1 : 64
C. foeniculacea (L.) Grev.	Kimmeridge, Dorset; July, 1976	–	–	–	1 : 4	1 : 1	–	–	–	–
	The Fleet, Dorset Oct., 1976	–	–	–	–	–	–	–	–	–
C. nodicaulis (With.) Roberts	Kimmeridge, Dorset; July, 1976	–	–	–	–	–	–	–	–	–
	Bembridge, I.O.W.; Oct., 1976	–	–	–	–	–	–	–	–	–

C. tamariscifolia (Huds.) Papenf.	Kimmeridge, Dorset; June, 1976	1 : 1	1 : 1	—	—	—	—	1 : 128	1 : 256	1 : 128
Halidrys siliquosa (L.) Lyngb.	Isle of Shuna, Argyll; July, 1975	1 : 8	1 : 8	1 : 8	1 : 8	1 : 8	1 : 8	1 : 1024	1 : 1024	1 : 512
	Plymouth, Devon; Feb., 1976	1 : 16	1 : 16	1 : 32	1 : 32	1 : 8	1 : 4	1 : 1024	1 : 1024	1 : 1024
Sargassaceae										
Sargassum muticum (Yendo) Fensholt	Bembridge, I.O.W., Sept., 1975	—	—	—	—	—	—	—	—	—
	Bembridge, I.O.W.; Oct., 1976	—	—	—	—	—	—	—	—	—
FLORIDEOPHYCEAE										
NEMALIALES										
Acrochaetiaceae										
Audouinella floridula (Dillw.) Woelkering	Isle of Shuna, Argyll; July, 1975	—	—	—	—	—	—	—	—	—
Gelidiaceae										
Gelidium latifolium (Grev.) Born. et Thur.	Kimmeridge, Dorset; April, 1976	—	—	—	—	—	—	—	—	—
G. pusillum (Stackh.) Le Jol.	Isle of Shuna, Argyll; July, 1975	—	—	—	—	—	—	—	—	—
	Bembridge, I.O.W.; June, 1976	—	—	—	—	—	—	—	—	—
Pterocladia capillacea (S. G mel.) Born. et Thur.	Loch Ine, Co. Cork; Sept., 1975	—	—	—	—	—	—	—	—	—
	Near Wembury, Devon; Nov., 1976	—	—	—	1 : 8	1 : 8	1 : 8	—	—	—

TABLE I—*continued*

Species	Collection area and date	Titres of extract with blood group								
		0·9% sodium chloride solution			albumin addition			papain pre-treatment of erythrocytes		
		A	B	O	A	B	O	A	B	O
Chaetangiaceae										
Scinaia forcellata Biv.	Bembridge, I.O.W.; Sept., 1975	—	—	—	—	—	—	—	—	—
GIGARTINALES										
Furcellariaceae										
Furcellaria lumbricalis (Huds.) Lamour.	Isle of Shuna, Argyll; July, 1975	—	—	—	—	—	—	—	—	—
	Shanklin, I.O.W.; March, 1976	—	—	—	—	—	—	—	—	—
Solieriaceae										
Solieria chordalis (C. Ag.) J. Ag.	Falmouth, Cornwall; April, 1976	—	—	—	1 : 2048	1 : 2048	1 : 1024	—	—	—
	The Fleet, Dorset; Aug., 1976	—	—	—	1 : 256	1 : 256	1 : 32	—	—	—
Polyideaceae										
Polyides rotundus (Huds.) Grev.	Kimmeridge, Dorset; May, 1976	1 : 8	1 : 8	1 : 4	1 : 16	1 : 16	1 : 64	1 : 32	1 : 16	1 : 32
Rhabdoniaceae										
Catenella caespitosa (With.) *in* Dixon et L. Irvine	Emsworth, Sussex; Oct., 1976	—	—	—	—	—	—	—	—	—

Rhodophyllidaceae										
Calliblepharis ciliata (Huds.) Kütz.	Kimmeridge, Dorset; April, 1976	–	–	–	–	–	–	–	–	–
	Bracklesham Bay, Sussex; May, 1976	–	–	–	–	–	–	–	–	–
C. jubata (Good. et Woodw.) Kütz.	Kimmeridge, Dorset; April, 1976	–	–	–	–	–	–	–	–	–
Cystoclonium purpureum (Huds.) Batt.	Southsea, Hants.; June, 1975	–	–	–	1 : 4	1 : 8	1 : 8	1 : 8	1 : 2	1 : 4
	Isle of Shuna, Argyll; July, 1975	–	–	–	1 : 8	1 : 2	1 : 8	–	–	–
	Kimmeridge, Dorset; June, 1976	–	–	–	1 : 1	1 : 1	1 : 1	–	–	–
Plocamiaceae										
Plocamium cartilagineum (L.) Dixon	Kimmeridge, Dorset; May, 1976	–	–	–	–	–	–	–	–	–
Sphaerococcaceae										
Sphaerococcus coronopifolius Stackh.	Bembridge, I.O.W.; Sept., 1975	–	–	–	–	–	–	1 : 8	1 : 8	1 : 8
Gracilariaceae										
Gracilaria bursa-pastoris (S. Gmel.) Silva	The Fleet, Dorset; July, 1976	–	–	–	1 : 8	1 : 8	1 : 32	1 : 32	1 : 32	1 : 32
G. verrucosa (Huds.) Papenf.	The Fleet, Dorset; July, 1976	–	–	–	–	–	–	1 : 8	1 : 8	1 : 8

Table I—*continued*

Species	Collection area and date	Titres of extract with blood group								
		0·9% sodium chloride solution			albumin addition			papain pre-treatment of erythrocytes		
		A	B	O	A	B	O	A	B	O
Phyllophoraceae *Ahnfeltia plicata* (Huds.) Fries	Kimmeridge, Dorset; April, 1976	—	—	—	—	—	—	—	—	—
Gymnogongrus crenulatus (Turn.) J. Ag. (= *G. norvegicus* (Gunn.) J. Ag.)	Bembridge, I.O.W.; July, 1976	—	—	—	—	1 : 4	—	1 : 4	1 : 64	1 : 2
Phyllophora crispa (Huds.) Dixon	Chapman's Pool, Dorset; June, 1976	—	—	—	1 : 1	—	1 : 2	—	—	—
	Kimmeridge, Dorset; April, 1976	—	—	—	—	—	—	—	—	—
P. pseudoceranoides (S. G. Gmel.) Newr. et A. R. A. Taylor	Chapman's Pool, Dorset; May, 1976	—	—	—	—	—	—	—	—	—
Gigartinaceae *Chondrus crispus* Stackh.	Southsea, Hants.; June, 1975	—	—	—	—	—	—	—	—	—
Gigartina stellata (Stackh. in With.) Batt.	Solent, Hants.; July, 1975	—	—	—	—	—	—	—	—	—
G. teedii (Roth) Lamour.	Near Wembury, Devon; Nov., 1976	—	—	—	1 : 2	1 : 2	1 : 2	—	—	—

CRYPTONEMIALES Corallinaceae *Corallina officinalis* L.	Isle of Shuna, Argyll; July, 1975	—	—	—	—	—	—	—	—	—
Jania rubens (L.) Lamour.	Bembridge, I.O.W.; Sept., 1975	—	—	—	—	—	—	—	—	—
Lithophyllum incrustans Phil.	Kimmeridge, Dorset; April, 1976	—	—	—	—	—	—	—	—	—
Phymatolithon calcareum (Pall.) Adey et McKibbin	Falmouth, Cornwall; April, 1976	—	—	—	—	—	—	—	—	—
Dumontiaceae *Dilsea carnosa* (Schmidel) O. Kuntze	Bracklesham Bay, Sussex; May, 1976	—	—	—	—	—	—	—	—	—
Dumontia incrassata (O. F. Müll.) Lamour.	Kimmeridge, Dorset; April, 1976	—	—	—	—	—	—	—	—	—
Cryptonemiaceae *Grateloupia doryphora* (Mont.) Howe	Emsworth, Hants.; Sept., 1976	—	—	—	—	—	—	—	—	—
G. filicina (Lamour.) C. Ag. var. *luxurians* A. et E. S. Gepp	Southsea, Hants.; May, 1975	—	—	—	—	—	—	—	—	—
Kallymeniaceae *Callophyllis* ? *flabellata* Crouan frat.	Solent, Hants.; May, 1976	—	—	—	—	—	—	—	—	—

Table I—*continued*

Species	Collection area and date	Titres of extract with blood group								
		0·9% sodium chloride solution			albumin addition			papain pre-treatment of erythrocytes		
		A	B	O	A	B	O	A	B	O
Kallymenia reniformis (Turn.) J. Ag.	Scilly Isles, Cornwall; July, 1976	—	—	—	—	—	—	—	—	—
PALMARIALES										
Palmariaceae										
Palmaria palmata (L.) O. Kuntze	Southsea, Hants.; June, 1975	—	—	—	—	—	—	1 : 2	1 : 2	1 : 2
	Bembridge, I.O.W.; Sept., 1975	—	—	—	—	—	—	—	1 : 2	1 : 2
RHODYMENTIALES										
Champiaceae										
Chylocladia verticillata (Lightf.) Bliding	Kimmeridge, Dorset; June, 1976	—	—	—	1 : 64	1 : 4	1 : 32	1 : 8	1 : 4	1 : 16
Gastroclonium ovatum (Huds.) Papenf.	Mewstone, Devon; Nov., 1976	—	—	—	—	—	—	—	—	—
Lomentaria articulata (Huds.) Lyngb.	Bembridge, I.O.W.; June, 1976	—	—	—	—	—	—	—	—	—
CERAMIALES										
Ceramiaceae										
Antithamnion cruciatum (C. Ag.) Näg.	Bembridge, I.O.W.; Oct., 1976	—	—	—	—	—	—	—	—	—

Callithamnion corymbosum (Sm.) Lyngb.	Portsmouth Harbour, Hants.; May, 1976	—	—	—	1 : 16	1 : 32	1 : 32	1 : 1	1 : 2	1 : 4
Ceramium echionotum J. Ag.	Scilly Isles, Cornwall; Oct., 1976	—	—	—	—	—	—	—	—	—
C. rubrum (Huds.) C. Ag.	Southsea, Hants.; June, 1975	—	—	—	—	—	—	—	—	—
	Bembridge, I.O.W.; Sept., 1975	—	—	—	—	—	—	—	—	—
Griffithsia corallinoides (L.) Batt.	Bembridge, I.O.W.; Sept., 1975	—	—	—	—	—	—	—	—	—
G. flosculosa (Ellis) Batt.	Southsea, Hants.; Aug., 1976	—	—	—	—	—	—	—	1 : 1	1 : 2
Halurus equisetifolius (Lightf.) Kütz.	Chapman's Pool, Dorset; May, 1976	—	—	—	—	—	—	—	—	—
Plumaria elegans (Bonnem.) Schmitz	Chapman's Pool, Dorset; March, 1975	1 : 2	—	1 : 2	1 : 64	1 : 64	1 : 64	1 : 32	1 : 64	1 : 64
Ptilota plumosa (Huds.) C. Ag.	Shetland Isles; July, 1973	—	1 : 16	—	1 : 2	1 : 64	1 : 2	1 : 2	1 : 128	1 : 2
	Near Oban, Argyll; Sept., 1974	—	1 : 4	—	1 : 8	1 : 256	1 : 2	—	1 : 128	—
	Bull Bay, Anglesey; Oct., 1975	1 : 4	1 : 64	1 : 16	1 : 32	1 : 1024	1 : 32	1 : 16	1 : 2048	1 : 64
	Aberffraw, Anglesey; Oct., 1975	—	1 : 64	1 : 4	1 : 4	1 : 512	1 : 8	1 : 8	1 : 2048	1 : 32
	Port Erin, I.O.M.; April, 1976	—	1 : 16	—	—	1 : 128	—	—	1 : 128	—

Table I—*continued*

Species	Collection area and date	Titres of extract with blood group								
		0·9% sodium chloride solution			albumin addition			papain pre-treatment of erythrocytes		
		A	B	O	A	B	O	A	B	O
Delesseriaceae										
Cryptopleura ramosa (Huds.) Kylin ex Newton	Bull Bay, Anglesey; Oct., 1975	—	—	—	—	—	—	—	—	—
Delesseria sanguinea (Huds.) Lamour.	Chapman's Pool, Dorset; May, 1975	—	—	—	—	1 : 1	1 : 1	1 : 2	1 : 2	1 : 2
Membranoptera alata (Huds.) Stackh.	Chapman's Pool, Dorset; June, 1975	—	—	—	1 : 4	1 : 4	1 : 8	—	1 : 1	—
	Bull Bay, Anglesey; Oct., 1975	1 : 4	1 : 4	1 : 4	1 : 16	1 : 16	1 : 16	1 : 4	1 : 32	1 : 2
Phycodrys rubens (L.) Batt.	Bull Bay, Anglesey; Oct., 1975	1 : 16	1 : 16	1 : 16	1 : 64	1 : 64	1 : 64	1 : 16	1 : 16	1 : 8
Polyneura gmelinii (Lamour.) Kylin	Solent, Hants.; June, 1976	—	—	—	—	—	—	1 : 2	1 : 2	1 : 4
P. hilliae (Grev.) Kylin	Scilly Isles, Cornwall; July, 1976	—	—	—	1 : 2	1 : 2	1 : 2	1 : 8	1 : 8	1 : 16
Rhodomelaceae										
Brongniartella byssoides (Good. et Woodw.) Schmitz	Solent, Hants.; June, 1976	—	—	—	—	—	—	—	—	—
Chondria coerulescens (J. Ag.) Falkenberg	Farlington Marsh, Hants.; Oct., 1975	—	—	—	—	—	—	1 : 8	1 : 4	—

C. dasyphylla (Woodw.) C. Ag.	Kimmeridge, Dorset; April, 1976	–	–	–	–	–	–	–	–	–
Halopitys incurvus (Huds.) Batt.	Bembridge, I.O.W.; Sept., 1975	–	–	–	–	–	–	–	–	–
Laurencia hybrida (DC.) Lenorm. ex Duby	Kimmeridge, Dorset; April, 1976	–	–	–	–	–	–	–	–	–
L. obtusa (Huds.) Lamour.	Bembridge, I.O.W.; Sept., 1975	–	–	–	–	–	–	–	–	–
L. pinnatifida (Huds.) Lamour.	Bembridge, I.O.W.; Sept., 1975	–	–	–	–	–	–	–	–	–
Polysiphonia elongata (Huds.) Spreng.	Near Wembury, Devon; Nov., 1976	–	–	–	–	–	–	–	–	–
P. lanosa (L.) Tandy	Bull Bay, Anglesey; Oct., 1975	1 : 1	–	1 : 1	–	–	–	–	–	1 : 1
Pterosiphonia complanata (Clem.) Falkenb.	Scilly Isles, Cornwall; July, 1976	–	–	–	–	–	–	–	–	–
Rhodomela confervoides (Huds.) Silva	Bembridge, I.O.W.; Oct., 1976	–	–	–	–	–	–	–	–	–
BANGIOPHYCEAE BANGIALES Bangiaceae *Porphyra linearis* Grev.	Kimmeridge, Dorset; April, 1976	–	–	–	–	–	–	–	–	–
P. umbilicalis (L.) J. Ag.	Southsea, Hants.; Oct., 1975	–	–	–	–	–	–	–	–	–

RESULTS AND DISCUSSION

As we reported earlier (Blunden *et al.*, 1975), the 0·9% sodium chloride extracts of most of the seaweeds tested did not show haemagglutinic activity. Of those that gave a positive result, the highest titres were achieved with *Ptilota plumosa* and *Codium fragile* subsp. *tomentosoides*, for which values of 1 : 64 were obtained. For the extracts of other species which were positive, low titres of 1 : 1 to 1 : 8 were most commonly recorded. The results achieved in this present survey are very similar to those of the earlier one, except that negative results were obtained this time with *Ulva lactuca, Chordaria flagelliformis, Desmarestia aculeata, D. ligulata, Dictyosiphon foeniculaceus, Fucus ceranoides* and *Furcellaria lumbricalis.* In the earlier survey, however, variable results were obtained with both *Ulva lactuca* and *Furcellaria lumbricalis* (as *F. fastigiata*) and only low activity was found in the other species, especially *Desmarestia aculeata, Dictyosiphon foeniculaceus* and *Fucus ceranoides.*

The species which did not show activity in the earlier screening, but for which 9·0% sodium chloride extracts produced haemagglutination in the present study are *Chaetomorpha capillaris, Cystoseira baccata, Halidrys siliquosa, Stictyosiphon tortilis, Membranoptera alata, Plumaria elegans* and *Polysiphonia lanosa.* However, from two collections of *M. alata*, the extract of one gave a positive result, but the other was negative. Of the other species, with the exception of *Halidrys siliquosa*, only very low titres of 1 : 1 or 1 : 2 were recorded.

As the *Polysiphonia lanosa* was found growing on *Ascophyllum nodosum*, it is possible that the very low activity of the extract of the former species may have been caused by a small quantity of *A. nodosum* being inadvertently retained in the sample.

The addition of 30% bovine albumin to the test medium increased the number of extracts that gave positive haemagglutination reactions and frequently significantly increased the titre of an extract in comparison to that obtained in 0·9% sodium chloride solution alone, for example with *Cystoseira baccata* (Table I).

The use in the haemagglutination test of erythrocytes that had previously been treated with papain also increased the number of algal species that gave positive reactions. The titres of extracts of species that had produced haemagglutination in the tests with untreated erythrocytes were usually higher when papain-treated cells were used. Particularly high titres were obtained with the extracts of *Codium fragile, Carpomitra costata, Cystoseira tamariscifolia, Fucus spiralis, Halidrys siliquosa, Pelvetia canaliculata, Spongonema tomentosum, Stictyosiphon tortilis* and *Ptilota plumosa.*

The extracts of a number of species agglutinated erythrocytes in the albumin medium alone and not in 0·9% sodium chloride solution either with or without papain treatment of the red cells. This effect was most pronounced with *Solieria chordalis,* which has recently been recorded for England by Farnham and Jephson (1977). The other species which produced similar results are *Cladophora rupestris, Cystoseira foeniculacea, Cystoclonium purpureum, Gigartina teedii, Phyllophora crispa* and *Pterocladia capillacea.* However, only weak activity was found in *Cystoseira foeniculacea, Gigartina teedii* and *Phyllophora crispa.* Moreover, with three different collections of *Cystoclonium purpureum,* the extracts of two agglutinated erythrocytes only in the albumin medium, whereas the extract of the third gave a positive reaction both in the albumin medium and, when papain-treated cells were used, in saline solution.

These haemagglutinins which react in albumin but fail to react with papain-treated cells are particularly interesting, since they may be reacting with erythrocyte-surface structures which are enzyme labile. These structures include the glycoprotein-carried blood-group antigens M, N, S, s, Fy^a and K (Issitt and Issitt, 1975; Judson and Anstee, 1977). This feature is currently under investigation.

Of the species whose extracts did not agglutinate erythrocytes in 0·9% sodium chloride solution in the present survey, but which had proved active in the earlier one, *Ulva lactuca, Desmarestia aculeata, Dictyosiphon foeniculaceus* and *Fucus ceranoides* all gave positive results when using one or both of the more sensitive procedures. In the present survey, the extract from one collection of *Membranoptera alata* caused haemagglutination in the 0·9% sodium chloride solution but the extract of a second collection did not. However, positive results were obtained with both extracts after the addition of bovine albumin or when papain-treated erythrocytes were used.

Most of the species found to cause haemagglutination appeared to be non-specific for human blood-group antigens, reacting equally with A_1 rr, B rr, O rr and O R_1R_2 erythrocytes. Exceptions to this were *Chara vulgaris* extract, which only reacted with papain-treated group A cells; *Chylocladia verticillata* extract, which appeared to have the unusual blood-group specificity anti A+H, as reported by Boyd *et al.* (1966) for some species of brown algae; *Ptilota plumosa* extract, which reacted preferentially with blood-group B erythrocytes; *Gymnogongrus crenulatus* extract, which also contained an agglutinin with B specificity. The blood-group specificity of *P. plumosa* has been examined by Blunden and Rogers (1976) and Rogers *et al.* (1977).

In this survey, both Rhesus (D) negative (rr) and Rhesus (D) positive (R_1R_2) erythrocytes were used. As expected, there was no significant difference in the haemagglutination produced with either the Rhesus (D) negative or the Rhesus

D positive erythrocytes by any of the algal extracts tested and hence the Rhesus characteristics of the red cells have not been listed in Table I.

The extracts of some species showed the phenomenon of prozoning, in which the level of agglutination increases as the extract is diluted, until a certain concentration is reached, after which the activity is reduced. This phenomenon was displayed by the extracts of *Cystoseira tamariscifolia*, *Halidrys siliquosa*, *Himanthalia elongata*, *Spongonema tomentosum* and *Stictyosiphon tortilis*. Prozoning was only observed when papain-treated erythrocytes were used.

The results presented in Table I are not by themselves of great taxonomic significance because there are no clear-cut patterns of haemagglutinic activity shown by most of the taxa tested. This is the case for the relatively few marine species of Chlorophyceae examined. Within the Ulvales, for example, *Enteromorpha compressa* gave negative results under all reaction conditions, as did *E. intestinalis*, except for low activity against group O erythrocytes after papain treatment. In addition, one out of two mixtures of the two species tested gave positive results, which were similar to, although much weaker than, those produced by two collections of *Ulva lactuca*. In the Cladophorales, *Chaetomorpha capillaris* showed weak, but fairly non-specific, haemagglutinic activity, whereas *C. linum* showed no activity. Of the two species of *Cladophora* tested, only *C. rupestris* gave a positive result and this was in the albumin medium alone. Extracts of *Codium fragile* (Codiales) agglutinated all three blood-groups, which is in agreement with the findings of Boyd *et al.* (1966) for *C. isthmocladum*. The two subspecies of *C. fragile* were tested separately, but there was no significant difference between the titres obtained.

Two species of Charophyceae were tested of which only the extract of *Chara vulgaris* produced haemagglutination. When papain-treated erythrocytes were used the extract appeared to have anti-A specificity, but further tests would be necessary to verify this.

There was generally a lack of haemagglutinic activity in extracts prepared from algae in the range of orders of Phaeophyceae examined. For some orders, for example Desmarestiales, only a few representatives could be tested and no reliable conclusions can be drawn. However, in other orders, more species were collected for testing.

The lower orders of brown algae cannot be separated on the basis of their haemagglutinic activity and so our results are not suggestive of any further reorganization of these groups, such as that recently carried out by Russell and Fletcher (1975) on the basis of a numerical taxonomic investigation. In the Ectocarpales, *Spongonema tomentosum* showed strong activity after papain treatment of the erythrocytes. Only *Stictyosiphon tortilis* showed a reaction under all test conditions; this was low, except after papain treatment of the erythrocytes. Apart from these two species and *Dictyosiphon foeniculaceus*, for which very low activity against papain-treated erythrocytes was recorded, all the species tested which are currently placed in the Ectocarpales gave negative results.

In the Dictyotales, only *Dictyopteris membranacea* extract produced haemagglutination, and that only after papain treatment of the erythrocytes, whereas Boyd *et al.* (1966) reported the unusual blood-group specificity A+H for five tropical members of this order.

Members of the Laminariales also showed no haemagglutinic activity, except for one collection out of three of *Laminaria saccharina*. In contrast, most members of the Fucales, which is the other advanced order of Phaeophyceae, showed marked haemagglutinic activity. This supports the generally held view that this order is well-defined and separate from the other orders of brown algae (Fritsch, 1945; Papenfuss, 1951; Nizamuddin, 1962). Haemagglutinic activity was most widespread in genera of the Fucaceae and in the one species of Himanthaliaceae; activity was more sporadic in members of the Cystoseiraceae. *Halidrys siliquosa* and *Cystoseira baccata* showed a general range of haemagglutinic activity like most of the Fucaceae, but *C. nodicaulis* and *C. foeniculacea* (two collections of each species) showed slight or no activity. *Sargassum muticum*, the only British representative of the Sargassaceae, was also inactive, although Boyd *et al.* (1966) found haemagglutinins in the cell sap of *S. natans* and *S. rigidulum*.

The increase in haemagglutinic activity shown by members of the Fucales when species in the Cystoseiraceae and Sargassaceae are compared with species in the Fucaceae provides additional confirmation to the contention of Manton (1964) and more recently Jensen (1974), that contrary to earlier opinion (Fritsch, 1943), the genus *Fucus* represents a specialized or derived type in the Fucales and that members of the Sargassaceae and Cystoseiraceae are the more primitive, since only the Fucaceae are characterized by general haemagglutinic activity, unlike virtually all the other brown algae.

As was suggested for the Phaeophyceae, the more advanced species of red algae appear to show more haemagglutinic activity. Only two species of Bangiophyceae were examined but these gave negative results, as did all the species of Nemaliales examined, except for one collection of *Pterocladia capillacea* and all Cryptonemiales (Florideophyceae). Haemagglutinic activity was found in some members of the remaining orders of Florideophyceae, which were Gigartinales, Palmariales, Rhodymeniales and Ceramiales; these can be regarded as being more advanced.

The Polyideaceae was originally placed by Kylin (1956) in the Cryptonemiales, but was later transferred by Papenfuss (1966) to the Gigartinales because of its undifferentiated auxiliary cell. The haemagglutinic reaction of *Polyides rotundus* is more consistent with other members of the Gigartinales, whereas no species of the Cryptonemiales tested showed any haemagglutinic activity.

Twenty-six species of Ceramiales were examined and some differences between the families can be discerned. Eleven species in seven genera of the Rhodomelaceae were tested and, apart from weak activity in extracts of *Chondria coerulescens* and *Polysiphonia lanosa*, negative results were obtained. However, of the six species of Delesseriaceae tested, only *Cryptopleura ramosa* showed no haemagglutinic activity.

Nine species of Ceramiaceae were tested and extracts of *Callithamnion corymbosum*, *Griffithsia flosculosa*, *Plumaria elegans* and *Ptilota plumosa* were shown to contain haemagglutinins. The strongest agglutination was produced by the extracts of *P. plumosa*. This species has often been confused in the past with *Plumaria elegans* (Cullinane and Murphy, 1976). However, the extracts of undried *Ptilota plumosa* preferentially agglutinate blood-group B erythrocytes, whereas the agglutination produced by the extract of undried *Plumaria elegans* was non-specific. The preferential reaction of *Ptilota plumosa* with blood-group B erythrocytes distinguishes this species from other members of the family. It would be of interest to examine other species of *Ptilota* and such related genera as *Neoptilota* for their haemagglutinic properties.

ACKNOWLEDGEMENTS

We thank Mr N. A. Jephson, Dr C. Barwell and Mr T. M. Blunden for the collection of some of the algal species tested. We thank also Dr J. M. Kain for the Port Erin collection of *Ptilota plumosa*.

REFERENCES

Barrett, J. T. and Howe, M. L. (1968). Haemagglutination and haemolysis by lichen extracts. *App. Microbiol.* **16**, 1137–1139.

Bird, G. W. G. (1959). Haemagglutinins in seeds. *Brit. med. Bull.* **15**, 165–168.

Blunden, G., Rogers, D. J. and Farnham, W. F. (1975). Survey of British seaweeds for hemagglutinins. *Lloydia* **38**, 162–168.

Blunden, G. and Rogers, D. J. (1976). An anti-B haemagglutinin from the red alga *Ptilota plumosa*. *In* "Food-Drugs From the Sea Proceedings 1974," pp. 252–257. Marine Technology Society, Washington.

Boyd, W. C., Almodóvar, L. R. and Boyd, L. C. (1966). Agglutinins in marine algae for human erythrocytes. *Transfusion (Philadelphia)* **6**, 82–83.

CULLINANE, J. P. and MURPHY, J. P. (1976). The distribution of *Ptilota plumosa* (Huds.) C. Ag. and *Plumaria elegans* (Bonnem.) Schm. in Ireland. *Scient. Proc. R. Dubl. Soc.* ser. A, **5**, 475–490.

FARNHAM, W. F. and JEPHSON, N. A. (1977). A survey of the maërl beds of Falmouth (Cornwall). *Br. phycol. J.* **12**, 119.

FRITSCH, F. E. (1943). Studies in the comparative morphology of the algae. III. Evolutionary tendencies and affinities among Phaeophyceae. *Ann. Bot.* **7**, 63–87.

FRITSCH, F. E. (1945). "The Structure and Reproduction of the Algae", Vol. II. Cambridge University Press, Cambridge.

GIBBS, R. D. (1974). "Chemotaxonomy of Flowering Plants", Vol. 1. McGill-Queens University Press, Montreal and London.

GOLD, E. R. and BALDING, P. (1975). "Receptor Specific Proteins, Plant and Animal Lectins". Excerpta Medica, Amsterdam.

ISSITT, P. D. and ISSITT, C. H. (1975). "Applied Blood-Group Serology", 2nd ed. Spectra Biochemicals, California.

JENSEN, J. B. (1974). Morphological studies in Cystoseiraceae and Sargassaceae (Phaeophyceae) with special reference to apical organisation. *Univ. Calif. Publs Bot.* **68**, 1–61.

JUDSON, P. A. and ANSTEE, D J. (1977) Comparative effect of trypsin and chymotrypsin on blood group antigens. *Med. Lab. Sci.*, **34**, 1–6.

KYLIN, H. (1956). "Die Gattungen der Rhodophyceen". CWK Gleerups Förlag, Lund.

MANTON, I. (1964). A contribution towards understanding of "the primitive fucoid". *New Phytol.* **63**, 244–254.

MOLLISON, P. L. (1972). "Blood Transfusion in Clinical Medicine", 5th edition. Blackwell Scientific Publications, Oxford.

NIZAMUDDIN, M. (1962). Classification and distribution of the Fucales. *Botanica mar.* **4**, 191–203.

PAPENFUSS, G. F. (1951). Problems in the classification of marine algae. *Svensk bot. Tidskr.* **45**, 4–11.

PAPENFUSS, G. F. (1966). A review of the present system of classification of the Florideophycidae. *Phycologia* **5**, 247–255.

PARKE, M. and DIXON, P. S. (1976). Check-list of British marine algae—third revision. *J. mar. biol. Ass. U.K.* **56**, 527–594.

ROGERS, D. J. (1974). "Investigations of haemagglutinins in seaweeds with special reference to *Ptilota plumosa*." M.Phil. thesis (C.N.A.A.), Portsmouth Polytechnic, 98 pp.

ROGERS, D. J., BLUNDEN, G. and EVANS, P. R. (1977). *Ptilota plumosa*, a new source of a blood-group B specific lectin. *Med. Lab. Sci.* **34**, 193–200.

RUSSELL, G. and FLETCHER, R. L. (1975). A numerical taxonomic study of the British Phaeophyta. *J. mar. biol. Ass. U.K.* **55**, 763–784.

TOMS, G. C. and WESTON, A. (1971). Phytohaemagglutinins. *In* "Chemotaxonomy of the Leguminosae" (J. B. Harborne, D. Boulter and B. L. Furner, eds), pp. 367–462. Academic Press, London and New York.

3 | Do the Polysaccharides of Brown and Red Seaweeds Ignore Taxonomy?

ELIZABETH PERCIVAL

Bourne Laboratory, Royal Holloway College, University of London, England

Abstract: The major polysaccharides of the marine Phaeophyceae and Rhodophyceae are complex heteropolymers often containing uronic acid residues and/or half ester sulphate groups. They consist of families of macromolecules; each type is built up on the same general structural plan but the extracts from different seaweeds often differ in the finer details of structure. These differences determine the physical properties such as gel strength and it is here that occasional generic and specific differences have been detected. However, some of the variations may be due to the age of tissue extracted, the preponderance of reproductive organs, environmental or seasonal conditions; all these factors are examined before taxonomic attributes are discussed.

The major families of polysaccharides are described, but detailed complex chemical formulae, representing for example the "fucans" and galactans, are avoided, and only significant variations are described. Proportionately few of the known species have been examined chemically and there is great need for more workers in this field.

INTRODUCTION

Both brown and red seaweeds synthesize complex heteropolysaccharides which often consist of a family of macromolecules all built up on the same general plan but differing in the finer details of structure. It is here that occasionally taxonomic differences emerge between genera and even species but care in the interpretation is necessary. As will be seen in the subsequent discussion of individual polysaccharides, not all the variations are taxonomic; some are due to seasonal changes, some to extremes of environmental conditions. In some instances a particular type of polysaccharide may fulfil one function in the vegetative part of the weed and another in the reproductive organs and this

---

Systematics Association Special Volume No. 10, "Modern Approaches to the Taxonomy of Red and Brown Algae", edited by D. E. G. Irvine and J. H. Price, 1978, pp. 47–62, Academic Press, London and New York.

involves a change in the fine structure. Slight changes in the overall structure cause considerable differences in the physical properties.

It must be emphasized that only in comparatively recent years have chemical techniques been available for the determination of the detailed structures of these polysaccharides; even then it involves lengthy and somewhat tedious investigation. Consequently the polysaccharides in relatively few species have been chemically studied in detail. Finally, in the majority of cases, the whole plant has been extracted so that any difference between the polysaccharides from the young growing tissues and those from the secondary growth or from the reproductive organs would not have been detected. This may explain in some instances why a whole family of molecules has been isolated.

The major polysaccharides, what is known about them, and what, if any, taxonomic conclusions may be drawn, will now be briefly considered (Percival and McDowell, 1967).

PHAEOPHYCEAE

Brown seaweeds all produce at least three types of polysaccharide. These are laminaran, a 1,3-linked β-glucan; alginic acid, a polymer of D-mannuronic and L-guluronic acids; and a so-called "fucan", a sulphated polymer which besides L-fucose almost invariably contains some D-xylose and D-glucuronic acid. In

TABLE I. Approximate percentage of "fucans" based on dry weight of weed

FUCALES		SPHACELARIALES	
Fucaceae		Cladostephaceae	
Ascophyllum nodosum	15[a]	*Cladostephus spongiosus*	2[b]
Fucus vesiculosus	9[a]	DICTYOTALES	
Fucus spiralis	10[a]	Dictyotaceae	
Pelvetia canaliculata	18–24	*Padina pavonia*	5
Himanthaliaceae		LAMINARIALES	
Himanthalia elongata (lorea)	18	Laminariaceae	
Cystoseiraceae		*Laminaria hyperborea*	6[a]
Birfurcaria bifurcata	17	DESMARESTIALES	
Durvilleaceae		Desmarestiaceae	
Durvillea potatorum[c]	0–2	*Desmarestia aculeata*	2–5
		Desmarestia ligulata	6
		Desmarestia firma	2

[a] Calculated from combined L-fucose.
[b] Calculated from combined L-fucose as the phenylhydrazone.
[c] Madgwick and Ralph, 1969.

certain species D-galactose and, in others, D-mannose may also be present in the "fucan". The relative proportions of the polysaccharides present vary from species to species, although changes in laminaran are mainly seasonal. The proportion of alginic acid is rarely less than 10% of the dry weight of the weed and may be as high as 50%. The proportion of the "fucans" appears to vary with the genus. Early studies, however, have shown that the amount decreases with the depth of immersion in which the weed normally grows. *Pelvetia canaliculata* has a high content, whereas in species of *Desmarestia* and *Durvillea*

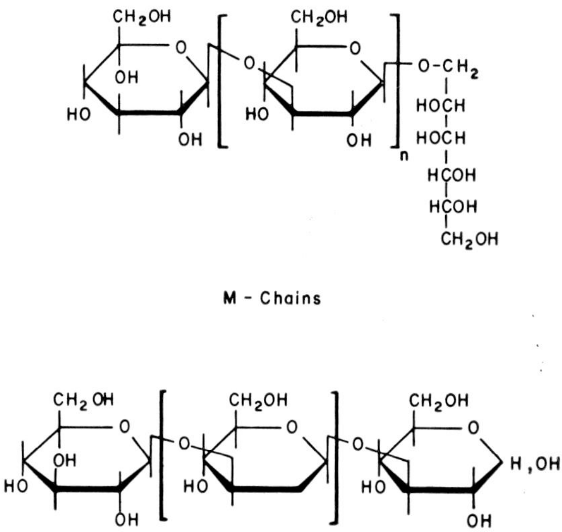

M - Chains

G - Chains

FIG. 1.

the amount of "fucan" is very small. The figures given in Table I can only be regarded as very approximate but they do show the wide variation in the amount between genera.

1. *Laminaran*

In the autumn and winter 20–30% of the dry weight of the frond of *Laminaria* can be laminaran but at other seasons of the year it is considerably less. It has been shown to be an active metabolite (Yamaguchi *et al.*, 1966; Bidwell, 1967). The only differences that have been detected in the laminaran from different species lie in the proportion of 1,6-linked glucose units which may be present and in the presence or absence of mannitol terminating some of the glucan chains at the reducing end (Fig. 1).

Laminaria digitata and *L. hyperborea* laminarans contain about 2–3% mannitol, indicating that about 40% of the chains are terminated in this way (these are known as M-chains). *Desmarestia aculeata* contains 1·7% of mannitol end groups or 30% of M-chains, and a high proportion of M-chains has been reported for the laminarans from some far eastern seaweeds (Elyakova and Zvyagintseva, 1974). The laminarans from species of *Laminaria* and *Fucus* also have a small proportion of 1,6-linked residues (Fig. 2), whereas about one-third of the

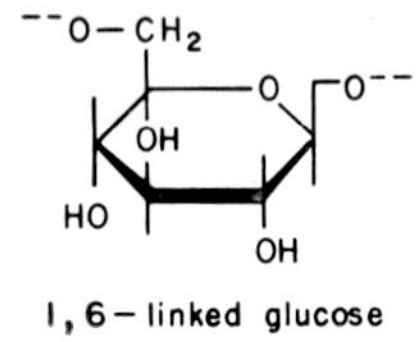

FIG. 2.

linkages are 1,6 in the laminaran from *Eisenia bicyclis*. The laminaran content of *Himanthalia elongata*, *Bifurcaria bifurcata* and *Padinia pavonia* is in each case less than 1% of the dry weight (Jabbar Mian and Percival, 1973); this was possibly due to the season of collection, late summer. In each case the laminaran was devoid of mannitol and contained a small proportion of 1,6-linked units.

2. *Alginic Acid*

This polymer consists of molecular chains constructed of blocks of D-mannuronic acid, blocks of L-guluronic acid and blocks of alternating units of the two acids

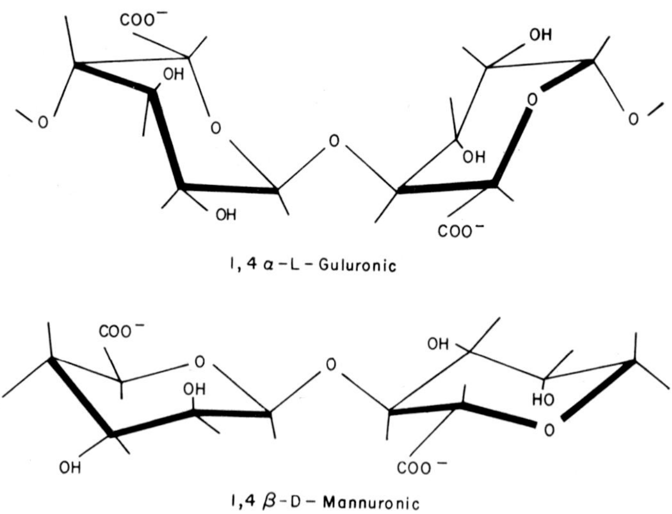

FIG. 3.

(Haug *et al.*, 1976) (Fig. 3). The proportions of the two acids have been found to vary in the extracts from different species of weeds (Haug *et al.*, 1974) (Table II). This does not appear to have any taxonomic significance since it has been shown (Haug *et al.*, 1969) that the intercellular alginates of the receptacle medullas of *Ascophyllum nodosum* and of *Fucus vesiculosus* are almost entirely composed of

TABLE II. The M/G ratios in alginates from different species

	M/G
Ectocarpales	
Pilayella	0·7
Chordariales	
Spermatochnus paradoxus	0·8
Dictyosiphonales	
Dictyosiphon foeniculaceus	0·7
Scytosiphon lomentaria	1·0
Desmarestiales	
Desmarestia aculeata	0·7
Desmarestia viridis	0·43
Desmarestia ligulata	0·47
Fucales	
Pelvetia canaliculata	1·3
Fucus serratus	1·2
Ascophyllum nodosum	1·7
Laminariales	
Laminaria digitata	1·5
Laminaria hyperborea	
fronds	1·3
stipes	0·5

polymannuronic acid, present as a voluminous soft gel. Furthermore, examination by the same authors of the different tissues in species of *Laminaria* and *Ascophyllum nodosum* reveals that the young tissue alginate is mannuronic acid rich, whilst that formed by secondary growth is relatively rich in guluronic acid residues. It is known that guluronic acid rich alginates have a higher affinity for Ca^{2+} ions which are mainly responsible for gel strength. The presence of guluronic acid gives sufficient affinity for Ca^{2+} and therefore a sufficiently strong gel in those parts of the plant where it is required, such as in the stipes of *L. hyperborea* where the guluronic acid content of the alginate is about 65%.

3. "Fucans"

The major difficulties in comparing these so-called "fucans" (otherwise fucoidan, glucuronoxylofucan) extracted from different species of seaweed are firstly that in the early studies the extracts were subjected to extensive purification procedures in an attempt to obtain a homopolysaccharide that contained only fucose, and secondly that much of the work described only the fractions of this complex family of heteropolymers that are readily removed from the weed with water (Fujikawa and Nakashima, 1975) or with dilute acid, leaving behind the more tightly bound molecules, which may differ in their fine structure.

TABLE III. Mole proportions of sugars in the different fractions of the glucuronoxylofucans[a]

Species	Fraction KCl	Fucose	Xylose	Glucuronic acid	Galactose
Himanthalia	0·3M	2·5	1·0	1·8	+[b]
	0·5M[c]	10·0	1·0	4·0	+++[b]
	1·0M[c]	14·0	1·0	2·0	++[b]
Bifurcaria	0·3M	2·8	1·0	2·4	+[b]
	0·5M	10·5	2·0	4·4	++[b]
	1·0M	12·0	1·0	2·3	++[b]
Padina	0·3M	2·0	1·0	2·0	+[b]
	0·5M[c]	9·0	3·0	3·6	++[b]
	1·0M[c]	12·5	1·0	3·0	+++[b]

[a] Jabbar Mian and Percival, 1973.
[b] Small; not determined quantitatively.
[c] Determined by g.l.c. as peak areas of derived alditols and L-gulonic acid.

Recent work has in fact shown that the "fucans" consist of a family of polysaccharides containing varying proportions of fucose, xylose and glucuronic acid; in certain species the mucilage also contains mannose and/or galactose, whilst half ester sulphate occurs in every species. It has been possible by column fractionation on DEAE-cellulose to separate from some extracts fractions containing a high proportion (22%) of uronic acid, with less fucose and sulphate than the original extract, and at the other extreme low uronic acid (3–4%) and high fucose and sulphate (Jabber Mian and Percival, 1973; Carlberg, 1976). All the intermediate type molecules between these two can be found in the other fractions (Table III). As shown earlier, the genera do differ in the proportion of "fucan" that is metabolized. In addition, the proportions of the different constituents vary and in Table IV some of the fucans that have been examined chemically and the relative proportions of sugars that have been found, are listed. Since the methods of extraction and determination are not identical, exact comparison is impossible. However, the difference in the proportion of galactose present is very striking. The macromolecules from species of *Sargassum*

TABLE IV. The approximate relative percentage of the individual sugars in the "fucans" from different genera

Species	Fuc.	Xyl.	Gal.	Man.	GlcUA
Fucaceae					
Ascophyllum nodosum[a]	49	10	4	t	12
Fucus vesiculosus[a]	54	t	t	a	?
Pelvetia sp.[b]	50	28	7	15	—
Himanthaliaceae					
Himanthalia elongata (lorea)	70	8	4	a	20
Sargassaceae					
Sargassum linifolium[c]	13	12	42	5	23
Sargassum pallidum[d]	x	x	x	x	x
Dictyotaceae					
Padina pavonia	64	10	6	t	20
Desmarestiaceae					
Desmarestia aculeata[g]	20	3	40	t	34
Desmarestia ligulata	45	3	—	2	10
Lessoniaceae					
Macrocystis pyrifera exudate[e]	67	1·2	5	t	a
Cystoseiraceae					
Bifurcaria bifurcata	55	11	6	a	26
Spermatochnaceae					
Nemacystus decipiens[f]	50	2	2	—	—

t = trace; a = absent; x = present but proportions not determined.
[a] Larsen *et al.*, 1970. [b] Pavelenko and Ovodov, 1974. [c] Abdel-Fattah *et al.*, 1974. [d] Ovodov *et al.*, 1973. [e] Schweiger, 1962. [f] Fujikawa and Wadi, 1975. [g] Percival and Young, 1974.

and from *Desmarestia aculeata* clearly contain a much higher proportion of galactose than the "fucans" from any other species so far investigated. Although this does not appear to be so for *D. ligulata*, it must be remembered that this weed contains a high concentration of sulphuric acid in its cells; although care was taken, it is possible that galactose was hydrolysed off the molecules before or during extraction. The presence of an appreciable amount of mannose in *Sargassum* spp. and *Desmarestia ligulata* (whereas in other genera it appears to be absent or present in only trace quantities) is another distinguishing feature. Uronic acid is also variable, the "fucan" from *Macrocystis* being devoid of this constituent.

RHODOPHYCEAE

1. *Galactans*

The major polysaccharides of the red weeds again consist of a family of polydisperse long chain molecules which, in contrast to the "fucans", are readily extracted with water from the seaweed. They all contain galactose, including modified or substituted galactose units. Research has shown (Rees, 1969) that they all appear to consist of chains of alternating units of 1,3-linked β-galactose

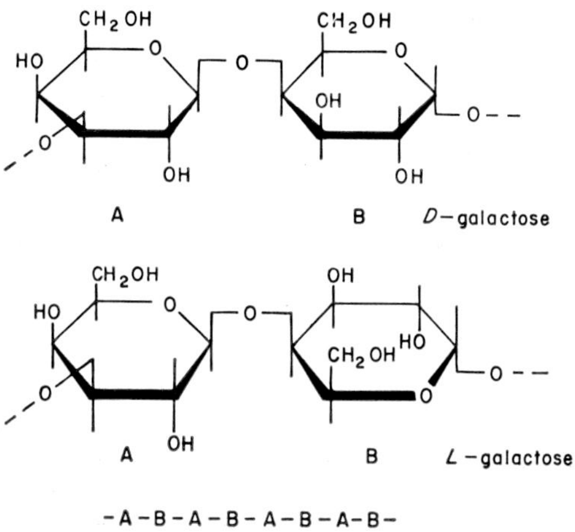

FIG. 4.

and 1,4-linked α-galactose, that is alternating A and B units (Fig. 4). However some of the units may be masked by modification or by substitution. Extracts from different weeds vary in the amount of D and of L-galactose, by the extent to which the units are changed to the 3,6-anhydro derivative (Fig. 5), by the extent and position of half ester sulphate and methoxyl groups and by substitution with pyruvic acid or glycerol (Fig. 6). Clearly there is a vast number of possible variations.

It is these differences in the fine structure that determine the conformation or shape of the individual molecular chains and hence the physical properties (Percival, 1972). There are extracts that give stiff gels in dilute solution—that is give gels which retain a definite shape even though they consist of 99·9% of water—and others that have no gelling properties at all. The presence of the 3,6-anhydro derivative induces gelling. Where this is replaced by galactose 6-sulphate the gelling power is considerably less, and gelation appears to be

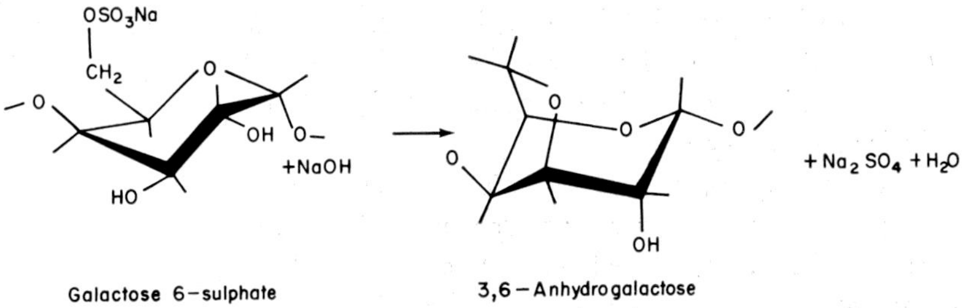

FIG. 5.

completely prevented by the presence of D-galactose 2,6-disulphate in place of the 3,6-anhydro sugar, or by the presence of 2-sulphate on the 1,3-linked galactose units. It is in these variations that differences in the galactans from different species have appeared. Again it should be emphasized that great care must be taken in deciding if the difference is really generic or specific, if it is a seasonal change, or a change due to extreme environmental conditions or to the presence of a higher proportion of reproductive organs on the plants which have been extracted. I should point out that these galactans have almost always been isolated from a mixture of whole plants of all ages. Although xylose and uronic acid have been reported in a number of galactans extracted from red seaweeds, no evidence has so far been advanced that glucuronic acid forms part of the galactan molecule. On the other hand, definite evidence of linkage of xylose to

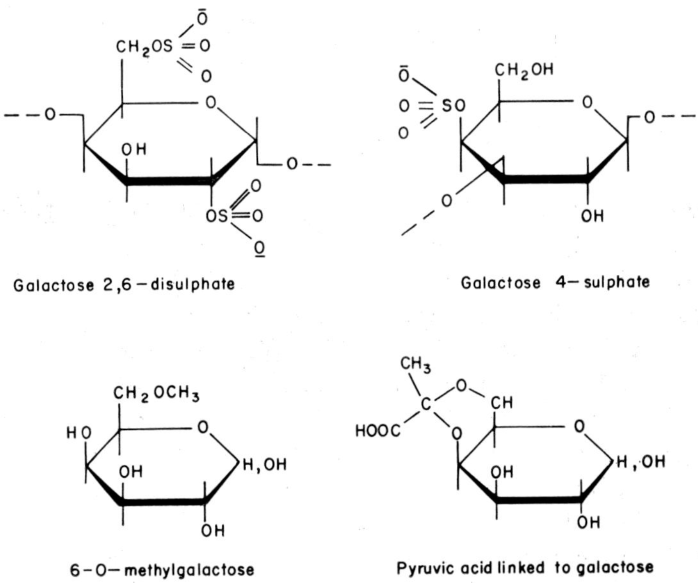

FIG. 6.

galactose has been provided for the galactan from *Laingia pacifica* (Kochetkov *et al.*, 1973). At the same time, several red weeds such as *Palmaria* (*Rhodymenia*) *palmata* and *Chaetangium erinaceum* (Nunn *et al.*, 1973a) have been shown to synthesize separate xylans.

From the polysaccharide angle, the Rhodophyceae can be divided into those genera that synthesize agar-type molecules—alternating 1,3-linked D-galactose units and 1,4-linked L-galactose units—and others that metabolize carrageenan-type molecules in which all the units are D-galactose and galactans which show affinities to agar and to carrageenan (Percival, 1972). In every case

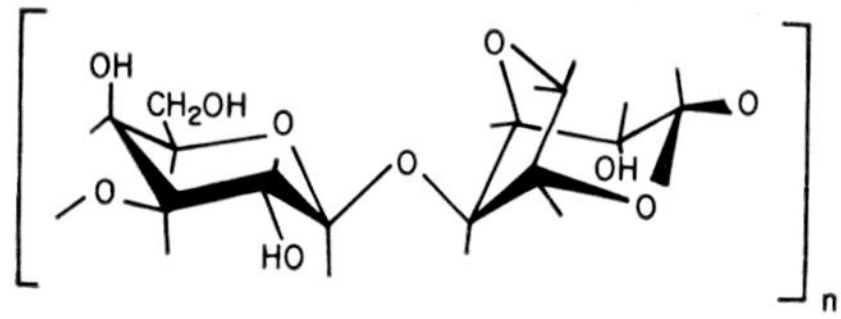

Agarose

FIG. 7.

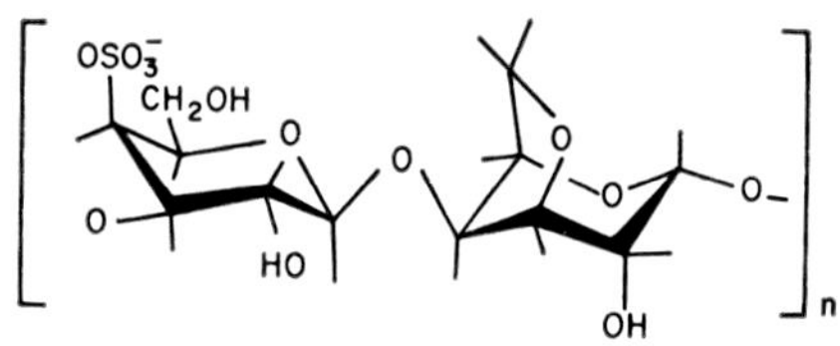

κ – Carrageenan

FIG. 8.

the extracts consist of a family of molecules which differ in the fine details of structure, as already indicated. It has been possible in some cases to separate the extremes of the structure. The separation (Araki, 1937) of agarose (Fig. 7), a non-sulphated fraction, from agar leaving behind agaropectin, a mixture of variously sulphated molecules (Izumi, 1971a, 1971b, 1972), and the precipitation by K ions of κ-carrageenan (Fig. 8), a sulphated gelling fraction from a whole extract of *Chondrus crispus* or *Gigartina* species, are examples of this fractionation In carrageenan the material left in solution, the non-gelling fraction known as λ-carrageenan, is a mixture of variously substituted molecules which can be further fractionated.

Apart from the extreme complexity of these extracts, identifying generic differences is further complicated by the facts that the early studies on these mucilages failed to establish the detailed structure, and that insufficient species have been examined by the methods now available. In addition, work by McCandless *et al.* (1973; 1975) in Canada on *Chondrus crispus* and *Iridaea cordata* extracts and by Bailey (Pickmere *et al.*, 1973) in New Zealand on *Gigartina* sp. extracts has shown that the gametophytic generation of these weeds synthesizes κ-carrageenan and its precursor, whereas the sporophytic generation produces mainly λ-carrageenan.

The presence of only D- or of D- and L-galactose units in these galactans does not appear to follow taxonomic classification since agar synthesizers are found in the Gracilariaceae and carrageenan synthesizers in the Gigartinaceae, both members of the Gigartinales. However, the two types of polysaccharide have never been found in different species belonging to the same genus and the distinction is in this sense a valid taxonomic criterion.

The polysaccharides from different species have been found to yield certain differences. Pernas *et al.* (1967) distinguished the galactans from *Gigartina stellata*, *Chondrus crispus* and *Gigartina skottsbergii* by the amounts precipitated by different concentrations of potassium chloride.

Enzymic analysis to characterize the types of agarose and agaropectin from *Gracilaria debilis*, *G. compressa*, *G. foliifera*, *G. domingensis*, *G. damaecornis* and *G. ferox* revealed similarities and differences between these six species (Hong *et al.*, 1969); it was concluded "that there appeared to be a correlation between the taxonomic grouping and the type of polysaccharide".

Other seaweeds produce galactans which have certain similarities to the agar-type polysaccharide whilst possessing structural features that resemble the carrageenans. For example, the galactan from *Porphyra umbilicalis*, a member of the Bangiaceae, is similar to agar in that it contains 3,6-anhydro-L-galactose, but 50% of these units are replaced by galactose 6-sulphate, a characteristic of the carrageenans. *Anatheca dentata* galactan has a backbone structure like agar but the high sulphate distinguishes it (Nunn *et al.*, 1973b). *Laurencia* (Bowker and Turvey, 1968) and several genera belonging to the Grateloupiaceae have C-2 of the D-galactose units substituted by methoxyl. In the extract from *Aeodes orbitosa* this is the major methyl sugar (Nunn and Parolis, 1968), comprising one in every six of the galactose residues. 4- and 6-O-methyl residues are also present, the former being on the L-galactose units. It should be mentioned that the last-named extract also contains glycerol. Methylated galactoses are characteristic of the Grateloupiaceae galactans. A unique feature of the galactan from *Aeodes ulvoidea* is the presence of single unit side chains of 4-O-methyl-L-galactose attached to C-6 of the 1,4-linked units (Allsobrook *et al.*, 1974). The galactan from *Odonthalia corymbifera* (Rhodomelaceae) contains a considerable proportion of 6-*O*-methyl-D-galactose (Usov and Kochetkov, 1968) and from 0·8–20·8% of this sugar is a constituent of agarose from different agarophytes (Araki, 1966). Carrageenan-type galactans so far appear to be devoid of methoxyl residues. Izumi (1972) considers that, although the macromolecules are all built up on the alternating 1,3- and 1,4-unit plan, their distribution in the agar and carrageenan families is peculiar to the species. He fractionated extracts and separated different proportions of the respective fractions.

Some galactans have been shown to contain pyruvic acid in 4,6-linkage to the galactose (Fig. 6). Species of *Gelidium*, *Gracilaria* (Young *et al.*, 1971), *Digenea simplex* (Rhodomelaceae) (Tomoda *et al.*, 1972), *Gigartina tenella* and *Grateloupia elliptica* (Hirase and Watanabe, 1972) all contain a small percentage of pyruvic

TABLE V. The presence of pyruvic acid in Agar

Agar source	% pyruvic acid	Agar source	% pyruvic acid
Gelidium amansii	1·3	*Digenea simplex*	0·11
Gelidium subcostatum	1·2	*Gelidiella acerosa*	0·12
Gelidium japonicum	nil	*Gracilaria verrucosa*	nil
Gelidium cartilagineum	1·29	*Gracilaria compressa*	2·92
Gelidium sesquipedale	0·04	*Gracilaria damaecornis*	0·17
Pterocladia tenuis	nil	*Gracilaria foliifera*	0·13
Pterocladia pumata	0·65	*Acanthopeltis japonica*	nil
Campylaephora hypnaeoides	nil	*Ceramium boydenii*	nil

acid linked to galactose, but other galactans, including that from *Gracilaria verrucosa*, were found (Araki, 1966) to be devoid of this acid (Table V).

In *Gloiopeltis furcata* and *G. tenax* galactans, much of the 3,6-anhydrogalactose of agar is replaced by L-galactose, only one in five of these being sulphated.

The carrageenan-type galactans, which have a higher sulphate content than the agars, differ from one another in the degree of sulphation and in the siting of these groups on the galactose residues, but with the present state of our knowledge it is not possible to apportion these differences to particular genera. *Eucheuma spinosum* and *Agardhiella tenera* galactans are distinguished as ι-carrageenans and consist of 1,3-linked galactose 6-sulphate and 1,4-linked 3,6-anhydrogalactose 2-sulphate with 10% of 1,4-linked galactose 2,6-disulphate (Anderson *et al.*, 1973). Differences in the siting of the sulphate groups distinguishes *Eucheuma spinosum* and *E. cottonii.*

These are some of the detected differences in the fine structure of the macromolecules of the galactans. While there are other differences between certain species, it is impossible to detail them without drawing very complex chemical formulae which would be inappropriate to this paper. Furthermore, there are many other species to be examined before final conclusions can be drawn.

EXTRACELLULAR MUCILAGES

Brief mention should be made of the unicellular microscopic red algae, *Rhodella maculata* (Evans *et al.*, 1974) and *Porphyridium* spp., which produce highly viscous extracellular polysaccharide mucilages. These differ from the galactans in that xylose and uronic acid are the major sugars, with less galactose and glucose. In *Rhodella* a small proportion of the galactose is modified to the 3,6-anhydro sugar and a small proportion of the xylose is present as the 3-O-methyl derivative. The mucilages of *Porphyridium cruentum* and *P. aerugineum* contain the same sugars as in *Rhodella* but in different proportions. The proportion of galactose and glucose is considerably higher and about half of the galactose is substituted by methoxyl groups (Percival and Foyle, unpubl.). *P. cruentum* mucilage contains only traces of 3-O-methylxylose, whereas in that from *P. aerugineum* the content is as high as 8–10%. Differences have also emerged in the stability of the sulphate groups in the two species.

CONCLUSIONS

The main conclusion that can be drawn from this paper is the remarkable degree to which characteristics of brown and red seaweed polysaccharides are in agreement with established classification. In some instances they provide generic and even specific distinctions which should be of help to the taxonomist where classification is doubtful on the basis of more usual criteria. Unfortunately, a vast amount of chemical work remains to be done on seaweed polysaccharides before a complete picture of possible generic and specific differences can be drawn. Not only do the polysaccharides of many more species need to be studied in detail but the new techniques available need to be applied to much of the earlier work. For the Phaeophyceae, new methods of extraction need to be devised, care being taken that the whole family of molecules is being investigated. Ideally, once the macromolecular structure of the total extract has been established, old and new tissues and reproductive organs should be extracted separately and the different extracts compared with that isolated from the whole plant. None of this is easy, and it is only through patient work by a large number of investigators that we shall finally obtain the required information.

REFERENCES

Abdel-Fattah, A. F., Hussein, M. M.-E. and Salem, H. M. (1974). Studies of the purification and some properties of sargassan; a sulphated heteropolysaccharide from *Sargassum linifolium*. *Carbohyd. Res.* **33**, 9–17.

Allsobrook, A. J. R., Nunn, J. R. and Parolis, H. (1974). The linkage of 4-O-methyl-L-galactose in the sulphated polysaccharide of *Aeodes ulvoidea*, *Carbohyd. Res.* **36**, 139–145.

Anderson, N. S., Dolan, T. C. S. and Rees, D. A. (1973). Carrageenans. Part VII. Polysaccharides from *Eucheuma spinosum* and *Eucheuma cottonii*. The covalent structure of ι-carrageenan. *J. Chem. Soc.* (C), 2173–2176.

Araki, C. (1937). Agar-agar III. Acetylation of the agar-like substance of *Gelidium amansii* L. *J. Chem. Soc., Japan*, **58**, 1338–1350.

Araki, C. (1966). Some recent studies on the polysaccharides of agarophytes. *Proc. int. Seaweed Symp.* **5**, 3–17.

Bidwell, R. G. S. (1967). Photosynthesis and metabolism in marine algae, VII. Products of photosynthesis in fronds of *Fucus vesiculosus* and their use in respiration. *Can. J. Bot.* **45**, 1557–1565.

Bowker, D. M. and Turvey, J. R. (1968). Water soluble polysaccharides of the red alga *Laurencia pinnatifida*, pt I. Constituent units. *J. Chem. Soc.* (C), 983–988.

Carlberg, G. E. (1976). Structural studies on the polysaccharides of the green seaweed *Urospora wormskioldii* and of the brown seaweeds *Desmarestia ligulata* and *D. firma*. Ph.D. Thesis (London).

Elyakova, L. A. and Zvyagintseva, T. N. (1974). A study of the laminarans of some far eastern brown seaweeds. *Carbohyd. Res.* **34**, 241–248.

Evans, L. V., Callow, M. E., Percival, E. and Fareed, V. (1974). Studies on the synthesis and composition of extracellular mucilage in the unicellular red alga *Rhodella*. *J. Cell. Sci.* **16**, 1–21.

Fujikawa, T. and Nakashima, K. (1975). Occurrence of fucoidan and fucoidan analogues in brown seaweed. *Nippon Nogei Kagaku Kaishi* **49**, 455–461.

Fujikawa, T. and Wada, M. (1975). Mucilage from the brown seaweed, *Nemacystus decipiens*. I. Constituents of the mucilage and their properties. *Agric. Biol. Chem.* **39**, 1109–1114.

Haug, A , Larsen, B. and Baardseth, E. (1969). Comparison of the constitution of alginates from different sources. *Proc. int. Seaweed Symp.* **6**, 443–451.

Haug, A., Larsen, B. and Smidsrod, O. (1974). Uronic acid sequence in alginate from different sources. *Carbohyd. Res.* **32**, 217–225.

Haug, A., Myklestad, S., Larsen, B. and Smidsrod, O. (1976). Studies on the sequence of uronic acid residues in alginic acid. *Acta Chem. Scand.* **21**, 691–704, 768–778.

Hirase, S. and Watanabe, K. (1972). The presence of pyruvate in λ-carrageenan and a similar polysaccharide. *Bull. Inst. Chem. Res., Kyoto University* **50**, 332–336.

Hong, K. C., Goldstein, M. E. and Yaphe, W. (1969). A chemical and enzymic analysis of the polysaccharides from *Gracilaria*. *Proc. int. Seaweed Symp.* **6**, 473–482.

Izumi, K. (1971a). Chemical heterogeneity of the agar from *Gelidium amansii*. *Carbohyd. Res.* **17**, 227–230.

Izumi, K. (1971b). Chemical heterogeneity in *Gloiopeltis furcata*. *Agric. Biol. Chem.* **35**, 651–655.

Izumi, K. (1972). Chemical heterogeneity of the agar from *Gracilaria verrucosa*. *J. Biochem. Tokyo* **72**, 135–140.

Jabbar Mian, A. and Percival, E. (1973). Carbohydrates of the brown seaweeds *Himanthalia lorea*, *Bifurcaria bifurcata* and *Padina pavonia*. Part I, Extraction and fractionation. *Carbohyd. Res.* **2** [= 26] **6**, 133–147.

Kochetkov, N. K., Usov, A. I., Miroshnikova, L. I. and Chizhov, O. S. (1973). The polysaccharides of algae. XII. Partial hydrolysis of a polysaccharide from *Laingia pacifica*. *Zh. Obsch. Khim.* **43**, 1832–1839.

Larsen, B., Haug, A. and Painter, T. (1970). Sulphated polysaccharides in brown algae. III. The native state of fucoidan in *Ascophyllum nodosum* and *Fucus vesiculosus*. *Acta Chem. Scand.* **24**, 3339–3352.

Madgwick, J. C. and Ralph, B. J. (1969). Chemical constituents of Australian bull kelp, *Durvillea potatorum*. *Proc. int. Seaweed Symp.* **6**, 539–544.

McCandless, E. L., Craigie, J. S. and Walter, J. A. (1973). Carrageenans in the gametophytic and sporophytic stages of *Chondrus crispus*. *Planta* **112**, 201–212.

McCandless, E. L., Craigie, J. S. and Hansen, J. E. (1975). Carrageenan of gametangial and tetrasporangial stages of *Iridaea cordata* (Gigartinaceae). *Can. J. Bot.* **53**, 2315–2318.

Nunn, J. R. and Parolis, H. (1968). A polysaccharide from *Aeodes orbitosa*. *Carbohyd. Res.* **6**, 1–11; **8**, 361–362.

Nunn, J. R., Parolis, H. and Russell, I. (1973a). Polysaccharides of the red alga, *Chaetangium erinaceum*. Part I. Isolation and characterisation of the water-soluble xylan. *Carbohyd. Res.* **26**, 169–180.

Nunn, J. R., Parolis, H. and Russell, I. (1973b). Sulphated polysaccharide of the Solieriaceae family. Part II. The acidic composition of the polysaccharide from the red alga, *Anatheca dentata*. *Carbohyd. Res.* **29**, 281–289.

Ovodov, Y. S., Khomenko, V. A. and Guseva, T. F. (1973). Polysaccharides of brown algae Sargassan from *Sargassum pallidum*. *Khim. Prir. Soedin* **9**, 107–108.

Pavelenko, A. F. and Ovodov, Y. S. (1974). Polysaccharides of brown algae. X. Structure of side chains of Pelvetian molecule. *Khim. Prir. Soedin.* **6**, 697–699.

Percival, E. (1972). Chemistry of agaroids, carrageenans and furcellarans. *J. Sci. Fd. Agric.* **23**, 933–940.

Percival, E. and McDowell, R. H. (1967). "Chemistry and Enzymology of Marine Algal Polysaccharides." Academic Press, London and New York (for most of the references prior to 1967).

Percival, E. and Young, M. (1974). Carbohydrates of the brown seaweeds. Part III. *Desmarestia aculeata*. *Carbohyd. Res.* **32**, 195–201.

Pernas, A. J., Smidsrod, O., Larsen, B. and Haug, A. (1967). Chemical heterogeneity of carrageenans as shown by fractional precipitation with potassium chloride. *Acta Chem. Scand.* **21**, 98–110.

Pickmere, S. E., Parsons, M. J. and Bailey, R. W. (1973). Composition of *Gigartina* carrageenan in relation to sporophyte and gametophyte stages of life cycle. *Phytochem.* **12**, 2441–2444.

Rees, D. A. (1969). Structure conformation and mechanism in the formation of polysaccharide gels and networks, *Advance Carbohyd. Chem. Biochem.* **24**, 267–332.

Schweiger, R. G. (1962). Methanolysis of Fucoidan. II. The presence of sugars other than L-fucose. *J. Org. Chem.* **27**, 4270–4272.

TOMODA, M., NAKATSUKA, S. and MINAMI, E. (1972). Plant mucilages. IV. Main structural features of the mucous polysaccharide isolated from *Digenea simplex*. *Chem. Pharm. Bull.* **20**, 935–937.

USOV, A. I. and KOCHETKOV, N. K. (1968). Polysaccharides of red algae. II. Polysaccharides of red alga *Odonthalia corymbifera*. Separation of 6-O-methyl-D-galactose. *Zh. obshch, khim.*, **38**, 234–238.

YAMAGUCHI, T., IKAWA, R. and NISIZAWA, K. (1966). Incorporation of radioactive carbon from $HC^{14}O_3$ into sugar constituents by brown alga, *Eisenia bicyclis* during photosynthesis and its fate in the dark. *Plant. Cell Physiol.* **7**, 217–225.

YOUNG, K., DUCKWORTH, M. and YAPHE, W. (1971). The structure of agar. Part III. Pyruvic acid a common feature of agars from different agarophytes. *Carbohyd. Res.* **16**, 446–448.

4 | The Importance of Cell Wall Constituents in Algal Taxonomy

E. L. McCANDLESS

Department of Biology, McMaster University, Hamilton, Ontario, Canada

Abstract: Enzymatic reactions involved in synthesis of cell wall polysaccharides are highly specific; consequently, even minor qualitative differences in polysaccharide composition imply genetic differences. In the case of certain siphoneous green algae, cell wall polysaccharide differences are proving to be taxonomically useful.

This paper considers the galactans of the red algae, agar- and carrageenan-type polysaccharides. Fine details of the structure of carrageenans, which have been demonstrated by immunochemistry and infrared spectroscopy as well as by classical biochemical methods, are discussed in terms of their possible use in the determination of phylogenetic relationships.

INTRODUCTION

Over a decade ago, Frei and Preston (1964a) stated very clearly the position which I shall take in this paper,

> "While the organization of the cell walls is only one of a number of taxonomically definitive characters, some considerable weight must be given to it . . . since the cell wall is the end product of a chain of biochemical reactions which must be fundamental."

There are really two questions to be asked. First, is the structure of the algal cell wall a good taxonomic character? If it is, are we in a position today to use this character in classification of red and brown algae? I will first discuss the former theoretical question and then venture some comments on the second one.

It is hardly necessary to say that a character, to be useful in taxonomy, should be an unvarying attribute of the organism, characteristic of the species or larger group. It must, on the other hand, be an attribute which displays differences between species, genera or larger groups, in order to be useful as a distinguishing

Systematics Association Special Volume No. 10, "Modern Approaches to the Taxonomy of Red and Brown Algae", edited by D. E. G. Irvine and J. H. Price, 1978, pp. 63–85, Academic Press, London and New York.

character. The attribute may be morphological, physiological or biochemical, but it should not be altered by environmental variations. Presumably it should have a genetic basis.

One might consider, as a taxonomic character, the physical organization of the cell wall as it is "seen" under the light or electron microscope. This is really the aspect of cell wall organization to which Frei and Preston (1964a) referred. Alternatively, one might consider the chemical composition of the cell wall. Does it contain a protein peculiar to the species, genus, family? What polysaccharides are found in the wall? How are they organized with respect to each other, or to other components of the wall? At a more complex level, how is the cell wall synthesized?

We have only fragmentary answers to some of these questions, analyses of a few species. It is important to keep in mind, as Lewin commented in his recent (1974) review, that we are all too ready to generalize on the basis of too few examples, when studying a character which may appear interesting.

At the time when the classifications were developed, the state of chemistry was inadequate to provide much information which might be used taxonomically. Consequently, as Heywood (1966) pointed out, characteristics were used which utilized man's visual abilities, i.e. morphological and, subsequently, histological characteristics. In the classification of angiosperms and in the algae as well, the emphasis has been on morphological characters, while in the younger field of microbiology, biochemical characters are employed (Heywood, 1967). The characteristics used in classification differ in part, then, because of their history. It is not necessary, however, that this remain the case, as newer methods of analysis come into use.

Every biologist knows that the nucleic acids, which constitute the basic genetic material that is passed on from one generation to another, determine the proteins which are synthesized. These proteins may be structural, as in the cell wall, or they may be enzymes which are involved in the synthesis of all the other biological molecules which constitute the organism, including such substances as cell wall polysaccharides. Although the relative abundance of these biological molecules may vary from individual to individual in a given species, many of them are characteristic of the species. This is the case with the cell wall polysaccharides of certain of the green algae, in which true cellulose appears to be absent. If different macromolecules are synthesized, then one must assume there are different enzymes present in the different species, which brings us back to the genetic basis of these differences.

Although we cannot yet describe how polysaccharides and other wall constituents are synthesized and laid down in the cell wall, we are gathering information on the enzymes which have particular rôles; this has been summarized by Karr (1976). We have a great deal of evidence that, for the incorporation of a monosaccharide into a polysaccharide, the monosaccharide must first be incorporated into a sugar nucleotide like UDP-glucose or GDP-glucose. There are specific epimerases, enzymes which convert one nucleotide sugar into another, epimeric, one. There is evidence that, in the cotton fibre at least, an acidic glycolipid plays a preliminary rôle in the formation of nucleotide sugars, in similar manner to that described for polysaccharides of the bacterial cell wall (Elbein and Forsee, 1973).

The enzymes which build the polysaccharides are very particular about the material they use for synthesis; there is specificity with respect to the sugar and also specificity with respect to the nucleotide phosphate. The enzyme which attaches another sugar unit on to the growing oligosaccharide chain is also selective with respect to the anomeric form of the sugar, whether the α or β configuration, and is also particular about the glycosidic linkages, such as 1,4 as 1,3, which it uses to attach that sugar. This means that numerous specific enzymes must be present and active for the synthesis of the complex polysaccharides found in the wall. The plant is somehow able to achieve complex co-ordination in space and time during this synthesis and incorporation in the wall which also allows for the changing size and orientation of the cells within the plant. The information is characteristic of the species and obviously has a genetic basis. If there are interspecific and intergeneric differences in cell wall structure and composition, then this structure, the cell wall, is an element well worth considering in taxonomic classification.

POLYSACCHARIDES OF ALGAL CELL WALLS

1. *The Siphoneous Green Algae*

True cellulose, the eucellulose of Myers and Preston (1959a), is a β-1,4-linked glucose polymer, the chains of which are arranged in a regular lattice, as indicated by the sharpness of the arcs in its X-ray diffraction diagrams; it is insoluble in alkali and in most other solvents. Unfortunately, in the literature the term "cellulose" has been used very loosely to mean an alkali-insoluble polysaccharide which yields β-glucose on hydrolysis and possibly other sugars as well. Because it is very important in the consideration of the cell wall as a taxonomic character, I shall use the term "cellulose" in its strict, biochemical sense, a polymer composed of β-1,4-linked glucose units only. True cellulose is of almost universal occurrence as a structural component of plant cell walls; hence its presence might not appear to be a useful feature for classification. However, there are exceptions.

Among the siphoneous green algae of the old order Siphonales, there exist plants which are devoid of cellulose. Instead, one or both of two other polysaccharides appear to play the structural rôle, β-1,4-linked mannan or β-1,3-linked xylan. In almost all instances, these unusual polysaccharides occur in crystalline form and confer the property of birefringence upon the cell wall. Although these algae are not members of either division considered in this symposium, I mention them briefly because they are usually discussed in any consideration of the relevance of cell wall composition as a taxonomic character. Then I shall consider the red algae, for which I think a similar case could be made.

Among the siphoneous green algae, there are some which contain cellulose as the chief structural polymer, others in which the chief cell wall polysaccharide is mannan or xylan, and still others in which the xylan exists in combination with a glucan. (For further information see Preston, 1974.) Then there is what I term the "*Derbesia–Halicystis–Bryopsis* complex", algae with heteromorphic life histories, of which gametophytes and sporophytes have only recently been recognized as belonging to the same species. In these the sporophytes, *Derbesia* species, contain only mannan and the gametophytes, either *Halicystis* or *Bryopsis*, contain xylan and a glucan (Preston, 1968; Wutz and Zetsche, 1970, 1976; Huizing and Rietema, 1975). Some botanists have declared a whole new order, Derbesiales, for these exceptional algae (Parker, 1970).

The "*Derbesia–Halicystis–Bryopsis* complex" constitutes a problem in interpreting the cell wall polysaccharide differences as a simple genetic difference, that is as an indication that different genes which specify polysaccharide synthesizing enzymes occur in the species which use xylan or mannan instead of cellulose in their cell walls. The different DNA sequences for all three enzymatic pathways must be present in the *Derbesia marina–Halicystis ovalis* combined species described by Kornmann (1938), in *Derbesia tenuissima–Halicystis parvula* described by Feldmann (1950), and in *Derbesia neglecta–Bryopsis halymeniae* described by Hustede (1964), because the abilities to produce the β-1,4-linked mannan, the β-1,3-linked xylan and a glucan, possibly cellulose, are all present, although they are not expressed at the same time in the life cycle (Preston, 1968; Wutz and Zetsche, 1970, 1976; Huizing and Rietema, 1975).

Are these differences in cell wall polysaccharides good taxonomic characters? Let us look at the criteria on which to base a decision. The fact that some algae make a different structural polymer is useful. In a given species, the polysaccharide which is used in making the cell wall is constant; it is not altered by environmental parameters and, most important, it requires a different set of enzymes to make xylan, or mannan, from that required to make cellulose.

2. The Red Algae

1. Cellulose

In the class generally considered to be the more primitive of the two which make up the Rhodophyta, the Bangiophyceae (cf. the Florideophyceae) cellulose is absent from the two species, *Porphyra umbilicalis* and *Bangia astropurpurea* (Frei and Preston, 1964b), for which evidence is available. Instead of cellulose, the skeletal cell wall polysaccharide appears to be a β-1,4-linked mannan, with some associated xylan. There may be glucan present, but it appears not to be cellulose, by any definition.

Whether true cellulose is present in the Florideophyceae or not has not been established. Frei and Preston (1961) stated that the X-ray diagrams of all the species they examined in this class were "consistent with the presence of cellulose 1 as the skeletal substance", reversing an earlier conclusion (Myers and Preston, 1959b). No evidence was presented, however. In the context of the present paper, the use of the term "cellulose I" is confusing; it has crystallographic rather than chemical meaning. Cronshaw *et al.* (1958) indicated that X-ray diffraction patterns of cell walls of several red algae were not identical, that several other sugars besides glucose were present in the wall polysaccharides in these species and that different polysaccharides might give similar patterns. In 1959 Myers and Preston (1959a) used the term cellulose to mean an alkali-insoluble polysaccharide which contained glucose. In the light of current information on enzyme specificity, we must be more careful in the use of the term. We certainly doubt the presence of true cellulose in the walls of one red alga, *Chondrus crispus*; all the birefringent material can be removed by hot water extraction, a most unlikely phenomenon if the birefringent material were cellulose (McCandless *et al.*, 1977). With respect to the X-ray patterns of other red algae already mentioned, Kreger (1962) commented that they never corresponded completely with those of cellulose II and pointed out that the presence, in reasonable amounts, of monosaccharides other than glucose means that the fibrils are not true cellulose. They should be called xyloglucans, or galactoglucans, or glucomannans. They may have taxonomic significance; we do not know.

2. Galactans

(*a*) *Agars and carrageenans.* The predominant polysaccharides of the red algae are galactans which constitute approximately 30–70% of the dry weight of the algae. These substances occur in both the cell walls and the intercellular matrix, and it is difficult to separate the two parts for biochemical analysis. With the development of histochemical and immunochemical techniques, however, it is becoming possible to dissect the plants biochemically and thus to determine whether cell walls and intercellular matrix polysaccharides are basically similar or not. In the meantime, as we are only just starting to do this, I shall discuss what we know about the total galactan composition of some of the red algae. Our immunochemical evidence (Gordon-Mills and McCandless, 1975) and the histochemical data of LaClaire and Dawes (1976), indicate quite clearly that carrageenans are located in the cell walls of *Chondrus crispus* and of *Eucheuma nudum*.

Those galactans which have been characterized seem to be made up of galactose units with alternating α-1,3- and β-1,4-linkages. The galactose units may be alternating D- and L-forms, as in the agar–porphyran series, or they may be D-galactose units only, as in the carrageenan–furcellaran series. To these basically similar skeletons a multiplicity of fine structural differences is added, for example, substituents of sulphate or methyl groups, or pyruvic acid, on various carbons, or anhydrogalactose instead of galactose. These constitute the differences which shall be considered, which may be specific, generic or familial.

It is possible to speculate from existing data that enzymes with the ability to link galactose units alternately α-1,3, β-1,4 are present in all members of the division, although the importance of the galactans is less in the Bangiophyceae, in which xylan is the chief polysaccharide of the cell wall (Frei and Preston, 1964b), than in the Florideophyceae. Some galactans of the Florideophyceae seem to be linked in this manner. To my knowledge this alternating linkage is not found in algae of other divisions. We do not know enough about its general occurrence in the Rhodophyta to use it as a taxonomic character.

Another possible splitting point for taxonomic purposes is the difference in the isomeric form of the galactose units, D- and L-galactose units in the agar series of polysaccharides, D-galactose only in the carrageenan series. This would seem to be an extremely important difference. It must be an integral part of the genetic programming of the alga which results in the production of both D- and L-galactose units and adds them alternately to a growing polysaccharide chain to make agar, as opposed to carrageenan in which D-galactose only is used. In 1957 Stoloff and Silva tried to set up a taxonomic system which divided the red algae on the basis of the galactans they produced, which they called agars, carrageenans or gelans. Unfortunately, the classification was based largely on physical properties of the polysaccharides and their gels, and/or the reactivity of carrageenan with casein or methylene blue; hence it was not sufficiently specific. These workers reported that all species of the same genus contained one or the other galactan but that their system broke down at the family level.

Two years later, Yaphe (1959) made a similar effort, using enzymes which digest agar or carrageenan. His method suffered by being too specific, as substituents on the galactose units which comprise the polymers interfered with the activity of both enzymes, and gave him false negative reactions. I have reproduced part of his Table 3 as my Table I.

TABLE I. The relation between taxonomic position and type of polysaccharide in Rhodophyta[a]

	Genus contains	
Order, Family	Agarose	or κ-carrageenan
NEMALIALES		
Gelidiaceae	*Gelidium*, 3 spp. *Pterocladia* *Suhria*	*Yatabella*
CRYPTONEMIALES		
Endocladiaceae	*Gloiopeltis*?	*Gloiopeltis*? *Endocladia*
GIGARTINALES		
Gracilariaceae	*Gracilaria*	
Furcellariaceae		*Furcellaria*
Hypneaceae		*Hypnea*
Phyllophoraceae	*Ahnfeltia*	
Gigartinaceae		*Cohndrus* *Iridaea* *Rhodoglossum* *Gigartina*, 6 spp.

[a] Modified from Yaphe, 1959.

The information which I was able to find about the types of galactan which are found in the Rhodophyta is summarized in Table II. For classification, I have used Parke and Dixon's (1976) checklist wherever possible, supplemented by Kylin's more comprehensive treatment (1956). The literature used in preparing this table is indicated in the footnote there. The evidence upon which this summary is based is of several types. At the very least (except for the Stoloff and Silva reference, or citation in Percival and McDowell, 1967) the literature mentioned the isolation of roughly equal amounts of D- and L-galactose, and the use of procedures such as methylation to indicate the position of substituents and glycosidic linkages. More specific evidence comes from the isolation of characteristic disaccharides (Nunn and Parolis, 1968, 1969; Nunn *et al.*, 1973; Allsobrook *et al.*, 1974; Peat *et al.*, 1961; Bowker and Turvey, 1968),

TABLE II. Type galactans produced in the Rhodophyta, in relation to taxonomic position

Order, Family	Genus contains: Agar-type	Genus contains: Carrageenan-type
BANGIOPHYCEAE		
BANGIALES		
Bangiaceae	*Bangia*[c]	
	Porphyra[d]	
FLORIDEOPHYCEAE		
NEMALIALES		
Gelidiaceae	*Gelidiella*[a]	*Yatabella*[b]
	Gelidium[b]	
	Pterocladia[b]	
	Suhria[b]	
	Acanthopeltis[a, e]	
CRYPTONEMIALES		
Endocladiaceae	*Gloiopeltis*[i]	*Endocladia*[b]
Corallinaceae	*Corallina*[e]	
Grateloupiaceae		*Aeodes*[g, l]
		Pachymenia[i]
		Phyllymenia[h]
Kallymeniaceae		*Erythrophyllum*[o]
GIGARTINALES		
Gracilariaceae	*Gracilaria*[b]	
	Corallopsis[a]	
Furcellariaceae		*Furcellaria*[b]
Solieriaceae	*Anatheca*[j]	*Agardhiella*[k]
		Eucheuma[i]
Polyideaceae		*Polyides*[e]
Hypneaceae		*Hypnea*[b]
Phyllophoraceae	*Phyllophora*[a, e]	*Gymnogongrus*[i]
Gigartinaceae	*Ahnfeltia*[b]	*Ahnfeltia*[i, m]
		Chondrus[b]
		Iridaea[b]
		Rhodoglossum[b]
		Gigartina[b]
		Petrocelis[n]
CERAMIALES		
Ceramiaceae	*Ceramium*[a, e]	
	Campylaephora[a, e]	
	Digenia[a]	
	Laurencia[f]	

References: [a] Stoloff and Silva (1957), [b] Yaphe (1959), [c] Wu and Ho (1959), [d] Peat *et al.* (1961), [e] Percival and McDowell (1967), [f] Bowker and Turvey (1968), [g] Nunn and Parolis (1968), [h] Nunn and Parolis (1969), [i] Stancioff and Stanley (1969), [j] Nunn *et al.* (1973), [k] Lawson *et al.* (1973), [l] Allsobrook *et al.* (1974), [m] Santos and Doty (1975), [n] McCandless (1978), [o] Craigie and McCandless (unpublished).

from the enzyme degradation studies of Yaphe (1959), from infrared spectral analysis (Stancioff and Stanley, 1969; Santos and Doty, 1975; McCandless, 1977; Craigie and McCandless, unpublished) and from our own immunochemical data.

Division on the basis of the constituent polysaccharides is either a taxonomist's nightmare, if one looks at the changes in classification which would have to result, or a biochemist's nightmare, if one looks at the existing classification and considers the analyses still to be done. Kylin (1956) estimated that there are 558 genera and 3740 species of red algae.

In the Bangiophyceae, species of two genera which have been examined, *Porphyra* and *Bangia*, make a galactose polymer called porphyran, which consists of alternating D- and L-galactose units. This would thus be a galactan of the agar family. Both species examined appear to make a similar polysaccharide but there are 22 other species in the subclass which have not been examined.

Among the Florideophyceae, the situation so far is clear in the Ceramiales; in other orders, the Nemaliales, Cryptonemiales and Gigartinales there are problems even at the family level; for example in the Endocladiaceae, *Gloiopeltis* and *Endocladia* make two different types of polymer.

One of the families of the Gigartinales, the Gigartinaceae, has been studied extensively. As Yaphe (1959) showed with his κ-carrageenase preparation, species in four of the five genera, *Chondrus*, *Iridaea*, *Rhodoglossum* and *Gigartina* are carrageenophytes. We have confirmed this with chemical and infrared spectral analysis as well as immunochemically, and added the genus *Petrocelis* which Parke and Dixon (1976) include in the family Gigartinaceae. Neither Yaphe (1959) nor we have analysed *Besa*, the small member of the family.

The Phyllophoraceae present a real problem. In addition to the fact that *Phyllophora* makes an agar-like polysaccharide and *Gymnogongrus* is reported to be a carrageenophyte, there is one genus, *Ahnfeltia*, the type-species of which (*A. plicata*) is an agarophyte and two others, *A. durvilliae* and *A. concinna*, make carrageenans (Stancioff and Stanley, 1969; Santos and Doty, 1975).

These are some of the problems which will have to be considered if the polysaccharides of the cell walls of red algae are to be used as taxonomic characters. Given the evidence we have today on genetic determination of enzyme synthesis which underlies the differences in polysaccharides present in various species, it would seem that if one is to reorder the red algae it would be important to consider these polysaccharides in the classification and the newer techniques make this a practicable possibility. I shall now discuss the carrageenans, and some of the studies with a possible bearing on taxonomic classification, which are being carried out in our laboratory.

(b) *Structural differences in the carrageenans as a taxonomic character.* Figure 1 shows the idealized formulae for repeating disaccharide units of several of the different forms of carrageenans, elucidated and proposed by Rees and co-workers at the University of Edinburgh (Rees, 1969). Two major differences in chemical structure are apparent, the presence or absence of anhydro-D-galactose

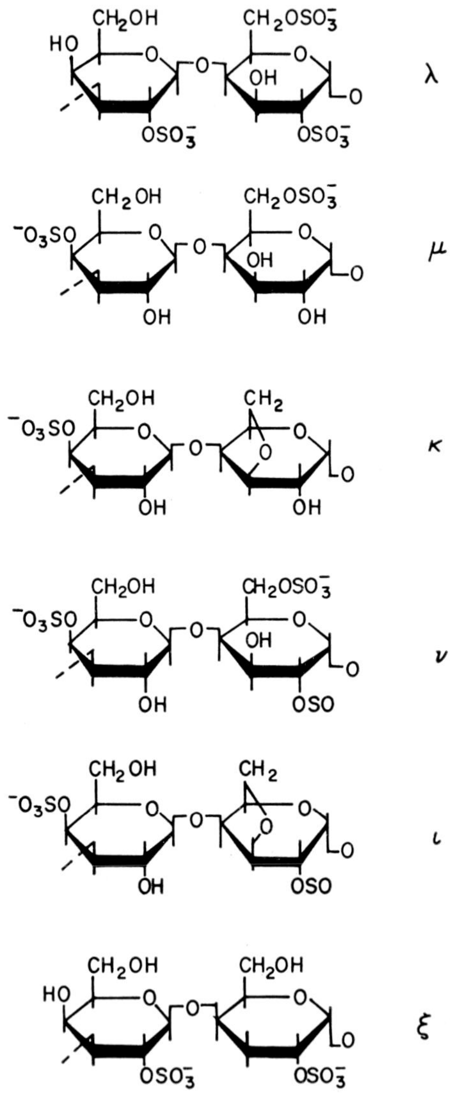

FIG. 1. Repeating disaccharide units of various carrageenan molecules (idealized).

(3,6 AG) in place of D-galactose as the 4-linked unit, and the number and position of sulphate groups on one or the other of the two units. λ-carrageenan, found in members of the family Gigartinaceae, contains essentially no 3,6 AG. The presence of 3,6 AG is characteristic of κ-carrageenan, which is found in the Gigartinaceae, in the same species as is λ-carrageenan, as well as in several other families. ι-carrageenan, found in, amongst others, certain *Eucheuma* species also contains 3,6 AG.

κ-carrageenan gels in the presence of potassium ions, whereas λ-carrageenan does not, but it has been somewhat complex to prepare one in the presence of the other, despite this property. Several years ago we discovered that the carrageenan present in the sporophyte of *Chondrus crispus* is λ-carrageenan, and that of the gametophyte is κ-carrageenan (McCandless *et al.*, 1973); this enabled the preparation of essentially pure κ- and λ-carrageenans. We have shown very clearly that the two types of carrageenan are sequestered in two different stages of the life cycle using fluorescent antibodies to κ- and λ-carrageenans (Gordon-Mills and McCandless, 1975). The same contrast is also seen in *Iridaea cordata*, the sporophytes producing λ- and the gametophytes κ-carrageenan (McCandless *et al.*, 1975). This means that the cell wall polysaccharides of the two stages are different, which is reminiscent of the situation in the "*Derbesia–Halicystis–Bryopsis* complex". (In the two species of red algae, the two stages are isomorphic, which makes it the more unusual.)

We demonstrated the differences in carrageenans extracted from the two stages of the life cycle of *C. crispus* by infrared spectral analysis, after we had noticed differences in anhydrogalactose content. This technique is based upon the fact that infrared radiation of specific frequencies is absorbed by specific chemical groups in the compound subjected to this radiation. The absorption peak which is due to 3,6 AG in carrageenan is found at about 940 cm^{-1}, and as can be seen in Fig. 2, it is prominent in the carrageenans from the gametophytic stage but not in those from the sporophytes. The peaks due to sulphate are also very informative; 840 cm^{-1} is due to sulphate on C4 of the 3-linked sugar, 830 cm^{-1} to sulphate on C2, 820 cm^{-1} to primary sulphate, on C6. The 840 peak is present in the spectra of the carrageenans from gametophytes, 830 and 820 in carrageenans from sporophytes. The carrageenans from *Iridaea cordata* gametophytes and sporophytes are similarly distinguished.

This means, then, that the mechanism for introducing an anhydro ring into the carrageenan molecule is either absent or not expressed in the sporophyte generation. It also means that the sulphotransferases, the enzymes which put sulphate on the galactose units, must be different in the two stages of the life cycle. These enzymes are specific as to where they put their sulphates; there may be as many as four different sulphotransferases involved in making these two carrageenans and not all of them are present, or expressed, in the two separate phases.

We have now analysed carrageenans which we have extracted from at least one species from every genus, except *Besa*, in the family Gigartinaceae (a total

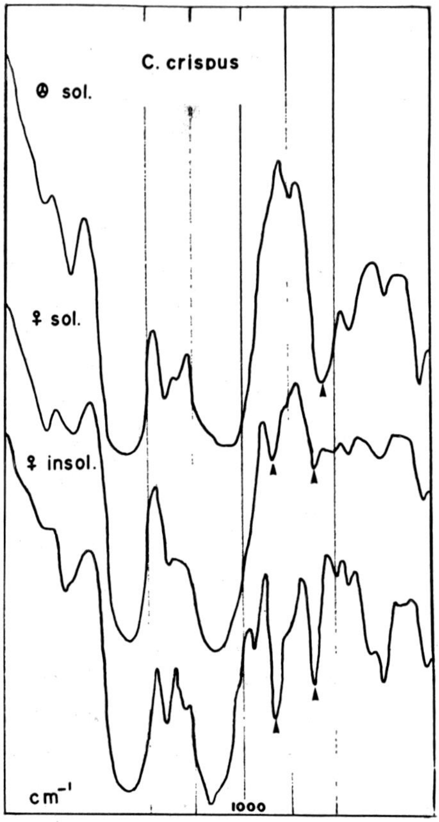

FIG. 2. Infrared absorption spectra of carrageenans from *C. crispus*, 600–1500 cm^{-1} (from McCandless, 1977).

of 18 species extracted). In each case the carrageenans from sporophytes contain little or no anhydrogalactose and have infrared spectra which resemble those of the λ-carrageenans of *C. crispus* and *I. cordata*. The carrageenans from the gametophytes, usually carposporic plants (although we have extracted male plants in a few cases), make κ-carrageenans. Representative data on 3,6 AG and sulphate contents are shown in Table III, carrageenans from *C. crispus*, *I.*

TABLE III. Biochemical analysis of carrageenans from algae of the family Gigartinaceae

Species	3,6 AG ♀ sol.	3,6 AG ♀ ins.	3,6 AG ⊕ sol.	SO_4 ♀ sol.	SO_4 ♀ ins.	SO_4 ⊕ sol.
Chondrus crispus[a]	11·6	25·1	0·7	26·9	25·6	32·8
Gigartina franciscana	12·3	21·7	3·1	21·5	22·3	29·0
Iridaea cordata[b]	12·7	21·7	0·6	28·3	22·1	32·8
Rhodoglossum californicum	10·4	17·9	1·3	23·6	22·2	35·7

Data expressed as % of dry defatted weight of alga.
Mean of 3–5 analyses.
sol. = KCl soluble.
ins. = KCl insoluble.
[a] From McCandless *et al.*, 1973.
[b] From McCandless *et al.*, 1975.

cordata, *Rhodoglossum californicum*, *Gigartina franciscana* (*exasperata*?), the four species in which we have analysed both phases.

We have also analysed carrageenans from both phases of the life cycle of a couple of carrageenophytes from families other than the Gigartinaceae (Table IV). It is interesting that no major difference between the carrageenans of the two phases has been found in *Hypnea musciformis*, both gametophyte and sporophyte phases of which make a κ-carrageenan, in *Furcellaria lumbricalis*, both phases of which make a poorly sulphated κ-type carrageenan called furcellaran, or in *Eucheuma nudum*, which makes an ι-type carrageenan. In all three the two stages make the same polymer. These plants then, do not demonstrate the *Chondrus crispus* phenomenon, which so far seems to be limited to the family Gigartinaceae. It may constitute a good taxonomic character for this family, although the classification of these algae as closely related species and genera

TABLE IV. Biochemical analysis of KC1 insoluble carrageenans from other algae

Species	3,6 AG ♀	3,6 AG ⊕	SO_4 ♀	SO_4 ⊕
Eucheuma nudum	16·8	14·7	25·5	26·7
Furcellaria lumbricalis	23·7	17·1	11·6	10·6
Hypnea musciformis[a]	24·0	22·8	20·4	21·0

[a] From McCandless, 1977.

was based on entirely different (morphological) grounds. However, the chemical morphology is also similar.

There is a disparity in the genus *Eucheuma* of the family Solieriaceae. *Eucheuma* species are well known as sources of a second gelling type of carrageenan, ι-carrageenan, which differs from κ-carrageenan in having an additional sulphate, on C2 of the 4-linked unit (Fig. 1). *Eucheuma spinosum* has been exploited for this polysaccharide, *E. uncinatum* and *E. isiforme* are also sources (Stancioff and Stanley, 1969). In the same genus, *Eucheuma cottonii* and *E. striatum* make κ-type rather than ι-carrageenans. It looks as though it might be important to re-evaluate the genus *Eucheuma* as to the relationship of its members. Stancioff and Stanley (1969) report that no members of the Solieriaceae, including *Eucheuma cottonii*, make λ-carrageenan, and Doty (personal communication) found κ-carrageenans in both stages of the life cycle in several other *Eucheuma* species, so that these should not be placed in the Gigartinaceae.

FINER DETAILS OF CARRAGEENAN STRUCTURE: IMMUNOCHEMISTRY AS A TOOL

An advantage which my entering the field of experimental botany with a background in animal physiology has conferred on our laboratory was experience in the use of animals as experimental tools. A few years ago my student K. H. Johnston demonstrated that when he injected carrageenans into rabbits, the rabbits responded by producing antibodies against the type of carrageenan used for injection (Johnston and McCandless, 1968). Using these antibodies as tools, he could show something about the type or types of carrageenans extracted from a number of carrageenophytes (Johnston and McCandless, 1969).

When the immunological defence mechanisms of an animal are stimulated by injection of an antigen such as a carrageenan, various structural details of the molecule, the antigenic determinants, cause an increase in cells which produce antibodies to those determinants. The combining sites of these antibodies are more or less specific for determinants of the antigens. We cannot specify, however, the exact numbers or types of antibodies that are produced; an animal may make antibodies to a number of different chemical aspects of the antigen, or, just as easily, may make antibody to a contaminant which happens to be present. Our early experiments in immunochemistry were impaired by the fact we did not have very pure λ- and κ-carrageenan preparations to inject into the animals. More recently, using the clue of the generation difference in *Chondrus crispus* to get a very pure λ-carrageenan from the sporophytes and similarly a very pure κ-carrageenan from the gametophytes, we prepared antibodies which were much more specific.

We used an anti-κ-carrageenan antibody which we prepared in a goat to test the gelling of KCl-insoluble κ-carrageenans obtained from a number of species of red algae (Hosford and McCandless, 1975). In all instances, the antiserum was precipitated by these κ-carrageenans from the gametophytes of *Chondrus crispus*, *Gigartina agardhii*, *G. papillata*, *G. stellata*, *Hypnea musciformis* and *Iridaea cordata*. With the exception of the polysaccharides from *H. musciformis* and *Furcellaria lumbricalis*, no preparations from sporophytic plants precipitated the anti-κ-carrageenan antibody. This agrees with the chemical data mentioned earlier. Figure 3 presents a visual display (left plate) of the precipitation reaction

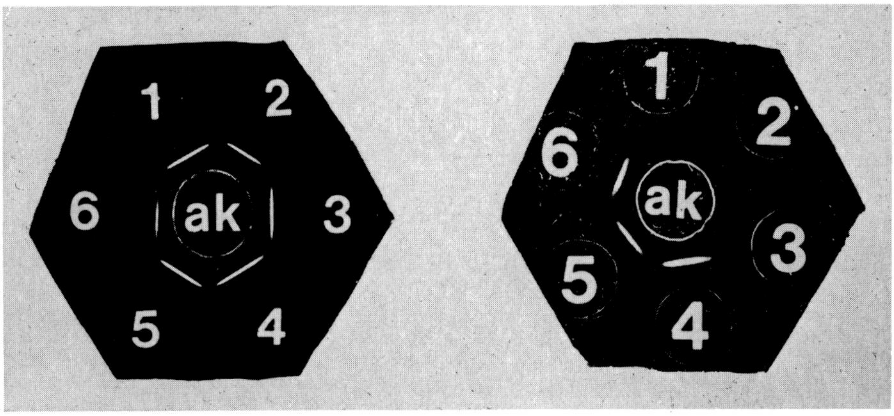

FIG. 3. Immunodiffusion plates, anti-κ-carrageenan in centre wells. Peripheral wells, left: κ-carrageenans from *C. crispus*, *G. papillata*, *G. agardhii*, *C. crispus*, *G. stellata*, *H. musciformis*; right, KCl-soluble carrageenans from tetrasporic *C. crispus*, *P. franciscana*, *Gigartina corymbifera*; gametophytic *C. crispus*, *G. agardhii*, *G. papillata* (from Hosford and McCandless, 1975).

on an agarose gel plate between anti-κ-carrageenan antibody in the centre well and various KCl-insoluble (kappa) carrageenans in the peripheral wells; on the right is shown the lack of reaction between the carrageenans from sporophytes and the reaction with KCl-soluble carrageenans from gametophytes. The precipitin lines suggest that identical determinants in the κ-carrageenans from the different species combined with the antibody, that is, they gave a reaction of identity. That does not mean that the κ-carrageenan molecules are identical. Quantitative immunoprecipitation reactions between the anti-κ-carrageenan antibody and various carrageenans (Fig. 4) showed that different amounts of antibody were precipitated at equivalence by the κ-carrageenan preparations from different species. These κ-carrageenan preparations gave qualitatively similar (although not quantitatively identical) infrared spectra (Fig. 5).

We had some difficulty with our anti-λ-carrageenan preparations, which were not as specific as we had hoped. Although the antibody reacted originally with all the λ-carrageenan preparations, it also reacted with κ- and ι-carrageenans. Eventually, we succeeded in making an antibody preparation which

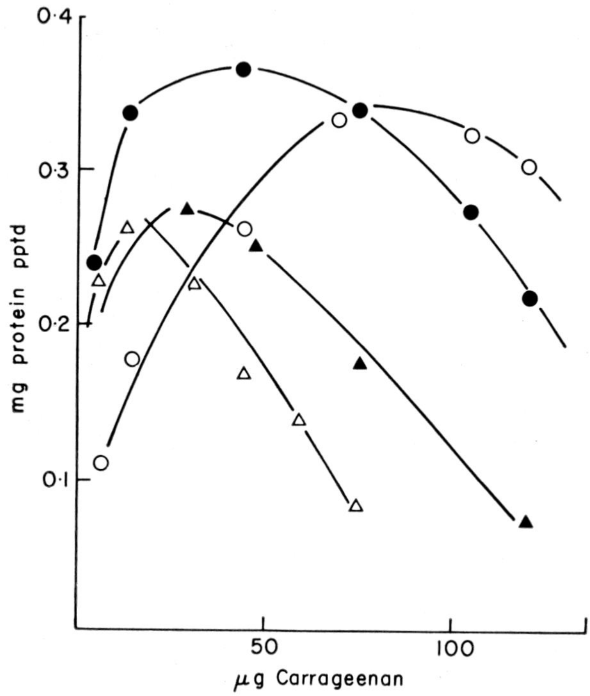

FIG. 4. Quantitative precipitin curves, anti-κ-carrageenan vs κ-carrageenans from *C. crispus, H. musciformis, G. stellata, G. papillata* (from Hosford and McCandless, 1975).

was quite specific for *Chondrus crispus* λ-carrageenan by absorbing out the antibodies which reacted with other substances. With this absorbed antibody, we found reactions with λ-carrageenans from *Chondrus crispus, Iridaea cordata, Rhodoglossum californicum* and *Petrocelis middendorfii*★ and to a slight extent *P.*

★ *Petrocelis franciscana* from western US coast, supplied 1973 by J. A. West; *P. middendorfii* from eastern Canadian coast, supplied 1976 by J. S. Craigie. Species names still in question.

franciscana but KCl-soluble carrageenans from the two *Gigartina* species did not react. (Figure 6 presents immunodiffusion data which shows the reaction of unabsorbed and absorbed anti-λ-carrageenan with various "λ"-carrageenans.)

The infrared spectra of the various "λ"-carrageenans are not as similar to each

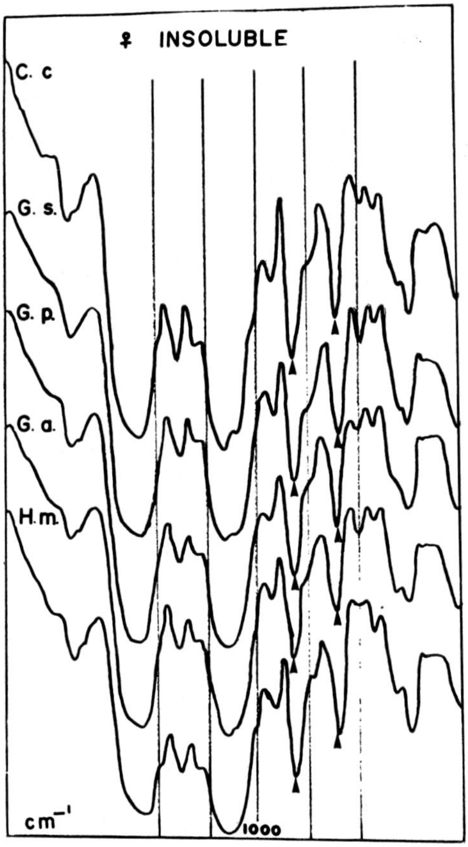

FIG. 5. Infrared spectra of KCl-insoluble (κ-) carrageenans from gametophytic *C. crispus*, *G. stellata*, *G. papillata*, *G. agardhii*, *H. musciformis*, 600–1500 cm^{-1} (from McCandless, 1977).

other as are those of the κ-carrageenans. In the region between 800 and 840 cm^{-1} where the characteristic absorption peaks for sulphates are found, a major difference is apparent between 800 and 820 cm^{-1} when the λ-carrageenans which reacted with our antibodies are compared with those which did not. In the latter category, the two *Gigartina* species contain carrageenans which show a sharp absorption peak at 830 (due to 2-sulphate) whereas in the carrageenans which react, this peak is broader seeming to include an 820 peak and possibly also absorptions at wave numbers approaching 800. The spectra of λ-carrageenan from *I. cordata* and of the carrageenan from *G. franciscana* are presented in Fig. 7. Absorption at 820 demonstrates the presence of a primary

FIG. 6. Immunodiffusion plates, KCl-soluble carrageenans from tetrasporic *C. crispus*, *P. middendorfii*, *G. corymbifera*, KCl-insoluble (κ-) from gametophytic *C. crispus*, KCl-insoluble from tetrasporic *G. corymbifera*, and KCl-soluble from tetrasporic *P. franciscana* in peripheral wells. Centre well, left, unabsorbed anti-λ-carrageenan, right, absorbed anti-λ-carrageenan.

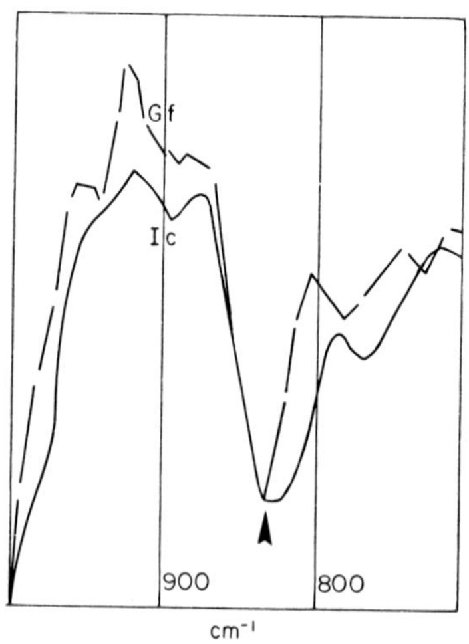

FIG. 7. Infrared spectra, 700–1000 cm^{-1}, of KCl-soluble carrageenans from tetrasporic *I. cordata* and *G. franciscana*.

sulphate (6-sulphate); thus, we can postulate that the *Gigartina* carrageenans are low in 6-sulphate, which is prominent in *C. crispus*, *I. cordata*, *R. californicum* and *P. middendorfii*. *P. franciscana*, which reacted weakly with our antiserum, occupies an intermediate position. The spectra of all seven carrageenans analysed are shown in Fig. 8.

FIG. 8. Infrared spectra 700–1000 cm^{-1}, of KCl-soluble carrageenans from *C. crispus*, *I. cordata*, *R. californicum*, *P. middendorfii*, *P. franciscana*, *G. corymbifera*, *G. franciscana*.

Treatment of carrageenans which bear a sulphate on C6 of the 4-linked galactose with alkali in the presence of borohydride eliminates that sulphate and yields an anhydro ring, 3,6 AG. Alkaline borohydride treated λ-carrageenan from *C. crispus* is characterized by a new peak at 940, the 3,6 AG peak, and by reduction or elimination of the 6-sulphate peak at 820 cm^{-1}. Similar changes are seen in the other λ-carrageenans from *I. cordata*, *R. californicum*, *P. middendorfii* and, to some extent, from *P. franciscana*. Absorption in the 820–830 region is somewhat resolved; the 830 peak becomes sharper. The 940 peak becomes more prominent. In the case of the two *Gigartina* species, very little change takes place in the 940 peak, or in the 820–830 region. Certainly these *Gigartina* carrageenans differ from those of *C. crispus* and the others. The relative absence of 6-sulphate suggests that they may contain a substantial amount of the ξ carrageenan described by Lawson *et al.* (1973) in *G. atropurpurea*, *G. chamissoi* and *G. canaliculata* (see Fig. 1). This suggests, to us at least, that these *Gigartina* species may not be so closely related to *C. crispus* as are the species from the other three genera.

A recent doctoral dissertation (Kim, 1976) has proposed the reclassification of members of the family Gigartinaceae, leaving only two genera, *Chondrus* and *Gigartina*. The genera *Iridaea* and *Rhodoglossum* are reduced to synonyms o *Gigartina*. The basis for this proposal is the presence of tissue enveloping the gonimoblast in *Gigartina*, *Iridaea* and *Rhodoglossum* and its absence in *Chondrus*. We suggest that the structure of the carrageenans of the cell walls, which links *Iridaea* and *Rhodoglossum* to *Chondrus*, is stronger evidence than the morphological detail used for this new taxonomic proposal.

This now leaves a question about *Petrocelis*. Both *P. franciscana* and *P. middendorfii* have been postulated to be sporophytic stages of the heteromorphic life cycles of two (or more) *Gigartina* species of that peculiar subgenus *Mastocarpus*, which have been recognized only as gametophytic plants (West, 1972). Both of the *Petrocelis* species produce λ-type carrageenans. The infrared spectra of these two are distinctive but in sulphation pattern *P. middendorfii* carrageenan at least bears a closer resemblance to that of *C. crispus*, as our absorbed anti-λ-carrageenan antibody recognized, than to *G. corymbifera* or *G. franciscana* carrageenans.

CONCLUSIONS

The anti-carrageenan antibodies and Vreeland's antibodies against alginate and sulphated fucans (Vreeland, 1977) demonstrate the usefulness of immunochemistry in classifying some of the polysaccharides of red and brown algae. To determine the presence of a particular polysaccharide, it is not even essential to purify the algal extract which contains it. To determine that the polysaccharide is part of the algal cell wall, the method is ideal when combined with appropriate histological techniques. For finer detail, however, which probably should be translated for use at the generic level, the method should be combined with an additional analytic tool such as infrared spectroscopy.

The analysis of cell wall components has just begun, when we recall Kylin's figure of 3740 species of red algae. I trust that this paper has provided evidence on the relevance of these analyses and an impetus to carry them out.

ACKNOWLEDGEMENTS

The author wishes to acknowledge the collaboration with Dr J. S. Craigie, Mr V. L. DiNinno, Mr M. E. Evelegh, Miss S. Jackson and Mrs C. M. Vollmer, as well as the research support from the National Research Council for the work reported in this paper.

REFERENCES

Allsobrook, A. J. R., Nunn, J. R. and Parolis, H. (1974). The linkage of 4-*O*-methyl-L-galactose in the sulphated polysaccharide of *Aeodes ulvoidea*. *Carbohyd. Res.* **36**, 139–145.

Bowker, D. M. and Turvey, J. R. (1968). Water-soluble polysaccharides of the red alga *Laurencia pinnatifida*. Part II. Methylation analysis of the galactan sulphate. *J. Chem. Soc., C*, **1968**, 989–992.

Cronshaw, J., Myers, A. and Preston, R. D. (1958). Chemical and physical investigation of the cell walls of some marine algae. *Biochim. Biophys. Acta* **27**, 89–103.

Elbein, A. D. and Forsee, W. T. (1973). Studies on the biosynthesis of cellulose. *In* "Biogenesis of Plant Cell Wall Polysaccharides" (F. Loewus, ed.), pp. 259–295. Academic Press, New York.

Feldmann, J. (1950). Sur l'existence d'une alternance de générations entre l'*Halicystis parvula* Schmitz et le *Derbesia tenuissima* (De Not.) Crn. *C.r. hebd. Séanc. Acad. Sci. Paris* **230**, 322–323.

Frei, E. and Preston, R. D. (1961). Variants in the structural polysaccharides of algal cell walls. *Nature, Lond.* **192**, 939–943.

Frei, E. and Preston, R. D. (1964a). Non-cellulosic structural polysaccharides in algal cell walls I. Xylan in siphoneous green algae. *Proc. Roy. Soc. Lond. B*, **160**, 293–313.

Frei, E. and Preston, R. D. (1964b). Non-cellulosic structural polysaccharides in algal cell walls II. Association of xylan and mannan in *Porphyra umbilicalis*. *Proc. Roy. Soc. Lond. B*, **160**, 314–327.

Gordon-Mills, E. M. and McCandless, E. L. (1975). Carrageenans in the cell walls of *Chondrus crispus* Stack. (Rhodophyceae, Gigartinales). I. Localization with fluorescent antibody. *Phycologia* **14**, 275–281.

Heywood, V. H. (1966). Phytochemistry and taxonomy. *In* "Comparative Phytochemistry" (T. Swain, ed.), pp. 1–20. Academic Press, London and New York.

Heywood, V. H. (1967). "Plant Taxonomy." St Martins Press, New York.

Hosford, S. P. C. and McCandless, E. L. (1975). Immunochemistry of carrageenans from gametophytes and sporophytes of certain red algae. *Can. J. Bot.* **53**, 2835–2841.

Huizing, H. J. and Rietema, H. (1975). Xylan and mannan as cell wall constituents of different stages in the life-histories of some siphoneous green algae. *Br. phycol. J.* **10**, 13–16.

Hustede, H. (1964). Entwicklungsphysiologische Untersuchungen über den Generationswechsel zwischen *Derbesia neglecta* Berth. und *Bryopsis Halymeniae* Berth. *Botanica mar.* **6**, 134–142.

Johnston, K. H. and McCandless, E. L. (1968). Immunologic response of rabbits to carrageenans sulfated galactans extracted from marine algae. *J. Immunol.* **101**, 556–562.

Johnston, K. H. and McCandless, E. L. (1969). Immunochemistry of carrageenans. *Proc. int. Seaweed Symp.* **6**, 507–519.

Karr, A. L. (1976). Cell wall biogenesis. *In* "Plant Biochemistry" (3rd edition) (J. Bonner and J. E. Varner, eds), pp. 405–426. Academic Press, New York.

Kim, D. H. (1976). A study of the development of cystocarps and tetrasporangial sori in figartinaceae (Rhodophyta, Gigartinales). *Nova Hedwigia* **27**, 1–146.

Kornmann, P. (1938). Zur Entwicklungsgeschichte von *Derbesia* und *Halicystis*. *Planta* **28**, 464–470.

Kreger, D. R. (1962). Cell walls. *In* "Physiology and Biochemistry of Algae" (R. Lewin, ed.), pp. 315–335. Academic Press, New York and London.

Kylin, H. (1956). "Die Gattungen der Rhodophyceen." C.W.K. Gleerups Förlag, Lund.

LaClaire, J. W. II and Dawes, C. J. (1976). An autoradiographic and histochemical localization of sulfated polysaccharides in *Eucheuma nudum* (Rhodophyta). *J. Phycol.* **12**, 368–375.

Lawson, C. J., Rees, D. A., Stancioff, D. J. and Stanley, N. F. (1973). Carrageenans. Part VIII. Repeating structures of galactan sulphates from *Furcellaria fastigiata, Gigartina canaliculata, Gigartina chamissoi, Gigartina atropurpurea, Ahnfeltia durvillaei, Gymnogongrus furcellatus, Eucheuma cottonii, Eucheuma spinosum, Eucheuma isiforme, Eucheuma uncinatum, Agardhiella tenera, Pachymenia hymantophora*, and *Gloiopeltis cervicornis*. *J. chem. Soc. Perkin I*, 2177–2182.

Lewin, R. A. (1974). Biochemical taxonomy. *In* "Algal Physiology and Biochemistry" (W. D. P. Stewart, ed.), pp. 1–39. Blackwell Scientific Publications, Oxford.

McCandless, E. L. (1977). Biological control of carrageenan structure: effects conferred by the phase of life cycle of the carrageenophyte. *Proc. int. Seaweed Symp.* **8**, (in press).

McCandless, E. L., Craigie, J. S. and Walter, J. A. (1973). Carrageenans in the gametophytic and sporophytic stages of *Chondrus crispus*. *Planta* **112**, 201–212.

McCandless, E. L., Craigie, J. S. and Hansen, J. E. (1975). Carrageenans of gametangial and tetrasporangial stages of *Iridaea cordata* (Gigartinaceae). *Can. J. Bot.* **53**, 2315–2318.

McCandless, E. L., Okada, W. T., Lott, J. N. A., Vollmer, C. M. and Gordon-Mills, E. M. (1977). Structural studies of *Chondrus crispus*: the effect of extraction of carrageenan. *Can. J. Bot.*, **55**, 2053-2064.

Myers, A. and Preston, R. D. (1959a). Fine structure in the red algae II. The structure of the cell wall of *Rhodymenia palmata*. *Proc. Roy. Soc. Lond. B*, **150**, 447–455.

Myers, A. and Preston, R. D. (1959b). Fine structure in the red algae III. A general survey of cell-wall structure in the red algae. *Proc. Roy. Soc. Lond. B*, **150**, 456–459.

Nunn, J. R. and Parolis, H. (1968). Sulphated polysaccharides of the Grateloupiaceae family. II. Isolation of 4-*O*-methyl-L-galactose, 6-*O*-methyl-D-galactose, and two disaccharides from hydrolyzates of aeodan. *Carbohyd. Res.* **8**, 361–362.

Nunn, J. R. and Parolis, H. (1969). Sulphated polysaccharides of the Grateloupiaceae family. III. Polysaccharides from *Phyllymenia cornea*. *Carbohyd. Res.* **9**, 265–276.

Nunn, J. R., Parolis, H. and Russell, I. (1973). Sulphated polysaccharides of the Solieraceae family. Part II. The acidic components of the polysaccharide from the red alga *Anatheca dentata*. *Carbohyd. Res.* **29**, 281-289.

Parke, M. and Dixon, P. S. (1976). Check-list of British marine algae—third revision. *J. mar. biol. Ass. U.K.* **56**, 527–594.

Parker, B. C. (1970). Significance of cell wall chemistry to phylogeny in the aglae. *Ann. N.Y. Acad. Sci.* **175**, 417–428.

PEAT, S., TURVEY, J. R. and REES, D. A. (1961). Carbohydrates of the red alga, *Porphyra umbilicalis*. *J. Chem. Soc.* **1961**, 1590–1595.

PERCIVAL, E. and McDOWELL, R. H. (1967). "Chemistry and Enzymology of Marine Algal Polysaccharides." Academic Press, London.

PRESTON, R. D. (1968). Plants without cellulose. *Scient. Amer.* **218**(6), 102–108.

PRESTON, R. D. (1974). "The Physical Biology of Plant Cell Walls." Chapman and Hall, London.

REES, D. A. (1969). Structure, conformation and mechanism in the formation of polysaccharide gels and networks. *Adv. Carbohyd. Chem. and Biochem.* **24**, 267–332.

SANTOS, G. A. and DOTY, M. S. (1975). IR studies on carrageenan of *Ahnfeltia concinna*, a marine red alga. *J. pharmaceut. Sci.* **64**, 1704–1706.

STANCIOFF, D. J. and STANLEY, N. F. (1969). Infrared and chemical studies on algal polysaccharides. *Proc. int. Seaweed Symp.* **6**, 595–609.

STOLOFF, L. and SILVA, P. (1957). An attempt to determine possible taxonomic significance of the properties of water extractable polysaccharides in red algae. *Econ. Bot.* **11**, 327–330.

VREELAND, V. (1978). Alginates and sulphated fucans in brown algal walls. *Proc. int. Seaweed Symp.* **8**, in press.

WEST, J. A. (1972). The life history of *Petrocelis franciscana*. *Br. phycol. J.* **7**, 299–308.

WU, Y. C. and HO, H. K. (1959). Water-soluble polysaccharides from red seaweeds (*Pterocladia tenuis* and *Bangia fusco-purpurea*). *J. Chinese Chem. Soc.* (*Taiwan*) **6**, 84–94.

WUTZ, M. and ZETSCHE, K. (1970). Biochemische Aspekte des heteromorphen Generationswechsels von *Halicystis-Derbesia*. *Ber. dt. bot. Ges.* **83**, 229–230.

WUTZ, M. and ZETSCHE, K. (1976). Zur Biochemie und Regulation des heteromorphen Generationswechsels der Grunalge *Derbesia-Halicystis*. *Planta* **129**, 211-216.

YAPHE, W. (1959). The determination of κ-carrageenin as a factor in the classification of the Rhodophyceae. *Can. J. Bot.* **37**, 751–757.

5 | Parasitic Red Algae: An Appraisal

L. V. EVANS, J. A. CALLOW and M. E. CALLOW

Department of Plant Sciences, University of Leeds, England

Abstract: A critical appraisal of the present state of knowledge on supposed red algal parasites shows that the taxonomy of these plants and the suggested taxonomic affinities they possess with their hosts are confused. Over-reliance on comparative structure and morphology and an almost complete absence of information on functional considerations must account, at least in part, for this situation. Extensive re-investigation by current microscopical and biochemical techniques is urgently required; to re-affirm separate red algal identity in many cases, to determine the reproductive behaviour, the degree of pigmentation and whether plastids are present, to obtain information on the host/parasite interface and the nature of the infection process and the nutritional relationship between host and parasite. Cross-inoculation experiments are suggested in those cases where phenotypic variability is suspected. Such improved information on the biology of these organisms will provide a better basis for subsequent taxonomic investigations. The feasibility of some modern techniques of biochemical taxonomy is briefly discussed.

Organisms which show the structural and reproductive characteristics of red algae but which lack pigments to a greater or lesser degree, have been known since the latter part of the nineteenth century. Since these algae also characteristically penetrate "host" tissues and show a reduction in thallus size, they are generally regarded as being nutritionally dependent upon their "hosts", i.e. are parasitic (Setchell, 1918).

Although over 40 such red algal parasites have been described, critical re-investigation by modern methods is required in most cases,

1. to re-affirm that the "parasite" has a distinct and separate red algal identity,
2. to establish their nutritional status.

In view of the wide spectrum of associations that occurs amongst seaweeds, ranging from cases of supposed "parasitism" to endophytic and epiphytic

---

Systematics Association Special Volume No. 10, "Modern Approaches to the Taxonomy of Red and Brown Algae", edited by D. E. G. Irvine and J. H. Price, 1978, pp. 87–109, Academic Press, London and New York.

relationships, it is initially important to establish a clear definition of the term "parasite". In this we are guided by the Federation of British Plant Pathologists (1973) which defines a parasite as "an organism or virus existing in intimate association with a living organism from which it derives an essential part of the material for its existence whilst conferring no benefit in return". This definition differs from that of Goff (1976) in that parasitism is clearly separated from pathogenesis, i.e. the ability of a parasite to cause disease in the host. The present authors are of the view that few, if any, of the effects of red algal parasites on the hosts can be described as disease (see later).

With three exceptions (see later), the nutritional relationship has not been the subject of extensive study by modern biochemical techniques. A consideration of Table I and some examples will illustrate the complexity of this subject. The presence or absence of photosynthetic pigments and plastids clearly regulates the potential for an autotrophic existence. As shown in Table I, pigmentation varies from totally absent to well developed. Within the same parasite genus, e.g. *Janczewskia*, both non-pigmented and pigmented species have been recorded. Even individual species, e.g. *Choreocolax polysiphoniae* (Table I), can be variable with respect to pigmentation, although Martin and Pocock (1953) suggest that in some cases pigmentation of otherwise colourless parasites might result from the leaching of host pigments into the parasite when removed from the natural marine environment. Clearly, in the total absence of pigmentation there must be a substantial dependence upon the host plant, but what of pigmented forms? *Schmitziella endophloea* is endophytic between cell wall layers of *Cladophora* spp. (Batters, 1892a) and has been referred to as a parasite (Adey and Sperapani, 1971), despite well-developed pigmentation. Furthermore, since the endophyte does not produce the ramifying intercellular filaments characteristic of most parasitic red algae, there appears to be no substantial reason to regard *S. endophloea* as anything but an autotrophic endophyte. However, it must be realized that nutritional dependence of a parasite on a host does not necessarily imply total dependence for all nutrient requirements. Court (1977) has suggested that the pigmented autotrophic "parasite" *Janczewskia gardneri*, growing on *Laurencia spectabilis*, is an obligate epiphyte since physiological experiments failed to reveal any net translocation of photosynthate from *Laurencia* to *Janczewskia*. However, such an association would still constitute parasitism if there was an obligate dependence on the part of one organism for minor metabolites, growth factors or vitamins supplied by the other partner. Terms implying varying degrees of nutritional dependence (hemi- or holoparasitic, partially parasitic) are best avoided until good quantitative data are available to justify their usage. Such data will be difficult to obtain in some cases since parasitism may vary with the stage of development, or may be restricted to certain regions of the thallus. Nonomura (personal communication) has shown that young stages of *Janczewskia morimotoi* growing on *Laurencia nipponica* are non-pigmented and contain proplastids. After three weeks' growth pigmentation develops and chloroplasts are present. In *Colacopsis* and *Choreonema* (Dixon, 1973), the vegetative thallus is reduced to an endophytic form, except for the reproductive stages which are borne externally to the host plant. The term "epiphyte" should be restricted to algae which are totally independent of their "hosts" for their nutrient requirements. In such associations the supporting alga is best referred to as the "basiphyte" (Linskens, 1963). Movement of assimilated radiocarbon has been shown not to occur between the basiphyte *Ascophyllum nodosum* and the epiphyte *Polysiphonia lanosa* (Harlin and Craigie, 1975; Turner and Evans, 1977), although *Polysiphonia* rhizoids penetrate deeply into the basiphyte tissue causing cellular disruption and destruction (Rawlence, 1972).

In a number of cases where parasites have been described, subsequent study failed to substantiate a separate red algal identity. Gall-like growths, the products of local tissue proliferation producing swellings or outgrowths commonly of characteristic shape but unlike any organ of the parent plant (FBPP, 1973) have been interpreted as parasites. Indeed, Goff (1977) has recently confirmed that tuberculate pustules on plants of *Neoagardhiella baileyi* are the result of localized parasitic infections by *Gardneriella tuberifera*. The reproductive pustules emerge through disrupted host cortex and are composed of both host and parasite tissues. Alternatively, systemic infection by rhizoidal filaments produces highly deformed plants. However, *Choreocolax cystoclonii* and *Lobocolax deformans* appear to be tumour-like proliferations of *Cystoclonium* (Dixon, 1973) and *Prionitis* (Chiang, 1970; McBride *et al.*, 1974) respectively. For reviews on algal galls see Tokida (1958) and Künzenbach and Brucker (1960). *Callocolax neglectus*, thought by Schmitz (1894) to be a parasite of *Callophyllis laciniata*, is suggested by Chemin (1937) after re-examination to be a tumour resulting from bacterial activity. The pale-brown parasite *Gonimophyllum buffhamii* (Batters, 1892b), forms leaflet-like outgrowths on *Cryptopleura ramosa*. Grubb (1925) has suggested that these leaflets represent abnormal outgrowths of the host possibly induced by the presence of an endophyte, since the development and structure of antheridia, procarps and cystocarps are almost identical in both. We would suggest, however, that similarity in reproductive behaviour between the host and parasite does not necessarily preclude parasitism (see later) and that the reduced pigmentation, the relatively minute size of the thallus and the presence of extensive endophytic filaments are cogent arguments for a separate identity of host and parasite (cf. Dixon, 1973). Globular bodies formed on *Phyllophora truncata* were considered to represent the parasite *Actinococcus subcutaneus* but were later shown to represent the much reduced sporophyte of *Phyllophora* growing on the gametophyte (Rosenvinge, 1931).

Parasitic red algae occur in four of the five major orders of the Florideophyceae (Parke and Dixon, 1976) but are particularly well-represented in the Gigartinales, Cryptonemiales and Ceramiales (Table I). Those parasites which are taxonomically closely related to their hosts, i.e. in the same family, are designated as adelphoparasites by Feldmann and Feldmann (1958), whilst those which are unrelated to their hosts, i.e. in a different family, are referred to as alloparasites. The majority of parasites listed in Table I are adelphoparasites and, commonly, secondary pit-connections may be observed between host and parasite in this group. Feldmann and Feldmann (1958) suggest that such connections are indicative of a strong biochemical affinity or compatibility between the cells of the two plants and that they are unlikely to be produced in cases of alloparasitism where a greater degree of incompatibility or even antagonism might be expected. However, secondary pit connections have also been reported in several of the alloparasitic genera so far investigated (Table I). Clearly, the presence of secondary pits is of little taxonomic significance and it is doubtful whether biochemical compatibility can be simply related to their presence or absence. Furthermore, adelpho- and alloparasites cannot be distinguished on the basis of any antagonistic effects or hypersensitive reactions. Spore germination and parasite development appear to proceed unimpaired in those adelpho- and alloparasites which have been studied in detail (Feldmann and Feldmann, 1958; Goff and Cole, 1976). Hypertrophy (excessive growth due to an enlargement of host cells), hyperplasia (growth by cell division) and small amounts of host cell destruction, have been reported in both alloparasites (Martin and Pocock, 1953; Kugrens and West, 1973b; Goff, 1976; Goff and Cole, 1976; Evans *et al.*, 1978) and adelphoparasites (Batters, 1892b; Kylin, 1956; Adey and Sperapani, 1971).

Modern microscopical and biochemical techniques have recently been applied to the re-investigation of alloparasites (Evans *et al.*, 1977, 1978; Goff, 1976; Goff and Cole, 1976; Kugrens and West, 1973b) and adelphoparasites (Kugrens, 1970; Kugrens and West, 1972a and b, 1973a; Court, 1977; Goff, 1977).

The alloparasite, *Holmsella pachyderma*, forms smooth white cushions, totally lacking in pigment, 1–2 mm in diameter, closely adpressed to the host, *Gracilaria verrucosa* (Fig. I). The thallus is enveloped in mucilage and is pseudoparenchymatous, being composed of a number of large-celled filaments

TABLE I. Taxonomic distribution of red algal parasites

Adelphoparasites							
Order and family	Parasite genus	Host genus	Parasite species	Host species	Pigmentation and presence of plastids	Secondary pits	References
GIGARTINALES							
Solieriaceae	*Gardneriella* Kylin, 1941	*Agardhiella* Schmitz	*G. tuberifera*	*A. coulteri*			Kylin, 1941
Rhabdoniaceae	*Catenellocolax* Weber van Bosse, 1928	*Catenella* Greville	*C. leeuwenii*	*C. impudica*			Kylin, 1956
Hypneaceae	*Hypneocolax* Børgesen, 1920	*Hypnea* Lamour.	*H. stellaris*	*H. nidifica* *H. cervicornis* *H. chordaceae* *H. spinella*	pink-red	+	Fan, 1961 Mshigeni, 1976
Plocamiaceae	*Plocamiocolax* Setchell, 1923	*Plocamium* Lamour.	*P. papenfussiana*	*P. corallorhiza*	white	+	Martin and Pocock, 1953
Gracilariaceae	*Gracilariophila* Setchell and Wilson in Wilson, 1910	*Gracilaria* Greville	*G. oryzoides*	*G. verrucosa*	white		Wilson, 1910
		Gracilariopsis Dawson *Corallopsis* Greville *Tylotus* J. Ag.		*Gracilariopsis sp.* *Corallopsis sp.* *Tylotus sp.*			
	Gracilariocolax Weber van Bosse, 1928	*Gracilaria* Greville	*G. henriettae*	*G. radicans*			Weber van Bosse, 1928
Phyllophoraceae	*Ceratocolax* Rosenvinge, 1898	*Phyllophora* Greville	*C. hartzii*	*P. truncata*[a]	pink-red	+	Rosenvinge, 1931
CRYPTONEMIALES							
Corallinaceae	*Chaetolithon* Foslie, 1898	*Corallina* L.	*C. deformans*	*Corallina spp.*			Kylin, 1956
	Choreonema Schmitz, 1889	*Corallina* L. *Jania* Lamour.	*C. thuretii*	*Corallina sp.* *J. rubens*	white		Schmitz, 1889

TABLE I—*continued*

Order and family	Parasite genus	Host genus	Parasite species	Host species	Pigmentation and presence of plastids	Secondary pits	References
Corallinaceae	*Kvaleya* Adey and Sperapani, 1971	*Leptophytum* Adey	*K. epilaeve*	*L. laeve*	white, no plastids	–	Adey and Sperapani, 1971
	Polyporolithon Mason, 1953	*Corallina* L. *Calliarthron* Manza	4 species[b]	*Corallina sp.* *Calliarthron sp.*	pink + chlorophyll		Mason, 1953
Cryoptonemiaceae	*Lobocolax* Howe, 1914	*Prionitis* J. Ag.	= host gall				McBride *et al.*,1974
Kallymeniaceae	*Callocolax* Schmitz, 1894	*Callophyllis* Kütz.	*C. neglectus*	*C. laciniata*	pink-red		Batters, 1895; Kylin, 1930
RHODYMENIALES							
Rhodymeniaceae	*Faucheocolax* Setchell, 1923	*Fauchea* Mont.	*F. attenuata*	*F. laciniata*/ *F. fryeria*	white-pink		Setchell, 1923
	Rhodymeniocolax Setchell, 1923	*Rhodymenia* Greville	*R. botryoidea*	*R. pseudopalmata*			Setchell, 1923
	Halosacciocolax Lund, 1959[c]	*Halosaccion* L.	*H. kjellmanii*	*H. ramentaceum*	yellow		Lund, 1959
	Gloiocolax Sparling, 1957	*Gloioderma* J. Ag.,	*G. novae-zelandiae*	*G. saccatum*	rose/brown	+	Sparling, 1957
CERAMIALES							
Delesseriaceae	*Asterocolax* J. and G. Feldmann, 1951	*Erythroglossum* J. Ag. *Phycodrys* Kütz. *Nienburgia* Kylin	*A. erythroglossii*	*E. sandrianum* *Phycodrys sp.* *Nienburgia sp.*	white + colourless plastids	+	Feldmann and Feldmann, 1958
	Gonimocolax Kylin, 1924	*Myriogramme* Kylin	*G. australe*	*M. livida*			
	Gonimophyllum Batters, 1892b	*Cryptopleura* Kütz *Botryoglossum* Kütz	*G. buffhamii*	*C. ramosa*[d] *Botryoglossum sp.*	pale brown		Batters, 1892b
		Acrosorium Zan ard.	*G. africanum*	*A. maculatum*			Martin and Pocock, 1953
	Polycoryne, Skottsberg in Kylin and Skottsberg, 1919	*Myriogramme* Kylin *Phycodrys* Kütz. *Nienburgia* Kylin	*P. radiata*	*Myriogramme sp.* *Phycodrys sp.* *Nienburgia sp.*			Kylin, 1956
	Loranthophycus Dawson, 1944	undescribed Delesseriacean, near *Platyclinia*	*L. californicus*	undescribed	rose-coloured		Dawson, 1944

Dasyaceae	*Colacodasya* Schmitz, 1897	*Heterosiphonia* Mont.	*C. inconspicua*	*Heterosiphonia sp.*			Schmitz and Hauptfleisch, 1897
D* Rhodomelaceae	*Antarctocolax* Skottsberg, 1953	*Picconiella* J. De Toni	*A. lambii*	*P. plumosa*			Kylin, 1956
	Benzaitenia Yendo, 1913; Morrill, 1976a	*Chondria* C. Ag.	*B. yenoshimensis*	*Chondria sp.*			Yendo, 1913 Kylin, 1956; Morrill, 1976a
		Laurencia Greville		*Laurencia sp.*			
	Calacopsis De Toni, 1903	*Lophurella* Schmitz	*C. pulvinata*	*L. hookeriana*	white		Kylin, 1956
		Vidalia Lamour.		*V. serrata*			Kylin, 1956
	Chamaethamnion Falkenberg in Schmitz and Falkenberg, 1897	*Polysiphonia* Greville	*C. schizandra*	*P. nigrita*			Kylin, 1956
	Janczewskia Solms-Laubach, 1877	*Laurencia* Lamour.	*J. verrucaeformis*	*L. obtusa*	white, plastids present	+	Feldmann and Feldmann, 1958
			J. solmsii	*L. subopposita*	white		Setchell, 1914
			J. gardneri	*L. pinnatifida*	pink		Setchell, 1914
			J. tasmanica	*L. forsteri*			Setchell, 1914
			J. meridionalis	*L. flexuosa*	white		Martin and Pocock, 1953
		Chondria C. Ag.	*J. moriformis*	*C. atropurpurea*	white		Setchell, 1914
			J. lappacea	*C. nidifica*	pink	+	Setchell, 1914
	Jantinella Kylin, 1941; Morrill, 1976b	*Chondria* C. Ag.	*J. verrucaeformis*[e]	*Chondria sp.*			Kylin, 1956; Morrill, 1976b
	Levringiella Kylin, 1956	*Pterosiphonia* Falkenberg	*L. gardneri*	*Pterosiphonia sp.*	white, reduced plastids	+	Kugrens, 1970; Kugrens and West, 1972; 1973a
	Melanocolax Martin and Pocock, 1953	*Rhodomelopsis* Pocock	*M. velutina*	*R. africana*	pink-dark red	+	Martin and Pocock, 1953
	Meridiocolax Morrill, 1976c	*Polysiphonia* Greville	*M. narcissus*	*P. ferulacea*			Morrill, 1976c
			M. polysiphoniae[f]	*P. denudata*			Morrill, 1976c
	Microcolax Schmitz, 1897	*Streblocladia* Schmitz	*M. africana*	*S. tenuissima*	red		Martin and Pocock, 1953
	Onychocolax Pocock, 1956	*Polysiphonia* Greville	*O. polysiphoniae*	*P. incompta*			Pocock, 1956

TABLE I—*continued*

Order and family	Parasite genus	Host genus	Parasite species	Host species	Pigmentation and presence of plastids	Secondary pits	References
Rhodomelaceae	*Ricardia*, Derb. et Solier, 1856	*Laurencia* Lamour.	*R. montagnei* *R. saccata*[g]	*Laurencia spp.*	red		Derbès, 1856; Børgesen, 1930
	Sporoglossum Kylin in Kylin and Skottsberg, 1919	*Lophurella* Schmitz	*S. lophurellae*	*L. hookeriana*			Kylin, 1956
	Stromatocarpus Falkenberg in Schmitz and Falkenberg, 1897	*Tayloriella* Kylin	*S. parasiticus*	*T. virgata*[h]	hyaline	+	Martin and Pocock, 1953
	Tylocolax Schmitz, 1897	*Lenormandia* Sonder	*T. microcarpus*	*L. spectabilis*			Kylin, 1956
	Pleurostichidium Heydrich, 1893	*Fucodium* J. Ag.	*P. falkenbergii*	*F. chondrophyllum*			Heydrich, 1893

Alloparasites

Parasite order and family	Parasite genus	Host order and family	Host genus	Parasite species	Host species	Pigmentation and presence of plastids	Secondary pits	References
CRYPTONEMIALES Choreocolacaceae	*Holmsella*, Sturch, 1926	GIGARTINALES Gracilariaceae	*Gracilaria* Greville	*H. pachyderma*	*G. verrucosa*	white, plastids reduced or absent	+	Sturch, 1924; Evans *et al.*, 1973; 1978
	Harveyella Schmitz and Reinke in Reinke, 1889	CERAMIALES Rhodomelaceae	*Rhodomela* C. Ag.	*H. mirabilis*	*R. confervoides*			Sturch, 1899, 1924
			Odonthalia Lyngbye		*O. floccosa*	white, no plastids	+	Goff and Cole, 1973; 1975; 1976. Goff, 1976

Choreocolacaceae	*Choreocolax* Reinsch, 1875		*Polysiphonia* Greville	*C. polysiphoniae*	*P. lanosa*	white-pink, reduced plastids	+	Sturch, 1926; Evans *et al.*, 1978; Callow *et al.*, 1977
			Pterosiphonia Falkenberg		*Pterosiphonia spp.*	white, plastids reduced	+	Kugrens and West, 1973b
			Odonthalia Lyngbye		*O. dentata*			Kylin, 1956
	Episporium Moebius, 1885		*Centroceras* Kütz.	*E. centroceratis*	*C. clavulatum*	white		Pocock, 1956
	Gelidiocolax Gardner, 1927	NEMALIALES Gelidiaceae	*Gelidium* Lamour.	*G. microsphaerica*	*G. pulchrum*	white		Gardner, 1927
			Pterocladia J. Ag.	*G. mammillata*	*Pterocladia sp.*		+	Fan and Papenfuss, 1959; Fan, 1961
			Suhria J. Ag.	*G. suhriae*[i]	*S. vittata*	pink-red		Martin and Pocock, 1953; Fan and Papenfuss, 1959
			Beckerella Kylin	*G. margaritoides*[k]	*B. pinnatifida*	white		Fan and Papenfuss, 1959; Martin and Pocock, 1953
			Gelidium Lamour.		*G. pinnatifidum*			
Pterocladio-philaceae	*Pterocladiophila* Fan and Papenfuss, 1959		*Pterocladia* J. Ag.	*P. hemisphaerica*	*P. lucida*		+	Fan and Papenfuss, 1959
CERAMIALES[m] Rhodomelaceae	*Syringocolax* Reinsch, 1875		*Gelidium* Lamour.	*S. macroblepharis*	*G. cartilagineum*	pale pink		Martin and Pocock, 1953
RHODYMENIALES Rhodymeniaceae	*Halosacciocolax* Lund,[c] 1959	PALMARIALES Palmariaceae	*Palmaria* Stackhouse	*H. lundii*	*P. palmata*[l]	white		Edelstein, 1972; Guiry, 1974
				H. kjellmanii		yellow		Guiry, 1975

[a] *P. brodiaei*; [b] *P. conchatum*, *P. patena*, *P. parcum*, *P. recinatum*; [c] *Halosacciocolax* Lund, 1959, is of uncertain taxonomic position; Cabioch and Guiry (1977) propose to place the genus in the Acrochaetiaceae (Nemaliales). *H. lundii* may be conspecific with *H. kjellmanii* (Guiry, 1974, 1975; Parke and Dixon, 1976); [d] *Nitophyllum laceratum*; [e] *Colacodasya verrucaeformis*; [f] formerly *Levringiella polysiphoniae*; [g] *Erythrocystis saccata*; [h] *Polysiphonia virgata* = *Carradoria virgata*; [i] *Choreocolax suhriae*; [k] *Choreocolax margaritoides*; [l] *Rhodymenia palmata*; [m] *Aiolocolax* Pocock 1956, parasitic on *Falkenbergiella caespitosa* (Rhodomelaceae), is of unknown systematic position, probably in the Ceramiales (Pocock, 1956).

forming a medulla from which arises a peripheral cortex of branched filaments (Fig. 2). Branched endophytic filaments penetrate between the cells of the host, but not, so far as is known, into them. Intracellular penetration has been recorded in *Ceratocolax* (Rosenvinge, 1931). Secondary pit connections are established between host cells and those of the endophytic filaments in *Holmsella* (Sturch, 1924). In the electron microscope, small proplastid-like structures have

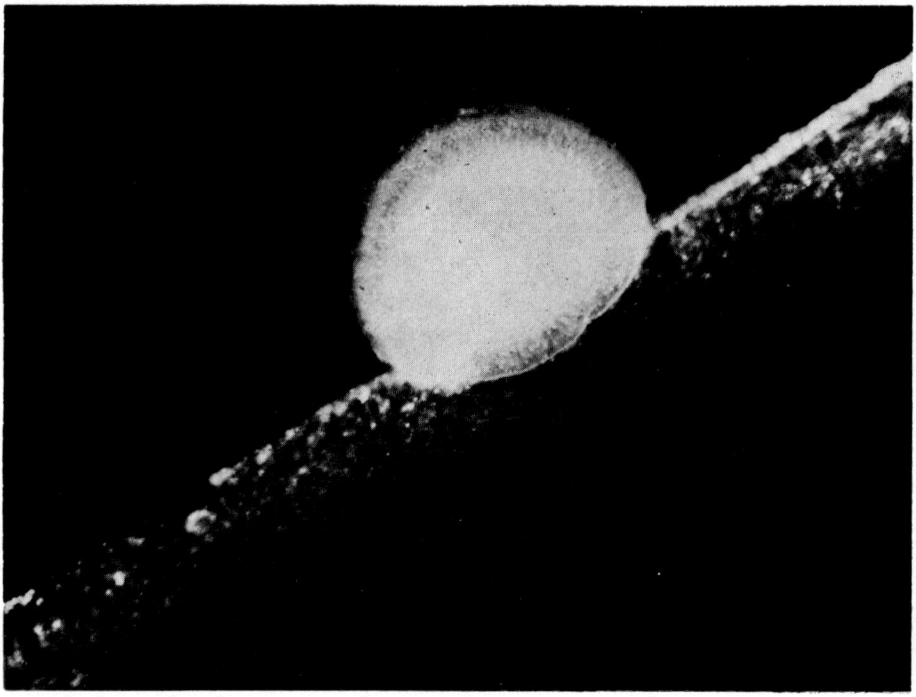

FIG. 1. A branch of *Gracilaria verrucosa* bearing a plant of *Holmsella pachyderma*. Living material, × 50. From Evans, *et al.*, 1973.

been seen in cortical cells and the medullary cells contain large numbers of starch grains.

On what nutritional grounds can *Holmsella* be termed a parasite? Segments of *Gracilaria*, each bearing a plant of *Holmsella*, were allowed to photosynthesize in ^{14}C-labelled bicarbonate in seawater. After two hours the segments were removed, washed and either dissected immediately into host and parasite regions, or allowed to photosynthesize for a further four or eight hours in unlabelled medium before dissection (Evans *et al.*, 1973).

It can be seen from Table II that the total amount of radioactivity in *Holmsella* increases with time, implying translocation of fixed carbon from host to parasite. The alcohol glycoside, floridoside (2-o-α-D-galactopyranosyl glycerol) (Colin and Guéguen, 1930; Putman and Hassid, 1954) is the major

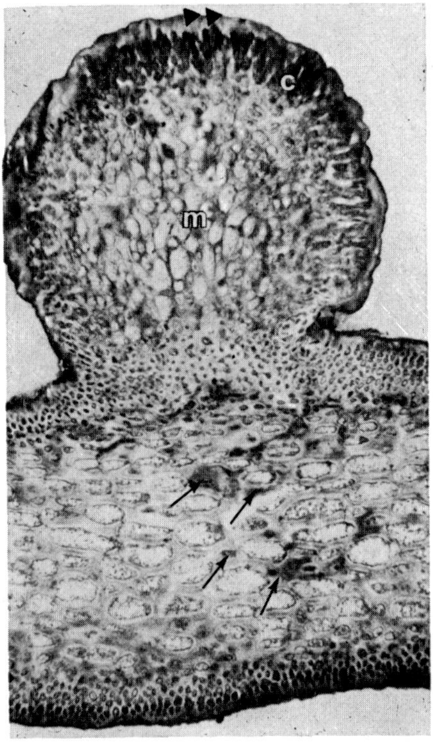

FIG. 2. Non-median transverse section through *Gracilaria/Holmsella* showing medulla (m), cortex (c), outer mucilage layer (arrow heads) and endophytic filaments of *Holmsella* in the extracellular mucilage of *Gracilaria* medullary cells (arrows). (Embedded in glycol methacrylate, 1 μm section stained in toluidine blue.) Light micrograph, × 100, from Evans, *et al.*, 1973.

labelled soluble carbohydrate in *Gracilaria* and radioactivity in this compound decreases with time, presumably due to respiration and conversion into in-soluble materials. In *Holmsella*, floridoside and the hexitol sugar mannitol were labelled and radioactivity in both compounds increased with time. The amount of radioactivity in starch also increased in both plants with time. It was proposed (Evans *et al.*, 1973) that carbon was translocated from host to parasite in the form of floridoside, which was then converted in the parasite to mannitol and

TABLE II. Total and percentage radioactivity in *Holmsella* and *Choreocolax* in different treatments

	Holmsella		*Choreocolax*	
Time	Total ^{14}C/unit dry wt ($min^{-1} \times 10^{-3}$ mg^{-1})	% total	Total ^{14}C/unit dry wt ($min^{-1} \times 10^{-3}$ mg^{-1})	% total
2 h	136	4·3	71	5
2+4 h	217	6·8	109	9
2+8 h	313	9·6	133	11
dark control[a]	6	13·9	14	0·5

[a] *Holmsella*: 2 h dark + 8 h dark; *choreocolax*: 2 h dark

starch. Since starch is insoluble, and mannitol is an unique parasite sugar, it was further proposed that these conversions would create a source–sink relationship promoting continued translocation from autotroph to heterotroph, as discussed by Smith *et al.* (1969).

Turning now to the second alloparasite examined in this laboratory, *Choreocolax polysiphoniae* forms white to pink cushions, up to 1 mm diameter on the surface of *Polysiphonia lanosa* (Fig. 3). In this case, a proportion of the pigment is due to the presence of minute epiphytes. The structure of the thallus is similar to that of *Holmsella* but the degree of contact with the host plant is considerably greater, since *Polysiphonia* cells are displaced deeply into the *Choreocolax* thallus (Sturch, 1926) and cells of the branched endophytic filaments are joined to host cells by secondary pit connections initiated by the host (Kugrens and West, 1973b). Although there are no reports that host plants are ever killed by the parasites, the two alloparasites *Harveyella* and *Choreocolax* do cause some degree of host disruption. In both cases, host cells undergo

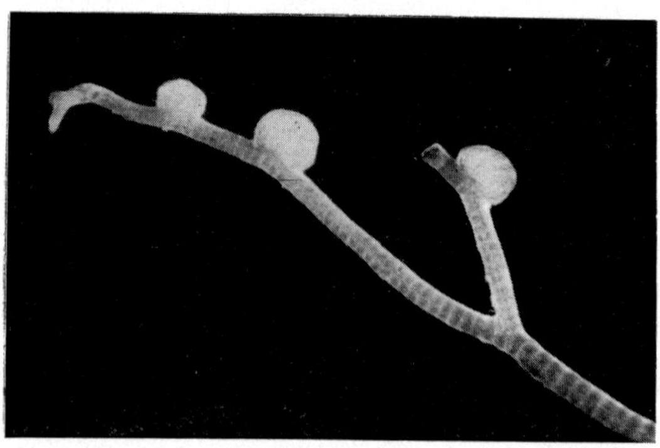

FIG. 3. A branch of *Polysiphonia lanosa* bearing plants of *Choreocolax polysiphoniae*. Living material, × 15.

hypertrophy, hyperplasia and cell destruction (Kugrens and West, 1973b; Goff, 1976; Callow *et al.*, 1978), although these processes are extremely localized and do not appear to have substantial deleterious effects on the hosts.

Within the cortical cells of *Choreocolax*, small plastids, each approximately 1 μm long, occur (Fig. 4). This is in agreement with the findings of Kugrens and West (1973b).

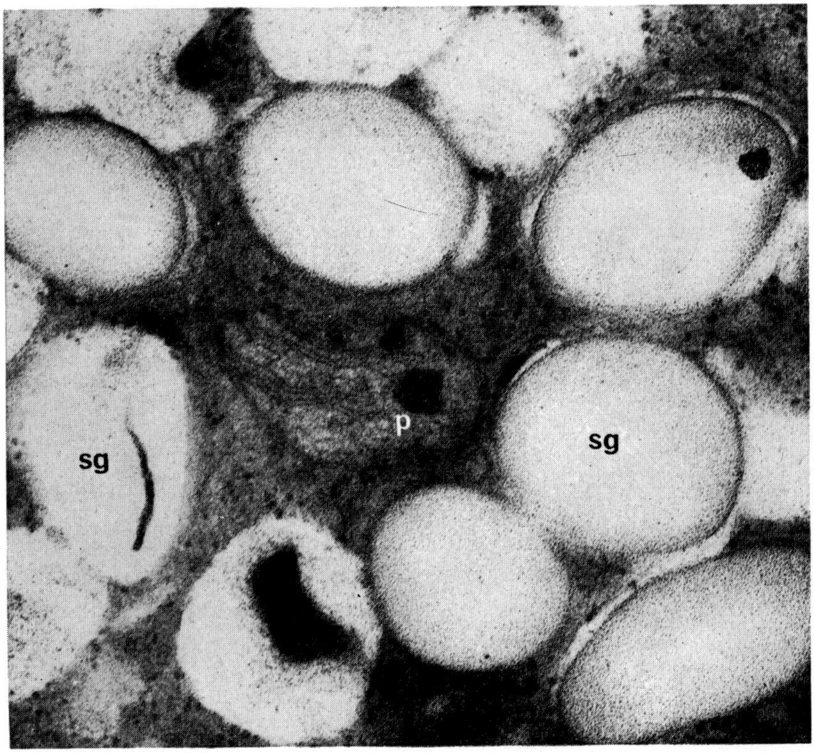

FIG. 4. Electron micrograph of a part of a cortical cell of *Choreocolax* showing a plastid (p), surrounded by floridean starch grains (sg). × 60 000.

To determine whether there is movement of carbon from host to parasite, translocation experiments were carried out as described for *Holmsella*. As can be seen from Table II, movement of labelled carbon occurs, the amount of radioactivity in the parasite increasing with time in the absence of external label. In the host, *Polysiphonia*, the major product of photosynthesis is another alcohol glycoside, sodium mannoglycerate, a compound which is restricted to the Ceramiales. In *Choreocolax*, most of the radioactivity in the soluble fraction is in two components, sodium mannoglycerate, which increases with time, and another as yet unidentified complex sugar in which radioactivity decreases with time. It is believed that the sodium mannoglycerate in the parasite represents contamination from the displaced *Polysiphonia* cells inside the *Choreocolax* thallus (Evans *et al.*, 1978), and that the as yet unidentified compound is unique to the *Choreocolax* and synthesized by it (see later). During translocation there is also a large accumulation of radioactivity into insoluble materials in both host and parasite.

Somewhat different results have been obtained by Court (1975, 1977) for mature plants of the pigmented adelphoparasite *Janczewskia gardneri* growing on *Laurencia spectabilis*. *J. gardneri* was shown to be autotrophic and no net translocation of ^{14}C-labelled assimilates to the "parasite" could be detected.

Interpretation of ^{14}C photosynthesis and translocation experiments is difficult if there is some degree of independent fixation by the parasite. Moreover, the metabolism of the host in the region of the parasite may be affected, especially in those situations where hypertrophy or hyperplasia create additional metabolic sinks. Although critical quantitative evidence is lacking it has been reported that *Odonthalia* tissue infected by *Harveyella* shows a greater incorporation of $^{14}CO_2$ than uninfected tissue, consistent with an enhanced rate of photosynthesis (Goff, 1976). Goff suggested that an increase in the number of plastids per cell, or an increase in photosynthetic tissue from induced cell division, could account for these results.

Reports of the occurrence of plastids in parasites are summarized in Table I. Although the majority of parasites lack pigmentation, many of these colourless forms appear to contain organelles resembling simple plastids. Thus, *Asterocolax* is colourless, or sometimes very faintly pink, but the cells contain numerous plastids (Feldmann and Feldmann, 1958). Examination with the electron microscope has revealed the presence of small reduced plastids containing 1–2 thylakoids in *Levringiella* (Kugrens and West, 1972a); and *Choreocolax* (Kugrens and West, 1973b; Callow *et al.*, 1978). More deeply pigmented parasites such as *Ricardia saccata* (= *Erythrocytis saccata*) contain chloroplasts of typical rhodophycean morphology suggesting a substantial degree of autotrophy. However, host tissues are still deeply penetrated by rhizoids and in the absence of any definitive evidence to the contrary, on the basis of criteria outlined earlier there appear to be no grounds for considering this organism as anything but a parasite.

The degree of pigmentation and photosynthetic capacity of parasite plastids may be different at various stages in the life history. Thus, pigmentation and differentiated chloroplasts only appear in *Janczewskia morimotoi* three weeks after infection. During development of tetraspores and carpospores of *Levringiella*, the numbers of plastids increase although they remain small and contain few thylakoids. Tetraspores of *Choreocolax* contain plastids of similar appearance to those of vegetative cells, although occasionally more extensive thylakoid development has been noticed. Furthermore, these plastids are photosynthetically active. The results of autoradiographic experiments in which tetrasporic plants were labelled for two hours in $NaH^{14}CO_3$ in the light, show that the tetraspores are able to fix carbon dioxide into stable, insoluble products in the absence of the host (Fig. 5). Spores of *Choreocolax* may thus

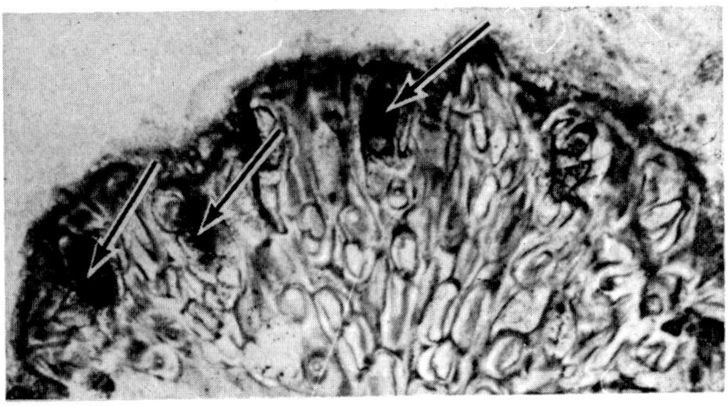

FIG. 5. Light microscope autoradiograph of *Choreocolax* after incubation in $NaH^{14}CO_3$ for 2 h in the light. Note the heavy accumulation of silver grains over tetrasporangia (arrows). (Embedded in glycol methacrylate, 1 μm section stained in toluidine blue, $\times 250$.)

represent an autotrophic stage of development which is repressed in the vegetative plant. The extent of carbon dioxide fixation by whole plants of *Choreocolax* dissected from the host thallus was investigated using $NaH^{14}CO_3$. Plants were incubated in sea-water in the light for varying periods of time and then allowed to assimilate $NaH^{14}CO_3$ for two and a half hours under the same conditions. From Table III it can be seen that there is a significant amount of carbon dioxide fixation in the light compared to dark controls, but in addition, the capacity for carbon dioxide fixation increases the longer the plants are incubated away from their host. The major soluble product of the neutral fraction synthesized by these plants is chromatographically identical with the unidentified compound found, after two hours incubation in $NaH^{14}CO_3$ in the light, in *Choreocolax* plants that are still attached to their hosts. These results suggest that the small reduced plastids present in the vegetative *Choreocolax* cells are potentially active in photosynthesis but that this capacity is normally repressed in the parasitic mode and may be released when translocation of

TABLE III. Photosynthesis by dissected plants of *Choreocolax*

Pre-treatment (h)	dpm mg^{-1}
3	16 202
18	30 635
29	41 725
66	206 053
3 dark	6 418
29 dark	6 742

Each sample contained 50 plants of *Choreocolax* and 25 μCi $H^{14}CO_3^-$. All incubations were performed at 15°C with a light intensity of 10 000 lux. Dark controls were also labelled in the dark.

carbon skeletons from the host ceases, as happens artificially following dissection, or more naturally when spores are released and germinate to form young plants. It would clearly be of interest in this context to examine the physiology and development of young plants before they become parasitic. If photosynthesis of attached *Choreocolax* plants is totally repressed, then the unidentified compound labelled after two hours incubation in the light must then be derived by translocation from *Polysiphonia*.

Examination of the parasite species listed in Table I shows that most species are extremely host-specific, usually attacking only one or two closely-related genera. Furthermore, adelphoparasitism is more common than alloparasitism. Martin and Pocock (1953) remark on this specificity and suggest that a close similarity between the metabolic products of the two plants ensures that the products of the host are more readily utilized by the parasite. In a similar way, Feldmann and Feldmann (1958) suggest, as indicated earlier, that the relative taxonomic positions of host and parasite will regulate or determine the formation of secondary pit connections and metabolic compatibility. These authors have clearly attempted to correlate the functional integrity of a host–parasite relationship with phylogeny and it is important to realize therefore, that no sound scientific basis for these generalizations in fact exists and, in particular, the continued use of the terms adelphoparasitism and alloparasitism in a functional context is of very limited value.

The point was made earlier that secondary pit connections may be present irrespective of the relative taxonomic position of host and parasite. Moreover, red algal pit connections are not thought to function in the movement of substances between cells (Pueschel, 1977) and since movement of materials between higher plant and fungal symbionts, for example, can occur in the absence of direct cellular connections, the presence of secondary pit connections must be considered to be of doubtful significance in red algal host–parasite relationships.

The elucidation of taxonomic affinities between host and parasite rests almost entirely on comparative morphology and in some cases, for example, *Halosacciocolax* (Edelstein, 1972; Guiry, 1974, 1975) and *Gelidiocolax* (Gardner, 1927; Fan and Papenfuss, 1959), a relationship may change from adelpho- to alloparasitism with changing systematic emphases. In addition, considerable differences in metabolism have been reported in algae, between different generations of the same species and different genera of the same family. Thus cell walls of the sporophytic and gametophytic generations of *Chondrus crispus* contain λ- and κ-carrageenen respectively (Chen *et al.*, 1973; McCandless *et al.* 1973; Gordon-Mills and McCandless, 1975, 1977). Cell wall composition could clearly affect the ability of sporeling germ-tubes to penetrate host tissue. Reference has already been made to the different soluble carbohydrate reserves of *Choreocolax* and *Holmsella*; in addition, *Bornetia*, *Callithamnion* and *Ceramium* (all Ceramiaceae) all possess different soluble reserves (see Craigie, 1974). In the wider context of parasitism and symbiosis it is also apparent that in many cases efficient carbohydrate translocation from autotroph to heterotroph is based upon the *dissimilarity* of host and parasite soluble reserves creating a strong source–sink relationship and minimizing reciprocal flow (Smith *et al.*, 1969). Equally it is evident that the comparative study of soluble reserves of red algal parasites is unlikely to provide any evidence of sound taxonomic value.

It is germane at this point to consider how parasites may have evolved. Setchell (1918) developed the hypothesis that parasites may have evolved through mutation of spores of a parent plant resulting perhaps in the total or partial loss of autotrophy, or in development of the ability to penetrate parent tissue, presumably through modified secretion of cell-wall degrading exo-enzymes. In support, Setchell cites the example of *Agardhiella tenera* where the tetraspores germinate *in situ* to produce a dwarf, pigmented epiphytic gametophyte with penetrating rhizoids that form secondary pit connections with the cells of the parent thallus (Osterhout, 1896; Fan, 1961). Osterhout records similar examples in other genera. In addition, Fan (1961) cites a number of examples of genera in which germination of carpospores within the parent thallus has been reported, although in such cases none of these germlings subsequently develop into whole dwarf plants. Feldmann and Feldmann (1958) suggest that circumstantial evidence supporting this hypothesis is the extensive distribution of adelphoparasitism and secondary pit connections although, as previously indicated, the functional significance of these characteristics is doubtful.

An alternative hypothesis is favoured by Sturch (1926) and Fan (1961). According to Sturch, parasites such as *Choreocolax* and *Harveyella* were originally epiphytes which developed a more aggressive habit, coming to represent one extreme of the whole spectrum of endophytic, parasitic and epiphytic relationships in the seaweeds. Although movement of nutrients between *Ascophyllum nodosum* and the epiphytic *Polysiphonia lanosa* has not been demonstrated (Harlin and Craigie, 1975; Turner and Evans, 1977) it is feasible that increased nutritional dependence could result by mutation.

It has been stated (Dixon, 1973) that the taxonomy of most algal groups is still in a pioneer phase. The aims of modern taxonomy have been outlined as follows (Dixon, 1973, after Davis and Heywood, 1963).

1. To provide a convenient method of identification and communication.
2. To provide a classification which as far as possible expresses the natural relationships of organisms.
3. To detect evolution at work, discovering its processes and interpreting its results.

From the foregoing account it is clear that with regard to the second and third of these three objectives, conventional systematic morphology has not provided any sound understanding of the evolution of red algal parasites, the relationships between different parasite genera, or their phylogenetic relationships with host plants.

Dixon (1973) has pleaded for more attention to be devoted to the species level, particularly with regard to phenotypic variability. Let us consider this in relation to the example of *Halosacciocolax*. Two species of *Halosacciocolax* have been described (Table I). The type species, *H. kjellmanii* grows primarily on *Halosaccion ramentaceum* (Edelstein and McLachlan, 1968). A different species, *H. lundii*, differing from *H. kjellmanii* in shape, size and colour (Table I), was recorded growing on *Palmaria palmata* (= *Rhodymenia palmata*) (Edelstein, 1972; Guiry, 1974). However, Guiry (1975) has also recorded *H. kjellmanii* growing on *P. palmata* and has suggested that the two forms may in fact be conspecific. This type of situation arising from phenotypic variability may be common, occurring in other parasite genera with more than one "species". Such situations may in fact be resolved by experimental studies on host specificity through cross-inoculation experiments using cultured material. The same type of experiment in a wider context would also test the functional significance of the allo- and adelphoparasite concepts.

A thorough appreciation of phylogenetic relationships between red algal parasites and their hosts will require more emphasis on criteria other than those of comparative morphology. One experimental approach has already been outlined. In other areas of plant taxonomy various biochemical criteria are employed. These include comparative serology, protein chemistry, carbohydrate chemistry and the chemistry of secondary plant products. Unfortunately it is impossible to predict whether such techniques will ever be applied to taxonomic studies of red algal parasites. Take, for example, the technique of comparative polyacrylamide gel electrophoresis. A rough calculation can show that enough soluble protein for the purposes of this technique could be obtained from some 1000 *Holmsella* plants, which requires some tedious but by no means impossible dissection. However, it is not yet known whether soluble proteins can be extracted from these organisms because the basic studies have not been carried out. Clearly the prime requirement for advancement in this area is a more extensive study of the basic biology of these organisms. Most of the well-known associations described before and soon after 1900 should be re-examined with modern techniques, particularly with respect to pigmentation, presence of plastids, the nature of the host–parasite interface and reproductive behaviour. Experimental studies on the nature of the infection process, enzyme secretion and the nutritional interrelationship should also be carried out. Attempts must be made to culture both host and parasite in the laboratory to obtain a more reliable and standardized source of material than that obtained from the wild. Only when such a body of information is available will it prove worth while to investigate and speculate upon the origins of these neglected plants.

REFERENCES

ADEY, W. H. and SPERAPANI, C. P. (1971). The biology of *Kvaleya epilaeve*, a new parasitic genus and species of Corallinaceae. *Phycologia* **10**, 29–42.

BATTERS, E. A. L. (1892a). On *Schmitziella*; a new genus of endophytic algae, belonging to the order Corallinaceae. *Ann. Bot.* **6**, 185–194.

BATTERS, E. A. L. (1892b). *Gonimophyllum buffhami*: a new marine alga. *J. Bot.* (*Lond.*) **30**, 65–67.

BATTERS, E. A. L. (1895). On some new British algae. *Ann. Bot.* **9**, 307–321.

BØRGESEN, F. (1920). The marine algae of the Danish West Indies, Vol. 2. Rhodophyceae with addenda to the Chlorophyceae, Phaeophyceae and Rhodophyceae. *Dansk bot. Ark.* **3**, 369–504.

BØRGESEN, F. (1930). Marine algae from the Canary Islands especially from Teneriffe and Gran Canaria III. Rhodophyceae Part III Ceramiales. *K. dansk. Vidensk. Selsk., Biol. Meddr* **9**(1), 1–159.

CABIOCH, J. and GUIRY, M. D. (1977). *Halosacciocolax kjellmanii* Lund, Rhodophycée parasite nouvelle pour les côtes de France. *Trav. Stat. biol. Roscoff,* (N.S.) **23**, 27–29.

CALLOW, J. A., CALLOW, M. E. and EVANS, L. V. (1978). Nutritional studies on the parasitic red alga *Choreocolax polysiphoniae* (in preparation).

Chemin, E. (1937). Role des bactéries dans la formation des galles chez les Floridées. *Annls Sci. Bot., nat. sér.* 10, **10**, 61–71.

Chen, L. [C.-M.], McLachlan, J., Neish, A. C. and Shacklock, P. F. (1973). The ratio of kappa- to lambda-carrageenan in nuclear phases of the rhodophycean algae *Chondrus crispus* and *Gigartina stellata*. *J. mar. biol. Ass. U.K.* **53**, 11–16.

Chiang, Y.-M. (1970). Morphological studies of red algae of the family Cryptonemiaceae. *Univ. Calif. Publs Bot.* **58**, 1–95.

Colin, H. and Guéguen, E. (1930). La constitution du principe sucré de *Rhodymenia palmata*. *C.r. hebd. Séanc. Acad. Sci., Paris*, **191**, 163–164.

Court, G. J. (1975). Photosynthesis and translocation studies of the symbiotic red algae *Janczewskia gardneri* and *Laurencia spectabilis* (Rhodophyceae, Ceramiales). *J. Phycol.* **11** (suppl.), 10 (abstr.).

Court, G. J. (1977). The symbiotic association of the red algae *Janczewskia gardneri* and *Laurencia spectabilis* (Ceramiales, Rhodophyceae). *J. Phycol.* **13** (suppl.), 14 (abstr.).

Craigie, J. S. (1974). Storage products. *In* "Algal Physiology and Biochemistry" (W. D. P. Stewart, ed.), pp. 206–235. Blackwell Scientific Publications, Oxford, London, Edinburgh, Melbourne.

Davis, P. H. and Heywood, V. H. (1963). "Principles of Angiosperm Taxonomy." Oliver and Boyd, Edinburgh and London.

Dawson, E. Y. (1944). A new parasitic red alga from southern California. *Bull. Torrey bot. Club* **71**, 655–657.

Derbès, F. (1856). Description d'une nouvelle espèce de Floridée, devant former un nouveau genre, et observations sur quelques algues. *Annls Sci. nat. Bot., sér.* 4, **5**, 209–220.

De Toni, G. B. (1903). "Sylloge Algarum . . .", Vol. IV, Sectio III. Sumptibus auctoris, Patavii.

Dixon, P. S. (1973). "Biology of the Rhodophyta." Oliver and Boyd, Edinburgh.

Edelstein, T. (1972). *Halosacciocolax lundii*, sp. nov., a new red alga parasitic on *Rhodymenia palmata* (L.) Grev. *Br. phycol. J.* **7**, 249–253.

Edelstein, T. and McLachlan, J. (1968). Investigations of the marine algae of Nova Scotia. V. Additional species new or rare to Nova Scotia. *Can. J. Bot.* **46**, 993–1003.

Evans, L. V., Callow, J. A. and Callow, M. E. (1973). Structural and physiological studies on the parasitic red alga *Holmsella*. *New Phytol.* **72**, 393–402.

Evans, L. V., Callow, M. E. and Callow, J. A. (1977). Host/parasite relationships in seaweeds. *Proc. int. Seaweed Symp.* **8** (in press).

Fan, K.-C. (1961). Studies on *Hypneocolax*, with a discussion on the origin of parasitic red algae. *Nova Hedwigia* **3**, 119–128.

Fan, K.-C. and Papenfuss, G. F. (1959). Red algal parasites occurring on members of the Gelidiales. *Madroño* **15**, 33–38.

Federation of British Plant Pathologists (1973). A guide to the use of terms in plant pathology. *Phytopath. Paper* No. 17.

Feldmann, J. and Feldmann, G. (1951). Sur quelques particularités cytologiques du *Janczewskia verrucaeformis* Solms-Laub., parasite du *Laurencia obtusa* (Huds.) Lamour. *C.r. hebd. Séanc. Acad. Sci., Paris* **233**, 1385–1386.

Feldmann, J. and Feldmann, G. (1958). Recherches sur quelques Floridées parasites. *Revue gén. Bot.* **65**, 49–128.

Foslie, M. (1898). Systematical survey of the Lithothamnia. *K. norsk. Vidensk. Selsk. Skr.* **1898**(2), 1–7.

Gardner, N. L. (1927). New Rhodophyceae from the Pacific coast of North America. III. *Univ. Calif. Publ Bot.* **13**, 333–368.

Goff, L. J. (1976). The biology of *Harveyella mirabilis* (Cryptonemiales; Rhodophyceae). V. Host responses to parasite infection. *J. Phycol.* **12**, 313–328.

Goff, L. J. (1977). Developmental studies of the red algal adelphoparasite *Gardneriella tuberifera* (Gigartinales). *J. Phycol.* **13** (suppl.), 24 (abstr.).

Goff, L. J. and Cole, K. (1973). The biology of *Harveyella mirabilis* (Cryptonemiales, Rhodophyceae). I. Cytological investigations of *Harveyella mirabilis* and its host, *Odonthalia floccosa*. *Phycologia* **12**, 237–245.

Goff, L. J. and Cole, K. (1975). The biology of *Harveyella mirabilis* (Cryptonemiales, Rhodophyceae). II. Carposporophyte development as related to the taxonomic affiliation of the parasitic red alga, *Harveyella mirabilis*. *Phycologia* **14**, 227–238.

Goff, L. J. and Cole, K. (1976). The biology of *Harveyella mirabilis* (Cryptonemiales, Rhodophyceae). III. Spore germination and subsequent development within the host *Odonthalia floccosa* (Ceramiales, Rhodophyceae). *Can. J. Bot.* **54**, 268–280.

Gordon-Mills, E. M. and McCandless, E. L. (1975). Carrageenans in the cell walls of *Chondrus crispus* Stack. (Rhodophyceae, Gigartinales). I. Localization with fluorescent antibody. *Phycologia* **14**, 275–281.

Gordon-Mills, E. M. and McCandless, E. L. (1977). Carrageenans in the cell walls of *Chondrus crispus* Stack. (Rhodophyceae, Gigartinales). II. Birefringence. *Phycologia* **16**, 169–176.

Grubb, V. M. (1925). The male organs of the Florideae. *J. Linn. Soc. Bot.* **47**, 177–255.

Guiry, M. D. (1974). The occurrence of the red algal parasite *Halosacciocolax lundii* Edelstein in Britain. *Br. phycol. J.* **9**, 31–35.

Guiry, M. D. (1975). *Halosacciocolax kjellmanii* Lund parasitic on *Palmaria palmata* forma *mollis* (S. *et* G.) Guiry in the eastern North Pacific. *Syesis* **8**, 113–117.

Harlin, M. M. and Craigie, J. S. (1975). The distribution of photosynthate in *Ascophyllum nodosum* as it relates to epiphytic *Polysiphonia lanosa*. *J. Phycol.* **11**, 109–113.

Heydrich, F. (1893). *Pleurostichidium*, ein neues Genus der Rhodomeleen. *Ber. dt. bot. Ges.* **11**, 344–348.

Howe, M. A. (1914). The marine algae of Peru. *Mem. Torrey bot. Club* **15**, 1–180.

Kugrens, P. (1970). Comparative ultrastructure of several parasitic red algae. *J. Phycol.* (Suppl.) **6**, 9.

Kugrens, P. and West, J. A. (1972a). Ultrastructure of spermatial development in the parasitic red algae *Levringiella gardneri* and *Erythrocystis saccata*. *J. Phycol.* **8**, 331–343.

Kugrens, P. and West, J. A. (1972b). Ultrastructure of tetrasporogenesis in the parasitic red alga *Levringiella gardneri* (Setchell) Kylin. *J. Phycol.* **8**, 370–383.

Kugrens, P. and West, J. A. (1973a). The ultrastructure of carpospore differentiation in the parasitic red alga *Levringiella gardneri* (Setch.) Kylin. *Phycologia* **12**, 163–173.

Kugrens, P. and West, J. A. (1973b). The ultrastructure of an alloparasitic red alga *Choreocolax polysiphoniae*. *Phycologia* **12**, 175–186.

Künzenbach, R. and Brucker, W. (1960). Zur Bildung von "Tumoren" an Meeresalgen II. *Ber. dt. bot. Ges.* **73**, 8–18.

Kylin, H. (1924). Studien über die Delesseriaceen. *Acta Univ. lund., N.F.*, Avd. 2, **20**(6), 1–111.

Kylin, H. (1930). Über die Entwicklungsgeschichte der Florideen. *Acta Univ. lund., N.F.*, Avd. 2, **26**(6), 1–104.

Kylin, H. (1941). Californische Rhodophyceen. *Acta Univ. lund. N.F. Avd. 2,* **37**(2), 1–51.

Kylin, H. (1956). "Die Gattungen der Rhodophyceen." CWK Gleerups Förlag, Lund.

Kylin, H. and Skottsberg, C. (1919). Zur Kenntnis der subantarctischen und antarctischen Meeresalgen II. Rhodophyceen. *In* "Wissenschaftliche Ergebnisse der schwedischen Südpolar-Expedition 1901–1093 . . ." (O. Nordenskjöld, ed.), Vol. 4, Lief. 15, pp. 1–88. Lithographisches Institut des Generalstabs, Stockholm.

Linskens, H. K. (1963). Beitrag zur Frage der Beziehungen zwischen Epiphyt und Basiphyt bei marinen Algen. *Pubbl. staz. zool. Napoli* **33**, 274–293.

Lund, S. (1959). The marine algae of East Greenland. I. Taxonomical part. *Meddr Grønland* **156**, 1–247.

Martin, M. T. and Pocock, M. A. (1953). South African parasitic Florideae and their hosts. 2. Some South African parasitic Florideae. *J. Linn. Soc. Bot.* **55**, 48–64.

Mason, L. R. (1953). The crustaceous coralline algae of the Pacific coast of the United States, Canada, and Alaska. *Univ. Calif. Publs Bot.* **26**, 313–390.

McBride, D. L., Kugrens, P. and West, J. A. (1974). Light and electron microscopic observations on red algal galls. *Protoplasma* **79**, 249–264.

McCandless, E. L., Craigie, J. S. and Walter, J. A. (1973). Carrageenans in the gametophytic and sporophytic stages of *Chondrus crispus*. *Planta* **112**, 201–212.

Moebius, M. (1885). Ueber eine neue epiphytische Floridee. *Ber. dt. bot. Ges.* **3**, 77–80.

Morrill, J. (1976a). Notes on parasitic Rhodomelaceae. I. The morphology and systematic position of *Benzaitenia yenoshimensis* Yendo, a parasitic red alga from Japan. *Proc. Acad. Sci. Phila.* **127**, 203–215.

Morrill, J. (1976b). Notes on parasitic Rhodomelaceae. II. The morphology and systematic position of the parasitic red alga *Jantinella verrucaeformis* (Setchell and McFadden) Kylin. *Proc. Acad. Sci. Phila.* **127**, 216-231.

Morrill, J. (1976c). Notes on parasitic Rhodomelaceae. III. *Meridiocolax narcissus*, a new red alga parasitic on *Polysiphonia ferulaceae* Suhr from the Florida Keys. *Proc. Acad. Sci. Phila.* **127**, 233–248.

Mshigeni, K. E. (1976). New records of *Hypneocolax stellaris* f. *orientalis* Weber-von-Bosse. A parasitic red alga. *Nova Hedwigia* **27**, 829–834.

Osterhout, W. J. V. (1896). On the life-history of *Rhabdonia tenera* J. Ag. *Ann. Bot.* **10,** 403-427.

Parke, M. and Dixon, P. S. (1976). Check-list of British marine algae—third revision. *J. mar. biol. Ass. U.K.* **56**, 527–594.

Pocock, M. A. (1956). South African parasitic Florideae and their hosts. 3. Four minute parasitic Florideae. *Proc. Linn. Soc. Lond.* **167**, 11–41.

Pueschel, C. M. (1977). A freeze-etch study of the ultrastructure of red algal pit plugs. *Protoplasma* **91**, 15–30.

Putman, E. W. and Hassid, W. Z. (1954). Structure of galactosylglycerol from *Iridaea laminarioides*. *J. Am. chem. Soc.* **76**, 2221–2223.

Rawlence, D. J. (1972). An ultrastructural study of the relationship between rhizoids of *Polysiphonia lanosa* (L.) Tandy (Rhodophyceae) and tissue of *Ascophyllum nodosum* (L.) Le Jolis (Phaeophyceae). *Phycologia* **11**, 279–290.

Reinke, J. (1889). "Algenflora der westlichen Ostsee deutschen Antheils. Eine systematisch-pflanzengeographische Studie." Paul Parey, Berlin.

Reinsch, P. F. (1875). "Contributiones ad algologiam et fungologiam", Vol. I. T.O. Weigel, Lipsiae.

Rosenvinge, L. K. (1898). Deuxieme mémoire sur les algues marines du Groenland. *Meddr Grønland* **20**, 1–125.

Rosenvinge, L. K. (1931). The marine algae of Denmark. Contributions to their natural history, Part IV. Rhodophyceae IV (Gigartinales, Rhodymeniales, Nemastomatales). *K. danske Vidensk. Selsk. Skr., 7. Raekke, Nat. Math. Afd.* **7**, 487–627.

Schmitz, F. (1889). Systematische Übersicht der bisher bekannten Gattungen der Florideen. *Flora* **72**, 435–456.

Schmitz, F. (1894). Kleinere Beiträge zur Kenntniss der Florideen. *Nuova Notarisia* **5**, 608–635.

Schmitz, F. and Falkenberg, P. (1897). Rhodomelaceae. *In* "Die natürlichen Pflanzenfamilien, . . ." (A. Engler and K. Prantl, eds), Teil 1, Abt. 2, pp. 421–480. Wilhelm Engelmann, Leipzig.

Schmitz, F. and Hauptfleisch, P. (1896–97). Rhodophyceae. *In* "Die natürlichen Pflanzenfamilian, . . .", (A. Engler and K. Prantl, eds), Teil 1, Abt. 2, pp. 298–420, 481–544. Wilhelm Engelmann, Leipzig.

Setchell, W. A. (1914). Parasitic Florideae, I. *Univ. Calif. Publs Bot.* **6**, 1–34.

Setchell, W. A. (1918). Parasitism among the red algae. *Proc. Am. phil. Soc.* **57**, 155–172.

Setchell, W. A. (1923). Parasitic Florideae, II. *Univ. Calif. Publs Bot.* **10**, 393–396.

Skottsberg, C. (1953). On two collections of Antarctic marine algae. *Arkiv. Bot.*, N.S. **2**, 531–566.

Smith, D., Muscatine, L. and Lewis, D. (1969). Carbohydrate movement from autotrophs to heterotrophs in parasitic and mutualistic symbiosis. *Biol. Rev.* **44**, 17–90.

Solms-Laubach, H. (1877). Note sur le *Janczewskia* Nouvelle Floridée parasite du *Chondria obtusa. Mém. Soc. Sci. nat. Cherbourg* **21** (sér. 3, Tom. 1), 209–224.

Sparling, S. R. (1957). The structure and reproduction of some members of the Rhodymeniaceae. *Univ. Calif. Publs Bot.* **29**, 319–396.

Sturch, H. H. (1899). *Harveyella mirabilis* (Schmitz and Reinke). *Ann. Bot.* **13**, 83–102.

Sturch, H. H. (1924). On the life-history of *Harveyella pachyderma* and *H. mirabilis. Ann. Bot.* **38**, 27–42.

Sturch, H. H. (1926). *Choreocolax polysiphoniae*, Reinsch. *Ann. Bot.* **40**, 585–605.

Tokida, J. (1958). A review on galls in seaweeds. *Bull. Jap. Soc. Phycol.* **6**, 93–99.

Turner, C. H. C. and Evans, L. V. (1977). Physiological studies on the relationship between *Ascophyllum nodosum* and *Polysiphonia lanosa. New Phytol.* **79**, 363–371.

Weber van Bosse, A. (1928). Liste des algues du Siboga. IV. Rhodophyceae Troisième partie Gigartinales et Rhodymeniales. . . . *Siboga-Exped. Monogr.* **59d**, 393–533.

Wilson, H. L. (1910). *Gracilariophila*, a new parasite on *Gracilaria confervoides. Univ. Calif. Publs Bot.* **4**, 75–84.

Yendo, K. (1913). Some new algae from Japan. *Nyt. Mag. Naturvid.* **51**, 275–288.

6 | The Importance of Sporangia in the Classification of the Florideophyceae

M. D. GUIRY

Department of Biological Sciences, Portsmouth Polytechnic, The Marine Laboratory, Hayling Island, England

Abstract: The ways in which sporangia may be categorized are discussed. The most important factor in determining the form and function of sporangia and spores is considered to be the presence or absence of meiosis. Accordingly sporangia are considered to be of two basic types: meiosporangia and mitosporangia. A literature survey suggests that most sporangia so far described (tetrasporangia, carposporangia, bisporangia, polysporangia and possibly monosporangia) have meiotic and mitotic homologues.

A detailed analysis of the manner in which spores are arranged within tetrasporangia is presented and this is considered to be of considerable importance in the classification of the Florideophyceae. The available evidence suggests that a cruciate mode of arrangement is the basic type from which the zonate and tetrahedral types have been developed. Additionally it is shown that there are two ways in which tetrasporangia may be cleaved; these are referred to as successive and simultaneous. Successive cruciate tetrasporangia are found in all the orders of the Florideophyceae; simultaneous cruciate tetrasporangia are probably only present in the Ceramiales. Successive zonate tetrasporangia seem to occur only in the Cryptonemiales and Gigartinales; simultaneous zonate tetrasporangia are characteristic of the Corallinaceae (Cryptonemiales). Successive tetrahedral tetrasporangia occur only in the Ceramiaceae (Ceramiales), while simultaneous tetrahedral tetrasporangia are found only in the Rhodymeniales and Ceramiales. A brief discussion of the manner of sporangial formation and the importance of generative sporangial mother cells (= stalk cells) in the classification of the Florideophyceae is presented.

The relevance of compound sporangia and pseudosporangia (vegetative reproductive organs resembling sporangia but not liberating spores) is discussed.

INTRODUCTION

The basis of the present ordinal classification for the Florideophyceae is, as proposed by Schmitz (1883, 1889) and modified by Oltmanns (1904) and Kylin (1928, 1930, 1931, 1932), the pre- and post-fertilization changes in, or

Systematics Association Special Volume No. 10, "Modern Approaches to the Taxonomy of Red and Brown Algae", edited by D. E. G. Irvine and J. H. Price, 1978, pp. 111–144, Academic Press, London and New York.

associated with, the female reproductive apparatus. Despite anomalies the system is widely accepted. Dixon (1961) and Papenfuss (1966) have reviewed the situation and have proposed acceptance of a system of five or six orders pending the acquisition of further information. Papenfuss (1966) recognized the Nemaliales, Gelidiales, Cryptonemiales, Gigartinales, Rhodymeniales and Ceramiales, whereas Dixon (1961, 1973) included the Gelidiales in the Nemaliales. Both Papenfuss and Dixon included the Acrochaetiales (Feldmann, 1953), the Bonnemaisoniales (Feldmann and Feldmann, 1942) and the Chaetangiales (Desikachary, 1964) in the Nemaliales. Recently Guiry and Irvine (1978) proposed a new order, the Palmariales, based largely on an unusual mode of sporangial formation. The purpose of the present paper is to consider the importance of sporangia in the classification of the Florideophyceae.

A spore is a cell which can independently reproduce a plant of the species, whilst a sporangium can be defined as a structure within which the spore arises. Sporangia and spores may be categorized in several different ways, including: the presence or absence of meiosis; the number of nuclei in each spore; the number of spores in a sporangium; the rôle of the spores in the life history; the position of the sporangia in the thallus and the presence or absence of a generative mother cell.

Meiosis typically occurs in a sporangium which is borne on a free-living diploid thallus. However, the evidence for meiosis in sporangia of the Florideophyceae is, for the vast majority, largely circumstantial. Although there are generally four spores formed, the number may be greater or smaller. Spores derived from sporangia which undergo meiotic divisions generally form haploid gametangial thalli but the assumption that all sporangia with four spores undergo meiosis has been shown to be erroneous. Sporangia with one or two spores have conversely been assumed to undergo mitosis whereas, as will be shown later, it is reasonable to assume that both mitoses and meioses may take place in such sporangia. The position of sporangia on the subtending filaments or on the surface of the thallus would seem to be of little major taxonomic importance and will not be considered further here. The formation of modified sporangial mother cells appears to be of sporadic occurrence but, as will be discussed later, may be of some taxonomic importance.

One may conclude, therefore, that meiosis is the most important factor governing the form and function of sporangia and the resultant spores. Sporangia may be divided into two general categories: meiosporangia (those in which meiosis takes place) and mitosporangia (those in which mitosis takes place). Meiospores form gametangial thalli while mitospores form thalli of a similar ploidy level to that of the parent thallus.

MEIOSPORANGIA

By far the commonest type of meiosporangium is the tetrasporangium, in which four spores are formed. Tetrasporangia are usually formed on diploid thalli; in a restricted number of instances tetrasporangia may be formed from cells derived from gonimoblast filaments, in which case they are referred to as carpotetrasporangia (Feldmann, 1939). Meiosis is also assumed to take place in polysporangia *sensu* Drew (1937, 1939) (sporangia with more than four spores, although formed in the position and manner of tetrasporangia but, as will be shown later, it is not unreasonable to assume that some polysporangia may form spores by mitosis. Some bisporangia have been shown to undergo meiosis in the formation of spores (Suneson, 1950) and such spores have two nuclei (i.e. four nuclei in a sporangium). There are some indications (West, 1969) that some monosporangia may have four nuclei in a single spore and it is possible that these are meiotically derived.

1. *Tetrasporangia*

Sporangia with three cleavages generally have four spores and the nuclei are then assumed to be meiotically derived. The formation of tetrasporangia has, particularly since the advent of precise culturing techniques, been shown to be an almost universal and basic feature of the Florideophyceae. However, it is interesting to note that tetrasporangia as such do not occur in the Bangiophyceae, a distinguishing feature of this class frequently not mentioned in text-books. Table I shows the occurrence of tetrasporangia in the families of the Florideophyceae and it can be seen that these are now known in almost every family except the freshwater representatives, which may all undergo somatic meiosis as reported by Mullahy (1952) for *Lemanea australis* Atk., by Magne (1964) for species of *Lemanea* and *Paralemanea*, and as indicated by Hurdlebrink and Schwantes (1972) for a species of *Batrachospermum*.

The manner in which the four spores are arranged clearly has some taxonomic significance. There are three basic types of arrangement (Fig. 1F–H): cruciate, tetrahedral and zonate. There would seem to be some confusion as to the precise orientation and timing of the cleavages which precede the formation of these three types and it is proposed to examine these in detail.

(*a*) *Cruciately divided tetrasporangia.* These are always ovate in shape and usually have a median division, typically along the short axis (Fig. 1F). The orientation of the other two cleavages, which typically take place along the long axis, is much more irregular. In some cruciately divided tetrasporangia the longitudinal cleavages are in a single plane but this may or may not be in the plane of view

TABLE I. Occurrence of various types of sporangia in the orders and families of the Florideophyceae

	Tetrasporangia										
	Cruciate	Zonate	Tetrahedral	Unknown	Irregularly cruciate	Irregularly zonate	Polysporangia	Parasporangia	Seirosporangia	Monosporangia	Bisporangia
NEMALIALES											
Acrochaetiaceae Fritsch[a]	+	.	.	.	.	.	.	.	+	+	+
Batrachospermaceae Fries	.	.	.	+	.	.	.	.	.	.	.
Lemaneaceae C. Agardh	.	.	.	+	.	.	.	.	.	.	.
Thoreaceae Hassall	.	.	.	+	.	.	.	.	.	.	.
Helminthocladiaceae J. Agardh	+	.	.	.	.	.	.	.	.	+	.
Nemaliaceae Pfitzer	+	.	.	.	.	.	.	.	.	+	.
Dermonemataceae Abbott	+	.	.	.	.	.	.	.	.	+?	.
Chaetangiaceae Kützing	+	.	.	.	.	.	.	.	.	+	.
Naccariaceae Kylin	+	.	.	.	.	.	.	.	.	.	.
Bonnemaisoniaceae Schmitz in Engler	+	.	.	.	+	.	.	.	.	.	.
Gelidiaceae Kützing	+	.	+?	.	+	.	.	.	.	.	.
Gelidiellaceae Fan	+	.	+?	.	+	.	.	.	.	.	.
Wurdemanniaceae Taylor	.	+	.	.	.	.	.	.	.	.	.
Gymnocodiaceae Elliott[b]	?	.	.	.	.	.	.	.	.	.	.
CRYPTONEMIALES											
Weeksiaceae Abbott	+	+	.	.	+	+	.	.	.	.	.
Dumontiaceae Bory	+	+	.	.	+	+	.	.	+	+	.
Choreocolacaceae Sturch	+	.	.	.	.	.	.	.	.	.	.
Cryptonemiaceae Harvey	+	.	.	.	.	.	.	.	.	+	.
Corynomorphaceae Balakrishnan	+	.	.	.	+	+	.	.	.	.	.
Pseudoanemoniaceae V. J. Chapman	.	.	.	+	.	.	.	.	.	.	.
Kallymeniaceae (J. Agardh) Kylin	+	.	.	.	.	.	.	.	.	.	.
Endocladiaceae (J. Agardh) Kylin	+	.	.	.	+	.	.	.	.	.	.
Crossocarpaceae Perestenko	+	.	.	.	.	.	.	.	.	.	.
Gloiosiphoniaceae Schmitz in Engler	+	+	.	.	+	+	.	.	.	.	.
Tichocarpaceae Kylin	+?	.	.	.	.	.	.	.	.	.	.
Hildenbrandiaceae Rabenhorst	+	+	.	.	+	+	.	.	.	.	.
Pterocladiophilaceae Fan and Papenfuss	.	+	.	.	.	.	.	.	.	.	.
Peyssonneliaceae Denizot	+	.	.	.	+	+?	.	.	.	.	.
Solenoporaceae Pia[b]	?	.	.	.	.	.	.	.	.	.	.
Corallinaceae Lamouroux	+	+	.	.	.	.	.	.	.	.	+
GIGARTINALES											
Gymnophlaeaceae Kützing	+	.	.	.	.	.	.	.	.	.	.
Gracilariaceae Nägeli	+	.	.	.	.	.	.	.	.	.	.
Sebdeniaceae Kylin	+	.	.	.	.	.	.	.	.	.	.
Calosiphoniaceae Kylin	+	.	.	.	.	.	.	.	.	.	.
Petrocelidaceae Denizot	+	.	.	.	+	.	.	.	.	.	.
Phyllophoraceae Nägeli	+	.	.	.	.	.	.	.	.	.	.
Gigartinaceae Bory	+	.	.	.	.	.	.	.	.	.	.
Chondriellaceae Levring	+	.	.	.	.	.	.	.	.	.	.
Polyideaceae Kylin	+	.	.	.	.	.	.	.	.	.	.
Nizymeniaceae Womersley	+	.	.	.	+	.	.	.	.	.	.
Rhizophyllidaceae (Montagne) Schmitz in Engler	+	+	.	.	+	+	.	.	.	.	.
Acrotylaceae Schmitz in Engler	.	+	.	.	.	.	.	.	.	.	.
Plocamiaceae Kützing	.	+	.	.	.	.	.	.	.	.	.
Sphaerococcaceae Dumortier	.	.	.	+	.	.	.	.	.	.	.
Phacelocarpaceae Searles	.	+	.	.	.	.	.	.	.	.	.
Sarcodiaceae Kylin	.	+	.	.	.	.	.	.	.	.	.
Furcellariaceae Greville	.	+	.	.	.	.	.	.	.	.	.
Solieriaceae Hauck	.	+	.	.	.	.	.	.	.	.	.
Hypneaceae J. Agardh	.	+	.	.	.	.	.	.	.	.	.

TABLE I—*continued*

	Tetrasporangia										
	Cruciate	Zonate	Tetrahedral	Unknown	Irregularly cruciate	Irregularly zonate	Polysporangia	Parasporangia	Seirosporangia	Monosporangia	Bisporangia
GIGARTINALES											
Rissoellaceae Kylin	.	+	.	.	.	.	.	.	.	.	.
Rhabdoniaceae Kylin	.	+	.	.	.	.	.	.	.	.	.
Cubiculosporaceae Kraft	.	+	.	.	.	.	.	.	.	.	.
Rhodophyllidaceae Schmitz in Engler	.	+	.	.	.	.	.	.	.	.	.
Mychodeaceae Kylin	.	+	.	.	.	.	.	.	.	.	.
Dicranemaceae Kylin	.	+	.	.	.	.	.	.	.	.	.
Cruoriaceae (J. Agardh) Kylin	.	+	.	.	.	.	.	.	.	.	.
Blinksiaceae Hollenberg & Abbott	.	+	.	.	.	.	.	.	.	.	.
RHODYMENIALES											
Rhodymeniaceae Harvey	+	.	+	.	.	.	.	.	.	.	.
Champiaceae Kützing	.	.	+	.	.	.	+	.	.	.	.
PALMARIALES											
Palmariaceae Guiry	+	.	.	.	.	.	.	.	.	.	.
CERAMIALES											
Ceramiaceae C. Agardh	+	.	+	.	.	.	+	+	+	+?	+
Delesseriaceae Bory	.	.	+	.	.	.	+	.	.	+?	.
Dasyaceae Kützing	.	.	+	.	.	.	.	.	.	.	.
Rhodomelaceae Harvey	.	.	+	.	.	.	.	.	.	+?	.

[a] Grateful acknowledgement is made to Dr P. C. Silva (Department of Botany, University of California, Berkeley) for providing the correct authorities for some of the families listed in this table.
[b] Fossil families for which sporangia are known but the mode of cleavage of the spores is not.

(Fig. 2C, D); if it is not, the sporangium appears to have only two spores (Fig. 2D). This may partly account for frequent reports of "bisporangia" being mixed with tetrasporangia. In the vast majority of cruciately divided tetrasporangia the median cleavage takes place before the longitudinal cleavages have been initiated (a situation here referred to as *successive* cleavage: Fig. 2B), but in a certain number of species referred to the Ceramiaceae (see p. 123) it has been reported that cruciate cleavages take place at the same time (here referred to as *simultaneous* cleavage: Fig. 4A, C). Cabioch (1972) indicated that cleavages in the sporangia of *Sporolithon erythraeum* (Rothpletz) Kylin (Corallinaceae; Mediterranean) are cruciate, and her illustration suggests that the cleavages may be simultaneous. Denizot (1968) also illustrated decussate cruciate tetrasporangia in *S. ptychoides* Heydr. (New Hebrides) but did not state whether or not the cleavages are successive or simultaneous. Rosenvinge (1931) reported that the cruciately divided tetrasporangia of *Harveyella mirabilis* (Reinsch) Reinke cleave simultaneously but Sturch (1924) and Goff (pers. comm.) reported that the cleavages are successive.

In a considerable number of red algae with cruciately divided tetrasporangia, cleavages take place which are not in the same plane (Fig. 2E–H) and the degree of divergence may vary through 180° (Guiry, 1976). Frequently one of the

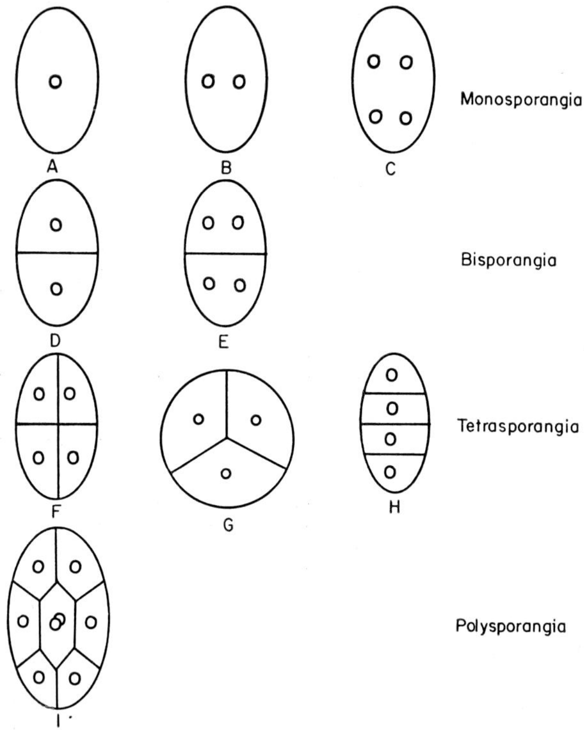

FIG. 1A–I. Diagrammatic representation of some sporangial types found in the Florideophyceae.

longitudinal cleavages may be at an angle of 90° to the other, which, when one or other of the cleavages is in the plane of view, gives the appearance of a sporangium with a single large spore and two smaller spores; this is referred to as decussate cruciate (Fig. 2E–H; cf. Denizot, 1968; Abbott and Hollenberg, 1976; Guiry, 1976, figs 15, 16). Sporangia have been reported where one of the longitudinal cleavages does not meet the median cleavage (Fig. 3C, D; Sjöstedt, 1926; Balakrishnan, 1962; Denizot, 1968) thus giving a tetrasporangium which is intermediate between cruciate and zonate. A further variation (which seems to be rather rare) is where two cleavages take place along the short axis and only one along the long axis, dividing the centre portion of the sporangium into two spores (Fig. 2I, K; Balakrishnan, 1962, fig. 25; Denizot, 1968, fig. 212; Rosenvinge, 1917, fig. 76A). In several species, e.g. *Cruoriopsis danica* Rosenv. and *C. hauckii* Batt. (Rosenvinge, 1917, fig. 107G, 109C); *Hildenbrandia rubra* (Sommerf.) Menegh. [= *H. prototypus* Nardo] (Rosenvinge,

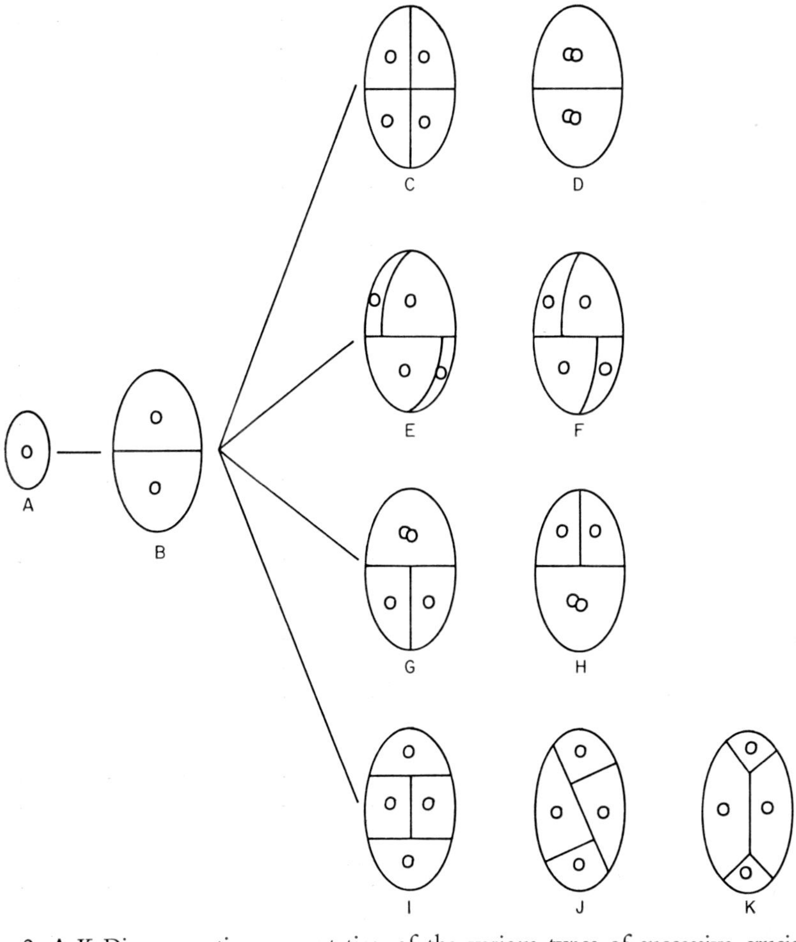

FIG. 2. A-K Diagrammatic representation of the various types of successive cruciate tetrasporangia.

1923–1924, fig. 125); *Petrocelis hennedyi* (Harv.) Batt. (Rosenvinge, 1917, fig. 96), sporangia have been recorded in which the median division is oblique (Fig. 2J) thus again very closely approaching a zonate tetrasporangium.

Cruciately divided tetrasporangia have been reported in all orders of the Florideophyceae (Table I) and would seem to be the most primitive of all the

TABLE II. Modes of cleavage of tetrasporangia in selected genera of the Cryptonemiales

	Cruciate	Zonate	Irregularly cruciate	Irregularly zonate	Unknown
Weeksiaceae					
Weeksia Setchell	+[a]	.	+[a]	.	.
Leptocladia J. Agardh	.	.	+[a]	.	.
Constantinea Postels and Ruprecht	.	.	+[a]	+[a]	.
Dumontiaceae					
Dumontia Lamouroux	+[d]	.	+[d]	.	.
Cryptosiphonia J. Agardh	.	.	+[c, l]	.	.
Farlowia J. Agardh*	.	.	+[b]	.	.
Dudresnaya P. and H. Crouan	.	+[d]	.	.	.
Pikea Harvey	+[e]	.	.	.	.
Dilsea Stackhouse	+[d]	.	+[n]	.	.
Neodilsea Tokida	.	.	+[o]	.	.
Acrosymphyton Sjöstedt*	.	.	+[g]	+[g]	.
Thuretellopsis Kylin	+[f]	.	.	.	.
Hyalosiphonia Okamura	.	.	+[j]	.	.
Dasyphloea Montague	.	+[j]	.	.	.
Gibsmithia Doty	+[p]	.	+[p]	.	.
Corynomorphaceae					
Corynomorpha J. Agardh	.	.	+[h]	+[h]	.
Gloiosiphoniaceae					
Gloiosiphonia Carmichael ex Berkeley*	.	.	+[i]	+[i]	.
Schimmelmannia Kützing*	+[b]	.	.	.	.
Thuretella Schmitz	.	.	.	.	+[j]
Endocladiaceae					
Endocladia J. Agardh	.	.	+[k]	.	.
Gloiopeltis J. Agardh	.	.	+[c]	+[c]	.
Hildenbrandiaceae					
Hildenbrandia Nardo	.	+[m]	+[m]	.	.
Peyssonneliaceae					
Peyssonnelia Decaisne	+[m]	+[m]	.	.	.
Cruoriella P. and H. Crouan	+[m]	.	.	.	.
Cruoriopsis Dufour	+[m]	.	+[m]	+[m]	.
Polystrata Heydrich	+[m]	.	.	.	.

* Where known; [a] Abbott, 1968; [b] Abbott and Hollenberg, 1976; [c] Sjöstedt, 1926; [d] Newton, 1931; [e] Scott and Dixon, 1972; [f] Dixon and Richardson, 1969; [g] Cortel-Breeman and Hoek, 1970; [h] Balakrishnan, 1962; [i] Edelstein, 1970; [j] Kylin, 1956; [k] Kylin, 1928; [l] Kylin, 1930; [m] Denizot, 1968; [n] Rosenvinge, 1917; [o] Tokida, 1943; [p] Doty, 1963.

sporangial types. In the Nemaliales, cruciately divided tetrasporangia occur in seven of the eight families for which tetrasporangia are known (Table I) and it is probable that all the marine Nemaliales will be shown to have successive cruciately divided tetrasporangia (see p. 123). In the Gigartinales cruciately divided tetrasporangia are known to characterize nine families of the twenty-five for which tetrasporangia are known (Table I) but the remaining families of the order are mostly characterized by regularly zonately divided tetrasporangia. All the cruciately and zonately divided tetrasporangia of the Gigartinales probably have successive cleavages. In the Cryptonemiales the distinction between cruciately divided tetrasporangia is often very difficult to make. In only three families (Cryptonemiaceae, Kallymeniaceae and Choreocolacaceae: Table I) is it possible to say that the species included have regularly cruciately divided tetrasporangia with any degree of consistency. In all the other families an admixture of types occurs, ranging from regularly cruciately divided to irregularly cruciately divided to irregularly zonately divided to regularly zonately divided (although as will be seen in the next section the last is only consistently found in members of the Corallinaceae). Table II outlines the occurrence of regularly and irregularly cruciately and zonately divided tetrasporangia in the genera of the Cryptonemiales and shows how variable is the mode of division within genera. It would seem probable, therefore, that the zonate mode of division is a relatively recent and haphazard development in the Cryptonemiales (except in the Corallinaceae) while in the Gigartinales the patterns seem better established. Irregularly divided tetrasporangia with a basically cruciate pattern have been reported in two genera of the Bonnemaisoniaceae (Nemaliales): *Delisea* and *Leptophyllis* (Levring, 1953). The tetrasporangia of *Leptophyllis* are stated by Kylin (1956) to be irregularly cruciately divided but the same author then refers to them as being irregularly zonate. The tetrasporangia of *Trailliella* and *Falkenbergia* (alternate phases in the life histories of algae of the Bonnemaisoniaceae) are rather irregularly successively cruciately divided (e.g. *Trailliella intricata*: Newton, 1931; Taylor, 1957) and not tetrahedrally divided as stated in some texts.

Cruciately divided tetrasporangia are characteristic of the subfamily Rhodymenioideae (Rhodymeniaceae, Rhodymeniales) where they are regularly and successively divided. The subfamily Hymenocladioideae (Rhodymeniaceae) and the family Champiaceae (Rhodymeniales) are both characterized by simultaneous tetrahedrally divided tetrasporangia (Sparling, 1957). Cruciately divided tetrasporangia are also found in some of the more primitive genera of the Ceramiaceae (Ceramiales) but most have tetrahedrally divided tetrasporangia (see Table III).

TABLE III. Occurrence of sporangial types and division of tetrasporangia of some of the tribes of the Ceramiaceae

	Tetrasporangia									
	"Oblique" *sensu* Wollaston 1968	Cruciate	Tetrahedral	Successive	Simultaneous	Parasporangia	Polysporangia	Seirosporangia	Bisporangia	Monosporangia
Crouanieae										
Crouania J. Agardh	.	+[d]	+[a]	.	.	.	.	.	+[d]	.
Euptilocladia Wollaston	.	.	+[a]	.	.	.	.	.	.	.
Gattya Harvey	.	.	+[a]	.	.	.	.	.	.	.
Gulsonia Harvey	.	.	+[a]	.	.	.	.	.	.	.
Gulsoniopsis Hommersand	.	+[c, e]	.	.	+[e]	.	.	.	.	.
Muellerena Schmitz in Schmitz and Hauptfleisch	.	.	+[m]	.	.	.	.	.	.	.
Ptilocladia Sonder	.	.	+[a]	.	.	.	.	.	.	.
Wrangelieae										
Wrangelia C. Agardh	.	.	+[d]	.	.	.	.	.	.	.
Callithamnieae										
Aristothamnion J. Agardh	.	.	.	.	.	.	+[r]	.	.	.
Callithamnion Lyngbye	.	.	+[d]	.	.	+[l]	.	.	+[l]	.
Thamnocarpus Harvey	unknown?		.	.	.	.	.	.	.	.
Seirospora Harvey	.	.	+[d]	.	.	.	.	+[d]	+[d]	.
Spyridieae										
Spyridia Harvey	.	.	+[c, k]	.	+[e]	.	.	.	.	.
Antithamnieae										
Acrothamnion J. Agardh	.	+[a]	.	.	.	.	.	.	.	.
Antithamnion Nägeli	.	+[a]	.	.	.	.	.	.	.	+[s]
Ballia Harvey	.	+[a]	.	.	.	.	.	.	.	.
Scagelonema Norris and Wynne	.	+[v]	.	+[v]	.	.	.	.	.	.
Bracebridgea J. Agardh	unknown[r]		.	.	.	.	.	.	.	.
Hollenbergia Wollaston	.	+[b]	.	+[b]	.	.	.	.	.	.
Macrothamnion Wollaston	+[a]	+[a]	.	.	.	.	.	.	.	.
Platythamnion J. Agardh	.	+[a]	.	+[a]	.	.	.	.	.	.
Warrenia Harvey	.	+[b]	.	.	.	.	.	.	.	.
Heterothamnieae										
Antithamnionella Lyle	.	+[b]	+[b]	.	.	.	.	.	.	.
Amoenothamnion Wollaston	+[a]	.	.	.	.	.	.	.	.	.
Heterothamnion J. Agardh	.	.	+[a]	.	.	.	.	.	.	.
Perithamnion J. Agardh	+[a]	.	.	.	.	.	.	.	.	.
Tetrathamnion Wollaston	+[a]	.	+[a]	.	.	.	.	.	.	.
Trithamnion Wollaston	.	.	+[a]	.	.	.	.	.	.	.
Scagelia Wollaston	.	+[b]	+[b]	.	.	.	.	.	.	.
Dohrnielleae										
Dohrniella Funk	.	+[d]	+[d]	.	.	.	.	+[d]	+[d]	.
Delesseriopsieae										
Delesseriopsis Okamura	.	+[h, q]	.	.	.	.	.	.	.	.
Balliella Itono and Tanaka	.	+[g, h]	.	+[g]	.	.	.	.	.	.
Spermothamnieae										
Bornetia Thuret	.	.	+[d]	.	.	.	.	.	.	.
Lejolisia Bornet	.	.	+[d]	.	.	.	.	.	.	.
Ptilothamnion Thuret	.	.	+[e]	.	.	.	+[e]	.	.	.
Spermothamnion Areschoug	.	.	+[e]	.	.	.	.	.	+[e]	.
Tiffaniella Doty and Meñez	.	.	+[e]	.	.	.	+[e, f]	.	.	.
Lomathamnion Gordon	.	.	+[e]	.	.	.	.	.	.	.
Interthamnion Gordon	.	.	+[e]	.	.	.	.	.	.	.
Ptilothamniopsis Dixon	.	.	+[o]	.	.	.	.	.	.	.

TABLE III—*continued*

	Tetrasporangia									
	"Oblique" *sensu* Wollaston 1968	Cruciate	Tetrahedral	Successive	Simultaneous	Parasporangia	Polysporangia	Seirosporangia	Bisporangia	Monosporangia
Sphondylothamnieae										
Diplothamnion Joly and Yamaguishi	.	.	+[e]	.	.	.	.	.	.	.
Drewiana Gordon	.	.	+[e]	.	.	.	.	.	.	.
Grallatoria Howe	.	+[u]	+[u]	.	.	.	.	.	.	.
Involucrana Baldock and Womersley	unknown[g]		.	.	.	.	+[e, g]	.	.	.
Medeiothamnion Pujals	.	.	+[w]	.	.	.	.	.	.	.
Shepleya Gordon	.	.	+[e]	.	.	.	.	.	.	.
Sphondylothamnion Nägeli	.	.	+[e]	.	.	.	.	.	.	.
Vickersia Karsakoff	.	.	+[d, e]	.	.	.	.	.	.	.
Wollastoniella Gordon	.	.	+[e]	.	.	.	.	.	.	.
Compsothamnieae										
Compsothamnion Nägeli	.	+[d]	+[d]	.	.	.	.	.	.	.
Dasythamnion J. Agardh	unknown[r]		.	.	.	.	.	.	.	.
Haloplegma Montagne	.	.	+[r]	.	.	.	.	.	.	.
Lophothamnion J. Agardh	unknown		.	.	.	.	+[r]	.	.	.
Mesothamnion Børgesen	.	.	+[p]	.	.	.	+[p]	.	.	.
Pleonosporium Nägeli	unknown		.	.	.	.	+[d, p]	.	.	.
Spongoclonium Sonder	.	.	+[r]	.	.	.	.	.	.	.
Mazoyerella Gordon-Mills and Womersley	unknown[i]		.	.	.	.	.	.	.	.
Griffithsieae										
Monosporus Solier	.	.	+[t]	.	.	.	.	.	.	.
Anotrichium Nägeli	.	.	+[j]	.	.	.	.	.	.	.
Griffithsia C. Agardh	.	.	+[j]	.	.	.	.	.	.	.
Halurus Kützing	.	.	+[j]	.	.	.	.	.	.	.

[a] Wollaston, 1968; [b] Wollaston, 1971a; [c] Hommersand, 1963; [d] Feldmann-Mazoyer, 1941; [e] Gordon, 1972; [f] Doty and Meñez, 1960; [g] Baldock and Womersley, 1968; [h] Itono and Tanaka, 1973; [i] Gordon-Mills and Womersley, 1974; [j] Baldock, 1976; [k] Womersley and Cartledge, 1975; [l] Rosenvinge, 1923–24; [m] Wollaston, 1972; [n] Wollaston, 1971b; [o] Dixon, 1971b; [p] Itono, 1971b; [q] Itono, 1971a; [r] Kylin, 1956; [s] Jaasund, 1965; [t] Bornet and Thuret, 1876; [u] Abbott, 1976; [v] Norris and Wynne, 1968; [w] Pujals, 1970.

(*b*) *Zonately divided tetrasporangia.* These tend to be narrower than cruciately divided tetrasporangia (except in the Corallinaceae). As Searles (1968) has pointed out, it is quite probable that zonately divided tetrasporangia were derived from cruciately divided tetrasporangia. Two distinct types of zonately divided tetrasporangia occur in the Florideophyceae. The first and most common is the successive type; here the divisions take place successively, a median division occurring initially, followed by two other divisions (Fig. 3E–G). In the second type (which, as far as can be ascertained, is only found in the Corallinaceae) the cleavages are all initiated simultaneously (Fig. 3J–L) and it seems logical to refer to this as simultaneous zonate. The Corallinaceae would appear to be unique amongst the families of the Cryptonemiales in the manner in which the tetrasporangia divide. This and other features of the majority of the Corallinaceae (e.g. consistent intercalary cell division, roofed conceptacles and, possibly, discharge of the spermatangia with the spermatia) suggest that this family constitutes a very distinct developmental line within the Cryptonemiales

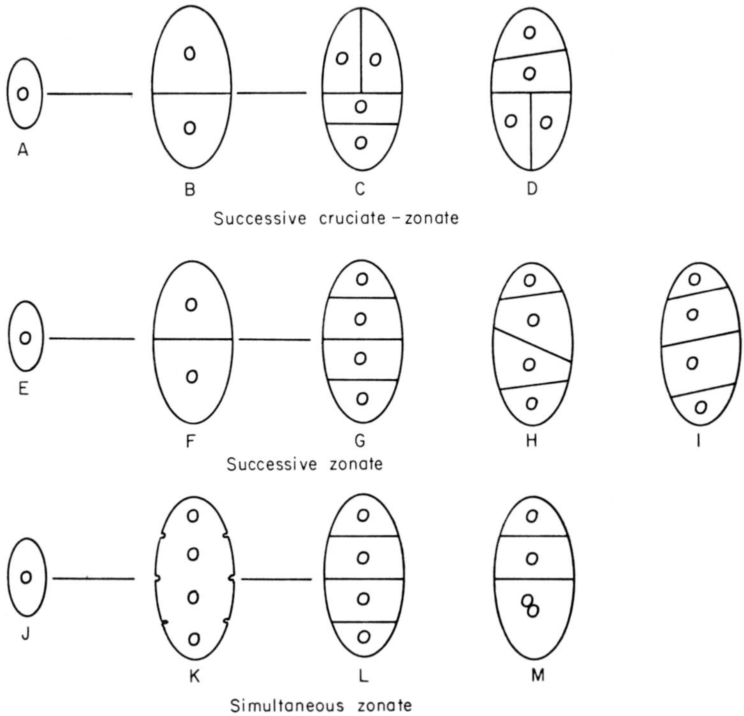

FIG. 3A–M. Diagrammatic representation of various types of zonate or near-zonate tetrasporangia.

as presently delimited. Lemoine (1971) has suggested that the genera *Sporolithon* (= *Archaeolithothamnium*), *Kymalithon* and *Hemiphyllum* (the last two being entirely fossil genera), all of which are known to lack tetrasporangial conceptacles, should possibly be referred to a new family intermediate between the Peyssonneliaceae and Corallinaceae.

The vast majority of the families of the Gigartinales have only zonately arranged regularly and successively divided tetrasporangia (Table I) and most of the others have regularly divided cruciate tetrasporangia only. The Wurdemanniaceae, described by Taylor (1960) with a single genus *Wurdemannia*, has been reported to have only zonate tetrasporangia and the gametangia and cystocarps are unknown (Kylin, 1956; Papenfuss, 1961). The genus is placed in the Nemaliales (or Gelidiales if this order is recognized) because of a supposed similarity in vegetative structure to the members of the Gelidiaceae, although zonate tetrasporangia do not otherwise occur in this group. Hence, as Papenfuss (1961) has pointed out, it is more likely that *Wurdemannia* belongs to the Gigartinales or Cryptonemiales, both of which do possess genera with zonate tetrasporangia.

Clearly, therefore, the zonate mode of division has arisen several times in the Florideophyceae. There is some evidence that it has arisen independently in a number of families of the Cryptonemiales, that a strong developmental line of families with zonate tetrasporangia has arisen in the Gigartinales, and that in the Corallinaceae a mode of division of the tetrasporangia has developed which is not found elsewhere in the Florideophyceae.

(*c*) *Tetrahedrally divided tetrasporangia.* These are usually spherical (Fig. 4D–I) and probably have been derived from cruciate tetrasporangia. The spatial arrangement of a tetrahedrally divided tetrasporangium may be visualized as three spheres of equal size arranged so that all are touching one another; a fourth is placed in the central depression so that the centres of the spheres are equidistant from one another. The spores, because they are enclosed in a bounding membrane, are closely adpressed to one another and all have an equivalent external surface area. Tetrahedrally divided tetrasporangia are generally regarded as undergoing simultaneous cleavages, while cruciately divided are regarded as undergoing two separate cleavages. Tetrasporangia which may appear tetrahedrally divided often do not have simultaneous cleavages (Fig. 4D–F) and sporangia in which the final arrangement is cruciate sometimes have simultaneous cleavages (Fig. 4A–C). Such variations occur most frequently in the Ceramiaceae. Gordon (1972) mentioned in her study of the Wrangelieae, Sphondylothamnieae and Spermothamnieae (Ceramiaceae) that in some species simultaneous cleavages take place and the sporangia appear cruciately divided whereas in others two successive divisions take place and the final orientation of the walls appears tetrahedral (Fig. 4D–F). Hommersand (1963) has also noted that in species of *Crouania*, *Gulsoniopsis* and possibly *Reinboldiella* (all Ceramiaceae) the tetrasporangia, although finally appearing cruciately divided, undergo simultaneous cleavages. Wollaston (1968) reported that in some species belonging to the Heterothamnieae (see Table III) successive divisions of the tetrasporangia occur but the final arrangement of the spores is similar to that of tetrahedral tetrasporangia, suggesting that nuclear division is completed in these sporangia before cleavage is initiated. Some further work is necessary to establish the precise time of nuclear division in these entities. In successively cruciately divided tetrasporangia the nuclei divide once before the transverse cleavage takes place and again after it has occurred (Fig. 1A–C). In simultaneously tetrahedrally divided tetrasporangia all the nuclear divisions seem to take place before, or while, cleavage is taking place (Fig. 4G–I). Tetrahedrally divided tetrasporangia occur in all four families of the Ceramiales but

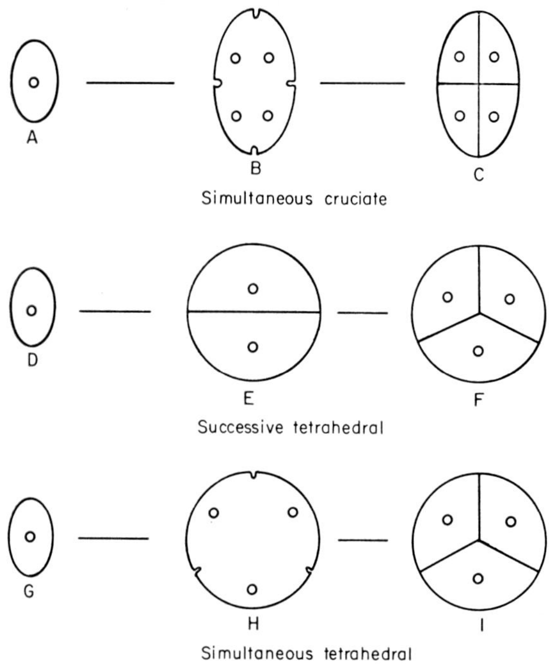

FIG. 4A–I. Diagrammatic representation of the different types of cleavage found in cruciate and tetrahedral tetrasporangia.

cruciately divided tetrasporangia are known only in a few tribes of the Ceramiaceae (see Table III).

Sporangia with tetrahedral divisions have also been reported in the Gelidiaceae and Gelidiellaceae (Nemaliales) and in the Rhodymeniaceae and Champiaceae (Rhodymeniales). In the former two families, both cruciately and tetrahedrally divided tetrasporangia have been reported in species of the same genus. It seems likely that the tetrahedrally divided tetrasporangia reported in these entities are really cruciately divided tetrasporangia with the spores arranged in such a manner that they appear to have tetrahedral divisions. This could, however, be regarded as successive tetrahedral cleavage since, in many of the illustrations of species of *Gelidiella*, sporangia can be seen which appear to have a single cleavage, suggesting successive divisions. In the Rhodymeniaceae intercalary tetrahedrally divided tetrasporangia with simultaneous cleavages are definitely only present in species of *Hymenocladia*, which genus is referred to a subfamily (Hymenocladioideae Sparling, 1957) distinct from the Rhodymenioideae, largely on the basis of this character. In the latter subfamily Joly (1957, 1965) reported tetrahedrally divided tetrasporangia in *Leptofauchea brasiliensis* Joly. Schneider and Searles (1976) in recording the species from North Carolina remarked that the tetrasporangia in their material were formed in superficial chains but made no mention of the mode of division. Further investigation of this species is clearly needed. In species referred to the Champiaceae simultaneously tetrahedrally divided tetrasporangia are the only type reported.

Tetrahedrally divided tetrasporangia with simultaneous cleavages are thus known definitely only in the Rhodymeniales and Ceramiales; they occur in conjunction with other features of structure and reproduction which are widely regarded as being advanced.

2. Carpotetrasporangia

In a very few instances quadripartite carposporangia are formed by the cleavage of the contents of cells derived from gonimoblast filaments; these are referred to as carpotetrasporangia. Such sporangia have been reported in at least five species of *Liagora* (*L. brachyclada* Decaisne (Levring, 1941); *L. harveyiana* Zeh (Womersley, 1965); *L. papenfussii* Abbott (Abbott, 1945); *L. pinnata* Harv. (Yamada, 1938) and *L. tetrasporifera* Børg. (Børgesen, 1927)), in at least two species of *Helminthocladia* (*H. agardhiana* Dixon (Feldmann, 1939); *H. senegalensis* Bodard (Bodard, 1971)), in *Yamadaella cenomyce* (Decaisne) Abbott (Abbott, 1970), in *Audouinella purpurea* (Lightf.) Woelk. (West, 1969) and in several species of *Phyllophora* and *Gymnogongrus* (see Schotter, 1968). The entity presently known as *Schizymenia epiphytica* (Setch. and Lawson) G. M. Smith and Hollenb. may also bear tetrasporangia derived from gonimoblast filaments (Smith and Hollenberg, 1943) but this requires further investigation (Abbott, 1967). A unique feature of the supposed carpotetrasporangia of *Schizymenia epiphytica* is that they are zonately divided (probably successively) whereas those of the other examples cited above are cruciately divided.

The formation of carpotetrasporangia is generally not accorded any taxonomic significance at the generic level, but species forming carpotetrasporangia are generally regarded as being distinct from similar species forming carposporangia with a single spore. Species forming carpotetrasporangia should probably be referred to separate genera from species with single-spored carposporangia. Schotter (1968) distinguished between a *Gymnogongrus* species (now known in the North Atlantic as *G. crenulatus* (Turn.) J. Ag.) which forms carpotetrasporangia, and a supposedly morphologically similar species (which he referred to as *G. devoniensis* (Grev.) Schotter) that formed, internally, carposporangia with a single spore. *G. devoniensis* has been included in synonymy with *G. crenulatus* by Parke and Dixon (1976) and by Dixon and Irvine (1977), a course of action which would not seem to be justified without evidence from culture; especially as there are ecological and morphologica- differences between these two species (Guiry and Morton, unpublished observal tions). It is probable that *Gymnogongrus* species with single-spored carpoporangia have a life history similar to that reported by DeCew and West (1977b) for *G. leptophyllus* J. Ag. (north-eastern Pacific).

3. Polysporangia

Drew (1937) showed that the polysporangia of *Tiffaniella snyderae* (Farlow) Abbott (as *Spermothamnion snyderae* Farlow) undergo meiosis in the formation of spores; she suggested that the term polysporangium be restricted to sporangia which form more than four spores meiotically and are borne in place of tetrasporangia. Such sporangia generally form spores arranged in a regular manner (Fig. 1I).

Polysporangia have been recorded in about eleven genera of the Ceramiaceae (*Mesothamnion, Lophothamnion, Involucrana, Tiffaniella, Ptilothamnion, Compsothamnion, Aristothamnion, Spermothamnion, Pleonosporium, Mazoyerella* and *Wollastoniella*; see Table III); in a single genus of the Delesseriaceae (a species of *Gonimophyllum*); and in two genera of the Champiaceae (*Gastroclonium* and *Coeloseira*; see p. 127). Polysporangia often have eight spores (Fig. 1I). In certain genera of the Ceramiaceae (listed previously) only polysporangia are formed (usually the spores are in multiples of four, suggesting a meiotic origin of the nuclei), but in other such genera both polysporangial and tetrasporangial species are found. In the type species of *Mazoyerella, M. arachnoidea* (Harv.) Gordon-Mills and Womers., Gordon-Mills and Womersley (1974) have reported that polyspores formed male and female plants in culture and that carpospores derived from the latter again formed polysporangial plants.

The occurrence of a single parasitic species of *Gonimophyllum* with polysporangia (*G. skottsbergii* Kylin: North Pacific) formed in place of tetrasporangia is very peculiar. The sporangia are reported to have 22–50 spores around a central body (Kylin, 1924; Wagner, 1954). Such an arrangement of spores is unique as there are no reports of such a central body in other polysporangia. An electron microscope investigation of the mode of cleavage of these sporangia would be valuable.

Polysporangia have been reported in three species of *Gastroclonium*: the lectotype species *G. ovatum* (Huds.) Papenf. (North Atlantic); *G. coulteri* (Harv.) Kylin and *G. reflexum* (Menegh. ex Zanard.) Kütz. (North Atlantic, Mediterranean). In the type species, polysporangia (with 16 spores) have only been reported once (Buffham, 1888) but repeated examination of populations in the British Isles (Guiry, unpubl.) has failed to reveal any further polysporangial plants. *Gastroclonium coulteri*, a species very similar to *G. ovatum* in outward appearance, frequently bears polysporangia with eight or more spores; these are most commonly found on plants from Monterey north to Northern California (North Pacific) and appear to be absent in the northern and southern parts of the range of the species (Guiry, unpubl. obs.) The polysporangial plants are frequently found with tetrasporangial plants and gametangial plants are virtually absent north of Point Conception (DeCew, pers. comm.). This species requires both cultural and cytological investigation. *Gastroclonium reflexum*, on the other hand, is found bearing only polysporangia with eight or more spores and tetrasporangial plants have not been reported; gametangial plants are always found with polysporangial plants in field populations in the British Isles.

The genus *Coeloseira* was erected by Hollenberg (1940) for two Champiaceae species which bore only polysporangia. In the holotype species *C. parva* Hollenb. the structure of the cystocarp is exactly like that of the genera *Chylocladia* and *Gastroclonium* and the vegetative morphology is very similar to that of *Gastroclonium ovatum*. The frequent occurrence of polysporangia in various species of *Gastroclonium* suggests that *Coeloseira parva* at least should be transferred to the genus *Gastroclonium*.

Polysporangia only occur in families where there are other entities with simultaneously tetrahedrally divided tetrasporangia and it is reasonable to suppose that polysporangia were derived from these.

4. *Bisporangia*

Sporangia with two spores (Fig. 1D, E) have been widely reported in the Corallinaceae, particularly in the crustose species, although Suneson (1937) and Johansen (1976) recorded bisporangia in *Amphiroa rigida* Lamour. and *A. zonata* Yendo respectively; Ganesan and Desikachary (1970) recorded bisporangia in *Lithothrix aspergillum* J. E. Gray and bisporangial plants are known for most of the species of *Bossiella* (Johansen, 1976). Bauch (1937) showed that there are two types of bisporangia; the first of these has binucleate sporangia (one nucleus in each spore: Fig. 1D) and the second has quadrinucleate sporangia (two nuclei in each spore: Fig. 1E). Suneson (1950) later reported that meiosis occurred in the formation of quadrinucleate bisporangia and did not occur in the binucleate bisporangia of the crustose coralline algae he examined. It seems clear from these and other studies that two widely different bisporangial types occur in the Florideophyceae; one is homologous with a meiotic tetrasporangium and the other homologous with a mitotic monosporangium. Only the former type will be treated in the present section.

TABLE IV. Occurrence of quadrinucleate bisporangia in the Corallinaceae.

Species	Author	Location
Phymatolithon polymorphum (L.) Fosl.	Bauch (1937)	Helgoland
Lithophyllum expansum Phil.	Bauch (1937)	Naples
Dermatolithon pustulatum (Lamour.) Fosl.	Suneson (1943)	Sweden
D. corallinae (P. and H. Crouan) Fosl.	Suneson (1950)	Sweden
Fosliella farinosa (Lamour.) Howe	Balakrishnan (1947)	S. India

Table IV lists the occurrence of quadrinucleate bisporangia in the Corallinaceae according to the literature; these are usually formed amongst normal tetrasporangia (Suneson, 1943, 1950) and are probably more frequent than the literature suggests. There are, however, many reports of bisporangia intermingled with cruciate tetrasporangia in many of the other Florideophyceae but such observations must be regarded as dubious particularly when the spores have not been observed discharging and when no comment is made on the number of nuclei present.

In the Nemaliales (including the Gelidiales) quadrinucleate bisporangia have been reported in *Suhria vittata* (L.) J. Ag. and *Gelidium pristoides* (Turn.) Kütz. from South Africa (Fan, 1961). In both species tetrasporangia are unknown and meiosis probably takes place in the bisporangial mother cells. Klavestad (1957), Knaggs (1967) and West (1969) recorded bisporangia in *Audouinella purpurea* (Acrochaetiaceae; Nemaliales). Knaggs (1967) found that the bisporangia on his material were quadrinucleate and the resulting bispores were binucleate but only one nucleus entered the germination tube. West (1969) also found that binucleate spores were liberated and that bisporangia and tetrasporangia were commonly derived from the same basal cell in a sporangial cluster (West, 1969). These data suggest that meiosis takes place in such bisporangia (Knaggs,

1967) but both authors were unable to grow the sporelings obtained from such spores to reproductive maturity.

The occurrence of quadrinucleate bisporangia seems to have little major taxonomic importance.

5. Monosporangia

Most monosporangia are formed by mitosis (see p. 132) but West (1969) reported that he observed occasional sporangia of *Audouinella purpurea* (Nemaliales) with four nuclei and no indications of cleavage (Fig. 1c). In an examination of 200 spores from this species collected in Washington and Alaska, he found that 70% were uninucleate, 23% binucleate, 6% trinucleate and less than 1% quadrinucleate. Despite being maintained in a variety of culture conditions for several months none of the quadrinucleate spores germinated. The possibility of the existence in nature of viable monospores with four meiotically derived nuclei, only one of which enters the germination tube, cannot be discounted.

MITOSPORANGIA

The commonest type of mitosporangium is the sporangium which forms a single uninucleate spore and these are referred to as monosporangia and carposporangia. Intrinsically there is little difference between these two types. Carposporangia are formed directly from gonimoblast filaments which may collectively be interpreted as a reduced diploid (or occasionally haploid) plant developing *in situ* on the female plants; whereas monosporangia are considered to be single-spored sporangia developing from vegetative cells of the thallus. It is possible therefore to regard the carposporophyte as a monosporangial generation which develops *in situ* on a female plant.

Bisporic mitosporangia forming uninucleate spores are also of common occurrence and it seems probable that most of these form further bisporangial generations (Chamberlain, see Chapter 10).

A considerable body of evidence now exists to show that in a number of species four uninucleate spores are formed by mitosis in a sporangium which looks exactly like a tetrasporangium.

1. Tetrasporangia

That spores derived from tetrasporangia can give rise to tetrasporangial plants in culture has now been demonstrated for a number of species (Table V). In one case (South and Whittick, 1976; for *Rhodophysema elegans* (P. and H. Crouan ex J. Ag.) Dixon from Newfoundland) cultural and cytological data are available which confirm that successive generations with the same chromosome number (30–35) are formed from tetraspores. Some problems of classification are presented by these algae as the female reproductive apparatus is unknown. In species of the genus *Hildenbrandia* only tetrasporangial plants are formed in the field and in culture (DeCew and West, 1977a); as *Hildenbrandia* is the type and only genus of the family Hildenbrandiaceae, that family must be classed as of unknown taxonomic position, since the orders of the Florideophyceae are distinguished solely on the basis of the pre- and post-fertilization changes in

TABLE V. Reported occurrences of tetrasporangial plants derived from tetraspores in culture

Entity	Author(s)
Callithamnion corymbosum (Sm.) Lyngb.	Hassinger-Huizinga (1952)
Antithamnion boreale (Gobi) Kjellm.	Sundene (1962)
A. defectum Kylin	West and Norris (1966)
Callithamnion sp.	West and Norris (1966)
Fauchea laciniata J. Ag.	West and Norris (1966)
Rhodochorton concrescens Drew	West (1970*a*)
Heterosiphonia densiuscula Kylin	West (1970b)
Acrochaetium proskaueri West	West (1972)
Pseudorhododiscus nipponicus Masuda	Masuda and Ohta (1975)
	Masuda (1976)
Rhodophysema georgii Batt.	Fletcher (1975)
R. elegans (P. and H. Crouan ex J. Ag.) Dixon	Ganesan and West (1975) South and Whittick (1976)
Audouinella membranacea (Magnus) Papenf.	West (1977)
Hildenbrandia occidentalis Setch.	DeCew and West (1977a)
H. prototypus Nardo	DeCew and West (1977a)

the female reproductive apparatus. The family is normally placed near the Corallinaceae because the tetrasporangia of *Hildenbrandia* species are formed in "conceptacles". The ontogeny and structure of these are quite unlike those of the conceptacles of the Corallinaceae. However, the basically zonate mode of division of the sporangia suggests that the Hildenbrandiaceae should be aligned with either the Cryptonemiales or Gigartinales and the irregular zonate division found in some species suggests a closer affinity with the Cryptonemiales (cf. Tables I and II).

A similar, but more intractable, problem is presented by the genera *Rhodophysema* and *Pseudorhododiscus* (Masuda, 1976) where female reproductive organs are also unknown. Several authors (see Table V) have shown that species of *Rhodophysema* form successive generations of tetrasporangial plants. However

Masuda and Ohta (1975) and Masuda (1976) have reported that tetraspores from *Pseudorhododiscus nipponicus* Masuda form plants which bear either tetrasporangia or spermatangia. The structure and reproduction of both these crustose genera suggest that they are closely related but in the absence of female reproductive organs they must be regarded as of uncertain taxonomic position. On the basis of present knowledge, they could be referred to the Cryptonemiales, Gigartinales, Rhodymeniales or Palmariales.

In the case of the other entities listed in Table V, features of organization of the vegetative filaments make their affinities somewhat clearer.

Dixon (1965) has observed that, in certain Florideophyceae, only tetrasporangial plants may be present near the limits of their geographical distribution. It is not known whether tetraspores from these plants recycle the tetrasporangial plants or whether they do not germinate in the wild and the populations maintain themselves vegetatively. Edwards (1973) cultured tetraspores from such plants but found that they formed gametangial plants, suggesting that the latter is the more likely explanation. A more extensive study may show that some species of Florideophyceae can form tetrasporangial plants from tetraspores at the limits or near-limits of geographical distribution.

2. *Polysporangia*

It is possible that some "polysporangia" do not undergo meiosis in the formation of spores, particularly where they occur towards the limits of distribution of the species, e.g. in *Gastroclonium coulteri* (p. 127).

3. *Bisporangia*

Mitotic bisporangia generally form two spores, each with a single nucleus (Fig. 1B). These are borne on thalli which bear only bisporangia. Such bisporangial plants are well developed in species belonging to the Corallinaceae at the limits of geographical distribution (particularly in colder waters: Foslie, 1905; Suneson, 1950) and may be the only reproductive organs found in populations at the northern limit of distribution of a species (Foslie, 1905).

In the Ceramiales bisporangia have been recorded sporadically in several genera, including *Callithamnion*, *Spermothamnion*, *Crouania* (Feldmann-Mazoyer, 1941), *Seirospora* (Feldmann-Mazoyer, 1941; Dixon, 1971a), *Dohrniella* (Mayoub, 1975) and *Antithamnion* (Feldmann-Mazoyer, 1941). In *Callithamnion byssoides* Arnott ex Harv. in Hook. (= *C. furcellariae* J. Ag.) bisporangia have frequently been recorded; they occur on separate plants (Rosenvinge, 1923–1924) and are binucleate. These are probably similar to the mitotic bisporangia of the Corallinaceae. The life histories of such plants are unknown and need further investigation; the practice of describing *formae* for them can only be deplored in the light of the present lack of data. Further, it is not always stated in the literature whether these bisporangia have two or four nuclei, information which would contribute somewhat towards a better understanding of their function. Generally the occurrence of sporangia with two uninucleate spores seems to have little overall taxonomic importance.

4. Monosporangia

Monosporangia form a single spore generally without meiotic division of the nucleus (but see p. 129). These spores are usually uninucleate but there are some reports of monospores which have a large number of nuclei. Monosporangia may be formed singly or in branched or unbranched series.

(a) Solitary monosporangia. Uninucleate solitary monosporangia may be found at any phase in the life history, probably duplicating the ploidy level of the parent thallus. Such monosporangia frequently occur in the Nemaliales (particularly the Acrochaetiaceae) where they are usually sessile structures formed singly from terminal cells of lateral branches. Recent reports of monosporangia in cultures of various algae of the Nemaliales and Cryptonemiales (*Pseudogloiophloea confusa* (Setch.) Levring: Ramus, 1969; *Halymenia floresia* (Clem.) C. Ag.: Hoek and Cortel-Breeman, 1970; *Pikea californica* Harv.: Scott and Dixon, 1972; *Thuretellopsis peggiana* Kylin: Richardson and Dixon, 1970) are interesting as these structures have not been detected in the wild. There is some variation in the phase upon which these structures are found (Dixon *et al.*, 1972) but the spores always seem to repeat the parent phase. Monosporangia have been reported in species of *Antithamnion* (Ceramiaceae) by Jaasund (1965) but nothing is known of the number of nuclei or of the phase reproduced by the spores. Also in the Ceramiales there are several questionable reports of small undivided sporangia on plants of *Polysiphonia urceolata* (Lightf. ex Dillw.) Grev. (Rosenvinge, 1923–1924), *P. flexicaulis* (Harv.) Collins (Yamanouchi, 1906) and *Griffithsia globulifera* Harv. (Lewis, 1909).

A very peculiar type of uninucleate monosporangium has been reported for *Ahnfeltia plicata* (Huds.) Fries (Gigartinales) where almost colourless sporangia are formed superficially on large pustules. Cystocarps have never been found on this species in the North Atlantic. Farnham and Fletcher (1976) indicated that a tetrasporangia-bearing alga (which resembles the crustose entity previously described as *Porphyrodiscus simulans* Batt.) occurs at the base of *Ahnfeltia plicata* in the north Atlantic and that this crustose alga showed a direct connection with the upright thallus. Chen (1977) has shown that monospores from the upright plants of this species (from Nova Scotia) formed crustose plants in culture; these formed tetrasporangia, tetraspores from which gave rise to further upright plants. It would seem likely that meiosis does not occur in tetrasporangia of this species and that male plants (as reported by Gregory, 1934) do not occur in the life-history. *Ahnfeltia plicata*, being the type species of the genus *Ahnfeltia*, therefore seems to present problems of classification. It is interesting to note that the tetrasporangia are zonately divided and are solitary whereas the tetrasporangia of the Phyllophoraceae (in which family the genus has been placed) are cruciate and borne in chains. There seems to be a case for describing a new family for *Ahnfeltia*.

Abbott (1969) recorded monosporangia borne terminally on the "nemathecial filaments" of *Ozophora latifolia* Abbott (Phyllophoraceae), but these probably represent undivided tetrasporangia (Abbott and Hollenberg, 1976).

Multinucleate monosporangia-like structures were reported by Svedelius (1914) in *Nitophyllum punctatum* (Stackh.) Grev. in close association with the procarps. These are thought (Fritsch, 1945, p. 723) to be a consequence of arrested development of these organs. Gordon-Mills and Womersley (1974) reported a similar development from unfertilized procarps of *Mazoyerella arachnoidea*, but did not indicate the number of nuclei present in the sporangia. In both of these cases release of the monospores was not observed, so that the viability of the "sporangia" is in question.

(*b*) *Monosporangia formed in chains.* Monosporangia formed in chains on thalli other than carposporophytes are generally called seirospores and are commonly formed on species of *Seirospora* (Ceramiaceae), where the chains are branched and terminal. The seirosporangia of *Dohrniella neapolitana* Funk (Mediterranean) are borne terminally in unbranched chains (Feldmann-Mazoyer, 1941; L'Hardy-Halos, 1966) on thalli that also bear tetrasporangia. Doty (1963) reported tetrasporangia and seirosporangia on the same thallus of an unnamed species of *Seirospora*. The seirosporangia of *Gibsmithia hawaiiensis* Doty (Doty, 1963) (Dumontiaceae, Cryptonemiales: Hawaii) are borne in unbranched chains at the apices of filaments of tetrasporangial plants. Seirosporangia are similar to solitary monosporangia (Dixon, 1973) in that they are formed on thalli bearing other reproductive organs. Certain sporangia have been reported on specimens of *Audouinella purpurea* (Nemaliales); these are borne terminally in unbranched chains (Klavestad, 1957) and are very similar to the seirosporangia of *Dohrniella neapolitana* and *Gibsmithia hawaiiensis*. However, it is not clear whether these sporangia are formed by the direct transformation of vegetative cells, or by division of a terminal cell.

5. Carposporangia

Carposporangia are similar to seirosporangia in that they are formed in branched or unbranched chains by the transformation of vegetative cells. The principal difference is that carpospores usually form a thallus (the tetrasporophyte) which is not like the parent carpasporophyte.

It has recently become apparent that not all carpospores are formed from diploid tissue resulting from the union of two gametes. It has been shown for a number of entities belonging to Gigartinales that sporangia developing from female reproductive organs may have the same ploidy level as the parent thallus (see Table VI). However, the vast majority of the Florideophyceae form carpospores which give rise to diploid tetrasporangia-bearing thalli.

TABLE VI. Reported occurrences of carposporangial plants derived from carpospores in culture

Entity	Authors	Location
Calosiphonia vermicularis (J. Ag.) Schmitz	Mayhoub (1973)	Syria
Gigartina stellata (Stackh. in With.) Batt.	Chen *et al.* (1974)	Nova Scotia
	Edelstein *et al.* (1974)	Nova Scotia
	West and Polanshek (1975a, b)	Scotland, Wales Maine, U.S.A.
G. papillata (C. Ag.) J. Ag.	West and Polanshek (1975b)	British Columbia Washington, Oregon
G. ochotensis (Rupr.) Rupr.	Masuda and Uchida (1976)	Japan
Gymnogongrus leptophyllus J. Ag.	DeCew and West (1977b)	California

PARASPORANGIA

Sporangia with more than four spores which reproduce the ploidy level of the parent thallus were referred to as parasporangia by Drew (1937, 1939). From a cytological investigation of *Plumaria elegans* (Bonnem.) Schmitz in 1939 she concluded that parasporangia were formed mitotically on triploid plants; Rueness 1968)) later found that paraspores from the same species formed parasporangial plants in culture but did not confirm the triploid nature of the plants. L'Hardy-Halos (1970) has suggested that parasporangia are derived from repeated mitotic divisions of a generative cell within an enclosing membrane; the pit connections between these cells can be clearly seen. Each of the cells then forms a sporangium with a single spore. What is classically referred to as a parasporangium is probably a group of monosporangia, each forming a single spore. Parasporangia (*sensu stricto*) have been reported only in species belonging to the Ceramiaceae.

According to Dixon (1960, 1973) they are frequently confused with bacteria galls, particularly in the genus *Ceramium*. Parasporangia are very similar in appearance to the cystocarps of the Ceramiaceae.

It is possible that parasporangia always reproduce thalli at a higher ploidy level than the tetrasporangial or gametangial plants, as they have been reported

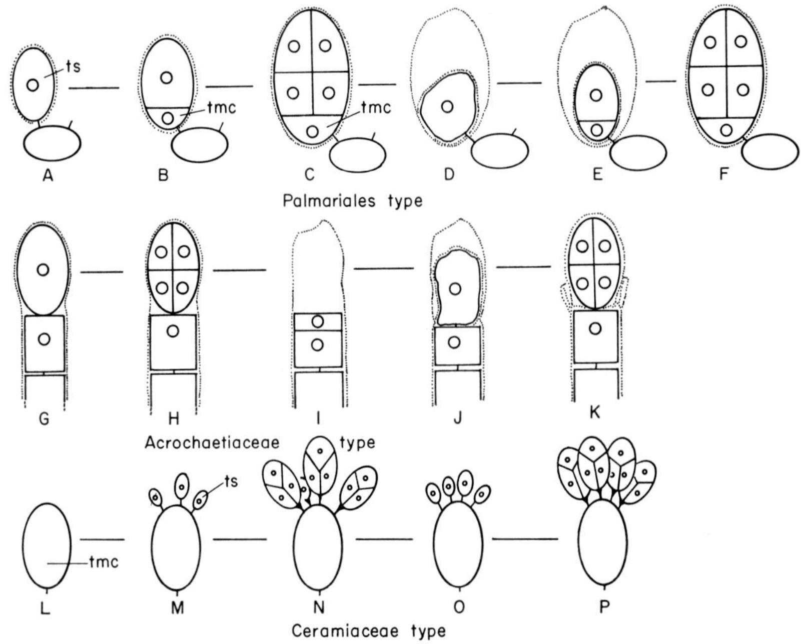

FIG. 5A–P. Diagrammatic representation of the various types of tetrasporangial mother cells found in the Florideophyceae. (ts: tetrasporangium; tmc: tetrasporangial mother cell.)

to do in *Plumaria elegans* (see Chapter 11 by Rueness). A fine structural investigation of the mode of formation of parasporangia might shed some further light on the nature of these sporangia.

STALKED TETRASPORANGIA

"Stalk cells' or "pedicels" have frequently been reported in the Ceramiaceae, where these cells correspond to the sub-terminal cells of the tetrasporangia-bearing filaments. Figure 5L–P shows a diagrammatic representation of such a cell where tetrasporangia may be repeatedly budded off from the apex of the cell. Such cells may bear a single tetrasporangium, or a number of tetrasporangia and may in fact be regarded as tetrasporangial mother cells. In the vast majority of the Florideophyceae only a single terminal or lateral tetrasporangium may be formed; the sub-terminal cell does not divide and produce another tetrasporangium. In these instances the sub-terminal cell is merely a vegetative cell which is non-generative.

"Stalked" tetrasporangia have been reported in the Antithamnieae (Wollaston, 1968, 1971a), the Dohrnielleae (Feldmann-Mazoyer, 1941), the Griffithsieae (Baldock, 1976), the Sphondylothamnieae, Wrangelieae and the Spermothamnieae (Gordon, 1972) tribes of the Ceramiaceae but it is not always clear whether these form one or several generations of tetrasporangia. In all these tribes the "stalk" cells may be present in some genera and species and absent in others.

In the Acrochaetiaceae (Nemaliales) the sub-terminal cell of a tetrasporangium-bearing branch may, on discharge of the tetraspores, divide and grow into the primary tetrasporangial sheath to form a secondary tetrasporangium (Fig. 5G–K). In the genus *Audouinella* (including *Rhodochorton* and *Acrochaetium*) such a process would seem to be particularly common to judge from the number of tetrasporangial membranes which may be observed in the vicinity of some tetrasporangia.

In the Cryptonemiales, Gigartinales and Rhodymeniales it would seem that such modified tetrasporangial mother cells are absent except perhaps in the Corallinaceae where sporangia commonly have a prominent basal cell which only seems to form one sporangium.

Guiry and Irvine (1978) have described a new order, the Palmariales* for the family Palmariaceae which includes the genera *Palmaria*, *Halosaccion* and *Leptosarca* (Guiry, 1974) formerly included in the Rhodymeniales. The main basis for the Palmariales is, in the absence of reports of female gametangia and carposporophytes, the consistent presence of a generative tetrasporangial mother cell. This cell is formed by the unequal division of a terminal cell of the lateral filaments of the thallus (Fig. 5A, B) to form a tetrasporangial mother cell (referred to as a "stalk cell" by Guiry, 1974) and a tetrasporangium. The tetrasporangium then divides to form four cruciately arranged tetraspores (Fig. 5C). On discharge of these the primary tetrasporangial mother cell expands into the vacated tetrasporangial sheath (Fig. 5D) and divides to form another (secondary) tetrasporangial mother cell (Fig. 5E) and a further tetrasporangium which again forms cruciately arranged tetraspores (Fig. 5F). This process may be repeated at least three times. Such a generative tetrasporangial mother cell is

---

* See note added in proof, p. 138.

formed in all the species examined of the three genera belonging to the Palmariaceae from both the northern and southern hemispheres.

The generative nature of the tetrasporangial mother cells of these species, and the complete absence of female gametangia and carposporophytes, despite the presence in the wild of relatively large numbers of spermatangial plants, led Guiry and Irvine (1978) to describe the Palmariales. These algae have no natural affinities with any one of the presently recognized orders of the Florideophyceae.

PSEUDOSPORANGIA

The "monosporangia" reported in *Monosporus pedicellatus* (J. E. Smith) Solier (Ceramiales; previously known as *Corynospora pedicellata*—see Baldock, 1976) are multinucleate and may have a short stalk (Feldmann-Mazoyer, 1941). Bornet and Thuret (1876), Feldmann-Mazoyer (1941) and L'Hardy-Halos (1970) all considered this structure to be a vegetative reproductive organ, as the whole sporangium is shed and the cell contents germinate without being released (L'Hardy-Halos, 1970). If this is correct then the genus is seen to be unfortunately named.

Gordon-Mills and Womersley (1974) described *Mazoyerella arachnoidea* as bearing what were described as multinucleate monosporangia on male, female or polysporangial plants (no tetrasporangial plants were found). It is stated that these repeated the parent phase but it is not clear whether the monospores were liberated or whether the structure is a vegetative reproductive organ similar to *Monosporus*.

Baldock (1976, p. 571) has suggested that the multinucleate monosporangia of *Monosporus* species may represent the culmination of a trend in the Griffithsieae tribe towards a reduction of numbers of tetrasporangia in a fertile whorl branchlet, culminating in *Anotrichium licmophorum* (Harv.) Baldock where only a single, terminal, pedicellate tetrasporangium is formed. This seems unlikely as, firstly, the monosporangium is multinucleate (similar to the vegetative cells of the thallus) and secondly, Bornet and Thuret (1876) and others report tetrasporangia on thalli of *Monosporus pedicellatus*. Additionally L'Hardy-Halos (1968) reported that the tetrasporangia of *M. pedicellatus* are sessile whereas those of *Griffithsia* and *Anotrichium* are pedicellate.

As the "monosporangia" of *Monosporus* do not liberate a monospore and the whole structure germinates on release, it is probably best to refer to them as propagules (Feldmann-Mazoyer, 1941). Cultural investigation of *Monosporus* species would be interesting.

CONCLUSIONS

It is clear from the above brief summary of form and function of sporangia that there is a great deal of variation that may be utilizable for classification purposes. For many years the ordinal classification of the Florideophyceae has been based on a single character, the pre- and post-fertilization changes in, or associated with, the female reproductive apparatus. This has proved somewhat unsatisfactory, particularly in the arrangement of the families of the Nemaliales, Gigartinales and Cryptonemiales. It is possible that in the future the characters associated with sporangia and spermatangia will prove to be of more use at the ordinal level of classification. The classification problems associated with algae that have been shown to lack female reproductive organs and carposporophytes will continue until some efforts are made to search for alternative characters.

NOTE ADDED IN PROOF

At going to press, Guiry and Irvine (1978) had not appeared. The following diagnosis is therefore required:

PALMARIALES GUIRY ET D. IRVINE, **ORD. NOV.**

Carpogonia et carposporophyta ignota. Tetrasporangia cruciata, magna, in cellula stipali portata. Spermatia in spermatangiis contiguis ex cellulis matricalibus spermatangialibus binatim evolutis singulatim facta. Familia typica: Palmariaceae Guiry.

Carposporophytes and carpogonia unknown. Tetrasporangia cruciate, large, borne on a stalk cell. Spermatia formed singly in contiguous spermatangia formed in pairs on the spermatangial mother cell. Type family: Palmariaceae Guiry.

ACKNOWLEDGEMENTS

I am particularly grateful to W. F. Farnham, Dr G. F. Papenfuss, Dr P. C. Silva and Dr J. A. West for their helpful comments on various portions of the manuscript. Completion of the manuscript was facilitated by N.A.T.O. grant no. 1130 to me and John West.

REFERENCES

Abbott, I. A. (1945). The genus *Liagora* (Rhodophyceae) in Hawaii. *Occ. Paps Bernice P. Bishop Mus.* **18**, 145–169.

Abbott, I. A. (1967). Studies in some foliose red algae of the Pacific coast II. *Schizymenia. Bull. So. Calif. Acad. Sci.* **66**, 161–174.

Abbott, I. A. (1968). Studies in some foliose red algae of the Pacific coast III. Dumontiaceae, Weeksiaceae, Kallymeniaceae. *J. Phycol.* **4**, 180–198.

Abbott, I. A. (1969). Some new species, new combinations and new records of red algae from the Pacific coast. *Madroño* **20**, 42–53.

Abbott, I. A. (1970). *Yamadaella*, a new genus in the Nemaliales (Rhodophyta). *Phycologia* **9**, 115–123.

Abbott, I. A. (1976). On the red algal genera *Grallatoria* Howe and *Callithamniella* Feldmann-Mazoyer (Ceramiales). *Br. phycol. J.* **11**, 143–149.

Abbott, I. A. and Hollenberg, G. J. (1976). "Marine Algae of California." Stanford University Press, Stanford.

Balakrishnan, M. S. (1947). The morphology and cytology of *Melobesia farinosa* Lamour. *J. Indian bot. Soc.* **1946**, 305–319.

Balakrishnan, M. S. (1962). Studies on Indian Cryptonemiales—II. *Corynomorpha* J. Ag. *Phytomorphology* **12**, 77–86.

BALDOCK, R. N. (1976). The Griffithsieae group of the Ceramiaceae (Rhodophyta) and its Southern Australian representatives. *Aust. J. Bot.* **24**, 509–593.

BALDOCK, R. N. and WOMERSLEY, H. B. S. (1968). The genus *Bornetia* (Rhodophyta, Ceramiaceae) and its Southern Australian representatives, with a description of *Involucrana* gen. nov. *Aust. J. Bot.* **16**, 197–216.

BAUCH, R. (1937). Die Entwicklung der Bisporen der Corallinaceen. *Planta* **26**, 365–390.

BODARD, M. (1971). Étude morphologique et cytologique d'*Helminthocladia senegalensis* (Rhodophycées), Némalionale nouvelle à carpotetraspores et à cycle haplodiplophasique. *Phycologia* **10**, 361–374.

BØRGESEN, F. (1927). Marine algae from the Canary Islands. III. Rhodophyceae. Part 1. Bangiales and Nemalionales. *K. danske Vidensk. Selsk. Skr., Biol. Medd.* **6**(6), 1–97.

BORNET, E. and THURET, G. (1876). "Notes algologiques,", Vol. 1. Masson, Paris.

BUFFHAM, T. H. (1888). On the reproductive organs, especially the antheridia of some of the Florideae. *J. Quekett micros. Club*, ser. 2, **3**, 257–266.

CABIOCH, J. (1972). Étude sur les Corallinacées. II. La morphogenèse: consequénces systématiques et phytogénétiques. *Cah. Biol. mar.* **13**, 137–288.

CHEN, L. C.-M. (1977). The sporophyte of *Ahnfeltia plicata* (Huds.) Fries (Rhodophyceae, Gigartinales) in culture. *Phycologia* **16**, 163–168.

CHEN, L. C.-M., EDELSTEIN, T. and MCLACHLAN, J. (1974). The life history of *Gigartina stellata* (Stackh.) Batt. (Rhodophyceae, Gigartinales) in culture. *Phycologia* **13**, 287–294.

CORTEL-BREEMAN, A. M. and VAN DEN HOEK, C. (1970). Life history studies on Rhodophyceae I. *Acrosymphyton purpuriferum* (J. Ag.) Kyl. *Acta Bot. neerl.* **19**, 265–285.

DECEW, T. C. and WEST, J. A. (1977a). Culture studies on the marine red alga *Hildenbrandia occidentalis* and *H. prototypus* (Cryptonemiales, Hildenbrandiaceae). *Bull. Jap. Soc. Phycol.* **25**, 31–34.

DECEW, T. C. and WEST, J. A. (1977b). A life history relationship between *Gymnogongrus leptophyllus* J. Ag. and *Erythrodermis* (= *Petrocelis*) *haematis* (Hollenberg) Denizot (Phyllophoraceae, Gigartinales). *Br. phycol. J.* **12**, 108.

DENIZOT, M. (1968). "Les algues Floridées encroûantes (à l'exclusion des Corallinacées)." Muséum National d'Historie Naturelle, Paris.

DESIKACHARY, T. V. (1964). Status of the order Chaetangiales (Rhodophyta). *J. Indian bot. Soc.* **42A**, 16–26.

DIXON, P. S. (1960). Studies on marine algae of the British Isles: the genus *Ceramium*. *J. mar. biol. Ass. U.K.* **39**, 331–374.

DIXON, P. S. (1961). On the classification of the Florideae with particular reference to the position of the Gelidiaceae. *Botanica mar.* **3**, 1–16.

DIXON, P. S. (1965). Perennation, vegetative propagation and algal life histories, with special reference to *Asparagopsis* and other Rhodophyta. *Botanica gothoburg.* **3**, 67–74.

DIXON, P. S. (1971a). Studies of the genus "Seirospora." *Botaniste* **54**, 35–48.

DIXON, P. S. (1971b). A study of *Callithamnion lejolisea* Farl. *J. Phycol.* **7**, 58–63.

DIXON, P. S. (1973). "Biology of the Rhodophyta." Oliver and Boyd, Edinburgh.

DIXON, P. S. and IRVINE, L. M. (1977). "Seaweeds of the British Isles." British Museum (Natural History), London.

DIXON, P. S., MURRAY, S. N., RICHARDSON, W. N. and SCOTT, J. L. (1972). Life history studies in genera of the Cryptonemiales. *Mém. Soc. bot. Fr.* **1972**, 323–332.

DIXON, P. S. and RICHARDSON, W. N. (1969). The life-history of *Thuretellopsis peggiana* Kylin. *Br. phycol. J.* **4**, 87–89.

DOTY, M. S. (1963). *Gibsmithia hawaiiensis* gen. n. et sp. n. *Pacif. Sci.* **17**, 458–465.

DOTY, M. S. and MEÑEZ, E. G. (1960). *Tiffaniella*, a new genus in the Ceramiales. *Trans. Am. microsc. Soc.* **79**, 135–144.

DREW, K. M. (1937). *Spermothamnion Snyderae* Farlow, a Floridean alga bearing polysporangia. *Ann. Bot.* N.S. **1**, 463–476.

DREW, K. M. (1939). An investigation of *Plumaria elegans* (Bonnem.) Schmitz with special reference to triploid plants bearing parasporangia. *Ann. Bot.* N.S. **3**, 347–367.

EDELSTEIN, T. (1970). The life history of *Gloiosiphonia capillaris* (Hudson) Carmichael. *Phycologia* **9**, 55–59.

EDELSTEIN, T., CHEN, L. C.-M. and MCLACHLAN, J. (1974). The reproductive structures of *Gigartina stellata* (Stackh.) Batt. (Gigartinales, Rhodophyceae) in nature and in culture. *Phycologia* **13**, 99–107.

EDWARDS, P. (1973). Life history studies of selected British *Ceramium* species. *J. Phycol.* **9**, 181–184.

FAN, K.-C. (1961). Morphological studies of the Gelidiales. *Univ. Calif. Publs Bot.* **32**, 315–368.

FARNHAM, W. F. and FLETCHER, R. L. (1976). The occurrence of a *Porphyrodiscus simulans* Batt. phase in the life history of *Ahnfeltia plicata* (Huds.) Fries. *Br. phycol. J.* **11**, 183–190.

FELDMANN, J. (1939). Une Némalionale à carpotétraspores: *Helminthocladia Hudsoni* (C. Ag.) J. Ag. *Bull. Soc. Hist. nat. Afr. N.* **30**, 87–97.

FELDMANN, J. (1953). L'évolution des organes femelles chez les Floridées. *Proc. int. Seaweed Symp.* **1**, 11–12.

FELDMANN, J. and FELDMANN, G. (1942). Recherches sur les Bonnemaisoniacées et leur alternance de générations. *Ann. Sci. nat. Bot.* sér. 11, **3**, 75–175.

FELDMANN-MAZOYER, G. (1941). "Recherches sur les Céramiacées de la Méditerranée occidentale." Minerva, Alger.

FLETCHER, R. L. (1975). The life history of *Rhodophysema georgii* in laboratory culture. *Mar. Biol. Berlin*, **31**, 299–304.

FOSLIE, M. (1905). Remarks on northern Lithothamnia. *K. norske Vidensk. Selsk. Skr.* **1905**(3), 1–138.

FRITSCH, F. E. (1945). "The Structure and Reproduction of the Algae", Vol. 2. Cambridge University Press, Cambridge.

GANESAN, E. K. and DESIKACHARY, T. V. (1970). Studies on the morphology and reproduction of the articulated corallines—V. *Lithothrix* Gray. *Phykos* **9**, 41–51.

GANESAN, E. K. and WEST, J. A. (1975). Culture studies on the marine red alga *Rhodophysema elegans* (Cryptonemiales, Peysonneliaceae). *Phycologia* **14**, 161–166.

GORDON, E. M. (1972). Comparative morphology and taxonomy of the Wrangelieae, Sphondylothamnieae, and Spermothamnieae (Ceramiaceae, Rhodophyta). *Aust. J. Bot.* Suppl. Ser. **4**, 1–180.

GORDON-MILLS, E. M. and WOMERSLEY, H. B. S. (1974). The morphology and life history of *Mazoyerella* gen. nov. (*M. arachnoidea* (Harvey) comb. nov.)—Rhodophyta, Ceramiaceae—from Southern Australia. *Br. phycol. J.* **9**, 127–137.

GREGORY, B. D. (1934). On the life history of *Gymnogongrus griffithsiae* and *Ahnfeltia plicata*. *J. Linn. Soc. Bot.* **49**, 531–551.

GUIRY, M. D. (1974). A preliminary consideration of the taxonomic position of *Palmaria palmata* (Linnaeus) Stackhouse = *Rhodymenia palmata* (Linnaeus) Greville. *J. mar. biol. Ass. U.K.* **54**, 509–528.

GUIRY, M. D. (1976). An assessment of *Palmaria palmata* forma *mollis* (S. et G.) comb, nov. (= *Rhodymenia palmata* forma *mollis* S. et G.) in the eastern north Pacific. *Syesis* **8**. 245–261, 1975.

GUIRY, M. D. and IRVINE, D. E. G. (1978). A critical assessment of infraordinal classification in the Rhodymeniales. *Proc. int. Seaweed Symp.* **8** (in press).

HASSINGER-HUIZINGA, H. (1952). Generationswechsel und Geschlechtsbestimmung die *Callithamnion corymbosum* (Sm.) Lyngb. *Arch. Protistenk.* **98**, 91–124.

HOEK, C. VAN DEN and CORTEL-BREEMAN, A. M. (1970). Life-history studies on Rhodophyceae. II. *Halymenia floresia* (Clem.) Ag. *Acta bot. neerl.* **19**, 341–362.

HOLLENBERG, G. J. (1940). New marine algae from southern California I. *Am. J. Bot.* **27**, 868–877.

HOMMERSAND, M. H. (1963). The morphology and classification of some Ceramiaceae and Rhodomelaceae. *Univ. Calif. Publs Bot.* **35**, 165–366.

HURDLEBRINK, L. and SCHWANTES, H. O. (1973). Sur le cycle de développement de *Batrachospermum*. *Mém. Soc. bot. Fr.* **1972**, 269–274.

ITONO, H. (1971a). On *Delesseriopsis elegans* Okamura (Ceramiaceae, Rhodophyta). *Bull. Jap. Soc. Phycol.* **19**, 1–4.

ITONO, H. (1971b). The genera *Callithamnion, Aglaothamnion, Seirospora, Pleonosporium* and *Mesothamnion* (Ceramiaceae, Rhodophyta) in southern Japan. *Mem. Fac. Fish. Kagoshima Univ.* **20**, 217–237.

ITONO, H. and TANAKA, T. (1973). *Balliella*, a new genus of Ceramiaceae (Rhodophyta). *Bot. Mag. Tokyo* **86**, 241–252.

JAASUND, E. (1965). Aspects of the marine algal vegetation of north Norway. *Botanica gothoburg.* **4**, 1–174.

JOHANSEN, H. W. (1976). Family Corallinaceae. *In* "Marine Algae of California" (I. Abbott and G. J. Hollenberg, eds), pp. 379–419. Stanford University Press, Stanford.

JOLY, A. B. (1957). Contribuição ao conhecimento da flora ficológica marinha da Baía de Santos e Arredores. *Bolm Fac. Filos. Ciênc. Univ. S. Paulo, Botânica* **14**, 1–196.

JOLY, A. B. (1965). Flora marinha do litoral norte do estado de São Paulo e regiões circunvizinhas. *Bolm Fac. Filos. Ciênc. Univ. S. Paulo, Botânica* **21**, 1–267.

KNAGGS, F. (1967). Modified sporangia on the tetrasporophyte of *Rhodochorton purpureum* (Lightf.) Rosenvinge. *Nova Hedwigia* **13**, 139–143.

KLAVESTAD, N. (1957). Paraspores in *Rhodochorton rothii*. *Nytt Mag. Bot.* **5**, 61–62.

KYLIN, H. (1924). Studien über die Delesseriaceen. *Acta Univ. lund.*, N.F., Avd. 2, **20**(6), 1–111.

KYLIN, H. (1928). Entwicklungsgeschichtliche Florideenstudien. *Acta Univ. lund.* N.F., Avd. 2, **24**(4), 1–127.

KYLIN, H. (1930). Über die Entwicklingsgeschichte der Florideen. *Acta Univ. lund.* N.F., Avd. 2, **26**(6), 1–104.

KYLIN, H. (1931). Die Florideenordnung Rhodymeniales. *Acta Univ. lund.* N.F., Avd. 2, **27**(11), 1–48.

KYLIN, H. (1932). Die Florideenordnung Gigartinales. *Acta Univ. lund.* N.F., Avd. 2, **28**(8), 1–88.

KYLIN, H. (1956). "Die Gattungen der Rhodophyceen." C. W. K. Glererups Förlag, Lund.

LEMOINE, M. (1971). Remarques sur la reproduction des algues calcaires fossiles Mélobésiées, la systématiques et la phylogénie. *Revue algol.* N.S. **2**, 152–161.

LEVRING, T. (1941). Die Meeresalgen der Juan Fernandez-Inseln. *In* "The Natural History of Juan Fernandez and Easter Island", Vol. 2, (C. Skottsberg, ed.), pp. 601–670. Almquist and Wiksells, Uppsala.

LEVRING, T. (1953). The marine algae of Australia I. Rhodophyta: Goniotrichales, Bangiales and Nemalionales. *Ann. Bot.*, ser. 2, **2**, 457–530.

LEWIS, I. F. (1909). The life history of *Griffithsia bornetiana*. *Ann. Bot.* **23**, 639–690.

L'HARDY-HALOS, M.-TH. (1966). Observations sur la morphologie et sur la position systématique de *Dohrniella neapolitana* Funk (Floridées-Céramiacées). *Bull. Soc. bot. Fr.* **113**, 295–304.

L'HARDY-HALOS, M.-TH. (1968). Observations sur la morphologie du *Neomonospora furcellata* (J. Ag.) G. Feldmann et Meslin (Rhodophyceae-Ceramiaceae) et sur sa position taxonomique. *Bull. Soc. bot. Fr.* **115**, 523–528.

L'HARDY-HALOS, M.-TH. (1970). Recherches sur les Céramiacées (Rhodophycées, Céramiales) et leur morphogénèse. I. Structure de l'appareil vegetatif et des organes reproducteurs. *Rev. gén. Bot.* **77**, 211–287.

MAGNE, F. (1964). Recherches caryologiques chez les Floridées (Rhodophycées). *Cah. Biol. mar.* **5**, 461–671.

MAGNE, F. (1972). Le cycle de développement des Rhodophycées et son évolution. *Mém. Soc. bot. Fr.* **1972**, 247–268.

MASUDA, M. (1976). On a new red algal genus *Pseudorhododiscus*. *Acta phytotax. geobot. Kyoto* **27**, 123–132.

MASUDA, M. and OHTA, M. (1975). The life history of *Rhodophysema georgii* Batters (Rhodophyta, Cryptonemiales). *J. Jap. Bot.* **50**, 1–10.

MASUDA, M. and UCHIDA, T. (1976). On the life history of *Gigartina ochotensis* (Rupr.) Rupr. from Muroran, Hokkaido. *Bull. Jap. Soc. Phycol.* **24**, 41–47 (in Japanese with an English summary).

MAYHOUB, H. (1973). Cycle du developpement de *Calosiphonia vermicularie* (J. Agardh) Schmitz (Rhodophycée, Gigartinale). *C.r. hebd. Séanc. Acad. Sci., Paris*, sér. D, **277**, 1137–1140.

MAYHOUB, H. (1975). Une nouvelle espèce de *Dohrniella* (Rhodophycées, Ceramiacées) de Méditerranée orientale. *Bull. Soc. phycol. Fr.* **20**, 7–11.

MULLAHY, J. H. (1952). The morphology and cytology of *Lemanea australis* Atk. *Bull. Torrey bot. Club* **79**, 393–406 and 471–484.

NEWTON, L. (1931). "A Handbook of the British Seaweeds." British Museum (Natural History), London.

NORRIS, R. E. and WYNNE, M. J. (1969) ["1968"]. Notes on marine algae of Washington and southern British Columbia, III. *Syesis* **1**, 133–146.

OLTMANNS, F. (1904). "Morphologie und Biologie der Algen." Part 1. Gustav Fischer, Jena.

PAPENFUSS, G. F. (1961). (Review of) "marine algae of the Eastern Tropical and Subtropical Coasts of the Americas". By W. R. Taylor. The University of Michigan Press, Ann Arbor. 870 pp. *Bull. Torrey bot. Club* **88**, 421–423.

PAPENFUSS, G. F. (1966). A review of the present system of classification in the Florideophycidae. *Phycologia* **5**, 247–255.

PARKE, M. and DIXON, P. S. (1976). Check-list of British marine algae—third revision *J. mar. biol. Ass. U.K.* **56**, 527–594.

Pujals, C. (1970). "*Medeiothamnion*" nuevo genero de Ceramiaceae. *Revta Mus. argent. Cienc. nat. Bernardino Rivadavia Inst. nac. Invest. Cienc. nat.* **3**, 287–299.

Ramus, J. (1969). The developmental sequence of the marine red alga *Pseudogloiophloea* in culture. *Univ. Calif. Publs Bot.* **52**, 1–28.

Richardson, W. N. and Dixon, P. S. (1970). Culture studies on *Thuretellopsis peggiana* Kylin. *J. Phycol.* **6**, 154–159.

Rosenvinge, L. K. (1917). The marine algae of Denmark. Contributions to their natural history Part II, Rhodophyceae II. (Cryptonemiales). *K. danske Vidensk. Selsk. Skr.*, 7. Raekke, Nat. Math. Afd. **7**, 155–283.

Rosenvinge, L. K. (1923–24). The marine algae of Denmark. Contributions to their natural history Part III, Rhodophyceae III. (Ceramiales) *K. danske Vidensk. Selsk. Skr.*, 7. Raekke, Nat. Math. Afd. **7**, 287–486.

Rosenvinge, L. K. (1931). The marine algae of Denmark. Contributions to their natural history Part IV, Rhodophyceae IV. (Gigartinales, Rhodymeniales, Nemastomatales) *K. danske Vidensk. Selsk. Skr.*, 7. Raekke, Nat. Math. Afd. **7**, 491–630.

Rueness, J. (1968). Paraspores from *Plumaria elegans* (Bonnem.) Schmitz in culture. *Nytt Mag. Bot.* **15**, 220–224.

Schmitz, F. (1883). Untersuchungen über die Befruchtung der Florideen. *Sber. Akad. Wiss.* **1883**, 215–258.

Schmitz, F. (1889). Systematische Übersicht der bisher bekannten Gattungen der Florideen. *Flora, Jena* **72**, 435–456.

Schneider, C. W. and Searles, R. B. (1976). North Carolina marine Algae. VII. New species of *Hypnea* and *Petroglossum* (Rhodophyta, Gigartinales) and additional records of other Rhodophyta. *Phycologia* **15**, 51–60.

Schotter, G. (1968). Recherches sur les Phyllophoracées. Notes posthumes publiées par Jean Feldmann et Marie-France Magne. *Bull. Inst. océanogr. Monaco* **67**(1383), 1–99.

Scott, J. L. and Dixon, P. S. (1972). The life history of *Pikea californica* Harv. *J. Phycol.* **7**, 295–300.

Searles, R. B. (1968). Morphological studies of red algae of the order Gigartinales. *Univ. Calif. Publs Bot.* **43**, 1–86.

Sjöstedt, L. G. (1926). Floridean studies. *Acta Univ. lund.* **22**(4), 1–94.

Smith, G. M. and Hollenberg, G. J. (1943). On some Rhodophyceae from the Monterey Peninsula, California. *Am. J. Bot.* **30**, 211–222.

South, G. R. and Whittick, A. (1976). Aspects of the life history of *Rhodophysema elegans* (Rhodophyta, Peyssonneliaceae). *Br. phycol. J.* **11**, 349–354.

Sparling, S. R. (1957). The structure and reproduction of some members of the Rhodymeniaceae. *Univ. Calif. Publs Bot.* **29**, 319–396.

Sturch, H. H. (1924). On the life-history of *Harveyella pachyderma* and *H. mirabilis*. *Ann. Bot.* **38**, 27–42.

Sundene, O. (1962). Reproduction and morphology in strains of *Antithamnion boreale* originating from Spitsbergen and Scandinavia. *K. norske Vidensk. Akad. Skr., Mat.-Nat. Kl.* **1962**(5), 1–19.

Suneson, S. (1937). Studien über die Entwicklungsgeschichte der Corallinaceen. *Acta Univ. lund.* N.F. Avd. 2, **33**(2), 1–102.

Suneson, S. (1943). The structure, life-history and taxonomy of the Swedish Corallinaceae. *Acta Univ. lund.* N.F. Avd. 2, **39**(9), 1–65.

SUNESON, S. (1950). The cytology of the bispore formation in two species of *Lithophyllum* and the significance of the bispores in the Corallinaceae. *Bot. Notiser* **103**, 429–450.

SVEDELIUS, N. E. (1914). Über Sporen an Geschlechtspflanzen von *Nitophyllum punctatum*; ein Beitrag zur Frage des Generationswechsels der Florideen. *Ber. dt. bot. Ges.* **32**, 106–116.

TAYLOR, W. R. (1957). "Marine Algae of the Northeastern Coast of North America", 2nd Edition. University of Michigan Press, Ann Arbor.

TAYLOR, W. R. (1960). "Marine Algae of the Eastern Tropical and Subtropical Coasts of the Americas." University of Michigan Press, Ann Arbor.

TOKIDA, J. (1943). On the so-called *Dilsea edulis* of Japan. *Bot. Mag. Tokyo* **47**, 93–97.

WAGNER, F. S. (1954). Contributions to the morphology of the Delesseriaceae. *Univ. Calif. Publs Bot.* **27**, 279–346.

WEST, J. A. (1969). The life histories of *Rhodochorton purpureum* and *R. tenue* in culture. *J. Phycol.* **5**, 12–21.

WEST, J. A. (1970a). The life history of *Rhodochorton concrescens* in culture. *Br. phycol. J.* **5**, 179–186.

WEST, J. A. (1970b). The conspecificity of *Heterosiphonia asymmetrica* and *H. densiuscula* and their life histories in culture. *Madroño* **20**, 313–319.

WEST, J. A. (1971). Environmental control of hair and sporangial formation in the marine red alga *Acrochaetium proskaueri* sp. nov. *Proc. int. Seaweed Symp.* **7**, 377–384.

WEST, J. A. (1978). The life history of *Rhodochorton membranaceum*, an endozoic red alga. *Proc. int. Seaweed Symp.* **8** (in press).

WEST, J. A. and NORRIS, R. E. (1966). Unusual phenomena in the life histories of Florideae in culture. *J. Phycol.* **2**, 54–57.

WEST, J. A. and POLANSHEK, A. R. (1975a). Culture studie son *Petrocelis cruenta* and *Gigartina stellata* (Rhodophyta) from the North Atlantic. *J. Phycol.* **11**, (suppl.), 17.

WEST, J. A. and POLANSHEK, A. R. (1975b). Genetic affinities among eastern Pacific populations of *Petrocelis* and *Gigartina* subgenus *Mastocarpus*. Paper presented at U.S.–Japan Cooperative Science Program Seminar, Kobe, Japan, 30 November–5 December 1975.

WOLLASTON, E. M. (1968). Morphology and taxonomy of Southern Australian genera of Crouanieae Schmitz (Ceramiaceae, Rhodophyta). *Aust. J. Bot.* **16**, 217–417.

WOLLASTON, E. M. (1971a). *Antithamnion* and related genera occurring on the Pacific coast of North America. *Syesis*, **4**, 73–92.

WOLLASTON, E. M. (1971b). The morphology and relationships of *Warrenia comosa* (Harvey) Kuetzing. *Phycologia*, **10**, 291–298.

WOLLASTON, E. M. (1972). The morphology and relationships of *Muellereana wattsii* (Harvey) Schmitz (Ceramiaceae: Rhodophyta). *Trans. R. Soc. S. Aust.* **96**, 119–124.

WOMERSLEY, H. B. S. (1965). The Helminthocladiaceae (Rhodophyta) of Southern Australia. *Aust. J. Bot.* **13**, 451–487.

WOMERSLEY, H. B. S. and CARTLEDGE, S. A. (1975). The Southern Australian species of *Spyridia* (Ceramiaceae: Rhodophyta). *Trans. R. Soc. S. Aust.* **99**, 221–234.

YAMADA, Y. (1938). The species of *Liagora* from Japan. *Sci. Paps Inst. algol. Res. Hokkaido Univ.* **2**, 1–34.

YAMANOUCHI, S. (1906). The Life-history of *Polysiphonia violacea*. *Bot. Gaz* **41**, 425–433; **42**, 401–449.

7 | The Application of Cytology to Taxonomy of Red and Brown Algae

F. MAGNE

Université Pierre et Marie Curie, Paris, France

Abstract: Cytology provides data of great value to the algal taxonomist. Both morphological and biochemical characteristics are manifest and these seem to be more reliable than characteristics of gross morphology. Interest in algal cytology increased dramatically when electron microscopy opened up the field of techniques. However, light microscopy still has an important part to play in preliminary observations, with the electron microscope being available to confirm and extend the observations.

Study of cell contents by light microscopy has contributed much towards fundamental distinction between the groups of multicellular algae on the basis of plastid characters and storage products. These contributions have been extended by ultrastructural and biochemical analysis.

Unusual cell types and special cellular behaviour can be used to characterize genera, or can promote very fruitful working hypotheses. Subcellular organelles studied in the light microscope can reveal much of importance.

Data on chromosome numbers permit better-founded taxonomic decisions. Unfortunately, to date, available information in the Phaeophyta is not of great use and in the Rhodophyta the situation is confused by many contradictory earlier data. Much more modern work is needed, since the existence of an unusual type of nuclear division in the Rhodophyta may prove a lead towards improvement of taxonomic knowledge of some orders.

With the light microscope, plastids as organelles have provided the greatest amount of taxonomically useful information.

In the brown algae, some parenchymatous heteroclitic genera were grouped into the order Scytosiphonales on the basis of their unusual possession of only one plastid, with pyrenoid, per cell. Traditionally, Phaeophyta possessing plastids with pyrenoids are considered the most primitive; the discovery in Fucales, where the plastids lack pyrenoids, of ultrastructures similar to pyrenoids within the plastids raises the need for a reappraisal of the situation.

In red algae, the morphology of plastidial apparatus is fairly uniform; in contrast,

Systematics Association Special Volume No. 10, "Modern Approaches to the Taxonomy of Red and Brown Algae", edited by D. E. G. Irvine and J. H. Price, 1978, pp. 145–156, Academic Press, London and New York.

there is great variety amongst the more primitive forms, so that more extensive utilization of this characteristic should lead to more accurate classification.

INTRODUCTION

The aim of taxonomy is the recognition of relationships between individuals by means of comparison of manifest characteristics. The more accessible and therefore most utilized characteristics derive from morphology. Cytology, the description of cell contents, is a part of morphology; in addition it is able to express some biochemical aspects. Experience proves that the characteristics it demonstrates are relatively stable and consistent, so that it can provide much valuable information for taxonomy. The electron microscope has led to wider use of cytological characteristics, having made possible the study of detailed fine structure of organelles, even at molecular level. The light microscope has nevertheless been used frequently in the detection and description of disposition and construction of cellular organelles; it has played a very important part in providing much needed data for red and brown algal taxonomy. The present paper primarily concerns these latter data.

CLASSIFICATION FEATURES

The major classification of the multicellular algae into several large groups is essentially based on their colour, following the early ideas of Lamouroux; other characteristics, especially reproductive, were added later. A group such as the Rhodophyta or Phaeophyta can be defined precisely only by light microscopic examination of the cell contents of individuals belonging to the group, clarified if necessary by the use of staining techniques. The distinguishing fundamental biochemical characteristics are expressed in the pigmentation—as shown by plastid colour—and in the nature and distribution of storage products. Starch is absent in both groups; the red algae store "floridean starch" grains outside the plastids, and the brown algae show no polysaccharide grains, only physodes, vesicles with contents which stain orthochromatically in vital stains.

These elementary but fundamental data provided by cell study have acquired their full value only with support provided by biochemistry to knowledge of cellular components. Electron microscopy, showing the ultrastructural differences among the plastids, has confirmed the major distinction among red, brown and green algae.

The presence of an unusual type of cell, defined by special characteristics of content or form, can be used to distinguish a species, genus, or even larger taxonomic group. Foslie (1909) was utilizing this fact at the beginning of this century, when he distinguished the genus *Porolithon* (Corallinaceae) on the existence of megacells. Such characteristics should not be employed in isolation or without an already detailed knowledge of the group concerned. Sauvageau (1898) grouped certain Myrionemataceae into the new genus *Ascocyclus*, distinguished on the basis of possession of ascocysts (unicellular paraphyses filled with physodes). It was later demonstrated (Loiseaux, 1967) that such paraphyses exist in several Phaeophycean families and that *Ascocyclus* cannot therefore be maintained.

1. *Secretory Cells*

In the Florideophycideae, so-called secretory cells, frequently with colourless, clear and highly refractive contents, are present in some families. In the Ceramiaceae, their importance in taxonomy seems to be limited to species level. However, in the Bonnemaisoniaceae they form an important family characteristic, having played a large part in the discovery that life-histories in that group involve morphologically very different gametophytes and tetrasporophytes. Feldmann and Feldmann (1942) demonstrated the life-history connections between *Asparagopsis armata* and *Falkenbergia rufolanosa* and between *Bonnemaisonia asparagoides* and *Hymenoclonium serpens*. This led them to hypothesize that *Trailliella intricata*, then referred to the Ceramiaceae but possessing unusually disposed secretory cells, was the tetrasporophyte missing from the life-history of the secretory-cell-bearing gametophyte *Bonnemaisonia hamifera*. This hypothesis was afterwards confirmed by the work of Harder and Koch (1949), in growing young *Bonnemaisonia hamifera* from *Trailliella* tetraspores. Thus a cytological character led to the identification of connections, hitherto unrecognized, between different morphological phases in the same life-histories; morphology alone would not have solved this complex taxonomic problem.

2. *Intercellular Connections*

If the existence of certain morphologically characteristic cells can eventually be utilized taxonomically, so can the existence of other cells with characteristic behaviour patterns. Anatomically more or less complex red algal species of the Cryptonemiales, Gigartinales, Ceramiales and Gelidiales, demonstrate three processes of secondary intercellular communication among vegetative cells. Most frequently, the phenomenon involves an acceptor cell, a donor cell and an intermediary cell, this latter arising from the donor cell after a division leaving a pit-connection between mother cell and product. The intermediary cell comes into contact with the acceptor cell, to which its contents pass during cell fusion. The primary pit-connection first set up between the intermediary cell and the acceptor cell, then constitutes an indirect secondary pit-connection. In other cases, communication is laterally developed between two adjacent cells belonging to two parallel rows by a more or less complete resorption of the cell walls separating them; such a case can be referred to as a cellular fusion. Finally, a pit-connection may be formed directly between two adjacent cells.

Only the latter two types of intercellular communication occur in the Corallinaceae. The members of this family constitute a coherent whole which is nevertheless taxonomically difficult. Since the different taxa cannot be distinguished by sexual reproductive characteristics, taxonomy was initially based on external morphology, principally by Areschoug (1852) and Foslie (1902). Following Lemoine (1911) anatomical characteristics were intensively used. More recently, distinctions have been based on both these major characteristics. Nevertheless, previous authors (Rosenvinge, 1917; Suneson, 1943) have pointed out the special taxonomic interest of some of the cytological characteristics and Johansen (1969), in a recent general treatment of the family, employed the presence or absence of secondary pit-connections as one of four major characteristics to support a new classification of this group.

Cabioch (1970) has shown that a comprehensive taxonomic treatment of the Corallinaceae is possible from cytological starting points. Studying distribution of secondary intercellular communications, she outlined the precise significance of this distribution, which expresses relationships and is of phylogenetic value. In *Sporolithon* (formerly *Archaeolithothamnium*), the best known fossil genus, both types of communication co-exist, which is considered to be an archaic character. In other genera, only one type exists, so that the group can be arranged into two natural series, usually in accord with some of the major trends of anatomical classification.

In conclusion, Cabioch (1971) proposed a new classification in which the major distinctions are based on cytology and defined by the distribution of the communications. The simultaneous existence of both communication types defines the subfamily Sporolithoideae; the exclusive occurrence of direct secondary pit-connections characterizes the Lithophylloideae; and the genera with only cellular fusions are grouped into the two subfamilies Corallinoideae and Lithothamnioideae, separated on the structure of their asexual conceptacles.

3. *The Nucleus*

Study of the nucleus can provide important information for taxonomy, especially through the use of the karyotype, one of the best characteristics of a species. In the Rhodophyta and Phaeophyta, data in this field can be used only in a very limited way, because they are fragmentary and frequently inconsistent; to a large extent this is a consequence of the great practical and technical difficulties frequently encountered during the study of many species. Moreover, some better known groups, such as the genus *Fucus* studied by Evans (1962), are distressingly uniform. I am convinced, however, that the position merits re-assessment; the study of some groups, especially Rhodophytan genera where species are numerous and nuclei and chromosomes reasonably sized, should provide interesting data.

Development of mitosis should be of particular taxonomic interest. During an extensive study of this phenomenon in Rhodophyta, I was able to point out (Magne, 1964a, 1964b) that, in some Gigartinales and Cryptonemiales, mitosis is particularly unusual. During the whole prophase, the nuclear vesicle is elongated between two poles, taking the form of a spindle; this phenomenon is especially obvious in *Calliblepharis*, hence the term "calliblepharidian division" was proposed. Little is known, even now, about the origin and significance of this peculiarity; however it seems consistent within both genus and family (Rhodophyllidaceae, Gigartinaceae). Logically, therefore, the phenomenon is considered fundamental and should be important in any reappraisal of the taxonomic treatment of Gigartinales and Cryptonemiales.

4. *Plastids*

The plastids, through their structure, number and disposition in the cell, are the organelles which, under the light microscope, have contributed most to algal taxonomy. Living cells provide the best conditions for their observation. Indeed, drying or the use of standard fixatives can produce artefacts such as aggregations of plastids or alterations in their structure. Variability of their characteristics according to the physiological activity of the cell has been suggested by some workers; however, it is usually possible to define without difficulty a structural type to be used as a standard against which comparisons can be made.

(*a*) *Phaeophyta.* In the Phaeophyta, Hamel—one of the first and more enthusiastic partisans of the use of cytological characteristics in taxonomy—proposed (1939) separation of the genera *Ectocarpus* and *Giffordia* chiefly on their cytology; in *Ectocarpus*, each cell contains only one ribbon-like plastid with numerous pyrenoids, while in *Giffordia* there are numerous plate-like plastids, each provided with a pyrenoid. Similarly, Bornet (1889) created the sub-genus *Bachelotia*—later elevated to generic level by Kuckuck (1929)—for *Pylaiella fulvescens* mainly on its peculiar cell structure with centrally located and stellate plastids, quite different from the other species of *Pylaiella*, in which each cell contains numerous plastids scattered throughout the parietal cytoplasm.

The stellate construction of the plastid, rare in the Phaeophyta, is actually known in only four genera. In *Bachelotia* it is the result of the disposition of numerous plastids converging, by the pole occupied by a pyrenoid, towards a focal point where all the pyrenoids are joined (Magne, 1976). In *Asteronema*, also of the Ectocarpales, it is similar (Delépine and Asensi, 1975) but not completely identical at ultrastructural level; in *Bachelotia*, the pyrenoid is terminal in each plastid (Magne, 1976), whereas in *Asteronema* (Delépine *et al*., 1976), it is lateral. In the two other genera where it is known, this disposition has a different origin; each cell has only one plastid of a stellate form, containing only a central pyrenoid. This was described in *Splachnidium* (Price and Ducker, 1966), and later in *Scytothamnus* (Asensi, 1975); confirmation was provided by electron microscopy. This characteristic, occurring in two genera actually placed in different orders (respectively Chordariales and Dictyosiphonales) is one of several reasons to associate them more closely (Delépine *et al*., 1976).

Among the polystichous Phaeophyta, members of the family Scytosiphonaceae contain only one parietal plastid in each cell and this bears a large hemispherical protruding pyrenoid in a lateral position. This characteristic makes them a distinct and homogeneous group, in spite of the very different thallus morphology from genus to genus. Reinke (1888), who first noted this homogeneity, considered it of systematic value. Feldmann (1949) founded the order of Scytosiphonales for this family and certain other genera chiefly on this characteristic; that order is nowadays usually accepted. Recently, in a systematic study based on numerical taxonomy, Russell and Fletcher (1975), however, employed some characteristics of the plastid, concluding that it is not convenient to separate the Scytosiphonaceae from Ralfsiaceae and Myrionemataceae.

Study of the life-history in members of the Scytosiphonales showed that there is potential for a discoid *Ralfsia*-like morphological phase (Dangeard, 1963; Nakamura, 1965; Tatewaki, 1966; and others . . .): sometimes this is referable to already-described species of the genera *Microspongium* (Lund, 1966) or *Ralfsia* (Fletcher, 1974), or to ectocarpoid forms such as *Streblonema anomalum* and *Compsonema sporangiferum* (Loiseaux, 1970). It is noteworthy that all the forms obtained in culture from members of the Scytosiphonaceae show, as with the parental material, only one plastid with pyrenoid in each cell. It will be necessary in future to reappraise the taxonomic position of all Phaeophyceae that possess only a single plastid per cell and to widen the concept of the Scytosiphonales, an order thus most certainly founded on a cytological character.

The distinction between Phaeophytan species with and without pyrenoids is usually in accordance with the grouping of the taxa into orders (Simon, 1954); the most primitive plants are provided with this organelle (Hamel, 1931–1939; Simon, 1954). Pyrenoids were thus never found in species of Fucales, considered from many characters to be an advanced order, until it was shown by Evans (1968) that the plastids of the eggs can contain tiny immersed pyrenoids; this was so in *Bifurcaria*, *Himanthalia* and *Cystoseira*, although not in *Fucus*, *Ascophyllum* or *Halidrys*. Such an important result demonstrates that pyrenoids are not always detectable with the light microscope and that no absolute rule can be made. On the other hand, this work has emphasized the value of the pyrenoid as an evolutionary criterion; according to Evans (1968) the very small size of these fucalean pyrenoids would indicate that they are in the process of disappearance, a state already achieved in those genera (which would be the more highly evolved ones) in which they are lacking. It is noticeable that Manton's (1964) spermatozoid study with the electron microscope led to a parallel conclusion.

The question is not yet entirely resolved, since these so-called "pyrenoids" are perhaps not true ones. For identification of phaeophycean pyrenoids in electron micrographs we possess only three criteria; their size, their protruded form and their apparent construction. The first and the second of these criteria are unusable in the Fucales. Utilizing the third, there is indeed a strong discrepancy between true and fucalean pyrenoids. As can be seen from pictures published by Evans himself (1968), material of true pyrenoids is homogeneous, dense and different from that of the stroma of the plastid; in fucalean pyrenoids this material is always similar to the stroma. A re-investigation is needed to elucidate the exact nature of the fucalean pyrenoids.

(*b*) *Rhodophyta* In most of the Rhodophyta, there is great uniformity in plastid construction and disposition (one or more parietal plastids, lacking pyrenoids). However, this is variable in the more primitive Rhodophyta, such as the Nemalionales and the whole of the Bangiophycideae.

In the Bangiophycideae, several types of plastid construction occur; the cell contains either one stellate axile plastid with a central pyrenoid (as in *Porphyra*), or one parietal plastid with (as in *Rhodosorus*) or without (as in *Porphyropsis*) a pyrenoid, or even several plastids without pyrenoids (as in *Phragmonema*).

The available classifications (Kylin, 1956; Dixon, 1973) are chiefly founded on frequently poor characters of vegetative morphology; cytology is discarded. However, the latter can provide characters that express relationships. In the Porphyridiales and Goniotrichales, which were suggested (Feldmann, 1955) for amalgamation, Feldmann (1967) has also suggested the gathering into one family (Phragmonemataceae) of all the genera in which the cells contain numerous plastids without pyrenoids.

Among the Bangiales, the Bangiaceae shows a striking homogeneity arising from cytological structure and reproductive characters. In contrast, the family Erythropeltidaceae as defined is deeply heterogeneous, the plastid construction varying from genus to genus and within genus from species to species (Kylin, 1956). The Erythropeltidaceae requires further scrutiny; cytological and other advances will perhaps permit reappraisal of generic, and possibly family, delimitation.

In the order Nemalionales (as defined by Kylin, 1956, but with the exclusion of the families Naccariaceae and Bonnemaisoniaceae) it is reassuring to note that the plastid construction is constant in most of the families, each cell containing either a single stellate plastid with central pyrenoid (in the Helminthocladiaceae) or several parietal plastids devoid of pyrenoids (in the Batrachospermaceae, Lemaneaceae and Thoreaceae). However, the situation differs in the Chaetangiaceae and Acrochaetiaceae.

In the Chaetangiaceae, both types of plastid construction noted above exist, leading to a division into two groups. The first, containing the genera *Galaxaura* and *Actinotrichia*, is characterized by a stellate plastid in each cell. The second, in which each cell contains several parietal plastids, includes the genera *Chaetangium*, *Scinaia*, *Pseudoscinaia*, *Gloiophloea*, *Pseudogloiophloea* and also *Whidbeyella*, a less well-known genus. This is an indication of some profound distinction that suggests the Chaetangiaceae is heterogeneous, its basic characteristic, the existence of a pericarp, being an artificial one. The genera *Galaxaura* and *Actinotrichia*, with multiaxial structure, single stellate plastid and terminal carpospores, must be considered closer to the Helminthocladiaceae, which possess the same characters and moreover show a definite trend toward the building of a cystocarpic wall. The taxonomic position of this family, which is then reduced to the second group listed above, will perhaps need revision. According to Feldmann (1959; 1977) the female organs in *Scinaia*, the simplicity of which has hitherto been considered primitive, are rather highly evolved, their simple form being secondarily acquired. Recently, the same conclusion was reached by Delépine and Lambert (1978) in *Chaetangium* and, after Ramus's study (1969), it is possibly similar in *Pseudogloiophloea*, where the female branch is identical to that of *Scinaia*. These female organs therefore do not entirely parallel the female organs in other Nemalionales. If, in the future, this point of view can be confirmed and extended to the whole of the genera, the Chaetangiaceae will have to be rejected from the Nemalionales.

The Acrochaetiaceae (as Chantransiaceae in Kylin, 1956) was and still is taxonomically very confused for historical and nomenclatural reasons. Our frequently scanty data on its members is one of the chief such reasons. Current belief is that very few genera are present in the family, either one (*Rhodochorton*: Drew, 1928; *Audouinella*: Parke and Dixon, 1976) or two (*Audouinella* and *Colaconema*: Woelkerling, 1971) being accepted. However, several authors, especially Rosenvinge (1909), Hamel (1927), Kylin (1944, 1956), Papenfuss (1945, 1947) and Feldmann (1962), have accepted a larger number of genera but not always with the same delimitation. The situation is therefore complex and confused. I cannot claim to provide a solution and cytology is certainly no universal remedy, but I do wish to emphasize the variety of plastid construction among the members of this family.

This very interesting characteristic has already been noticed frequently in the past and used (especially by Papenfuss and Feldmann) more than in any other group of Rhodophyta, to define several genera. Moreover, Feldmann created the order Acrochaetiales in which the major divisions are based on cytological characters. In the genus *Acrochaetium* itself, the plastid construction can be of single stellate and axial plastid with a central pyrenoid that gives the definition of the genus *Chromastrum* Papenfuss (as in *Acrochaetium parvulum*), of the type with single discoid parietal plastid with pyrenoid (as in *Acrochaetium asparagopsidis*, *Acrochaetium subtilissimum* and *Acrochaetium thuretii*), or of the type with single parietal plastid devoid of pyrenoid (as in *Acrochaetium infestans*).

For the past years, data on the members of this family have been accumulating from monographs, reviews and (especially fruitful) culture programmes initiated for developmental purposes (Knaggs and Conway, 1964; West, 1968, 1969, 1970, 1972; Feldmann, 1972; Boillot and Magne, 1973; Stegenga and Borsje, 1976, Stegenga and Vroman, 1976; Magne, 1977) or from experimental work (Knaggs, 1966; White and Boney, 1969). This will continue and I hope that reappraisal of the taxonomic treatment of this family will be possible, with the participation, perhaps major, of cytology.

ACKNOWLEDGEMENTS

The author is greatly indebted to Mr René Delépine for various data, and especially to Professor Jean Feldmann for many valuable discussions.

REFERENCES

Areschoug, J. E. (1852). Ordo XII. Corallineae. *In* "Species genera et ordines algarum . . ." (J. G. Agardh) **2**(2), 506–576. C. W. K. Gleerup, Lundae.

Asensi, A. O. (1975). La estructura, la distribución y el cultivo de *Scytothamnus fasciculatus* (Hook et Harv.) Cotton (Phaeophyta). *Physis*, sect. A **34**, 269–282.

Boillot, A. and Magne, F. (1973). La cycle biologique de *Kylinia rosulata* Rosenvinge (Rhodophycées, Acrochaetiales). *Bull. Soc. phycol. Fr.* **18**, 47–53.

Bornet, E. (1889). Note sur l'*Ectocarpus* (*Pylaiella*) *fulvescens* Thuret. *Rev. gén. Bot.* **1**, 5–10.

Cabioch, J. (1970). Sur l'importance des phénomènes cytologiques pour la systématique et la phylogénie des Corallinacées (Rhodophycées, Cryptonemiales). *C.r. hebd. Séanc. Acad. Sci., Paris* **271**(D), 296–299.

Cabioch, J. (1971). Essai d'une nouvelle classification des Corallinacées actuelles. *C.r. hebd. Séanc. Acad. Sci., Paris* **272**(D), 1616–1619.

Dangeard, P. (1963). Recherches sur le cycle évolutif de quelques Scytosiphonacées. *Botaniste* **46**, 5–132.

Delépine, R. and Asensi, A. (1975). *Asteronema* nov. gen., nouveau genre de Phéophycée australe. *Bull. Soc. bot. France* **122**, 295–304.

Delépine, R., Asensi, A. and Guglielmi, G. (1976). Nouveaux types d'ultrastructure plastidiale chez les Phéophycées. *Phycologia* **15**, 425–434.

Delépine, R. and Lambert, C. (1978). Observations sur le *Chaetangium fastigiatum* de la nature et son cycle de vie. *Bull. Soc. bot. France* **124** (in press).

Dixon, P. S. (1973). "Biology of the Rhodophyta." Oliver and Boyd, Edinburgh.

Drew, K. M. (1928). A revision of the genera *Chantransia, Rhodochorton*, and *Acrochaetium* with description of the marine species of *Rhodochorton* (Naeg.) *gen. emend.* on the Pacific coast of North America. *Univ. Calif. Publs Bot.* **14**, 139–224.

Evans, L. V. (1962). Cytological studies in the genus *Fucus. Ann. Bot.* N.S. **26**, 345–360.

Evans, L. V. (1968). Chloroplast morphology and fine structure in British fucoids. *New Phytol.* **67**, 173–178.

Feldmann, J. (1949). L'ordre des scytosiphonales. *Mém. Soc. Hist. nat. Afr. N.* hors-série **2**, 103–115.

Feldmann, J. (1955). Un nouveau genre de Protofloridée: *Calodictyon, nov. gen. Bull. Soc. bot. France* **102**, 23–28.

Feldmann, J. (1959). A propos de la phylogénie des Floridées (Floridéophycidae). *Proc. int. bot. Congr.* **9**(2), 113.

Feldmann, J. (1962). Rhodophyta order Acrochaetiales and its classification. *Proc. Pacific Sci. Congr.* **9** (4, Botany), 219–221.

Feldmann, J. (1967). Sur une Bangiophycée endozoique (*Neevea repens* Batters) et ses affinités (Rhodophyta). *Blumea* **15**, 25–29.

Feldmann, J. (1972). Les problèmes actuels de l'alternance de générations chez les Algues. *Mém. Soc. bot. France* **1972**, 7–38.

Feldmann, J. (1977). (Communication personnelle.)

Feldmann, J. and Feldmann, G. (1942). Recherches sur les Bonnemaisoniacées et leur alternance de générations. *Ann. Sci. nat., Bot.* **11**(3), 75–175.

Fletcher, R. L. (1974). Studies on the brown algal families Ralfsiaceae and Scytosiphonaceae. *Br. phycol. J.* **9**, 218.

Foslie, M. (1902). Den botaniske samling. *K. norske Vidensk. Selsk. Aarsb.* **1902**, 23–25.

Foslie, M. (1909). Algologiske Notiser VI. *K. norske Vidensk. Selsk. Skr.* **1909**(2), 1–63.

Hamel, G. (1927). Thèse, 117 p., Jacqueline, imp., Saint-Lô.

Hamel, G. (1931–1939). "Phéophycées de France." Paris.

HAMEL, G. (1939). Sur la classification des Ectocarpales. *Bot. Notiser* **1939**, 65–70.

HARDER, R. and KOCH, W. (1949). Life-history of *Bonnemaisonia hamifera* (*Trailliella intricata*). *Nature, Lond.* **163**, 106.

JOHANSEN, H. W. (1969). Morphology and systematics of coralline algae with special reference to *Calliarthron*. *Univ. Calif. Publs Bot.* **49**, 1–78.

KNAGGS, F. W. (1966). *Rhodochorton purpureum* (Lightf.) Rosenvinge. Observations on the relationship between reproduction and environment. I. The relationship between the energy of incident light and tetrasporangium production. *Nova Hedwigia* **11**, 405–411.

KNAGGS, F. W. and CONWAY, E. (1964). The life-history of *Rhodochorton floridulum* (Dillw.) Näg. I. Spore germination and the form of the sporelings. *Br. phycol. Bull.* **2**, 339–341.

KUCKUCK, P. (1930). Fragmente einer Monographie der Phaeosporeen. *Helgoländer wiss. Meeresunters.* **17**(4), 1–93. 1929–1930.

KYLIN, H. (1944). Die Rhodophyceen der schwedischen Westküste. *Lunds Univ. Årsskr.*, N.F. Avd. 2, **40**(2), 1–104.

KYLIN, H. (1956). "Die Gattungen der Rhodophyceen." C. W. K. Gleerups Förlag, Lund.

LEMOINE, P. (1911). Structure anatomique des Mélobésiées. *Ann. Inst. ocean. Monaco*, **2**, 1–215.

LOISEAUX, S. (1967). Morphologie et cytologie des Myrionémacées Critères taxonomiques. *Rev. gén. Bot.* **74**, 329–347.

LOISEAUX, S. (1970). *Streblonema anomalum* S. and G. and *Compsonema sporangiiferum* S. and G., stages in the life history of a minute *Scytosiphon*. *Phycologia* **9**, 185–191.

LUND, S. (1966). On a sporangia-bearing microthallus of *Scytosiphon lomentaria* from nature. *Phycologia* **6**, 67–78.

MAGNE, F. (1964a). La mitose calliblépharidienne de certaines Rhodophycées. *C.r. hebd. Séanc. Acad. Sci., Paris* **259**, 3811–3812.

MAGNE, F. (1964b). Recherches caryologiques chez les Floridées (Rhodophycées). *Cah. Biol. mar.* **5**, 461–671.

MAGNE, F. (1976). Quelques caractères cytologiques particuliers du *Bachelotia antillarum* (Phéophycées, Ectocarpales). *Phycologia* **15**, 309–319.

MAGNE, F. (1977). La reproduction sexuée chez l'*Acrochaetium asparagopsidis* (Rhodophycées). *Rev. algol.* N.S. **12**, 61–72.

MANTON, I. (1964). A contribution towards understanding of the primitive fucoid. *New Phytol.* **63**, 244–254.

NAKAMURA, Y. (1965). Development of zoospores in *Ralfsia*-like thallus, with special reference to the life cycle of the Scytosiphonales. *Bot. Mag. Tokyo* **78**, 109–110.

PAPENFUSS, G. F. (1945). Review of the *Acrochaetium–Rhodochorton* complex of the red algae. *Univ. Calif. Publs Bot.* **18**, 299–334.

PAPENFUSS, G. F. (1947). Further contributions toward an understanding of the *Acrochaetium–Rhodochorton* complex of the red algae. *Univ. Calif. Publs Bot.* **18**, 433–447.

PARKE, M. and DIXON, P. S. (1976). Check-list of British marine algae—third revision. *J. mar. biol. Ass. U.K.* **56**, 527–594.

PRICE, I. R. and DUCKER, S. C. (1966). The life history of the brown alga *Splachnidium rugosum*. *Phycologia* **5**, 261–273.

Ramus, J. (1969). The developmental sequence of the marine red alga *Pseudogloiophloea* in culture. *Univ. Calif. Publs Bot.* **52**, 1–42.

Reinke, J. (1888). Ueber die Gestalt der Chromatophoren bei einigen Phäosporen. *Ber. dt. bot. Ges.* **6**(6), 213–217.

Rosenvinge, L. K. (1909). The marine algae of Denmark. Contributions to their natural history. Part I. Introduction. Rhodophyceae I. (Bangiales and Nemalionales). *K. dansk. Vidensk. Selsk. Skr.*, 7. Raekke, Nat. Math. Afd. **7**(1), 1–154.

Rosenvinge, L. K. (1917). The marine algae of Denmark. Contributions to their natural history. Part II. Rhodophyceae II. (Cryptonemiales). *K. dansk. Vidensk. Selsk. Skr.*, 7. Raekke, Nat. Math. Afd. **7**(2), 155–283.

Russell, G. and Fletcher, R. L. (1975). A numerical taxonomic study of the British Phaeophyta. *J. mar. biol. Ass. U.K.* **55**, 763–783.

Sauvageau, C. (1898). Sur quelques Myrionémacées (premier mémoire). *Ann. Sci. nat. Bot.*, sér. 8, **5**, 161–288. 1897.

Simon, M.-F. (1954). Recherches sur les pyrénoïdes des Phéophycées. *Rev. Cytol. Biol. vég.* **15**, 74–106.

Stegenga, H. and Borsje, W. J. (1976). The morphology and life history of *Acrochaetium dasyae* Collins (Rhodophyta, Nemaliales). *Acta Bot. neerl.* **25**, 15–29.

Stegenga, H. and Vroman, M. (1976). The morphology and life history of *Acrochaetium densum* (Drew) Papenfuss (Rhodophyta, Nemaliales). *Acta Bot. neerl.* **25**, 257–280.

Suneson, S. (1943). The structure, life-history and taxonomy of the Swedish Corallinaceae. *Lunds Univ. Årssk.*, N.F. Avd. 2, **39**(9), 1–66.

Tatewaki, M. (1966). Formation of a crustaceous sporophyte with unilocular sporangia in *Scytosiphon lomentaria*. *Phycologia* **6**, 62–66.

West, J. A. (1968). Morphology and reproduction of the red alga *Acrochaetium pectinatum* in culture. *J. Phycol.* **4**, 89–99.

West, J. A. (1969). The life histories of *Rhodochorton purpureum* and *R. tenue* in culture. *J. Phycol.* **5**, 12–21.

West, J. A. (1970). The life history of *Rhodochorton concrescens* in culture. *Br. phycol. J.* **5**, 179–186.

West, J. A. (1972). Environmental regulation of reproduction in *Rhodochorton purpureum*. *In* "Contributions to the Systematics of Benthic Marine Algae of the North Pacific", (I. A. Abbott and M. Kurogi, eds), pp. 213–230. Japanese Society of Phycology, Kobe, Japan.

White, E. B. and Boney, A. D. (1969). Experiments with some endophytic and endozoic *Acrochaetium* species. *J. exp. mar. Biol. Ecol.* **3**, 246–274.

Woelkerling, W. J. (1971). Morphology and taxonomy of the *Audouinella* complex (Rhodophyta) in Southern Australia. *Aust. J. Bot.*, suppl. ser. 1, 1–91.

8 | The Rôle of Transmission Electron Microscopy in Elucidating the Taxonomy and Phylogeny of the Rhodophyta

J. G. DUCKETT

School of Plant Biology, University College of North Wales, Wales

and

M. C. PEEL

*Department of Plant Biology, University of Birmingham, England**

Abstract: Ultrastructural studies have permitted more precise definition of the Rhodophyta and have revealed new perspectives on possible relationships between red algae and other divisions; the impact on the systematics within the group has been more limited. This review outlines some of the unique or more unusual attributes of red algal cells, speculates on features of ultrastructure that may eventually prove taxonomically useful and re-examines phylogenetic arguments in the light of information from transmission electron microscopy (TEM).

INTRODUCTION

The advent of the transmission electron microscope (TEM) and its use in biology in the early 1950s provided a revolutionary new tool for algal systematics. Not only did it present the means by which a wealth of new characters could be recognized and thus assist in the specific classification of micro-organisms but it has also led to a clearer understanding and sometimes a re-assessment of the higher taxonomic hierarchies. Manton's (1965) classic ultrastructural studies on scales and flagella in the first decade of biological electron microscopy are a landmark in algal systematics. It is salutary to reflect that they followed immediately after her treatise on the chromosomes of pteridophytes, in which optical microscopy had been pushed to its extreme limits (Manton,

Systematics Association Special Volume No. 10, "Modern Approaches to the Taxonomy of Red and Brown Algae", edited by D. E. G. Irvine and J. H. Price, 1978, pp. 157–204. Academic Press, London and New York.

* Present address: Department of Biology, Queen Elizabeth College, London, England.

1950). More recently, ultrastructural investigations of the microtubule systems of the motile cells, the mitotic apparatus and cytokinesis of green algae have led to a complete reappraisal of the algal ancestry of land plants (Pickett-Heaps, 1975; Stewart and Mattox, 1975).

In striking contrast, it is probably true at the present time to say that the TEM has been of less use in red algal systematics than in similar studies of any other algal group. Interestingly in the context of this symposium, much the same applies to the Phaeophyta, probably for the same reasons. These are as follows. Firstly, in groups which are predominantly macrophytic, fixing and embedding material for TEM is a laborious way of finding new characters: a primary consideration for the taxonomist is convenience. However, already the TEM has provided (albeit largely by serendipity in the case of the red algae) data useful in classifying organisms which at other levels appear very similar. For example, the unicells *Rhodella* and *Porphyridium* exhibit fundamental differences in plastid structure. Similarly, the corallines *Jania* and *Corallina*, sometimes placed in a single genus (Solms-Laubach, 1881), clearly differ in the peri-nuclear morphology of the spores and in the nature of the spore cell wall. Secondly, red algae are inconvenient organisms for TEM use: they are difficult to culture and are nearly all marine. Both these features are reflected in the fact that nearly all ultrastructural studies have been performed in laboratories with easy access to the coast. Thirdly, in nearly all the known examples in which the TEM has been used extensively, its employment has been to resolve details of "cell walls" (e.g. scales, loricae, frustules) which are easy to examine, merely requiring the production of direct preparations. There appears to be no such use to which the TEM could be put in red algal taxonomy. Fourthly, the majority of ultrastructural studies on red algae have been developmental and any taxonomic inferences have been largely incidental. Finally, most electron microscopists are not algal taxonomists, and vice versa. Field detection of particular taxa, not to mention identification of critical phases in the life-histories, often requires long experience which all too often seems incompatible with the practice of laboratory-based electron microscopy.

Thus, it is not currently possible to produce an incisive review of systematic data derived from TEM studies on red algae. Information is available for far too few taxa; much of that information relates to different stages in the life-history and is therefore not strictly comparable. In this paper we shall use the existing data, patchy though it is, as a framework for speculation as to which features of Rhodophytan cytology may perhaps prove useful in the future. Even if many of our suggestions are subsequently found naïve or incorrect, their very negation will prove that the neglect (in our view unwarranted in the light of some of the remarkable information already available) of this approach to study in the group has been effectively redressed.

However, although the TEM has hitherto provided information of limited systematic value within the red algae, in a wider context its rôle has been dramatic. Not only has ultrastructural study led to a more precise definition of the group but it also continues to reveal new perspectives on the possible relationship between red algae and other groups. Thus, we shall outline the contribution of transmission electron microscopy to the argument that the red algae are probably the most primitive eukaryotes and shall document recent cytological evidence relevant to the century-old (but none the less controversial) notion of a phyletic relationship between red algae and the fungi.

In our experience, perhaps the most remarkable feature of red algae at the cytological level is that they display many features much more reminiscent of animal cells than of other algal groups. We disagree with the recent statement that the basic cell structure is simple (Dodge, 1973). For example, red algae synthesize diverse secretory and metabolic products ultrastructurally manifest in a variety of membrane-bound vesicles (indeed a new type was described by Wetherbee and West, 1977). They also display complex interactions between Golgi and endoplasmic reticulum, both of which exhibit a high level of structural differentiation. Nuclear/cytoplasmic interactions similar to those found in embryonic and rapidly differentiating animal cells are being discovered at critical phases in the life-histories of red algae. Many phenomena in the group appear to take place with remarkable speed, a fact that may explain why no definitive account of the entire mitotic cycle exists. Even in synchronous and logarithmic phase cultures of *Porphyridium*, this process has not been followed by electron microscopists. It is clearly absurd to suggest that such similarities are in any way indicative of phyletic affinities; our reasons for emphasizing them under a systematic mantle are twofold. Firstly, they constitute a possible aid to future workers; we have found that reports on animal cytology often prove far more helpful in interpreting red algal cytology than do those on other algae. Secondly, these parallels underline the fact that cytological similarities are just as likely to reflect similar subcellular phenomena as phyletic relationship. There is no *a priori* reason why ultrastructural similarities and differences should be more important than those at any other level.

MATERIALS AND METHODS

Porphyridium purpureum (Bory) Drew et Ross (strain 1380/1A) was grown in Erdschreiber sea-water under a 12 hour light/dark regime provided by a bank of "daylight" fluorescent tubes. After gentle centrifugation of the cultures, the resultant pellet was resuspended in a drop of culture medium and fixed for two hours at room temperature in 4% glutaraldehyde in 0·1 M sodium cacodylate buffer, pH 7·0. After a double rinse in buffer the material was post-fixed in 1% osmium tetroxide in 0·1 M buffer, rinsed again three times, then embedded in Epon via propylene oxide and a graded alcohol series. Collection data and full details of the fixation procedures for the remaining taxa in the micrographs illustrating this review can be found as follows: *Corallina officinalis* L., in Peel *et al.* (1973); *Furcellaria lumbricalis* (Huds.) Lamour., in Peel and Duckett (1975); *Jania rubens* Lamour., in Peel *et al.* (1973); *Nemalion helminthoides* (Vell. in With.) Batt., in Duckett *et al.* (1974) and *Nostoc sphaericum* Vauch. ex Born. et Flah., in Duckett *et al.* (1977).

GENERAL CYTOLOGY

The first major contribution of the TEM to the study of red algae was that it forever removed all uncertainty as to the true eukaryotic nature of the group. Although the red algae were generally placed in the Eukaryota, prior to 1960 there remained lingering doubts that they are true eukaryotes: certain membrane-bound organelles (e.g. mitochondria and Golgi bodies) had not been demonstrated. For this reason it was suggested that the red algae might occupy a position somewhere intermediate between the Cyanophyta and the remainder of the algae; this led Dougherty and Allen (1960) to coin the term Mesoprotista. However, it should be remembered that Fritsch (1945) thought that the red and blue–green algae were not closely related, a judgment based principally on the lack of sexual reproduction in the latter.

In cell structure red algae conform to the basic eukaryotic plan with double membrane-bound nuclei, mitochondria and plastids lying in cytoplasm containing endoplasmic reticulum and the Golgi bodies commonly ascribed to this level of organization (Fig. 1). Another feature of eukaryotic organization, the microtubule, is now being discovered in an increasing number of red algal cell types (see McBride and Cole, 1972a; McDonald, 1972; Goff, 1976) in much the same manner as was the case in higher plants following the introduction of aldehyde fixatives in the early 1960s. TEM has confirmed both the diagnostic absence of flagella from all stages of the life-history (Young, 1977) and, not surprisingly, the complete lack of centrioles. Interestingly in this context, *Chlorella* (Chlorophyta) appears to be one of the very few examples of an organism possessing centrioles but lacking flagellated stages (Atkinson *et al.*, 1971).

Despite their lack of flagella, it is erroneous to describe red algae as non-motile. Amoeboid shape changes have been described many times for released spermatia (Dixon, 1973) and Rosenvinge (1927) reported that mono-, tetra- and carpospores perform gliding movements. Recently, the motile cells of *Porphyridium purpureum* (Bory) Drew et Ross have been investigated ultrastructurally (Lin *et al.*, 1975). Moving cells have a smooth and blunt anterior end and an elongate tail-like posterior. Large vesicles in the tail contain fibrillar material similar to that in the extracellular sheath. In the light of Ramus's (1972) demonstration that Golgi-derived vesicles containing polysaccharides are transported externally, it is tentatively suggested that polar expulsion of mucilaginous material might account for the motility of *Porphyridium*. It would now be particularly interesting to see if this highly unusual method of propulsion also occurs in the spores of multicellular taxa and to explore possible parallels between the mechanism of movement in red algae; motility in diatoms (Drum and Hopkins, 1966; Harper and Harper, 1967; Hopkins and Drum, 1976); and the gliding movements in blue–green algae (Fogg *et al.*, 1973) and many Protista (Jarosch, 1962). Mucilage secretions appear to be involved in all the latter examples. It should also be noted that microfibrils, implicated in amoeboid movement in animals and in cytoplasmic streaming in higher plants (Gunning and Steere, 1975; Parthasarathy and Mühlethaler, 1972) have yet to be detected in red algae.

THE CELL WALL

In electron micrographs, the walls of red algal cells typically have a layered fibrillar appearance (Fig. 2), generally interpreted as a cellulose framework in a mucopolysaccharide matrix. Extensive survey of published micrographs clearly indicates that the considerable differences which exist in both the thickness of the layers and the density and disposition of the fibrils is as great between different parts of the same thallus as between taxa. Moreover, the exact appearance of the wall seems, to some extent, to be dependent on fixation procedures (Gordon and McCandless, 1973; McCandless, pers. comm.). At first sight, therefore, red algal walls seem to have little to offer the taxonomist. However, extensive observations by one of us (M. C. P.) on *Corallina* and *Jania* suggest that useful characters may come to light when critical comparisons are made. A characteristic feature of the developing spore walls in *Jania* is the presence of numerous well defined particles immediately outside the plasma membrane (Fig. 3). The absence of this feature in *Corallina* (Fig. 4), the inner spore-wall layers of which invariably have an amorphous appearance, is a clear-cut distinction that adds weight to generic separation.

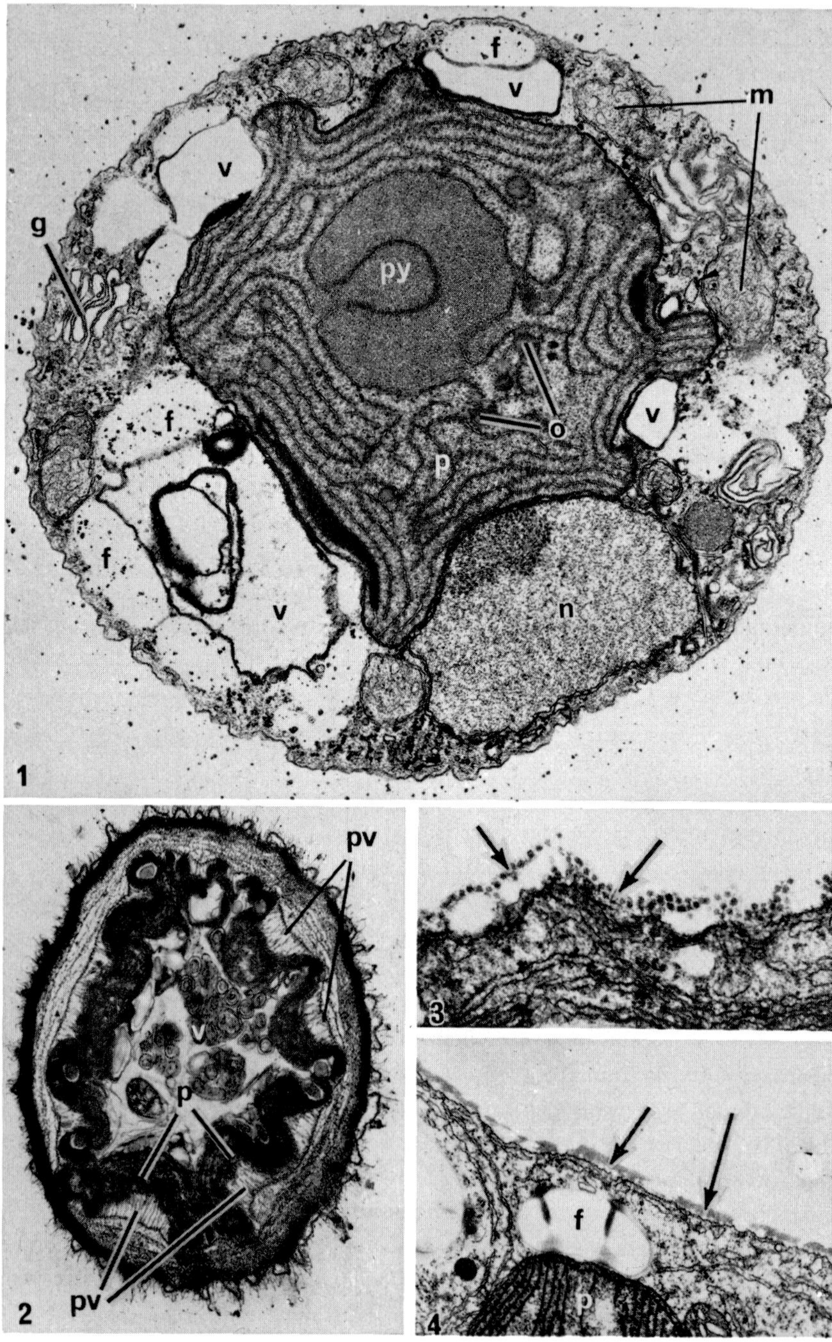

Figs 1-4. Captions on opposite page.

1. Pit-Connections

The one feature of wall ultrastructure which has attracted systematic investigation is the morphology of pit-connections, the lenticular plugs with multi-layered caps which occupy a central position in the septa between adjacent cells. However it must be pointed out that, although normally treated as wall components (Dixon, 1973; Dodge, 1973), they may not be strictly extracellular. Development takes place within the cytoplasm (Ramus, 1969a, b) and at maturity the cell membrane is continuous from cell to cell between the septal wall and the plug, the last feature clearly demonstrating that red algae possess the symplastic organization characteristic of other multicellular algal phyla, fungi and land plants (Marchant, 1976). However in a very recent freeze-etching study Pueschel (1977) states that, as a result of the junction of the plasmalemma and cap membranes, the plug itself (together with the cap layer on either face of the core) is entirely membrane-bounded and essentially extracellular.

The higher magnification of the TEM immediately settled the controversy, resulting from light microscope observations, that a fundamental difference between Bangiophyceae and Florideophyceae lay in the absence of pit-connections in the former. Pit-connections with exactly the same basic morphology as in the Florideophyceae are now well documented for the conchocelis phases of *Bangia* and *Porphyra* (Bourne *et al.*, 1970; Cole and Conway, 1975; Lee, 1971). However, in our view, speculation as to whether this last feature is a primitive or derived condition (Lee, 1971) is best left until more is understood

---

FIG. 1. Median section of *Porphyridium* showing various organelles in the peripheral cytoplasm around the central chloroplast. Note the absence of an inner limiting disc from the chloroplast. × 20 000.

FIG. 2. Transverse section of a medullary cell of *Nemalion* showing several layers of fibrillar wall material. × 13 000.

FIG. 3. *Jania*; developing carpospore showing particulate wall material (arrowed) adjacent to the plasma membrane. × 29 000.

FIG. 4. *Corallina*; developing carpospore showing amorphous wall material (arrowed) adjacent to the plasma membrane. × 19 000.

KEY TO ABBREVIATIONS USED IN FIGURES

b. bacteria. c. plug cap. ep. electron-dense perinuclear material. ern. nuclear endoplasmic reticulum. ers. swollen endoplasmic reticulum. f. floridean starch. ff. forming face of Golgi body. g. Golgi body. il. inner limiting disc. k. kinetochore. m. mitochondrion. n. nucleus. ne. nuclear envelope. o. osmiophilic globuli. p. plastid. pa. arms of stellate plastid. pc. pit-connection. pf. fibrillar vesicle derived from plastid. pv. plasmalemmavilli. py. pyrenoid. r. ribosomes. s. spermatangium. sm. spermatangial mother cell. sv. spermatial vesicle. v. vacuole. ve. electron-dense spheroidal vesicle.

about the function of pit-connections and about the physiological consequences of their absence in macrophytic Bangiophyceae.

As in the above case, TEM has also settled all arguments as to whether there exists a single type or two different pit structures within the Florideophyceae (Dixon, 1973). Following from the original demonstration of biconvex plugs in *Lomentaria* by Bouck (1962), subsequent investigations have all confirmed that open communications do not occur between red algal cells.

Despite these obvious systematic clarifications which TEM has brought to pit-connection distribution and structure at the class level, the usefulness at lower taxonomic levels remains debatable. Although gradations in the granularity in the plug and the presence or absence of material condensed on the cap are considered significant by Lee (1971) in the grouping of genera (e.g. the separation of *Bonnemaisonia* from the remainder of the Nemalionales), a recent study of *Nemalion* (Duckett *et al.*, 1974) indicates that the diversity in plug or pit form between different parts of the same thallus may be as great as that between different and often only apparently distantly related taxa. Thus, if pit-connection substructure is to be used taxonomically in the future it is essential that direct comparisons should be restricted to analogous cells of different organisms. Examples of the diversity in pit-connection morphology are illustrated in Figs 5–8, 12. In general, pits in spermatial branches are smaller (usually less than 0·3 μm diameter) and less narrowly waisted than those between vegetative cells. At the other extreme, they may reach 30–40 μm in mature axial filaments (e.g. in *Ptilota*). Similarly, shape varies from disc-like to almost spherical, whilst in the hair-base septa of *Nemalion* asymmetrical pit-connections are found. A comparative study of the latter thus appears particularly inviting.

A further aspect of pit-connections which might well have far-reaching systematic significance and may shed light on the enigma of their function is their rôle in the liberation of the cytoplasmic contents of some specialized cells. There is now good evidence from several taxa that the liberation of spermatia (Duckett *et al.*, 1974), tetraspores and carpospores (Chamberlain and Evans, 1973) is via apertures, from which pit connections have been displaced, in the walls of the parent plants. Immediately after release, these cells are apparently naked protoplasts. The absence of well-developed waists in these pit-plugs may be an adaptation facilitating their dislocation. In contrast, liberation of the hair cytoplasm in *Nemalion* does not involve removal of the plugs, since these are retained in the walls of the old hairs (Fig. 11). Similarly in spores or spermatia, where these are apical cells not produced by percurrent proliferation, discharge simply cannot involve displacement of pit-connections since these will only be present in the basal septa. There are thus two different release mechanisms, only one of which involves the pit-connections.

Percurrent proliferation (the successive development of new structures within existing walls) is a widespread phenomenon in red algae. Although visible under the light microscope, such proliferation requires TEM for detailed analysis of its frequency. Consequently, it appears to have been largely neglected by systematists. Percurrent proliferation commonly occurs on spermatial branches throughout red algae (Grubb, 1925; Hughes, 1971) but development of successive generations of carpospores and tetraspores in this manner appears to be much more infrequent in the specialized Florideophyceae (Fritsch, 1945). If detailed TEM studies were to confirm such a restricted distribution in what are generally considered to be advanced taxa this might be a finding of considerable evolutionary significance. Similarly, it would be of interest to discover whether percurrent proliferation of hairs is commoner in the lower Florideophyceae. Certainly it is extremely prolific in this situation in *Nemalion* (Fig. 11).

2. Plasma Membrane

A final noteworthy feature related to red algal walls is the sometimes highly structured nature of the plasma membrane. In 1971, Cottler described regular villus-like outgrowths in the outer medulla of *Chondrus*. It was suggested that these "plasmalemmavilli", which are quite unlike anything previously described in the plant kingdom, might perhaps be related to cell wall formation. It is therefore particularly interesting that the same structures should be detected in *Nemalion*, a genus widely separated from *Chondrus*. Here also they are most conspicuous in the outer medulla, reaching maximal development in bays overlying arms of the plastids and associated with extra layers of fibrillar wall material (Figs 9, 10). Although subsequent work on *Chondrus* (Gordon and McCandless, 1973; McCandless, pers. comm.) suggests that these villus-like outgrowths may be fixation or embedding artefacts, in *Nemalion* their form if unaffected by different buffers and embedding schedules. Should further investigations reveal plasmalemmavilli to be widespread in red algae, then they might well become an important diagnostic feature of the group. Fine extensions of the plasma membrane are clearly visible in many published micrographs of red algae (e.g. in Bisalputra *et al.*, 1967; McBride and Cole, 1969; Scott and Dixon, 1973a, b).

From the foregoing discussion, it will be clear that red algal wall ultrastructure offers diverse possibilities for taxonomic studies. One of the chies advantages is that many features may be observed even when fixation is poor. For example, details of percurrent proliferations and pit-connections can still be resolved if materials initially stored in formaldehyde are subsequently embedded.

3. Fungi and Red Algae

The other intriguing aspect of red algal walls is the number of features they share with those of higher fungi. Not surprisingly, these features figure strongly in speculations about a possible red algal ancestry for the Ascomycetes and Basidiomycetes (Denison and Carroll, 1966; Hughes, 1971; Lee, 1971; Demoulin, 1974; Pueschel, 1977). Most frequently cited is the resemblance between red algal pit-connections and septal plugs in fungi. Although in overall structure pit-connections more closely resemble the pulley wheels in Ascomycetes, Uredinales and Ustilaginales than they do the dolipores of Homobasidiomycetes (Bracker, 1967; Littlefield and Bracker, 1971), ontogenetic, biochemical and functional homologies have yet to be demonstrated. As Pueschel (1977) concludes, the resemblance between red algal pit plugs and fungal septa pore plugs is largely superficial and of little value in resolving the question of a possible phylogenetic affinity between the two groups. Hughes (1971) makes a careful comparison between spermantagial ontogeny in red

---

FIGS 5–7, 9–12. *Nemalion*. FIG. 8. *Corallina*.

FIG. 5. Median longitudinal section of a pit-connection between cells of the middle cortex, showing details of the substructure of the plug cap and the continuity of the plasma membrane between the cells (arrowed). × 57 000.

FIG. 6. Spermatial branch. Note the very small, scarcely waisted pit-connections in the septa between the spermatangial mother cell and the spaces left after the liberation of the spermatia. The four sets of cell walls delimiting these spaces indicate that four generations of spermatia have been produced. × 8700

FIG. 7. Large, flat pit-connection in the septum between a fusion cell and the parent gametophyte × 22 000.

FIG. 8. Almost spherical pit-connection in the carposporophyte. × 11 000.

FIG. 9. Longitudinal section of plasmalemmavilli in an inner cortical cell. Fine arms of cytoplasm extend for a considerable distance into the loosely fibrillar wall. The villi are most conspicuous outside the arms of the stellate plastid. × 48 000.

FIG. 10. Transverse section of plasmalemmavilli illustrating that these form tubules of a diameter 40–60 nm rather than baffles. × 27 000.

FIG. 11. Longitudinal section of a hair. Remains of seven previous hairs are visible surrounding the one most recently produced. Numerous bacteria are trapped between the walls of the old hairs. A pit-connection from the septum of a previous hair is arrowed. × 2800.

FIG. 12. Median longitudinal section of an asymmetrical pit-connection at the base of a hair. × 66 000.

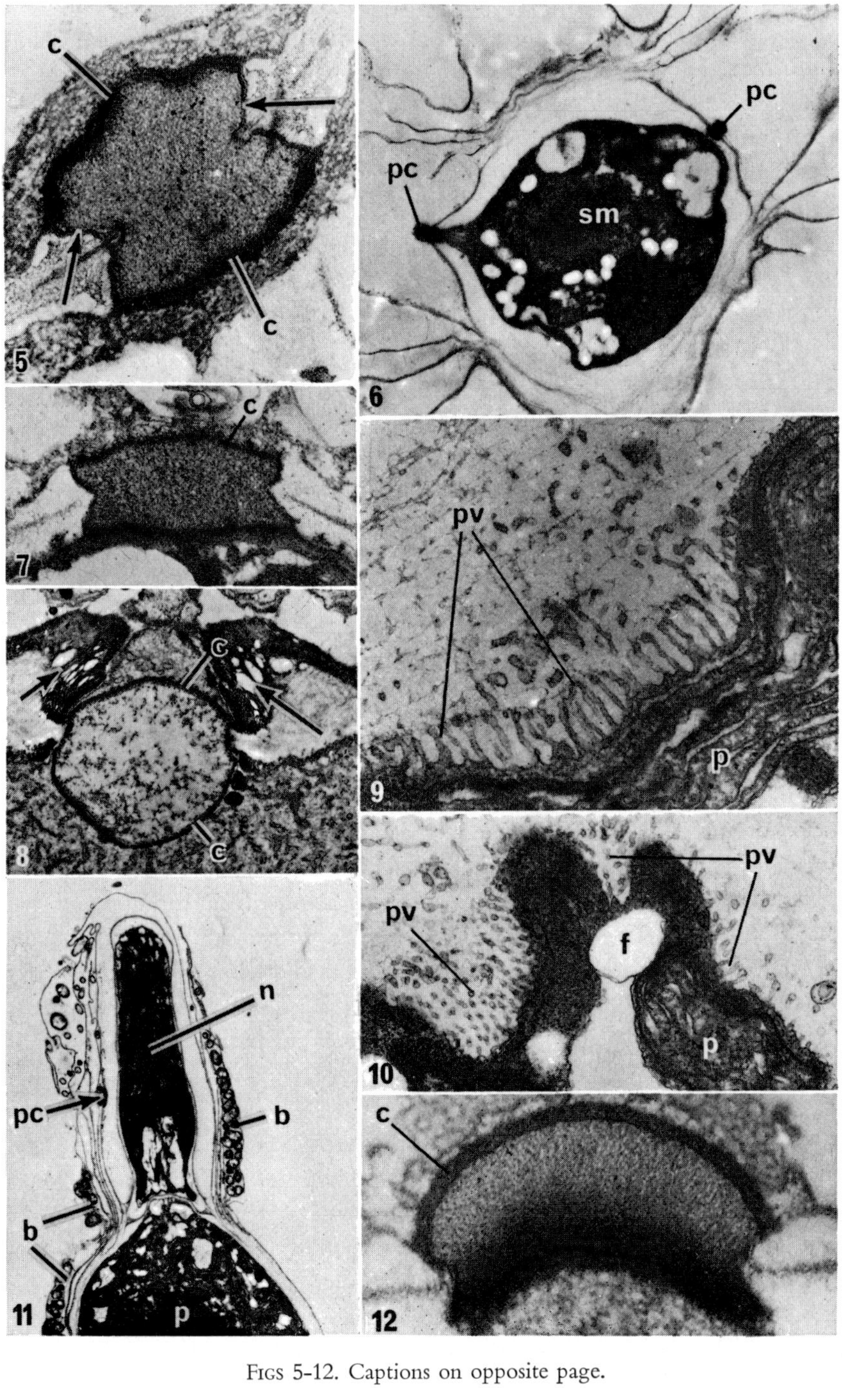

FIGS 5-12. Captions on opposite page.

algae and phialide development in some of the Ascomycetes and Fungi Imperfecti. Hughes avoids firm commitment to the algal ancestry theory, as does Pueschel, concluding that the production of spermatia singly in percurrently-produced spermatangia indicates considerable differences between these and the phialospores and phialides of those Ascomycetes whose sexual apparatus has been held to be homologous with that of red algae. We agree, on present evidence, that ultrastructural similarity between red algae and fungi based on evidence from pit-connections and percurrent proliferations is probably as much a consequence of similar filamentous organization as of any fundamental phylogenetic relationship.

In our view, the best cytological evidence that can be presented in support of the red algal ancestry of higher fungi is the production of apoplastic cells by a process which completely excludes all plastid material from specific cells at cytokinesis. It occurs during hair formation in *Nemalion* (Duckett *et al.*, 1974) and during spermatangial development in several genera (Peel and Duckett, 1975; Young, 1977). Apart from in the red algae, the only other well-documented examples are in the hair cells of the filamentous green alga *Bulbochaete* (Frazer and Gunning, 1973) and in the spermatozoids of *Biddulphia* (Heath and Darley, 1972). Absence of plastids elsewhere in the plant kingdom is brought about either by degeneration (e.g. during rhizoid formation in ferns) or sequestration of the plastid-containing part of the cell cytoplasm (e.g. in the motile spermatozoids of archegoniate plants; Duckett, 1975). Proplastids are invariably present in colourless and heterotrophic members of all the other major groups of plants, even when such plants have been rendered permanently apochlorotic by major experimental manipulation (Fraser and Gunning, 1973). The origin of higher fungi is generally supposed to be from parasitic red algae that, losing their plastids in the process, gave rise to parasitic fungi (Demoulin, 1974). It is unfortunate for protagonists of such algal origin that proplastids have now been discovered in *Holmsella* (Evans *et al.*, 1978, this symposium), a parasite formerly thought to be completely devoid of plastids (Evans *et al.*, 1973).

THE GOLGI APPARATUS

In marked contrast to fungi (Bracker, 1967; Morré *et al.*, 1971), one of the most striking features of red algal cells is the high level of structural organization of their Golgi bodies. To date there are probably more published electron micrographs of the red algal Golgi than of any other organelle, with the possible exception of plastids. Extraordinary changes in Golgi morphology have been described in developing spores of diverse taxa and several types of Golgi-derived vesicles have been implicated in both cell wall formation and the production of various internal storage products (Peyrière, 1970; Kugrens and West, 1972a, 1973; Ramus, 1972; Chamberlain and Evans, 1973; Evans *et al.*, 1974; Konrad Hawkins, 1974a, b; Tripodi, 1974; Ramus and Robins, 1975). Two particularly common types are the fibrillar and cored vesicles thought to contain respectively wall mucilages and adhesive materials (Figs 13, 14). As can be seen in Fig. 13, the mature carpospores of *Nemalion* are literally packed with cored vesicles. On the other hand, some types of Golgi-derived vesicles, such as the striated vesicles seen during carpospore formation in *Polysiphonia* (Wetherbee and West, 1976, 1977) and during tetrasporogenesis in *Levringiella* (Kugrens and West, 1972c) and the electron-dense spheroidal vesicles (Figs 15, 16) appearing during tetrasporogenesis in *Corallina* (Peel *et al.*, 1973), have a much more restricted distribution. However, this remarkable diversity in Golgi morphology appears to be of little use for the taxonomist since it mainly reflects particular subcellular processes rather than taxonomic differences. For example, many of the ultrastructural features accompanying sporogenesis are extremely similar in all species investigated to date.

However, careful scrutiny of published micrographs coupled with our own observations on *Corallina* and *Jania* indicate that some red algal Golgi possess a unique attribute which may have far-reaching systematic significance, distinguishing the group from all other algal divisions. It is particularly noteworthy that an article reporting the independent recognition of this same feature in tetrasporangia of *Polysiphonia* appeared in print (Alley and Scott, 1977) during the preparation of this paper after the Symposium was held. However, these authors do not fully explore the taxonomic implications of their discovery.

As in other divisions, the red algal Golgi body consists basically of a stack of cisternae with peripheral fenestrations, which gives way in turn to anastomosing tubules and vesicles. Each stack is clearly polarized, with distal and proximal faces associated respectively with large irregular and smaller more uniform vesicles. The latter are often of the coated type and transition elements between them and endoplasmic reticulum can often be recognized (Peel *et al.*, 1973; Peel and Duckett, 1975). As illustrated in Figs 15 and 16, in *Corallina* and *Jania* the unique feature of well-developed Golgi bodies is the close apposition of the membranes of adjacent cisternae. Here the intercisternal space is obliterated by darkly staining material. As in *Polysiphonia* (Alley and Scott, 1977), the cisternal membranes are about 6 nm thick and the overall dimension of the laminated regions 15–17 nm. Thus, the intercisternal space must be considerably narrower than the usual dimension of about 10 nm given by Gunning and Steere (1975). It is almost as if fusion has taken place between the outer portion of the adjacent

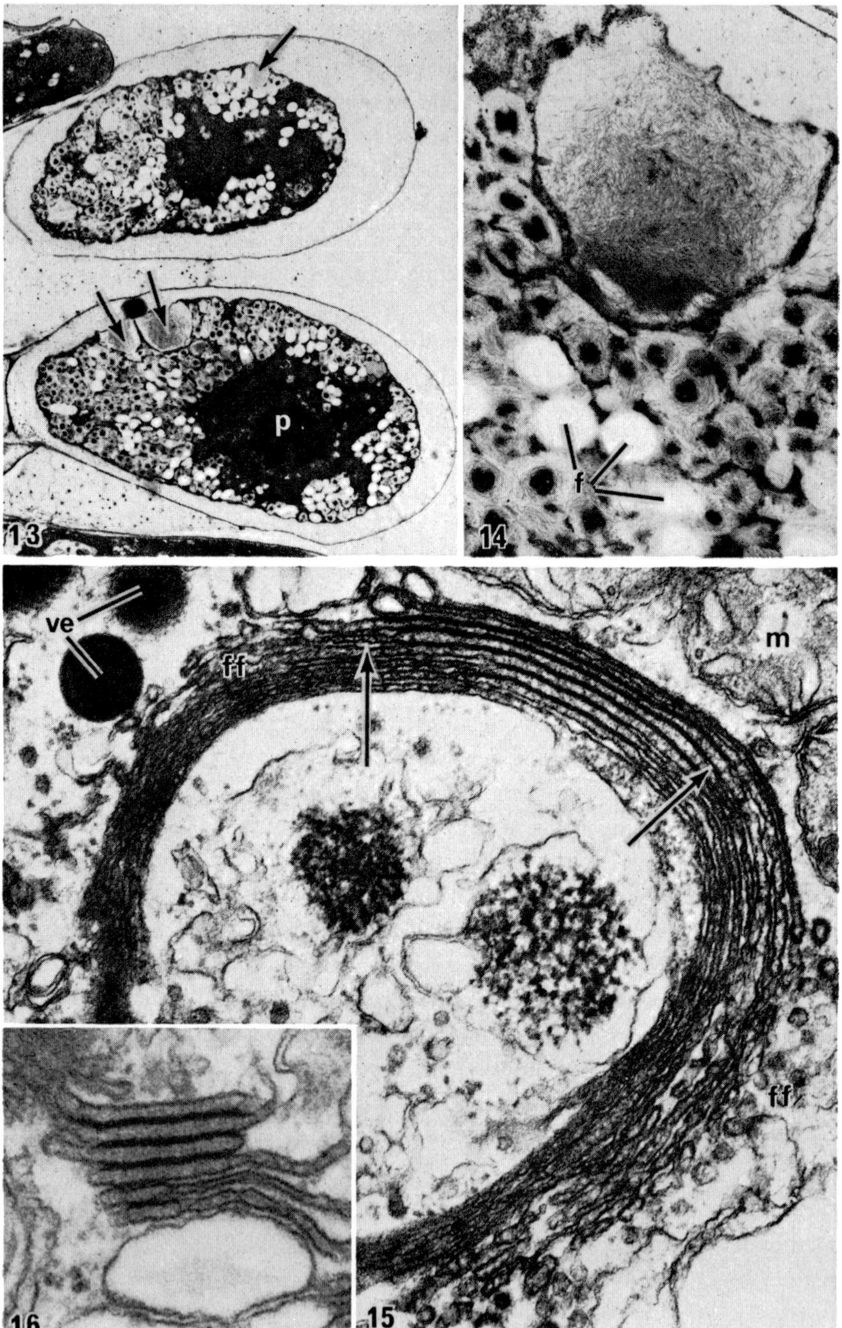

Figs 13-16. Captions on opposite page.

cisternal membranes. This is quite unlike the electron-opaque intercisternal rods or cross-bridges which are considered to provide membrane–to–membrane connections between adjacent cisternae in other plant and animal Golgi bodies (Franke *et al.*, 1972) Such structural links are thought to be responsible for maintaining the shape of the central parts of the cisternae and the coherence of the cisternae within the stack. Although the laminated regions probably perform the same function in red algae, morphologically they have more in common with granal thylakoids and tight junctions between animal cells (Novikoff and Holtzman, 1970).

This unique modification of the red algal Golgi body has now been illustrated, although not necessarily commented upon in the sporangia of the following genera: *Callithamnion* (Konrad Hawkins, 1974a); *Ceramium* (Chamberlain and Evans, 1973); *Corallina* (Duckett and Peel, this book); *Griffithsia* (Peyrière, 1970); *Jania* (this book); *Levringiella* (Kugrens and West, 1972a, 1973); three species of *Polysiphonia* (Tripodi, 1974; Alley and Scott, 1977; Wetherbee and West, 1977); *Pterosiphonia* (Tripodi, 1974) and *Ptilota* (Scott and Dixon, 1973a). It thus appears to be widespread in the "higher" orders of Floridophyceae but has yet to be detected in the Bangiophyceae or the Nemalionales. If subsequent studies show this distribution pattern to be correct, it could be an important new character for use in delimiting the Nemalionales and the Bangiophyceae.

PLASTIDS

Perhaps the single most important criterion used to delimit the Rhodophyta as a group is the nature of their photosynthetic apparatus. The phycobilin pigment balance (see Dixon, 1973, for review) is universally accepted as the primary feature used to establish the division and to link the Florideophyceae and Bangiophyceae (Christensen, 1962; Klein and Cronquist, 1967); this is supported by the ubiquitous occurrence, in the cytoplasm of group members, of the same

---

FIGS 13, 14. Mature carposporangia of *Nemalion*.

FIG. 13. Two mature carposporangia whose cytoplasm is packed with cored and fibrillar (arrowed) vesicles and starch spherules. × 3750.

FIG. 14. Details of the two types of carpospore vesicle. × 21 000.

FIG. 15. *Corallina*. Golgi body in differentiating carpospore. Note the electron-dense spheroidal vesicles outside the forming face (ff) and the larger central vesicles with coarsely granular contents in the centre. The region of cisternal fusion is arrowed. × 65 000.

FIG. 16. *Jania*. Portion of a Golgi body in a differentiating carpospore showing fusion of membranes of adjacent cisternae. × 92 000.

storage polysaccharide, floridean starch. It was therefore a discovery of paramount importance, made possible only with the TEM, that the chloroplasts of red algae have several common features unique to the group. So distinctive is chloroplast ultrastructure that it has rapidly assumed status equal to that of pigmentation and other biochemical criteria in the diagnosis of the Rhodophyta.

The outstanding attributes of red algal chloroplasts are:

1. The thylakoids are not stacked.
2. The accessory, water soluble photosynthetic pigments, phycocyanin and phycoerythrin, are located in granules called phycobilisomes, measuring about 35 nm in diameter. These are distributed in a regular manner on the outer faces of the thylakoids (Gantt and Lipschultz, 1972; Koller *et al.*, 1977).
3. The double membrane comprising the chloroplast envelope is not associated with endoplasmic reticulum (cf. the Chromophyta: Gibbs, 1970; Greenwood, 1974).
4. Physical continuity between outer and inner membranes of the envelope, endoplasmic reticulum, and thylakoids respectively has yet to be conclusively demonstrated (Bisalputra, 1974).

Other chloroplast features of red algae, such as DNA fibrils in the stroma (Bisalputra and Bisalputra, 1967), which also contains abundant ribosomes (presumably 70s, although we are not aware that this has been elucidated), are also common to all other eukaryotic algae (Bisalputra, 1974). The particulate nature of the freeze-fractured thylakoid faces is very similar to that in other algal groups, including the Cyanophyta (Neushul, 1971).

Despite these common basic features, many reports exist of considerable diversity in plastid ultrastructure; probably even more still await discovery. It was nonetheless an unpleasant surprise to discover (during the preparation of this review) that many apparently definitive generalizations about plastid number and morphology in relation to taxonomy, in particular to the higher categories, are based on totally inadequate observations derived from far too few taxa. Perhaps the most outstanding lacuna is that there has hitherto been little attempt to standardize observations. Before plastid characteristics can be fully utilized in red algal systematics, the following procedures will, we believe, need to be adopted:

1. Mature chloroplasts should be compared with other mature chloroplasts from the same parts of plants of different taxa. Features such as thylakoid distribution and frequency, overall shape, and the morphology of pyrenoids (including presence or absence), are greatly dependent on age and location.

2. Precise details of plastid ontogeny should be carefully documented in each taxon investigated, in order to discover possible instances of parallel evolution. For example, the multiplastidic condition may be achieved and maintained in two fundamentally different ways: either by fragmentation of a single multilobed plastid or by partitioning of a population of simple plastids at cytokinesis. Discrimination between these two phenomena may well have an important bearing on the taxonomy of the Bangiophyceae and the Nemalionales.
3. Data on plastid numbers and shape should be based on electron microscopy, preferably with serial sectioning or stereology. In our opinion, it is almost impossible to decide from light microscopy alone whether arms of a single multilobed plastid are becoming completely detached as cells age (as for example is reported to occur in more advanced Florideophyceae; see Dixon, 1973). It would not surprise us if undivided plastids in cells of red algae turn out to be far more frequent than has hitherto been assumed. Critical serial sectioning of several unicellular algae and fungi has led in the last decade to the unsuspected discovery (see Duckett and Toth, 1977, for review) that these contain just a single much-branched mitochondrion.

Although the available data are not wholly satisfactory, we feel that it is still worthwhile at the present time to attempt to distinguish major trends in red algal plastid structure; these may have systematic and evolutionary significance and may indicate some features worthy of special attention.

The simple ovoid and spherical plastids of *Corallina* and *Jania* (illustrated in Figs 17–19) are typical of many members of the Florideophyceae with several plastids per cell. In addition to the unstacked thylakoids, a further conspicuous feature is the outermost thylakoid which runs parallel to the chloroplast envelope. In some cases, more than one of these peripheral thylakoids are present (Bisalputra, 1974; Hara and Chihara, 1974). In this account, we shall follow Bouck (1962) who originally called this structure the "inner limiting disc" in the first detailed account of red algal plastid ultrastructure in *Lomentaria*. This disc is thought to be of major importance to the origin of the other thylakoids, which most likely arise as invaginations from it and not from the plastid envelope as in other divisions. There is also limited evidence that the disc itself may arise from the chloroplast envelope (e.g. Bouck, 1962) but this question needs careful re-investigation, as does the presence or absence of the disc throughout the Rhodophyta. Although disc presence is considered to be useful in subdividing the Florideophyceae and the Erythropeltidaceae from the rest of the Bangiophyceae (McBride and Cole, 1972b), the statement must be viewed with some caution. The demonstration by Cole and Conway (1975) of

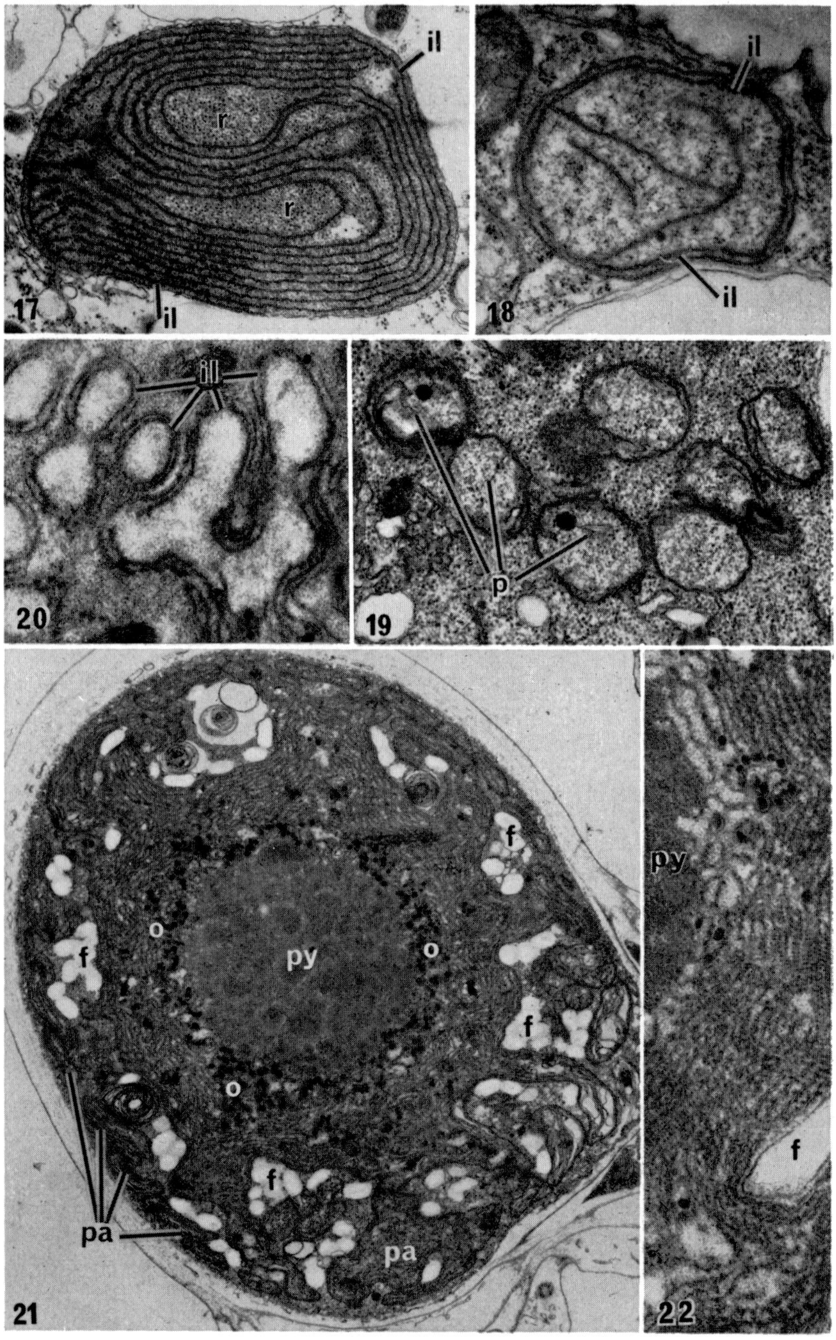

FIGS 17-22. Captions on opposite page.

disc presence in the conchocelis stages of *Bangia* and *Porphyra* and of its absence from the macroscopic phases of these genera and elsewhere in the Bangiophyceae (with the exception of the Erythropeltidaceae) casts doubt on whether the two classes are as distinct as previous authors (e.g. Dixon, 1973) have suggested. Here we can only add more fuel to the argument: no distinct inner limiting disc is visible in any of Lichtlé's (1973a) micrographs of *Rhodochorton* and *Rhodothamniella* (both now considered to fall within *Audouinella*, Acrochaetiaceae, Nemalionales). There is little evidence of the structural integrity of the disc in the labyrinthine proplastid of the fusion cell of *Nemalion* (Figs 23, 24), although it is invariably present elsewhere in the life-history of that plant (Fig. 20). Similarly, the disc is absent from the spermatial mother cell plastids of *Corallina* (Peel and Duckett, 1975) (Fig. 19) and *Levringiella*, although it is still recognizable in spermatia of *Erythrocystis* (Kugrens and West, 1972a) and in other highly specialized cells such as tetrasporangia and carposporangia (Fig. 18). All trace of internal thylakoids appears to have vanished from the plastids of the parasite *Choreocolax* (Kugrens and West, 1973).

It will be apparent that the precise pattern of thylakoid ontogeny may vary according to whether an inner limiting disc is present or absent. In the latter case, thylakoids probably derive from bifurcation of pre-existing thylakoids; in the former, they may arise either in a similar matter (e.g., in *Lemanea*; Bergfeld, 1970) or from the inner limiting disc. It would be very interesting to investigate whether any correlation exists between the presence of thylakoid bifurcation profiles and the absence of an inner limiting disc.

It is also noteworthy that cells lacking extensive thylakoid systems, usually

---

FIGS 17–19. *Corallina*. FIGS 20–22. *Nemalion*.

FIG. 17. Mature chloroplast in carposporophyte. Note the numerous ribosomes in the central stroma and the larger (35 nm) phycobilisomes attached to the single thylakoids. × 23 000.

FIG. 18. Proplastid in a tetrasporangium. Only three thylakoids are visible inside the inner limiting disc. × 25 000.

FIG. 19. Degenerating plastids in a spermatangial mother cell. The inner limiting disc is no longer visible. × 27 000.

FIG. 20. Arms of a proplastid in an apical cell of the thallus showing a distinct inner limiting disc but no central thylakoids. × 23 000.

FIG. 21. Mature chloroplast in an outer medullary cell. Note the very numerous peripheral arms between which lie numerous starch spherules. The central pyrenoid is surrounded by a zone of abundant osmiophilic globuli. × 7600.

FIG. 22. Detail of mature chloroplast showing the single thylakoids studded with abundant phycobilisomes. × 21 000.

assumed to be incapable of photosynthesis, still retain their capacity to form starch (Peel and Duckett, 1975). In the light of this, the novel suggestion by Tripodi (1971) that starch formation in red algae may be the result of Golgi activity clearly invites further investigation. However, Golgi bodies have not yet been implicated in starch formation in fungi (McCracken *et al.*, 1973). Possible interrelationships between photosynthetic activity, starch formation and phycobilisome frequency in red algae have not yet been explored; it is not clear whether the absence of phycobilisomes from some electron micrographs (e.g. of *Batrachospermum*, in Bergfeld, 1970) is a real taxonomic feature or merely the result of poor fixation.

The single, centrally located, chloroplasts surrounded by small peripheral vacuoles in *Porphyridium* (Fig. 1) and in the cortex of *Nemalion* (Fig. 21) illustrate the very different morphology of these organelles in the simpler Nemalionales, in macrophytic phases in the life-history of Bangiophyceae and in unicells in the latter class. The most notable difference (in internal organization) from simple Florideophycean plastids is the presence of the pyrenoid, a feature whose variability in the red algae has scarcely been explored by systematists (Griffiths, 1970; Dodge, 1973).

In both *Porphyridium* and *Nemalion*, the pyrenoid is central and numbers of thylakoids follow a sinuous course through its matrix. Osmiophilic droplets are conspicuous around the periphery. Other pyrenoid types described in red algae include a simple form in the *Porphyra* conchocelis (Cole and Conway, 1975) and a labyrinthine pyrenoid, whose interstices contain projections of nuclear material, in *Rhodella* (Evans, 1970). In neither of these cases do thylakoids enter the pyrenoid. Another variation is the crystalline rather than granular matrix in the germinating monospores of *Smithora* (McBride and Cole, 1972a). With the exception of the conchocelis of *Porphyra* and the trichogyne of *Nemalion* (Duckett, unpublished observations) which each have two pyrenoids, there appears to be only one in red algal plastids. It is likely that both of these examples are instances of division phases, which also perhaps indicates that the pyrenoids usually derive from division of pre-existing ones. Figure 27, of *Porphyridium*, may be interpreted as a midway stage in such a replication process. However, that pyrenoids can also arise *de novo* may be deduced from the fusion cell proplastid of *Nemalion* (Figs 23, 24), which lacks a pyrenoid; the derivatives of this plastid give rise to the pyrenoid-containing plastids in the future carpospores.

Figure 1 illustrates the unitary nature of the stellate chloroplast in *Porphyridum* but owing to the extreme attentuation of the arms in *Nemalion* (Fig. 21) serial sectioning is required to establish whether all the ultimate ramifications do actually belong to the same organelle. This figure clearly demonstrates how difficult it can be to draw meaningful conclusions about plastid development when the only evidence is that derived from light microscopy.

For an even clearer case, we refer the reader to *Rhodella* (Evans, 1970) where, in our view, it would have been impossible to deduce from light microscopy that the multiple parental lobes of the plastid are all connected to the same central pyrenoid by narrow isthmuses, the latter devoid of thylakoids and thus non-pigmented. Indeed, this singular morphology led to the establishment of *Rhodella* as a new genus.

1. *Plastid Replication*

The question of the possible systematic value of maturational fragmentation of chloroplasts becomes even more complex when considered in the wider context of the replication of Rhodophycean plastids in general. As with the replication of the thylakoids and pyrenoids, there is a similar lack of definitive information on the multiplication of the entire plastid. However, it already seems clear that continuity may be maintained in several ways. In *Antithamnion* (Burton, 1971; Bisalputra, 1974), chloroplasts of vegetative cells multiply by elongation, followed by formation of a central constriction. Median separation of the inner thylakoids precedes an internal division of the inner limiting disc. Consequently, half of the umbo-shaped plastid contains a separate disc delimiting separate sets of internal thylakoids. The final phase of division is the separation of the chloroplast envelopes. These events may perhaps represent the most typical multiplication behaviour in higher Florideophyceae and can equally well cover all ontogenetic stages from proplastids (Fig. 18) to mature chloroplasts (Fig. 17).

Division by constriction in *Porphyridium* (Fig. 27) involves median separation of the thylakoids and pyrenoid but at no stage is the stroma separated into two distinct halves since the inner limiting disc is absent. Similarly, in the vegetative thallus of *Nemalion*, plastid replication follows a similar course and there is no involvement of the inner limiting disc in an intermediary partitioning of the central stroma and internal thylakoids.

Although at first sight these different division patterns appear to underline the other structural differences between the chloroplasts, before they can be incorporated into the systematics of red algae some note must be taken of the other methods of replication which can occur in the same taxa. In the tetraspore mother cells of *Antithamnion* (Burton, 1971; Bisalputra, 1974), proplastids consisting of an envelope and an inner limiting disc, but with few or no internal thylakoids, arise directly from mature chloroplasts by a blebbing or budding process. Subsequently, the detached proplastids are differentiated into mature chloroplasts as thylakoids develop from invaginations of the disc. In *Nemalion*. separation of the lobes of the labyrinthine fusion cell proplastid (Figs 23, 24), which subsequently supplies the carposporophyte with chloroplasts, is a similar phenomenon. In both the above cases, the rapid multiplication of plastids brought about by budding may be related to the critical stages at which they occur in the life-history. Moreover, since the proplastid buds are initially achlorophyllous this process would not be visible under the light microscope. It thus appears to be a phenomenon somewhat different from the fragmentation of a mature chloroplast in a highly differentiated thallus cell. Since the latter would not normally undergo further development or division, these changes in their plastids might well be irreversible, with the detached lobes being incapable of further replication (and thus very different from the derivatives of budding processes which ensure the continuity of plastids in meristematic tissues). This may perhaps be the case in the inner cortex of *Nemalion* (Figs 25, 26). Prominent in many of these cells are large vesicles filled with fine whorls of fibrillar material. Similar material also occurs in the arms of the plastid where its presence seems to indicate a degenerative change in the thylakoids. We therefore suggest that the large vesicle perhaps arose by detachment of the arms containing the disorganized thylakoids. Since these events in *Nemalion* are very different from plastid degeneration in *Rhodochorton* (*Andouinella*), also in the Nemalionales (Lichtlé, 1973b), which involves accumulation of osmiophilic globuli between the thylakoids (a change more reminiscent of the chloroplast–chromoplast transformation in higher plants), comparative studies on plastid ageing processes may have considerable taxonomic potential.

From these diverse examples, it should now be clear that TEM is essential for meaningful systematic studies on plastids of red algae. Given the various ontogenetic complications we have enumerated, the taxonomic conclusions on Rhodophycean plastids may be summarized as follows. Single stellate or lobed chloroplasts containing a single pyrenoid but lacking an inner limiting disc are an important distinguishing feature of the lower Bangiophyceae. However, higher Florideophyceae have several discoid chloroplasts per cell; an inner limiting disc is present but pyrenoids are absent. Two exceptions to these triple character associations occur in *Nemalion* and *Smithora* (McBride and Cole, 1969), where both pyrenoids and an inner limiting disc are present. Furthermore, the inner limiting disc does not figure prominently in the replication process in *Nemalion* as it does in the higher Florideophyceae. Detailed comparative studies on further genera of the Bangiales and Nemalionales thus appear particularly inviting.

2. Plastids and Phylogeny

Although in common with other authors (e.g. Dixon, 1973) we have alluded to the Florideophyceae as "higher", any attempt at an evolutionary interpretation of the variations in plastid ultrastructure would be pure speculation. A supposed phylogenetic trend from the complex Bangiophycean plastids to the simple discs in the Florideophyceae could be read the other way equally well. Even when far more ontogenetic data are available for many more taxa, it will be difficult to visualize what key pieces of information might unravel this dilemma.

Regardless of whether plastids ultimately prove to be of paramount importance in Rhodophycean systematics, there can be little doubt that together with the absence of flagella they feature strongly in overall speculative schemes of algal phylogeny (for reviews see Christensen, 1962; Klein and Cronquist, 1967; Gibbs, 1970; Klein, 1970; Lee, 1972; Fogg *et al.*, 1973; Greenwood, 1974). Since the arguments produced are chiefly based on biochemical criteria and are discussed in detail by these authors, it is unnecessary here to do more than outline the major points of contention. Most frequently presented are the numerous biochemical and structural similarities in the photosynthetic apparatus (Fig. 28) of red and blue–green algae. It should be stressed that, in view of the fundamental differences between eukaryotic and prokaryotic levels of organization, such parallels should not be interpreted too strongly in terms of phylogeny. As Fogg *et al.* (1973) state, it would be naïve to suppose that present day red algae are derived directly from an ancestral blue–green alga. A less contentious working hypothesis is to regard the two groups as successive stages of evolution. Nevertheless, the derivation of the red algal plastid from an endosymbiotic cyanophyte is an extremely attractive notion, even though much of its appeal almost certainly derives from the fact that it is unlikely ever to be "proven". On the other hand, the idea has led to extensive investigations of possible missing links with the result that cyanelle-containing organisms are now far better understood.

Although, from an ultrastructural standpoint, the gulf between other eukaryotic algae and the Rhodophyta is undoubtedly large, suggestions of possible links between the latter and the cryptomonads are perhaps the most interesting (Greenwood, 1974). However, here again comparisons are better expressed as possible stages in the evolution of the photosynthetic apparatus rather than in terms of phylogenetic relationship. The Rhodophyta and the Cryptophyta are the only eukaryotic algae possessing phycobiliproteins. In the cryptomonads these, together with the other accessory photosynthetic pigments, are probably located either within the thylakoid membranes or within the

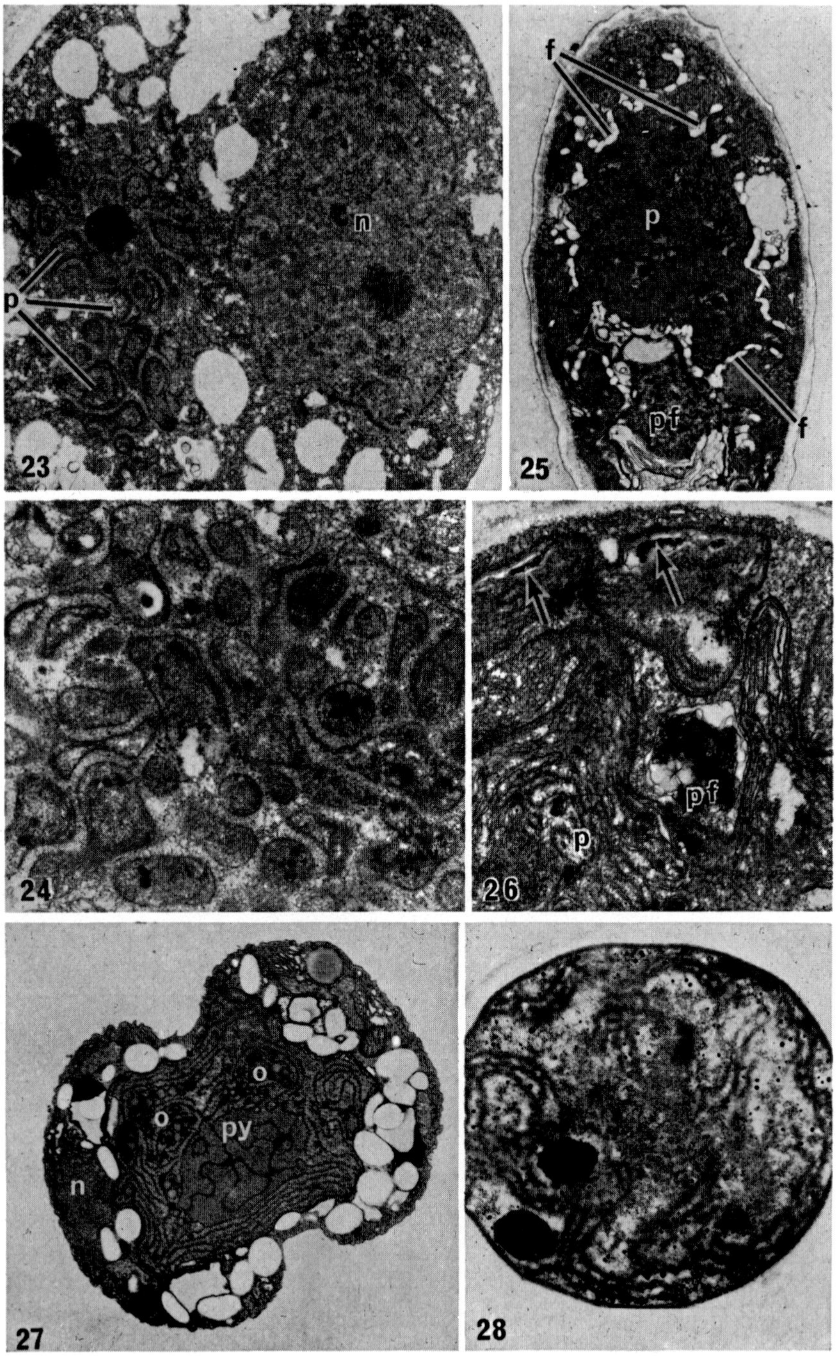

Figs 23–28. Captions on opposite page.

lumina rather than superficially as phycobilisomes. Greenwood (1974) presents a highly cogent argument that replacement of the water-soluble and superficially attached pigments by lipidic chlorophylls and carotenoids (characteristic of the Chromophyta), within the thylakoid membranes or in the lumina, is a major evolutionary step which permits elaborate stacking of the thylakoids in close apposition. Amongst eukaryotes only the Rhodophyta retain the original photosynthetic equipment comparable with that which initially appeared at the prokaryotic level of evolution. Nevertheless it should be pointed out that, although stacking is tacitly assumed advantageous, nowhere can we find a definitive statement on what exactly the benefits might be. Investigation of photosynthetic efficiency in terms of unit area of thylakoid membrane in different divisions might be particularly rewarding.

The other notable similarity between red algae and cryptomonads, also distinguishing them from the Chromophyta, is that they both produce starch grains. In the latter, these lie within an additional cytoplasmic compartment delimited by the extension of the perinuclear cisterna between nucleus and plastid envelope (chloroplast endoplasmic reticulum), rather than freely in the cytoplasm as in red algae and dinoflagellates (Dodge, 1973). Greenwood suggests that the ability to form starch in the Cryptophyta may perhaps be related to the presence of a "crypto-nucleus" (complete with double membranes, pores and nucleolar-like regions) in this compartment, and correlates

---

FIGS 23–26. *Nemalion*.

FIG. 23. Fusion cell showing numerous irregular profiles of the labyrinthine proplastid adjacent to the large nucleus. × 9000.

FIG. 24. Detail of the fusion cell proplastid. Osmiophilic globuli are common in many of the profiles but an inner limiting disc is absent. × 18 000.

FIG. 25. Mature chloroplast in a cortical cell. The arms of the plastid stand out against the surrounding zone of cytoplasm which contains abundant starch. This cell contains a large vesicle of fibrillar material probably derived from the sequestration of some of the arms of the plastid. × 3800.

FIG. 26. Detail of the arms of a mature chloroplast showing accumulations of fibrillar material apparently derived from the breakdown of the thylakoids. A small vesicle containing identical material is prominent between two of the arms. × 13 000.

FIG. 27. *Porphyridium*. Umbo-shaped cell interpreted as a division profile. Note that the central chloroplast still retains numerous thylakoids and the central pyrenoid. × 10 000.

FIG. 28. Mature vegetative cell of *Nostoc*. Note the superficial resemblance to the red algal chloroplasts. Loose whorls of single peripheral thylakoids bear phycobilisomes and the central cytoplasm is packed with ribosomes. × 21 000.

the emptiness of the latter with the incapacity to form starch in other Chromophytes. The cryptomonads may represent a primitive condition retaining within this compartment components which were lost in later evolution. These characteristics of the Cryptophyta suggest that the group is better placed at the base of the Chromophyta, between them and the Rhodophyta, rather than at the base of, or ancestral to, the latter (Lee, 1972).

THE MALE GAMETES

Although bereft of all the adornments such as flagellar roots, cytoskeletal arrays of microtubules or specialized mitochondria associated with motility, which render the zoospores and spermatozoids of other algal divisions and of fungi so attractive under the TEM, the male gametes of Rhodophyta are no less interesting in the cytological phenomena which they exhibit. It is indeed remarkable that with only about a dozen taxa yet investigated ultrastructurally (the first published accounts of the spermatial ultrastructure did not appear until 1971), several very different types of spermatium have already been discovered and features, previously unknown in any other plant gametes, described in what (compared with their motile counterparts elsewhere) appear superficially to be unrewarding material for TEM. Although Grubb's (1925) monograph is an exemplary light microscopical study (and forms a useful starting point for subsequent ultrastructural investigations) because of the minuteness of the cells many questions could not be resolved. Amongst the more important were the condition of the chromatin, the state of the nuclear envelope, the presence or absence of plastids, the nature of the bounding layer(s) and the liberation mechanism. Not only has the TEM enabled such problems to be critically re-investigated but the variations already revealed also clearly indicate that spermatia have considerable systematic potential. It would nevertheless be premature to suggest, on present information, that wide ranging studies on spermatia might lead to such a fundamental reappraisal of the taxonomy and phylogenetic relationships of the Rhodophyta as has occurred in the Chlorophyta and the Archegoniatae.

1. Chromatin

A particularly distinctive spermatium is that found in *Corallina* (Peel and Duckett, 1975) and *Jania* (Peel, unpublished observations). Under the light microscope, the nuclei of the mature gametes contain several well-defined, deeply staining bodies alluded to by earlier workers (Grubb, 1925, Suneson, 1937) as Körnchen, Karyosomes or chromatin granules. Since their number appeared to be constant for any particular taxon it was suggested that they might be true chromosomes in a condensed state and that the nucleus was in a state of division. Figure 30 illustrates that in *Corallina* the Körnchen of light microscopists appear to correspond to discrete, electron-dense, precisely aligned intranuclear bodies. Sections in different planes indicate that the nucleus is a short cylindroid with these bodies arranged in two rows parallel to either end of the cylindroid (a situation confirmed from Feulgen stained resin sections under the light microscope; Peel and Duckett, 1975). Grubb (1925) considered the Körnchen to be chromosomes at prophase or metaphase but in *Corallina* and *Jania* their alignment is much more typical of anaphase plates, thus these bodies are chromatids rather than chromosomes. Moreover, other features of the nucleus indicate that the mitosis is highly unusual.

Firstly, the chromatids are made up of two closely associated but structurally distinct bodies. The larger type, about 0·75 μm in diameter, has a granular texture typical of condensed chromatin and is sometimes closely associated with the nuclear envelope. The other, 0·125 μm in diameter, is more compact and of greater electron density. It occurs on the periphery of the larger bodies. Material of this nature has not been seen in the nuclei of *Corallina* and *Jania* elsewhere in the life-history nor, as far as we are aware, in any other red algae. Secondly, the larger bodies are associated with what are most readily interpreted as structured kinetochores. However, in contrast to those associated with chromosomes and chromatids in normally dividing cells of Rhodophyta (McDonald, 1972) and other algal divisions, those in the spermatia have no preferred orientation and are not associated with microtubules. Since, in our initial observations on spermatium differentiation in *Corallina* (Peel and Duckett, 1975), microtubules were never encountered we were uncertain as to whether the latter situation was real or merely a fixation artefact. The subsequent discovery of microtubules adjacent to the nuclear envelope, although they are still absent from the association with kinetochores, tends to support the former notion. Thirdly, we have been unable to detect any stages later than anaphase in nuclear division, although a large number of developmental stages in the formation of the Körnchen can readily be recognized.

Thus, we have concluded that the nuclear condition in *Corallina* and *Jania* is best described as an arrested mitosis. The closest parallel which we have been able to discover is in the eggs of some marine animals such as annelids, ascidians and molluscs, which at maturity are in prometaphase or metaphase I of meiosis (Crosby-Longwell and Stiles, 1968, 1970). They remain in this condition until the spermatozoid enters the egg, when the division goes to completion. Presumably the same occurs immediately the spermatia of *Corallina* and *Jania* make contact with the trichogyne, with the advanced mitotic state of their nuclei strongly suggesting that cytological changes take place very rapidly at this time. Unfortunately, no attempt has yet been made to follow fertilization in the Rhodophyta at the ultrastructural level.

In no other spermatia hitherto studied by TEM does the nuclear condition approach that in Corallinaceae. Indeed, perhaps the only feature common to the nuclei is the apparent disappearance of their nucleoli during the maturation process. However, some care must be exercised when making comparative observations on nuclear cytology: in some genera (e.g. *Janczewskia*; Kugrens, 1974) the spermatia may be released at any stage during their development, making recognition of the mature condition difficult. The nuclei of *Rhodomela* (Peyrière, 1974) bear the closest resemblance to those of *Corallina* and *Jania*. In the liberated spermatia of *Rhodomela*, the nucleus has shallow depressions at either end which contain randomly orientated cytoplasmic microtubules. Microtubules are also present within the nucleus, adjacent to fenestrae on the centre of the polar depressions but these do not appear to extend into the chromatin which lies in two distinct bands parallel to the polar depression. Thus, these nuclei strikingly resemble the prophase condition described in spermatangial mother cells of *Griffithsia* (Peyrière, 1971) and in young germlings of *Membranoptera* (McDonald, 1972), but, as in *Corallina* and *Jania*, microtubules are not associated with the chromatin bands. It is just possible therefore that the mitosis may again be arrested in *Rhodomela*, albeit at an earlier stage since the chromatin pattern is almost identical to that in young spermatia of *Corallina* (Peel and Duckett, 1975). Apart from cup-shaped nuclei in *Ptilota* (Scott and Dixon, 1973b) those of all other spermatia investigated are approximately spherical at maturity. Although Young (1977) describes the chromatin in *Bonnemaisonia* spermatia as condensed, this forms an electron-dense granular reticulum with no indication that the nuclei are in a state of division. Similarly, in *Erythrocystis* (Kugrens and West, 1972a) and *Ptilota* (Scott and Dixon, 1973b) dense masses of chromatin are in evidence adjacent to the nuclear envelope. There is some suggestion in Fig. 14 of Kugrens and West (1972a) that the chromatin forms discrete bodies in *Levringiella* but this feature is not remarked upon by the authors. In *Nemalion* (Fig. 34) and *Janczewskia* (Kugrens, 1974) the chromatin is uniformly dispersed throughout the nucleus in the mature gametes.

2. *Nuclear Envelope*

A similar picture of ultrastructural diversity is found in the appearance of the nuclear envelope in the mature spermatia. During development in *Corallina* and *Jania* pores are very numerous; although still present at maturity, they are far less prominent, so that it is then probably more correct to describe the envelope as containing larger irregular discontinuities (Fig. 30). Pores are very numerous in the polar depression of *Rhodomela* but rare elsewhere; this latter is the case for the entire nuclear envelope in *Levringiella*, *Erythrocystis* (Kugrens and West, 1972a), *Bonnemaisonia* (Young, 1977) and *Janczewskia* (Kugrens, 1974). In contrast to all these taxa with intact envelopes, Scott and Dixon (1973b) described the disappearance of the nuclear envelope in the liberated spermatia of *Ptilota*, stating this to be a confirmation of earlier light microscope observations (Grubb, 1925). Careful re-examination of their micrographs, however, suggests to us that this may be partly due to problems of fixation and partly to the fact that the envelope is obscured by a dense external coat of attached ribosomes, also a feature of the mature spermatium of *Nemalion* (Fig. 34).

A final noteworthy feature of *Corallina* and *Jania* nuclei (Fig. 30) not yet observed in other taxa are the caps, comprising an inner amorphous and an outer granular region, which form a layer over both ends of the nucleus parallel to the Körnchen.

3. *Spermatial Plastids*

As with the nuclei there are considerable variations, of possible taxonomic import, in the fate of the spermatial plastids. Perhaps the most common situation is typified by *Corallina*, where there is progressive degeneration, including the loss of the inner limiting disc, in the spermatangial mother cells; only rarely are remnants of plastids encountered in the spermatia. Similarly, in *Bonnemaisonia nootkana* (Young, 1977) and *Furcellaria* (Duckett, unpublished observations) plastids are present in the spermatangial mother cells but are not distributed to the spermatangia during the final cytokinesis. This exclusion process is clearly illustrated in Fig. 35 (*Nemalion*) where no trace of the prominent mother cell plastid enters the developing gametes. The presence of spermatial plastids in *Polysiphonia*, *Rhodomela*, *Laurencia* and *Nitophyllum* is not mentioned by Peyrière (1974), nor are they present in any of the micrographs. However, reduced chloroplasts have been reported in the spermatia of *Levringiella*, *Erythrocystis* (Kugrens and West, 1972a), *Janczewskia* (Kugrens, 1974) and *Ptilota* (Scott and Dixon, 1973b), although in the last case there is no published micrograph showing a recognizable spermatial plastid. Functional chloroplasts are apparently retained in *Bonnemaisonia hamifera* (Simon-Bichard-Bréaud, 1971) (an important interspecific difference from *B. nootkana*, if substantiated) and *Griffithsia* (Peyrière, 1971).

Thus there appears to be little uniformity in plastid behaviour. Nevertheless, the trend towards reduction and exclusion of the plastids from the male line during sexual reproduction in other divisions (see Kugrens, 1974; Bell and Duckett, 1976, for reviews) is also clearly manifest in the Rhodophyta.

4. The Spermatial Vesicle

A prominent feature of the majority of mature spermatia (Fig. 33) is the so-called spermatial or spermatangial vesicle (Kugrens and West, 1972a; Scott and Dixon, 1973b; Kugrens, 1974) which can often occupy at least half the volume of the gamete. These vesicles are regarded as essential in release from the mother cell and the spermatangium. Although some workers (Peyrière, 1971, 1974; Simon-Bichard-Bréaud, 1972a; Scott and Dixon, 1973b) have suggested that formation of the spermatial vesicle is due entirely to Golgi activity, more detailed developmental studies by Kugrens and West (1973a), Kugrens (1974) and Young (1977) indicate that it arises initially by swelling of ER cisternae. Vesicles (derived from Golgi bodies which become conspicuous later in spermatial development) contribute, with additional material from the ER, to the enlarging spermatial vesicles. Concomitant with the involvement of the Golgi, several changes take place in the contents of the vesicles. Ultimately, these contents may be coarsely fibrillar, as in *Ptilota* (Scott and Dixon, 1973b) and *Furcellaria* (Fig. 32 and Peel and Duckett, 1975), or granular as in *Erythrocystis* (Kugrens and West, 1972a) and *Nemalion* (Fig. 33).

Of potential taxonomic interest is the fact that the vesicles vary in number and position between genera. Most commonly (in relation to the mother cell) there is a single proximal vesicle (Peyrière, 1971, 1974; Kugrens and West, 1972a; Scott and Dixon, 1973b; Kugrens, 1974) derived from coalescence of several vesicles. In *Bonnemaisonia* and *Petrocelis* (Simon-Bichard-Bréaud, 1971;

FIG. 29. *Corallina.* Light micrograph of released spermatium showing the extremely long posterior appendage (over 90 μm). × 1000.

FIG. 30. *Corallina.* Longitudinal section of a mature spermatangium. Note the alignment of the intranuclear electron-dense spheroidal bodies and the accumulations of granular electron-dense material outside the nucleus. Kinetochores are just visible adjacent to three of the intranuclear dense bodies which are themselves of two types. Profiles of the smaller, more finely textured type are arrowed. The spermatial coat is indistinct and poorly defined. × 30 000.

FIG. 31. *Corallina.* Detail of mature spermatium showing cytopalsmic microtubules (arrowed) adjacent to the nuclear envelope. × 70 000.

FIG. 32. Section of a spermatangial region of *Furcellaria.* The spermatia contain large vesicles of fibrillar material and are deeply embedded in the thick confluent walls of the gametophyte. Egress is presumably via narrow channels through this extracellular material (arrowed). × 2800.

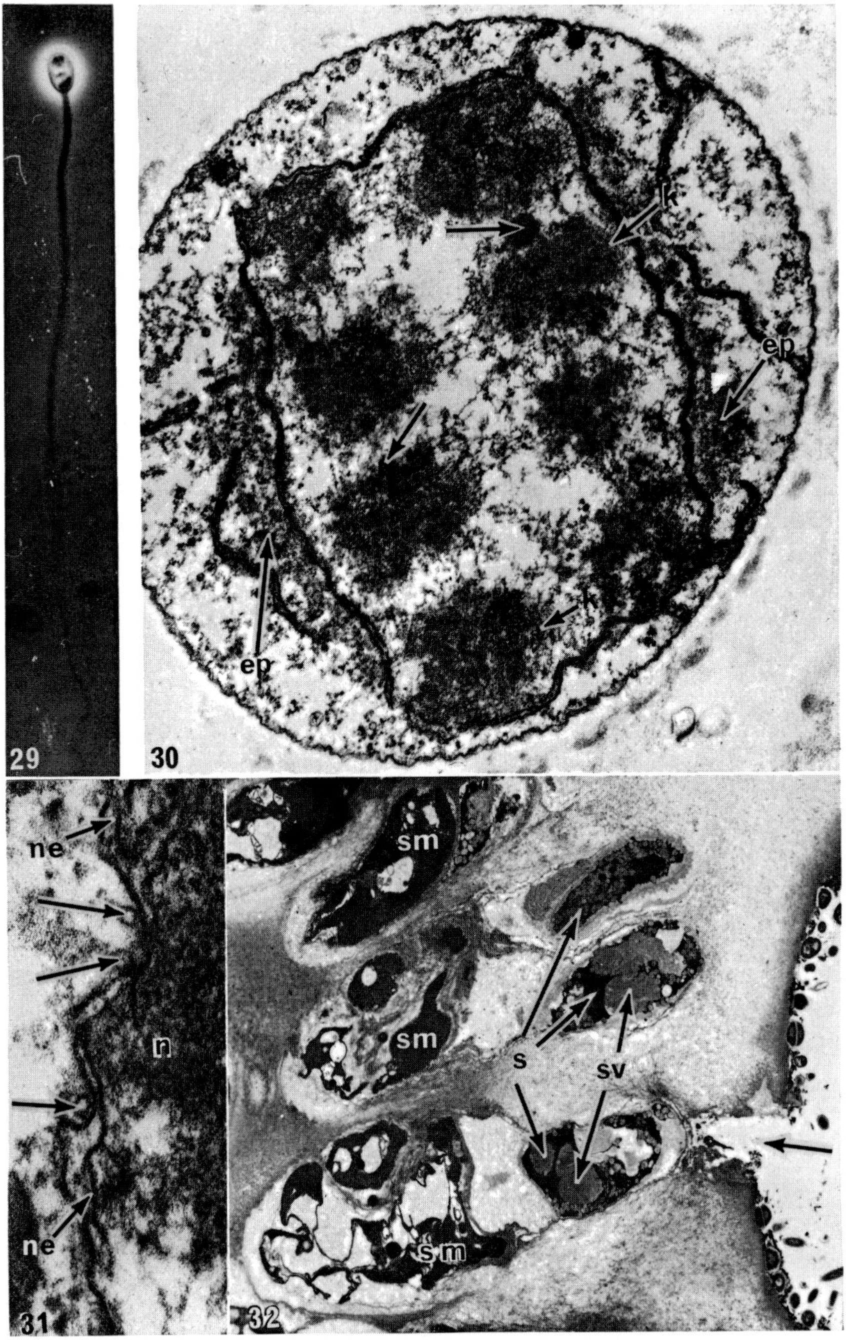

FIGS 29–32. Captions on opposite page.

Young, 1977), however, each spermatium produces one distal and one proximal vesicle lying on either side of the central nucleus. The number and position of the vesicles is variable in *Furcellaria* (Fig. 32 and Peel and Duckett, 1975), and in *Nemalion* (Fig. 33) two vesicles lie side-by-side in a proximal position.

At maturity the contents of the vesicle are extruded from the spermatium, laterally in *Bonnemaisonia* and *Furcellaria* (Peel and Duckett, 1975; Young, 1977) but basally in other genera. Although the exact nature of the extrusion mechanism is unknown, it has been suggested that the force responsible for release of the gametes from the spermatangium is generated by hydration of polysaccharides within the vesicle during maturation. Lending support to this idea is the absence of spermatial vesicles from the mature spermatia of *Corallina* and *Jania*. These complete their development lying free within the conceptacle and therefore neither require a mechanism to sever connections with the mother cell or to release themselves from the spermatangium. However, the immature elongate attached spermatia possess basal vesicles of electron-dense fibrillar material which at the time of release appear to be extruded from the cytoplasm. Our studies are currently being extended to the genus *Gracilaria*, where the spermatia also develop within conceptacles (Newton, 1931), to establish whether precocious formation of vesicles coincides with early release.

Apart from their contribution to the spermatial vesicle, the Golgi also play a prominent part in the formation of the spermatial coat. Irrespective of whether discrete spermatangial walls are separate from adjacent cells and thus clearly distinct (e.g. *Nemalion*, Figs 6, 33–35); are indistinguishable within the confluent outer gelatinous coat to the thallus (e.g. *Furcellaria*, Fig. 32); or are absent (e.g. *Corallina*, Peel and Duckett, 1975); the spermatia are surrounded by extracellular material of a nature rather different from that comprising the walls of

---

Spermatangia of *Nemalion*.

FIG. 33. Spermatangial branches showing three spermatia and a spermatangial mother cell. The prominent chloroplast in the latter is absent from the spermatangia which contain large spermatial vesicles and numerous smaller cored vesicles (arrowed). Remains of old walls around the spermatangia indicate that several generations of spermatia have been produced from the same mother cell. ×11 000.

FIG. 34. Section of mature spermatangium showing the nucleus containing homogeneous dispersed chromatin. The nuclear envelope is coated with abundant ribosomes. Both the gamete and its nucleus are much smaller than in *Corallina* (cf. Fig. 30). ×24 000.

FIG. 35. Longitudinal section of two spermatangial branches illustrating the exclusion of plastid material from the male gametes. ×5800.

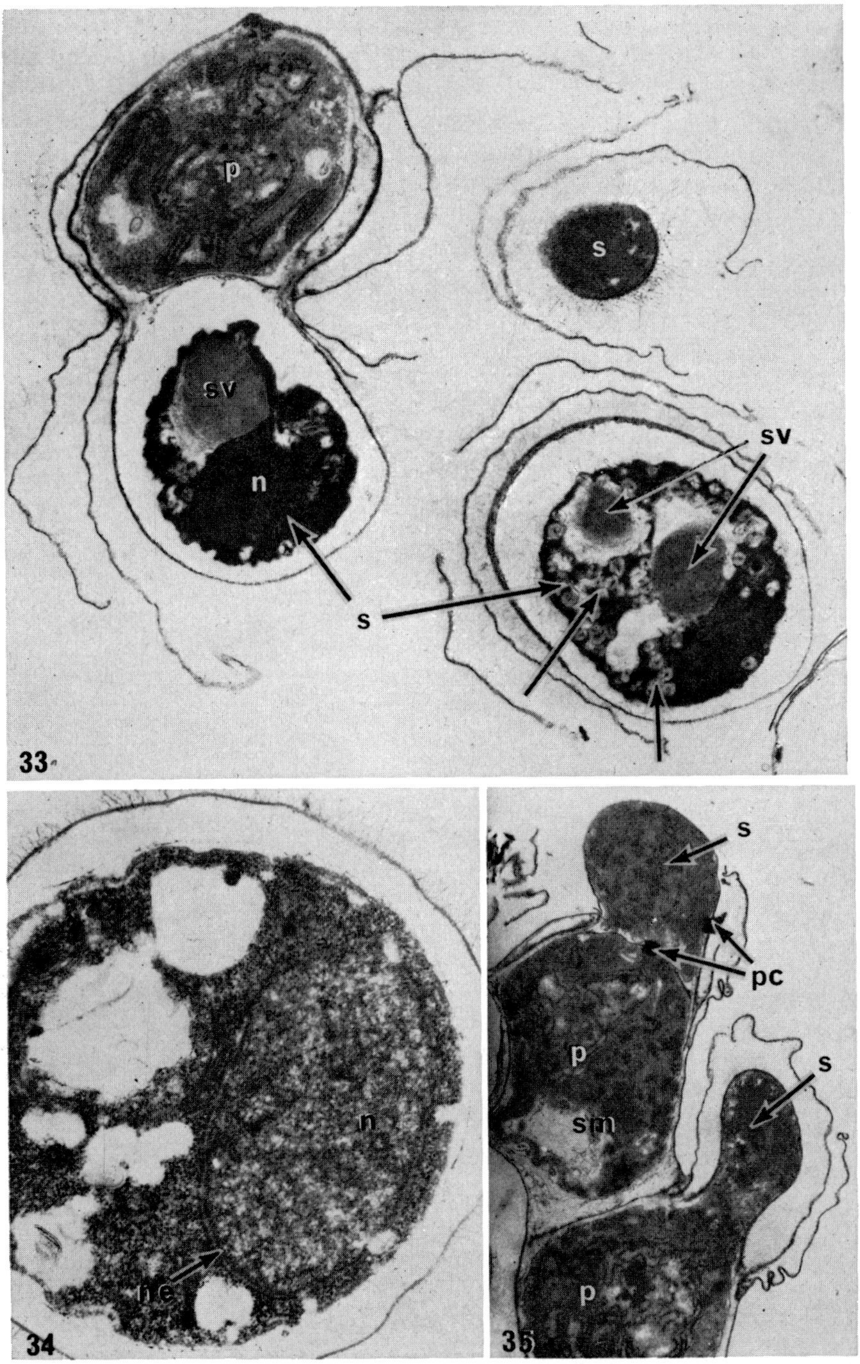

FIGS 33–35. Captions on opposite page.

G*

cells of vegetative plants. Although spermatangial coats are described as being composed of variable numbers of layers according to species (e.g. three in *Levringiella* but only one in *Erythrocystis*), developmental studies on *Janczewskia* (Kugrens, 1974) suggest that these differences are more likely to relate to observations on spermatia at different stages of maturity, thus being of little systematic value. After release, definite stratification is lost and the spermatia are naked except for a non-laminated coat of mucilage. This apparently originates from gelatinization of the contents of the spermatial vesicle around the gametes, together with additional mucilage from the spermatangial wall and spermatial coat. Specialized secretory vesicles from the Golgi probably also contribute to this mucilage (Kugrens, 1974; Peel and Duckett, 1975) but vesicles still present after release (e.g. the cored vesicles in *Nemalion*, Fig. 33) perhaps contain adhesive materials important in attachment to the trichogyne.

By far the most remarkable feature of the spermatial coat is its behaviour in *Corallina* (Peel and Duckett, 1975) and *Jania* (Peel, unpublished observations). After liberation from the mother cell, excess coat material forms a peculiar posterior appendage or "tail" which superficially resembles a flagellum. Thus, under the light microscope (Fig. 29), these cells bear an uncanny similarity to an animal spermatozoid. Although the function of this appendage is at present unknown (it would certainly considerably enhance the chances of contact with the trichogyne), it cannot be overstressed that it neither behaves like a flagellum nor has the same construction. Similarly the claim by Simon-Bichard-Bréaud (1971, 1972b) that the spermatia of *Bonnemaisonia* also possess a flagellum appears to have no foundation whatsoever (see Kugrens and West, 1972a; Peel and Duckett, 1975; Young, 1977 for discussion). Her "crypte flagellaire" is little more than dilated ER cisternae (Young, 1977) and the "basal bodies" are probably nuclear pores (Kugrens and West, 1972a). It is unfortunate that these gross misinterpretations have attracted such wide attention.

THE NUCLEUS

At present, plastids have made far more impact on Rhodophytan systematics and phylogeny than have all other ultrastructural features; the nucleus, however, holds the greatest potential for the future. Definitive information on the place of meiosis is essential for an understanding of any life-history. In red algae, the position is particularly perplexing; according to Dixon (1973), widely accepted data are frequently of doubtful validity and many unverified assumptions are regarded as definitive proof (a situation exemplified by the fact that different phases of the same organism have, on numerous occasions, initially been classified as distantly related plants). The site of meiosis has yet to be demonstrated in those orders of the Bangiophyceae that exhibit sexual reproduction; even in the higher Florideophyceae, its general occurrence in the tetrasporangia is not proven. Apart from Austin's (1960) study of *Furcellaria*, conclusive photomicrographic (or experimental) evidence of meiosis is lacking due to the small size of the chromosomes, if indeed they actually exist throughout the red algae as the readily visualized discrete entities seen in higher plants.

1. *Synapsis*

A major breakthrough was achieved when Kugrens and West (1972b) announced the discovery of the synaptonemal complex in the tetraspore mother cells of the four Florideophycean genera *Janczewskia*, *Levringiella*, *Gonimophyllum* and *Polycoryne*; synaptonemal complexes were first described in brown algal unilocular sporangia in the following year (Toth and Markey, 1973). The demonstration of this feature, generally regarded as reliable ultrastructural evidence for meiosis throughout green plants, fungi and animals, has far-reaching systematic significance. Identification of the complex in other taxa (e.g. Bangiophyceae and the Nemalionales) is the key to further elucidation of their life-histories. The presence or absence of the complex should provide more information on taxa which lack tetrasporophytes or for which only this phase is known. It may prove indispensible in distinguishing polysporangia (the result of meiosis) from purely mitotic parasporangia. Finally, counting the number of synaptonemal complexes within a nucleus, by serial sectioning, may be the only way of obtaining accurate data on chromosome numbers, since each bivalent has only a single complex. However, the task of locating complexes is far from easy; they are not only transient and restricted to a few cells but might also occur where not expected. Given such a situation, all those embarking on fine structural studies of red algae would be well advised to familiarize themselves with the appearance of the complex.

2. *Nuclear Division*

Photosynthetic pigments and plastid ultrastructure serve to link Cyanophyta with Rhodophyta and the total absence of flagellated cells from all stages in the life-history serves to separate the two divisions from all other eukaryotic algae. Even the most complex multicellular algae of other groups, almost without exception, return to the flagellated condition for sexual reproduction. From the various ultrastructural studies on spermatia, it now seems established that this is not the case in red algae. The essential progenitor of the flagellum is the centriole, which figures prominently in the mitotic apparatus of many divisions of eukaryotic algae (see Pickett-Heaps, 1971, for a discussion of exceptions).

Thus it becomes of paramount importance to the theory that red algae diverged from the main line of eukaryotic evolution at a very early stage and are not descended from ancestors with motile cells, to establish whether or not centrioles are present at nuclear division (Leedale, 1970). Despite repeated attempts at the TEM level, and many confused data from light microscope studies (due, in our view, mainly to over-extrapolation from other divisions, but see Dixon, 1973, for review), there are only five accounts of the ultrastructure of nuclear division that go any way towards resolving this problem (Peyrière, 1971, 1974; McDonald, 1972; Peel and Duckett, 1975; Scott and Thomas, 1975). For ultrastructural studies on mitosis in Phaeophyta we refer the reader to Brawley *et al.* (1977).

The most extensive study is that by McDonald (1972) who described all stages of mitosis in *Membranoptera*. Centrioles were not detected. Instead, structures termed polar rings (which are probably equivalent to the centrosomes, polar formations and polar bodies of light microscopists) occupy each spindle pole at prophase. The "corpuscule polaire dense" of Peyrière (1971), associated with the poles of prophase nuclei in spermatangial mother cells of *Griffithsia* and the extra nuclear polar bodies in *Porphyridium* (Bronchart and Demoulin, 1977, Fig. 2), might well be non-median sections of the same structure whose true nature was not revealed due to the paucity of favourable sections. Neither polar rings nor centrioles are mentioned in the other accounts of dividing cells. Since the polar rings are associated with perinuclear microtubules in *Membranoptera*, McDonald (1972) suggested that they are some sort of microtubule organizing centre (Pickett-Heaps, 1969). Certainly their general texture and electron opacity is very similar to other non-centriolar polar foci for extra-nuclear mitotic microtubule systems (see Heath, 1974, for review) observed in a few algae (e.g. the rhizoplast of *Ochromonas*) and more especially fungi (e.g. the spindle plaques in many Ascomycetes).

The second important discovery from these ultrastructural studies is the absence, in every case, of "open" spindles. Except for polar fenestrae in *Porphyridium* (Bronchart and Demoulin, 1977), *Membranoptera* (McDonald, 1972), *Griffithsia* (Peyrière, 1971) and *Rhodomela* (Peyrière, 1974), the nuclear envelope remains intact throughout mitosis. Although, in view of the variability in the behaviour of the nuclear envelope in other divisions (Leedale, 1970; Heath, 1974), it would be extremely unwise to imply that all red algal mitoses are of this type, it is clear that ultrastructure does not support light microscope observations on the breakdown of the envelope. Whilst the fundamental phylogenetic significance of closed versus open spindles remains highly speculative, all available evidence from other groups indicates that the former is the primitive condition (Leedale, 1970; Heath, 1974). However, possession of polar fenestrae, a character shared with some green algae and fungi, means that red algal mitosis is not completely intranuclear. Other supposedly primitive features (Leedale, 1970), such as chromosome segregation by attachment to the envelope, and absence of well ordered spindle microtubules, are not seen in red algae. The mitotic apparatus in *Membranoptera* comprises three distinct microtubule systems: continuous pole–to–pole tubules (also seen at metaphase in *Porphyridium*) which extend through the polar fenestrae, perinuclear tubules, and a third group running from the chromosomes to the poles. The last are attached by structured kinetochores, an uncommon feature in plants (Heath, 1974). Interestingly, these kinetochores were not detected by Bronchart and Demoulin (1977) in *Porphyridium*, which also apparently lacks discrete condensed chromosomes at mitosis. A unique feature of *Porphyridium*, however, is the presence of a single large microbody at each spindle pole.

The final study (Scott and Thomas, 1975), although restricted to telophase II of meiosis in *Polysiphonia*, is nevertheless extremely interesting in that it reveals a form of nuclear division only previously described in the two fungi *Saccharomyces* (Moens and Rapport, 1971) and *Saprolegnia* (Howard and Moore, 1970). Three nuclei are shown joined together at telophase II with the clear implication that the second meiotic division has immediately followed the first and that the four nuclei do not separate until the end of the process. The functional advantage of this uninuclear meiosis remains obscure (Heath, 1974).

3. *Nuclear–Cytoplasmic Interactions*

Although comparative data on division have not yet been obtained, ultratructural studies are now revealing some remarkable forms of nuclear behaviour during sporogenesis in Rhodophyta. These events may be best described as visible manifestations of intense transcriptional activity and reorganization of the cytoplasm at critical stages in the life-history; collectively, they are termed nucleo–cytoplasmic interactions (Kessel, 1973; Peel *et al.*, 1973; Franke, 1974; Franke *et al.*, 1974; Spring *et al.*, 1974; Bell and Duckett, 1976; Dickinson and Heslop-Harrison, 1977).

In post-meiotic tetraspore mother cells of *Corallina*, each nucleus becomes ensheathed by a complex of precisely orientated ER (Figs 36, 38). In regions between this nuclear endoplasmic reticulum (NER) and the nuclear envelope, electron-dense material (EDM) accumulates. Adjacent to the nuclei, the NER comprises sheets aligned parallel to the nuclear envelope, while further away these sheets frequently diverge and project away from the nucleus. The NER is of two types, one smooth, proximal to the nucleus where EDM is present and the other studded with ribosomes distally where the EDM becomes discon-

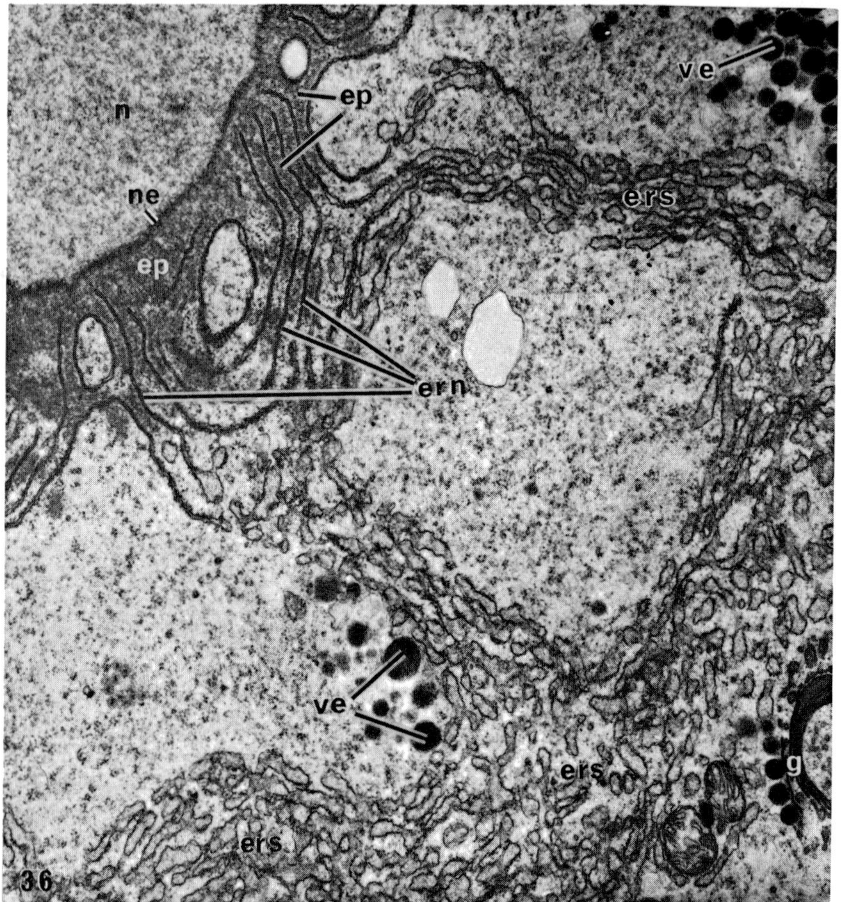

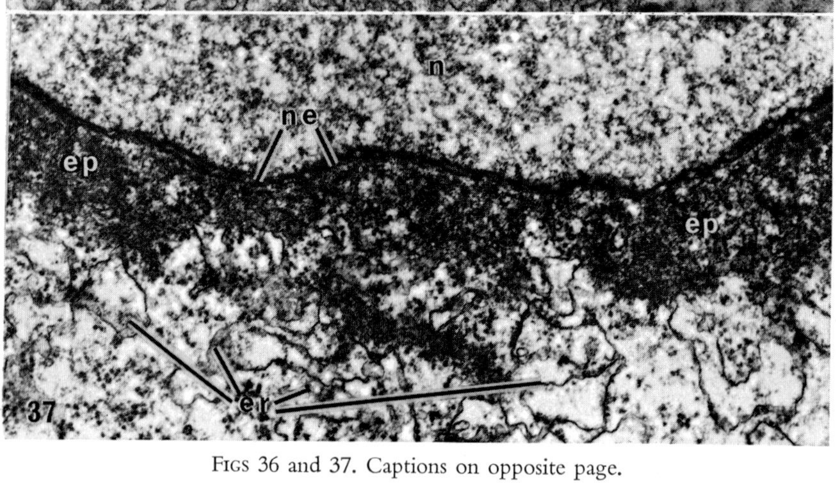

FIGS 36 and 37. Captions on opposite page.

tinuous. The NER appears to be active in the production of smaller swollen cisternal elements which form anastomosing reticular tracts between adjacent nuclei. It is suggested that the EDM is involved in the assembly of ribosomes and that the material within the swollen ER cisternae is transported via vesicles to the proximal face of the Golgi bodies (Peel *et al.*, 1973).

EDM similar to that in *Corallina* also accumulates around the nuclei of post-meiotic tetraspore mother cells in the related genus *Jania*. However, here it lies in cytoplasmic pockets formed by regular convolutions of the nuclear envelope and is not associated with precisely orientated ER (Fig. 39). No closely comparable perinuclear complexes appear to accompany tetrasporogenesis in *Polysiphonia* (Scott and Thomas, 1975; Alley and Scott, 1977), *Levringiella* (Kugrens and West, 1972c) or *Ptilota* (Scott and Dixon, 1973a), the only other genera investigated ultrastructurally. However, in every case the nucleoli are particularly prominent and highly structured, and pores are especially numerous in the nuclear envelope. Both these features are strongly indicative of hyper-activity. Moreover, extensive evaginations of the nuclear envelope are produced, a situation reminiscent of the nuclear–cytoplasmic interaction in differentiating fern eggs (Bell and Duckett, 1976).

By inference from other systems, such as *Acetabularia* (Franke *et al.*, 1974; Spring *et al.*, 1974), *Fucus* zygotes (Brawley *et al.*, 1977), eggs of archegoniate plants (Bell and Duckett, 1976) and animal oocytes (Kessel, 1973; Franke, 1974), the other stage in the red algal life-history, which might be expected to reveal examples of profound cytoplasmic reorganization at the ultrastructural level is that associated with sexual reproduction. Extensive blebbing of the nuclear envelope occurs in developing carpospores of *Levringiella* (Kugrens and West, 1973), *Polysiphonia* and *Pterosiphonia* (Tripodi, 1974), where it is accompanied by radical changes in the internal structure of the mitochondria and plastids (a striking parallel with fern eggs, Bell and Duckett, 1976; and with meiotic tissues of angiosperms, Dickinson and Heslop-Harrison, 1977). In *Polysiphonia* and *Pterosiphonia* (Wetherbee *et al.*, 1974) intranuclear and cytoplasmic annulate lamellae (AL) are present (ER with pore complexes similar to those of the

---

FIGS 36, 37. *Corallina*.

FIG. 36. Post-meiotic tetraspore mother cell showing electron-dense perinuclear material between sheets of precisely orientated endoplasmic reticulum. Away from the nucleus the latter give way to extensive tracts of swollen endoplasmic reticulum. × 18 000.

FIG. 37. Developing carpospore showing an extensive accumulation of electron-dense perinuclear material into which extend several profiles of swollen endoplasmic reticulum. × 33 000.

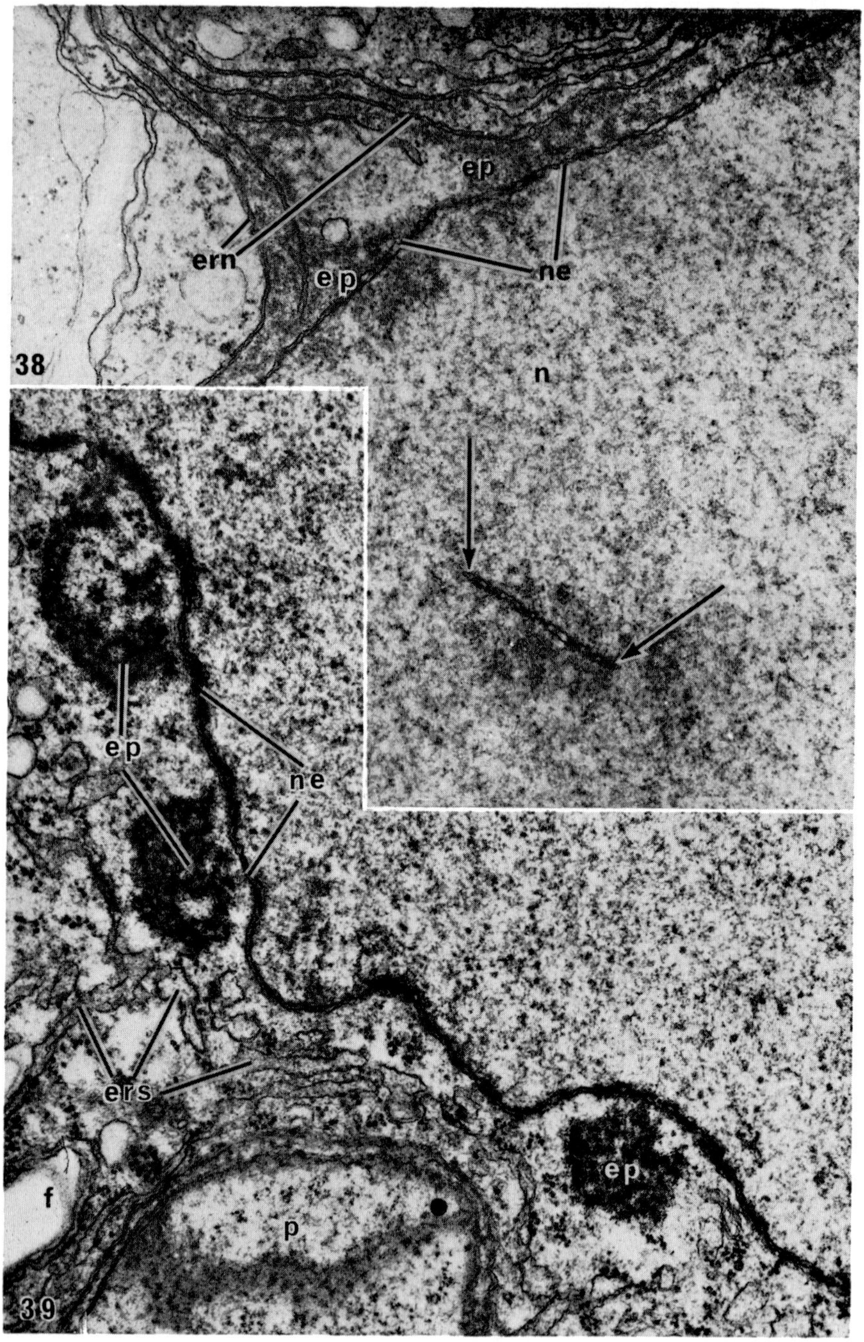

Figs. 38 and 39. Captions on opposite page.

nuclear envelope (Kessel, 1971; Franke, 1974)). The former are single and occur in haploid nuclei of supporting cells of the carposporangia. Cytoplasmic AL, on the other hand, only become apparent after the formation of the fusion cell, in which they proliferate throughout the cytoplasm as single cisternae or elaborate stacks with the pores in register across them.

Although common in rapidly differentiating germ cells (of both sexes) and in embryonic and tumour cells of animals, in plants AL has been described elsewhere only in angiosperm pollen; in eggs of the heterosporous fern *Marsilea* (D. Gold-Myles, personal communication); in nuclei of the tetraspore mother cells of *Corallina* (Fig. 38) and in the cytoplasm, during prophase, of the first mitotic division of *Fucus* zygotes (Brawley *et al.*, 1977).

Neither AL nor nuclear evaginations have been detected during carposporogenesis in *Jania* or *Corallina*. Instead, perinuclear EDM like that in the tetraspore mother cells is present. In *Jania* this again lies in pockets of the nuclear envelope (Fig. 39) but in *Corallina* it forms a continuous layer not associated with precisely orientated ER (Fig. 37). Osmiophilic granular material very similar to the EDM is associated with the AL in *Polysiphonia* (Wetherbee *et al.*, 1974) and surrounds the developing carpospore nucleus in *Nemalion* (Duckett, unpublished observations).

Undoubtedly these nuclear–cytoplasmic interactions will become increasingly prominent in causal analyses of red algal life-histories. Their interest, however, is far from solely morphogenetic. It should be clear from the above examples that dramatic differences (particularly striking between *Corallina* and *Jania*) have already come to light amongst the mere handful of taxa yet investigated. Thus, nvaluable systematic information may derive from sources both unusual and, since they are intimately related to fundamental events in the life-history, somewhat unexpected. Clearly, many more surprises await the enterprising cytologist amidst the Rhodophyta.

ACKNOWLEDGEMENTS

We are indebted to Mr D. Andrew Davies and Mrs Sheila Walker (Bangor) for skilled technical assistance and to A. D. Greenwood (Imperial College, London) for many

---

FIG. 38. *Corallina*. Post-meiotic tetraspore mother cell showing sheets of nuclear endoplasmic reticulum and perinuclear electron-dense material. Similar material is also visible in the peripheral nucleoplasm which also contains annulate lamellae (arrowed). × 30 000.

FIG. 39. *Jania*. Developing carpospore, showing electron-dense perinuclear material localized in cytoplasmic pockets formed by convolutions of the nuclear envelope. × 43 000.

hours of stimulating discussion on algal cytology. M. C. P. gratefully acknowledges the financial support provided by a Natural Environment Research Council Studentship and the facilities and assistance made available at the N.E.R.C., Unit of Marine Invertebrate Biology, Marine Science Laboratories, Menai Bridge, North Wales. J. G. D. thanks Drs J. S. Buchanan and M. T. Martin for initially arousing his interest in the cytology of *Nemalion*.

REFERENCES

ALLEY, C. D. and SCOTT, J. L. (1977). Unusual dictyosome morphology and vesicle formation in tetrasporangia of the marine red alga *Polysiphonia denudata*. *J. Ultrastruct. Res.* **58**, 289–298.

ATKINSON, A. W., GUNNING, B. E. S., JOHN, P. C. L. and MCCULLOUGH, W. (1971). Centrioles and microtubules in *Chlorella*. *Nature New Biol., Lond.* **234**, 24–25.

AUSTIN, A. P. (1960). Life history and reproduction of *Furcellaria fastigiata* (L.) Lam. 2. The tetrasporophyte and reduction division in the tetrasporangium. *Ann. Bot.* **24**, 296–310.

BELL, P. R. and DUCKETT, J. G. (1976). Gametogenesis and fertilization in *Pteridium*. *In* "The Biology of Bracken" (F. H. Perring and B. G. Gardiner, eds). *Bot. J. Linn. Soc.* **73**, 47–78. Academic Press, London and New York.

BERGFELD, R. (1970). Die Feinstruktur und Entwicklung der Plastiden einiger Sußwasser-rotalgen. *Cytobiologie* **1**, 411–419.

BISALPUTRA, T. (1974). Plastids. *In* "Algal Physiology and Biochemistry, Botanical Monographs 10" (W. D. P. Stewart, ed.), pp. 124–160. Blackwell, Oxford.

BISALPUTRA, T. and BISALPUTRA, A.-A. (1967). The occurrence of DNA fibrils in chloroplasts of *Laurencia spectabilis*. *J. Ultrastruct. Res.* **17**, 14–22.

BISALPUTRA, T., RUSANOWSKI, P. C. and WALKER, W. S. (1967). Surface activity, cell wall, and fine structure of pit connection in the red alga *Laurencia spectabilis*. *J. Ultrastruct. Res.* **20**, 277–289.

BOUCK, G. B. (1962). Chromatophore development, pits, and other fine structure in the red alga, *Lomentaria baileyana* (Harv.) Farlow. *J. Cell Biol.* **12**, 553–569.

BOURNE, V. L., CONWAY, E. and COLE, K. (1970). On the ultrastructure of pit connections in the conchocelis phase of the red alga *Porphyra perforata*. J. Ag. *Phycologia* **9**, 79–81.

BRACKER, C. E. (1967). Ultrastructure of fungi. *A. Rev. Phytopath.* **5**, 343–374.

BRAWLEY, S. H., QUATRANO, R. S. and WETHERBEE, R. (1977). Fine-structural studies of the gametes and embryo of *Fucus vesiculosus* L. (Phaeophyta). III. Cytokinesis and the multicellular embryo. *J. Cell Sci.* **24**, 275–294.

BRONCHART, R. and DEMOULIN, V. (1977). Unusual mitosis in the red alga *Porphyridium purpureum*. *Nature, Lond.* **268**, 80–81.

BURTON, A. H. S. (1971). "Chloroplast Continuity during the Formation of the Tetraspore in *Antithamnion subulatum*". M.Sc. Thesis, University of British Columbia, Canada.

CHAMBERLAIN, A. H. L. and EVANS, L. V. (1973). Aspects of spore production in the red alga *Ceramium*. *Protoplasma* **76**, 139–159.

CHRISTENSEN, T. (1962). Alger. *In* "Botanik", Vol. 2, no. 2 (T. W. Bocher, M. Lange and T. Sorensen, eds). Munksgaard, Copenhagen.

COLE, K. and CONWAY, E. (1975). Phenetic implications of structural features of the perennating phase in the life history of *Porphyra* and *Bangia* (Bangiophyceae, Rhodophyta). *Phycologia* **14**, 239–245.

COTTLER, M. H. (1971). Plasmalemmal extensions in *Chondrus crispus* (L.) Stackh. *J. Ultrastruct. Res.* **37**, 31–36.

CROSBY-LONGWELL, A. and STILES, S. S. (1968). Fertilization and completion of meiosis in spawned eggs of the American oyster, *Crassostrea virginica* Gmelin. *Caryologia* **21**, 65–73.

CROSBY-LONGWELL, A. and STILES, S. S. (1970). The genetic system and breeding potential of the commercial American oyster. *Endeavour* **29**, 94–99.

DEMOULIN, V. (1974). The origin of ascomycetes and basidiomycetes. The case for a red algal ancestry. *Bot. Rev.* **40**, 315–344.

DENISON, W. C. and CARROLL, G. C. (1966). The primitive ascomycete: a new look at an old problem. *Mycologia* **58**, 249–269.

DICKINSON, H. G. and HESLOP-HARRISON, J. (1977). Ribosomes, membranes and organelles during meiosis in angiosperms. *Phil. Trans. R. Soc. Lond. B*, **277**, 327–342.

DIXON, P. S. (1973). "Biology of the Rhodophyta." Oliver and Boyd, Edinburgh.

DODGE, J. D. (1973). "The Fine Structure of Algal Cells." Academic Press, London and New York.

DOUGHERTY, E. C. and ALLEN, M. B. (1960). Is pigmentation a clue to protistan phylogeny? *In* "Comparative Biochemistry of Photoreactive Systems" (M. B. Allen, ed.), pp. 129–144. Academic Press, New York and London.

DRUM, R. W. and HOPKINS, J. T. (1966). Diatom locomotion: an explanation. *Protoplasma* **62**, 1–33.

DUCKETT, J. G. (1975). Spermatogenesis in pteridophytes. *In* "The Biology of the Male Gamete" (J. G. Duckett and P. A. Racey, eds). *Biol. J. Linn. Soc.* **7**, Suppl. 1, 97–127. Academic Press, London and New York.

DUCKETT, J. G., BUCHANAN, J. S., PEEL, M. C. and MARTIN, M. T. (1974). An ultrastructural study of pit connections and percurrent proliferations in the red alga. *Nemalion helminthoides* (Vell. in With.) Batt. *New Phytol.* **73**, 497–507.

DUCKETT, J. G. and TOTH, R. (1977). Giant mitochondria in a periclinal chimera of *Ficus elastica* Roxb. forma *variegata*. *Ann. Bot.* **41**, 903–912.

DUCKETT, J. G., PRASAD, A. K. S. K., DAVIES, D. A. and WALKER, S. (1977). A cytological analysis of the *Nostoc*-bryophyte relationship. *New Phytol.* **79**, 349–362.

EVANS, L. V. (1970). Electron microscopical observations on a new red algal unicell, *Rhodella maculata* gen. nov., sp. nov. *Br. phycol. J.* **5**, 1–13.

EVANS, L. V., CALLOW, J. A. and CALLOW, M. E. (1973). Structural and physiological studies on the parasitic red alga *Holmsella*. *New Phytol.* **72**, 393–402.

EVANS, L. V., CALLOW, M. E., PERCIVAL, E. and FAREED, V. (1974). Studies on the synthesis and composition of extracellular mucilage in the unicellular red alga *Rhodella*. *J. Cell Sci.* **16**, 1–21.

FOGG, G. E., STEWART, W. D. P., FAY, P. and WALSBY, A. E. (1973). "The Blue-Green Algae." Academic Press, London and New York.

FRANKE, W. W. (1974). Structure, biochemistry and functions of the nuclear envelope. *Int. Rev. Cytol.* Suppl. **4**, 71–236.

FRANKE, W. W., BERGER, S., FALK, H., SPRING, H., SCHEER, U., HERTH, W., TRENDELBURG, M. F. and SCHWEIGER, H. G. (1974). Morphology of the nucleo-cytoplasmic interactions during the development of *Acetabularia* cells. I. The vegetative phase. *Protoplasma* **82**, 249–282.

FRANKE, W. W., KARTENBECK, J., KRIEN, S., VANDERWOUDE, W. J., SCHEER, U. and MORRÉ, D. J. (1972). Inter- and intra-cisternal elements of the Golgi apparatus. *Z. Zellforsch. mikrosk. Anat.* **132**, 365–380.

FRAZER, T. W. and GUNNING, B. E. S. (1973). Ultrastructure of the hairs of the filamentous green alga *Bulbochaete hiloensis* (Nordst.) Tiffany: an apoplastidic plant cell with a well developed Golgi apparatus. *Planta* **113**, 1–19.

FRITSCH, F. E. (1945). "Structure and Reproduction of the Algae", Vol. 2. Cambridge University Press, Cambridge.

GANTT, E. and LIPSCHULTZ, C. A. (1972). Phycobilisomes of *Porphyridium cruentum*. I. Isolation. *J. Cell Biol.* **54**, 313–324.

GIBBS, S. P. (1970). The comparative ultrastructure of the algal chloroplast. *Ann. N.Y. Acad. Sci.* **175**, 454–473.

GOFF, L. J. (1976). The biology of *Harveyella mirabilis* (Cryptonemiales; Rhodophyceae). V. Host responses to parasite infection. *J. Phycol.* **12**, 313–328.

GORDON, E. M. and MCCANDLESS, E. L. (1973). Ultrastructure and histochemistry of *Chondrus crispus* Stackhouse. *Proc. Nova Scotia Inst. Sci.* **27**, Suppl. 111–133.

GREENWOOD, A. D. (1974). The Cryptophyta in relation to phylogeny and photosynthesis. *Proc. int. Congr. el. Micr. Canberra* **8**, 566–567.

GRIFFITHS, D. J. (1970). The pyrenoid. *Bot. Rev.* **36**, 29–58.

GRUBB, V. M. (1925). The male organs of the Florideae. *J. Linn. Soc.* (*Bot.*), **47**, 177–255.

GUNNING, B. E. S. and STEERE, M. W. (1975). "Ultrastructure and the Biology of Plants Cells." Edward Arnold, London.

HARA, Y. and CHIHARA, M. (1974). Comparative studies on the chloroplast ultrastructure in the Rhodophyta, with special reference to their taxonomic significance. *Sci. Rep. Tokyo Kyoiku Daigaku*, ser. B, **15**, 209–235.

HARPER, M. A. and HARPER, J. F. (1967). Measurements of diatom adhesion and their relationship with movement. *Br. phycol. Bull.* **3**, 195–207.

HEATH, I. B. (1974). Genome separation mechanisms in prokaryotes, algae and fungi. *In* "The Cell Nucleus", Vol. II (H. Busch, ed.), pp. 487–515. Academic Press, New York and London.

HEATH, I. B. and DARLEY, W. M. (1972). Observations on the ultrastructure of the male gametes of *Biddulphia laevis* Ehr. *J. Phycol.* **8**, 51–59.

HOPKINS, J. T. and DRUM, R. W. (1966). Diatom motility: an explanation and a problem. *Br. phycol. Bull.* **1**, 63–67.

HOWARD, K. L. and MOORE, R. T. (1970). Ultrastructure of oogenesis in *Saprolegnia terrestris*. *Bot. Gaz.* **131**, 311–336.

HUGHES, S. J. (1971). Percurrent proliferations in fungi, algae and mosses. *Can. J. Bot.* **49**, 215–231.

JAROSCH, H. (1962). Gliding. *In* "Physiology and Biochemistry of Algae" (R. A. Lewin, ed.), pp. 573–581. Academic Press, New York and London.

KESSEL, R. G. (1968). Annulate lamellae. *J. Ultrastruct. Res.* Suppl. **10**, 1–82.

KESSEL, R. G. (1973). Structure and function of the nuclear envelope and related cytomembranes. *In* "Progress in Surface and Membrane Science. Vol. 6" (J. F. Danielli, M. D. Rosenberg and D. A. Cadenhead, eds), pp. 243–329. Academic Press, London and New York.

KLEIN, R. M. (1970). Relationships between blue–green and red algae. *Ann. N.Y. Acad. Sci.* **175**, 623–633.

KLEIN, R. M. and CRONQUIST, A. (1967). A consideration of the evolutionary and taxonomic significance of some biochemical, micromorphological and physiological characters in the Thallophytes. *Quart. Rev. Biol.* **42**, 105–295.

KOLLER, K. P., WEHRMEYER, W. and SCHNEIDER, H. (1977). Isolation and characterization of disc-shaped phycobilisomes from the red alga *Rhodella violacea. Arch. Mikrobiol.* **112**, 61–68.

KONRAD HAWKINS, E. (1974a). Golgi vesicles of uncommon morphology and wall formation in the red alga, *Polysiphonia. Protoplasma* **80**, 1–14.

KONRAD HAWKINS, E. (1974b). Growth and differentiation of the Golgi apparatus in the red alga, *Callithamnion roseum. J. Cell Sci.* **14**, 633–655.

KUGRENS, P. (1974). Light and electron microscopic studies on the development and liberation of *Janczewskia gardneri* Setch. spermatia (Rhodophyta). *Phycologia* **13**, 295–306.

KUGRENS, P. and WEST, J. A. (1972a). Ultrastructure of spermatial development in the parasitic red algae *Levringiella gardneri* and *Erythrocystis saccata. J. Phycol.* **8**, 331–343.

KUGRENS, P. and WEST, J. A. (1972b). Synaptonemal complexes in red algae. *J. Phycol.* **8**, 187–191.

KUGRENS, P. and WEST, J. A. (1972c). Ultrastructure of tetrasporogenesis in the parasitic red alga *Levringiella gardneri* (Setchell) Kylin. *J. Phycol.* **8**, 370–383.

KUGRENS, P. and WEST, J. A. (1973). The ultrastructure of carpospore differentiation in the parasitic red alga *Levringiella gardneri* (Setch.) Kylin. *Phycologia* **12**, 163–173.

LEE, R. E. (1971). The pit connections of some lower red algae: ultrastructure and phylogenetic significance. *Br. phycol. J.* **6**, 29–38.

LEE, R. E. (1972). Origin of plastids and the phylogeny of algae. *Nature, Lond.* **237**, 44–46.

LEEDALE, G. F. (1970). Phylogenetic aspects of nuclear cytology in the algae. *Ann. N.Y. Acad. Sci.* **175**, 429–453.

LICHTLÉ, C. (1973a). Ultrastructure du plaste de deux Rhodophycées: *Rhodochorton purpureum* (Lighft. [*sic*]) Rosenvinge et *Rhodothamniella floridula* (Dillwyn) J. Feldmann. *C.r. hebd. Séanc. Acad. Sci. Paris*, sér. D, **277**, 1865–1868.

LICHTLÉ, C. (1973b). Dégénérescence du plaste du *Rhodochorton purpureum* (Lighft. [*sic*]) Rosenvinge. Rhodophycée. Acrochaétiale. *C.r. hebd. Séanc. Acad. Sci. Paris*, sér. D, **277**, 2341–2344.

LIN, H., SOMMERFELD, M. R. and SWAFFORD, J. R. (1975). Light and electron microscope observations on motile cells of *Porphyridium purpureum* (Rhodophyta). *J. Phycol.* **11**, 452–457.

LITTLEFIELD, L. J. and BRACKER, C. E. (1971). Ultrastructure of septa in *Melampsori lini. Trans. Br. mycol. Soc.* **56**, 181–188.

MANTON, I. (1950). "Problems of Cytology and Evolution in Pteridophyta." Cambridge University Press, Cambridge.

MANTON, I. (1965). Some phyletic implications of flagellar structure in plants. *Adv. bot. Res.* **2**, 1–34.

MARCHANT, H. J. (1976). Plasmodesmata in algae and fungi. *In* "Intercellular Communication in Plants: Studies on Plasmodesmata" (B. E. S. Gunning and A. W. Robards, eds), pp. 59–80. Springer-Verlag, Berlin, Heidelberg and New York.

McBRIDE, D. L. and COLE, K. (1969). Ultrastructural characteristics of the vegetative cell of *Smithora naiadum* (Rhodophyta). *Phycologia* **8**, 177–186.

McBRIDE, D. L. and COLE, K. (1972a). Ultrastructural observations on germinating monospores in *Smithora naiadum* (Rhodophyceae, Bangiophycidae). *Phycologia* **11**, 181–191.

McBRIDE, D. L. and COLE, K. (1972b). Fine structural studies on several Pacific coast representatives of the Erythropeltidaceae (Rhodophyceae). *Proc. int. Seaweed Symp.* **7**, 159–164.

McCRACKEN, D. A., NADAKAVUKAREN, M. J. and DODD, J. D. (1973). Starch in fungi. II. Induction of amyloidity in members of diverse genera. *Am. J. Bot.* **60**, 940–943.

McDONALD, K. (1972). The ultrastructure of mitosis in the marine red alga *Membranoptera platyphylla*. *J. Phycol.* **8**, 156–166.

MOENS, P. B. and RAPPORT, E. (1971). Spindles, spindle plaques, and meiosis in the yeast *Saccharomyces cerevisiae* (Hansen). *J. Cell Biol.* **50**, 344–361.

MORRÉ, D. J., MOLLENHAUER, H. H. and BRACKER, C. E. (1971). Origin and continuity of Golgi apparatus. *In* "Results and Problems in Cell Differentiation. 2. Origin and Continuity of Cell Organelles" (J. Reinert and H. Urspring, eds), pp. 82–126. Springer-Verlag, Berlin, Heidelberg and New York.

NEUSHUL, M. (1971). Uniformity of thylakoid structure in a red, a brown and two blue-green algae. *J. Ultrastruct. Res.* **37**, 532–543.

NEWTON, L. (1931). "A Handbook of the British Seaweeds." British Museum (Natural History), London.

NOVIKOFF, A. B. and HOLTZMAN, E. (1970). "Cells and Organelles." Holt, Rinehart and Winston, London, New York, Sydney and Toronto.

PARTHASARATHY, M. V. and MÜHLETHALER, K. (1972). Cytoplasmic microfilaments in plant cells. *J. Ultrastruct. Res.* **38**, 46–62.

PEEL, M. C. and DUCKETT, J. G. (1975). Studies of spermatogenesis in the Rhodophyta. *In* "The Biology of the Male Gamete" (J. G. Duckett and P. A. Racey, eds). *Biol. J. Linn. Soc.* **7**, Suppl. 1, 1–13. Academic Press, London and New York.

PEEL, M. C., LUCAS, I. A. N., DUCKETT, J. G. and GREENWOOD, A. D. (1973). Studies of sporogenesis in the Rhodophyta. I. An association of the nuclei with endoplasmic reticulum in post-meiotic tetraspore mother cells of *Corallina officinalis* L. *Z. Zellforsch. mikrosk. Anat.* **147**, 59–74.

PEYRIÈRE, M. (1970). Evolution de l'appareil de Golgi au cours de la tétrasporogenèse de *Griffithsia flosculosa* (Rhodophycée). *C.r. hebd. Séanc. Acad. Sci. Paris*, sér. D, **270**, 2071–2074.

PEYRIÈRE, M. (1971). Etude infrastructurale des spermatocystes du *Griffithsia flosculosa* (Rhodophycée). *C.r. hebd. Séanc. Acad. Sci. Paris*, sér. D, **273**, 2071–2074.

PEYRIÈRE, M. (1974). Etude infrastructurale des spermatocystes et spermaties de différentes Rhodophycées floridées. *C.r. hebd. Séanc. Acad. Sci. Paris*, sér. D, **278**, 1019–1022.

Pickett-Heaps, J. D. (1969). The evolution of the mitotic apparatus: an attempt at comparative ultrastructural cytology in dividing plant cells. *Cytobois* **1**, 257–270.

Pickett-Heaps, J. D. (1971). The autonomy of the centriole: fact or fallacy? *Cytobios* **3**, 205–214.

Pickett-Heaps, J. D. (1975). "Green Algae, Structure, Reproduction and Evolution in Selected Genera." Sinauer Assoc. Inc. Sunderland, Mass.

Pueschel, C. M. (1977). A freeze-etch study of the ultrastructure of red algal pit plugs. *Protoplasma* **91**, 15–30.

Ramus, J. (1969a). Pit connection formation in the red alga *Pseudogloiophloea*. *J. Phycol.* **5**, 57–63.

Ramus, J. (1969b). Dimorphic pit connections in the red alga *Pseudogloiophloea*. *J. Cell Biol.* **41**, 340–345.

Ramus, J. (1972). The production of extracellular polysaccharide by the unicellular red alga *Porphyridium aeruginosum*. *J. Phycol.* **8**, 97–111.

Ramus, J. and Robins, D. M. (1975). The correlation of Golgi activity and polysaccharide secretion in *Porphyridium*. *J. Phycol.* **11**, 70–74.

Rosenvinge, L. K. (1927). On mobility in the reproductive cells of the Rhodophyceae. *Bot. Tidsskr.* **40**, 72–79.

Scott, J. L. and Dixon, P. S. (1973a). Ultrastructure of tetrasporogenesis in the marine red alga *Ptilota hypnoides*. *J. Phycol.* **9**, 29–46.

Scott, J. L. and Dixon, P. S. (1973b). Ultrastructure of spermatium liberation in the marine red alga *Ptilota densa*. *J. Phycol.* **9**, 85–91.

Scott, J. L. and Thomas, J. P. (1975). Electron microscope observations of telophase II in the Florideophyceae. *J. Phycol.* **11**, 474–476.

Simon-Bichard-Bréaud, J. (1971). Un appareil cinétique dans les gamétocystes mâles d'une Rhodophycée: *Bonnemaisonia hamifera* Hariot. *C.r. hebd. Séanc. Acad. Sci. Paris*, sér. D, **273**, 1272–1275.

Simon-Bichard-Bréaud, J. (1972a). Origine et devenir des vacuoles à polysaccharides des gamétocystes mâles de *Bonnemaisonia hamifera* Hariot (Rhodophycée). *C.r. hebd. Séanc. Acad. Sci. Paris*, sér. D, **274**, 1485–1488.

Simon-Bichard-Bréaud, J. (1972b). Formation de la crypte flagellaire et évolution de son contenu au cours de la gamétogenèse male chez *Bonnemaisonia hamifera* Hariot (Rhodophycée). *C.r. hebd. Séanc. Acad. Sci. Paris*, sér. D, **274**, 1796–1799.

Solms-Laubach, H. (1881). Die Corallinenalgen des Golfes von Neapel under der angrenzenden Meeres-Abschnitte. *Fauna Flora Golf Neapel* **4**, 1–64.

Spring, H., Trendelburg, M. F., Scheer, U., Franke, W. W. and Herth, W. (1974). Structural and biochemical studies of the primary nucleus of two green algal species, *Acetabularia mediterranea* and *Acetabularia major*. *Cytobiologie* **10**, 1–65.

Stewart, K. D. and Mattox, K. R. (1975). Comparative cytology, evolution and classification of the green algae with some considerations of the origin of other organisms with chlorophylls A and B. *Bot. Rev.* **41**, 104–135.

Suneson, S. (1937). Studien über die Entwicklungsgeschichte der Corallinaceen. *Acta Univ. lund* **33**, 1–101.

Toth, R. and Markey, D. R. (1973). Synaptonemal complexes in brown algae. *Nature, Lond.* **243**, 236–237.

TRIPODI, G. (1971). The fine structure of the cystocarp in the red alga *Polysiphonia sertularioides* (Grat.) J. Ag. *J. Submicr. Cytol.* **3**, 71–79.

TRIPODI, G. (1974). Ultrastructural changes during carpospore formation in the red alga *Polysiphonia. J. Submicr. Cytol.* **6**, 275–286.

WETHERBEE, R. and WEST, J. A. (1976). Unique Golgi apparatus and vesicle formation in a red alga. *Nature, Lond.* **259**, 566–567.

WETHERBEE, R and WEST, J. A. (1977). Golgi apparatus of unique morphology during early carposporogenesis in a red alga. *J. Ultrastruct. Res.* **58**, 119–133.

WETHERBEE, R., WEST, J. A. and WYNNE, M. J. (1974). Annulate lamellae in post-fertilization development in *Polysiphonia novae-angliae* (Rhodophyta). *J. Ultrastruct. Res.* **49**, 401–404.

YOUNG, D. N. (1977). A note on the absence of flagellar structures in spermatia of *Bonnemaisonia. Phycologia* **16**, 219–222.

9 | An Introduction to the Scanning Electron Microscopy of Red Algae

D. J. GARBARY

Hartley Botanical Laboratories, University of Liverpool, England

Abstract: The surface structure of sixteen species in nine genera of the Corallinaceae and nine species of Acrochaetiaceae was investigated using the scanning electron microscope (SEM). Using both quantitative and qualitative characters, most of the genera and species of Corallinaceae can be distinguished and a key to the species investigated is presented using characters generated from the SEM. In addition to several patterns of surface structure of vegetative cells, *Fosliella minutula*, *F. farinosa* and *Dermatolithon litorale* showed different patterns of calcification of the conceptacle surface. By use of a number of characters the Corallinaceae can be divided into two groups; *Jania* spp., *Corallina officinalis*, *Mesophyllum lichenoides* and *Phymatolithon calcareum* are separated from *Dermatolithon* spp., *Fosliella* spp. and *Melobesia membranacea*, with *Lithophyllum incrustans* and *Lithothamnium* spp. as intermediates. In species of Acrochaetiaceae, four types of surface were present on vegetative cells and at least three surface types were distinguished on monosporangia and tetrasporangia.

INTRODUCTION

Scanning electron microscopy has been little used in the study of seaweeds, except for investigations of generic concepts (Bailey *et al.*, 1976a) and morphology (Berger *et al.*, 1974; Schweiger *et al.*, 1974; Schweiger and Berger, 1975; Bailey *et al.*, 1976b) in the Dasycladaceae (Chlorophyta). In non-taxonomic studies, Moss (1974, 1975, 1977) used the SEM to investigate the attachment and development of rhizoids and holdfasts in genera of the Fucales and Laminariales. Borowitzka *et al.* (1974) used the SEM to study the structure and organization of calcium carbonate deposits in a number of algae, including red and brown seaweeds.

The red algae were initially studied with the SEM by Bailey and Bisalputra

Systematics Association Special Volume No. 10, "Modern Approaches to the Taxonomy of Red and Brown Algae", edited by D. E. G. Irvine and J. H. Price, 1978, pp. 205–222, Academic Press, London and New York.

(1970) when investigating *Corallina* and *Calliarthron*. More recently, Conway and Cole (1977), studying the conchocelis phase of *Porphyra* and *Bangia*, showed differences in the surface structure of monosporangia and conchosporangia. Variation in surface structure between the species they investigated, however, was not discussed. Thus the potential of the SEM as a tool for investigating characters of systematic importance has not been shown for the Rhodophyta.

A survey of the Florideae using the SEM has not been attempted; instead, several species and genera were chosen from two families of red algae for observation of the kinds of taxonomically valuable characters that could be derived for each group. The two families chosen were the Corallinaceae and the Acrochaetiaceae.

CORALLINACEAE

The confusion in generic delimitation in the Corallinaceae is apparent from a comparison of the treatments this family received in recent checklists of British marine algae (Parke and Dixon, 1968, 1976). This points out the need for different characters to improve current generic and specific delimitation in the family. In this study, sixteen species in nine genera (see Table I) were observed and a number of qualitative and quantitative characters interpreted from the electron micrographs. Nomenclature as in Parke and Dixon (1976) is used in all cases, unless noted otherwise in the text.

The preparation of coralline algae for the SEM was straightforward. Freshly collected plants were air dried, coated with gold–palladium and viewed in a Cambridge Instruments stereoscan microscope. Herbarium material was coated directly, whereas plants preserved in formalin were rinsed with distilled water prior to air drying.

The scanning electron micrographs of the Corallinaceae were suitable for quantitative analysis. Enlarged photographs were traced on to a polyester film of consistent weight and area. From the tracings, maximum epithallial concavity size (see below for explanation of terms) and thickness of calcification between epithallial concavities were measured. By cutting out the epithallial concavities from the tracings and weighing all the fragments, the percentage of the surface area occupied by the epithallial concavities was obtained. Despite the number of concavity and interconcavity measurements (for most specimens $n = 25$), they are based at the most on several views at varying magnifications of one or two specimens. As a result, considerable intraspecific variation has been overlooked. As more detailed studies are made on individual species and genera, these figures will have to be revised.

For descriptive purposes, variation in the Corallinaceae can be discussed in terms of several aspects of surface structure, including structure of vegetative surfaces, differentiation and structure of conceptacle surfaces and quantitative characters. These aspects will be discussed in turn. A key to the species included in this study is also presented.

TABLE I. List of members of the Corallinaceae examined, with values for quantitative characters. Figures indicate mean values ± standard error

Species	Epithallial concavity size (μm)	Distance between concavities (μm)	Percentage of surface composed of epithallial cells
Corallina officinalis	9·8 ± 0·4	1·7 ± 0·2	65
Dermatolithon corallinae	6·9 ± 0·2	4·2 ± 0·3	27, 36
	5·6 ± 0·2	5·2 ± 0·5	14
D. pustulatum	5·8 ± 0·2	6·0 ± 0·2	16
	7·9 ± 0·3	4·6 ± 0·5	30
D. litorale		no quantitative data taken	
Fosliella farinosa	5·4 ± 0·2	3·4 ± 0·4	19, 34
	6·1 ± 0·2	5·4 ± 0·2	25
F. lejolisii	3·9 ± 0·2	4·3 ± 0·3	12, 15
F. limitata	4·3 ± 0·1	4·6 ± 0·5	17, 28
F. minutula	5·1 ± 0·2	5·5 ± 0·2	22
Jania corniculata	13·8 ± 0·7	1·8 ± 0·3	65
J. rubens	15·4 ± 0·8	1·8 ± 0·2	63
	10·1 ± 0·5	1·5 ± 0·2	66
Lithothamnium corallioides	6·7 ± 0·2	2·4 ± 0·1	36, 37
L. glaciale	5·9 ± 0·2	1·7 ± 0·1	45, 48
Lithophyllum incrustans	5·2 ± 0·1	3·4 ± 1·0	30, 32
	5·4 ± 0·2	4·0 ± 0·2	23, 25
Melobesia membranacea	5·5 ± 0·1	4·2 ± 0·2	30, 37
Mesophyllum lichenoides	8·2 ± 0·2	1·7 ± 0·2	50, 65
Phymatolithon calcareum	8·6 ± 0·2	1·7 ± 0·1	58, 59

1. *Structure of Vegetative Surfaces*

The surface of most of the Corallinaceae showed an aspect of a series of pores separated by calcified regions (Fig. 10). The uncalcified regions (pores) represent the spaces formerly occupied by epithallial cells. During processing the epithallial cells disintegrate, leaving what may be referred to as epithallial concavities. This structure is contrasted with the condition found in three species in *Corallina* and *Jania*. In these species (*C. officinalis*, *J. corniculata*, *J. rubens*) the appearance was one of a calcium carbonate surface on which ridges were formed (Figs 1–4). There was an apparent lack of primary pit-connections on the outer surface which would preclude the presence of epithallial cells in the surface depressions. A sectional view of a broken branch (Fig. 3) agreed with this interpretation, all of the cells at the periphery of the plant possessing a calcium carbonate "roof". This regular surface was broken in *Jania* spp. by a number of raised thickenings each with a definite pore (Fig. 2). It is suggested that these structures represent the positions of hairs or trichocytes.

In the remaining species, definite epithallial concavities were present and occasionally primary pit-connections were visible. There was, however, considerable variation in the structure and placement of cell walls and in the shape of epithallial concavities. Three main patterns of epithallial concavity structure were recognized:

1. Round to irregular concavities scattered on the surface (Figs 22–24);
2. Rectangular to rounded concavities in rows separated by grooves (Figs 13, 15).
3. Concavities with greater calcification on one side than another (Figs 27, 28).

Types one and two were related in that calcification formed a flat surface. The latter two patterns were similar in that the basic unit of surface structure was rectangular.

On the basis of these characters, *Melobesia membranacea* was distinct, being the only species whose surface corresponded to type three. All species of *Fosliella*, and *Dermatolithon litorale*, exhibited type two epithallial structure. *D. pustulatum* and *D. corallinae* were intermediate between the first and second type in that grooves (ridges in *D. corallinae*) between the rows of concavities were absent except at the margins (Figs 6, 11). In addition, in older tissues it was often difficult to discern that the epithallial concavities were indeed in rows. All other taxa investigated had epithallial structure corresponding to type one. *Lithophyllum incrustans*, however, was anomalous in that concavities of type two were seen on the vertical face of the growing edge (Fig. 26).

Plants possessing type one surface can be separated into four groups on the basis of the differentiation of the calcium carbonate walls. The first state is undifferentiated, the calcium carbonate matrix separating the cells appearing smooth and flat. *Mesophyllum lichenoides* and *Lithothamnium glaciale* (Figs 23, 29) are representative of this group. The second state is characterized by the presence of a narrow ridge on the surface separating the epithallial concavities (Fig. 22). This pattern characterizes *L. corallioides*. In the third state, calcium carbonate is laid down in a rosette around the epithallial concavities (Fig. 31). This pattern has been demonstrated in this study for *Lithophyllum incrustans* and by Borowitzka *et al.* (1974) for two additional species of the genus. Thus the

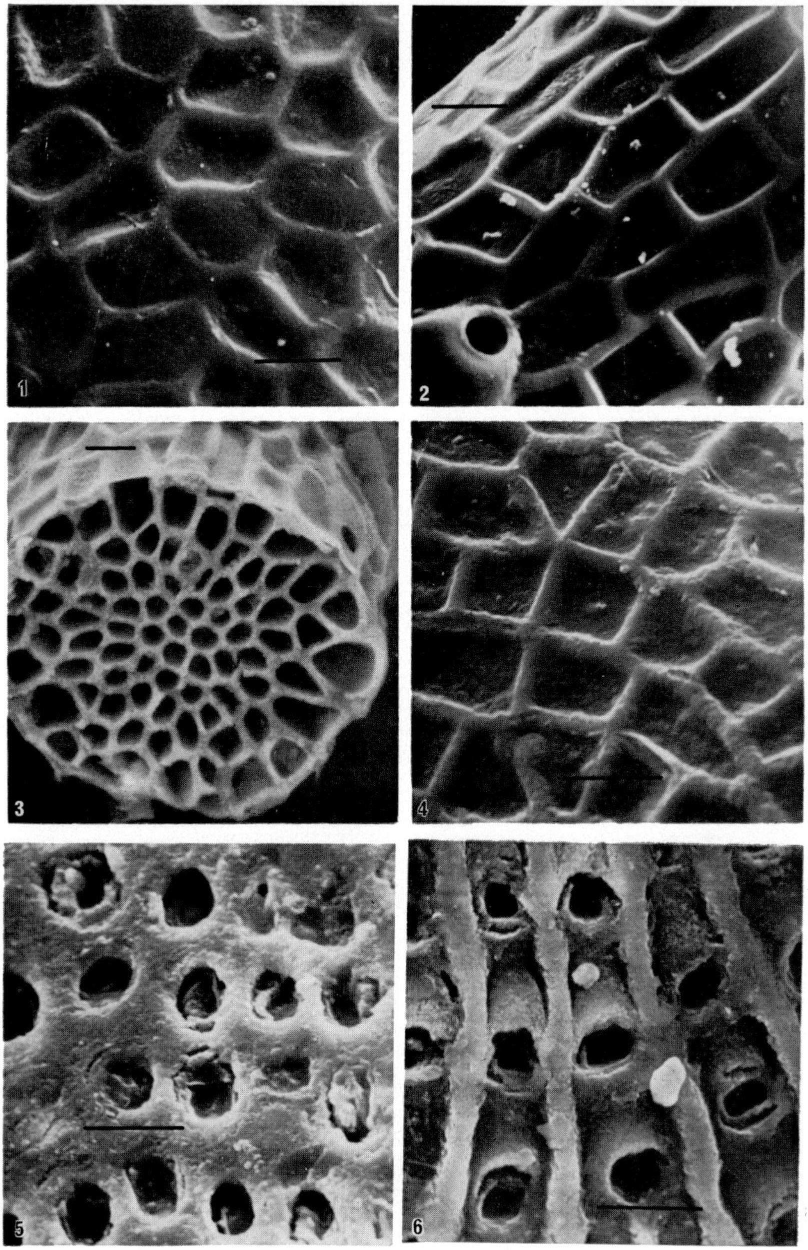

FIGS 1–6. All scales 10 μm.

FIG. 1. Surface of *Corallina officinalis*.

FIG. 2. Surface of *Jania corniculata* with trichocyte.

FIG. 3. Sectional view of broken branch of *J. corniculata*.

FIG. 4. Surface of *J. rubens*.

FIG. 5. Surface of *Dermatolithon corallinae* in older part of crust.

FIG. 6. Surface of *D. corallinae* near edge of crust with ridges.

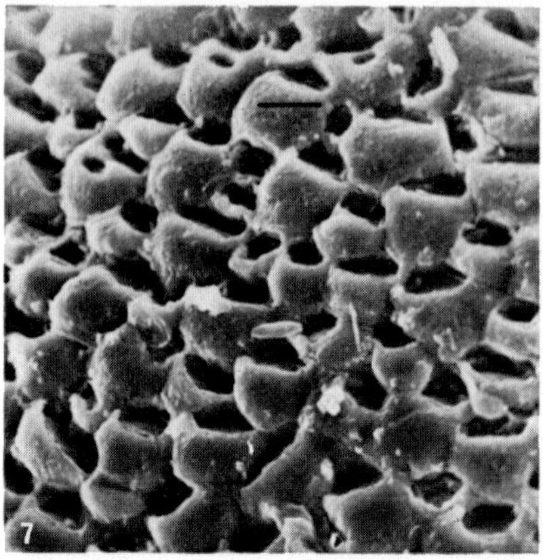

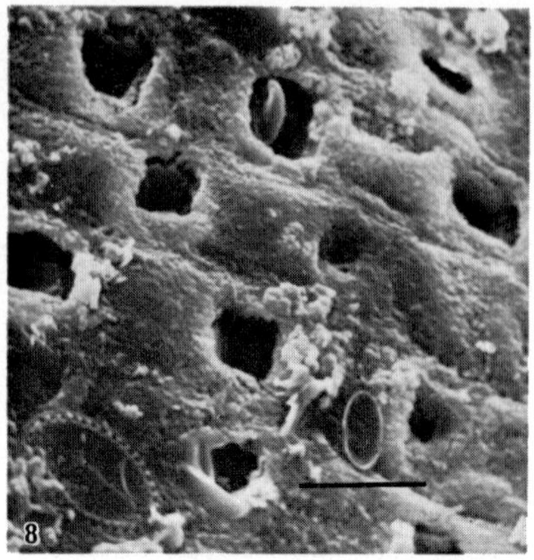

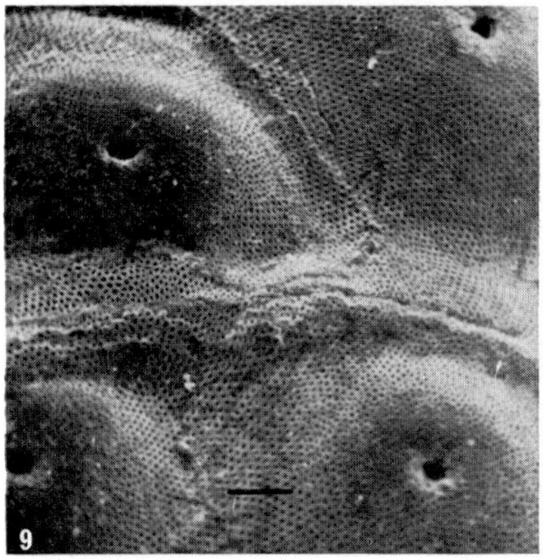

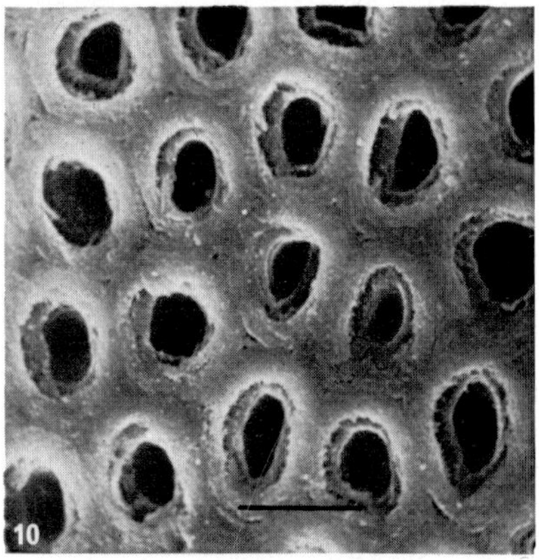

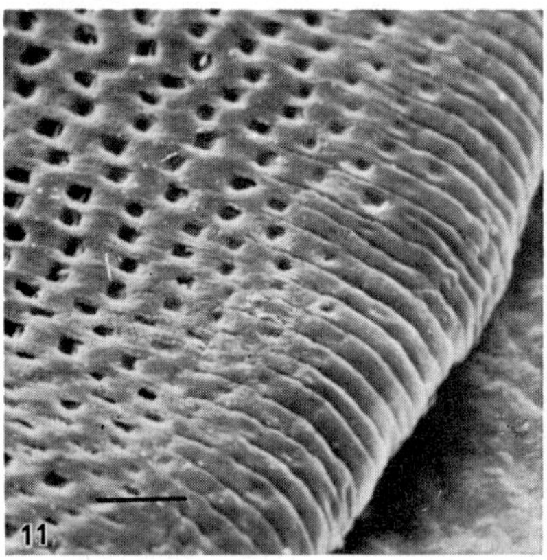

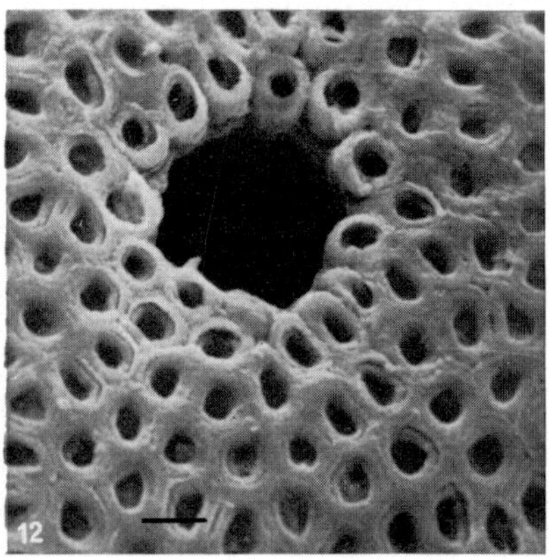

FIG. 7. *Dematolithon litorale*—side view of conceptacle with cup-shaped cells. Scale 10 μm.

FIG. 8. *D. litorale*—surface view of vegetative tissue with epiphytes and debris. Scale 10 μm.

FIG. 9. General view of *D. pustulatum* with conceptacles. Scale 100 μm.

FIG. 10. *D. pustulatum*—detail of cells on conceptacle. Scale 10 μm.

FIG. 11. *D. pustulatum*—edge of crust with groves and cells in rows. Scale 25 μm.

FIG. 12. *D. pustalatum*—detail of cells around conceptacle pore. Scale 10 μm.

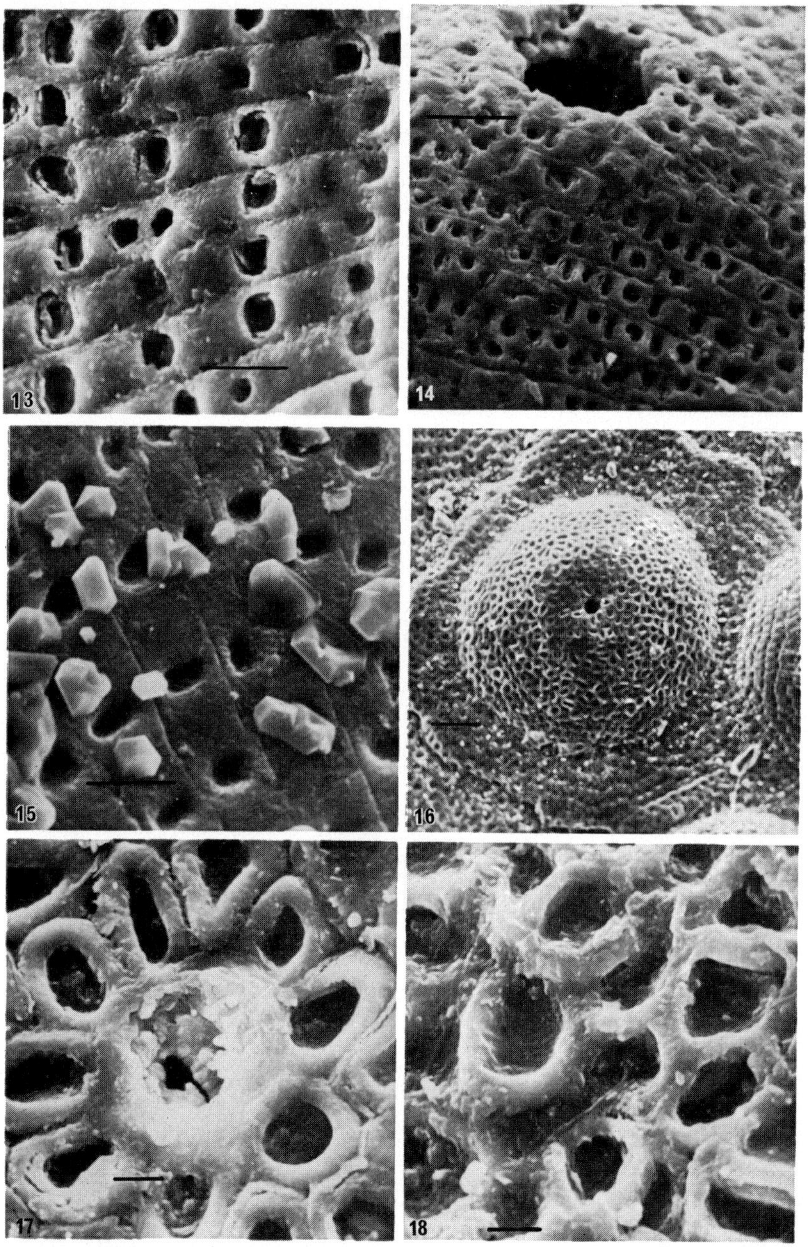

FIG. 13. *Fosliella limitata*—surface of crust with cells in rows separated by grooves. Scale 10 μm.

FIG. 14. *F. limitata*—surface of conceptacle. Scale 20 μm.

FIG. 15. *F. farinosa*—surface of crust with accumulated debris. Scale 10 μm.

FIG. 16. *F. farinosa*—general view of crust with conceptacle. Scale 50 μm.

FIG. 17. Detail of cells around conceptacle pore in *F. farinosa*. Scale 5 μm.

FIG. 18. Detail of cells on side of conceptacle. Scale 5 μm.

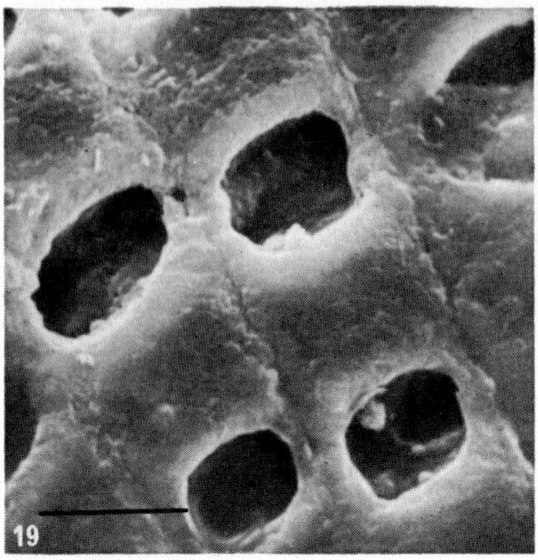
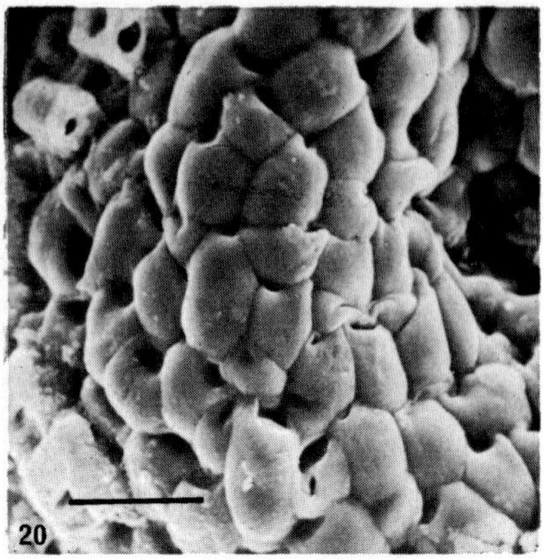
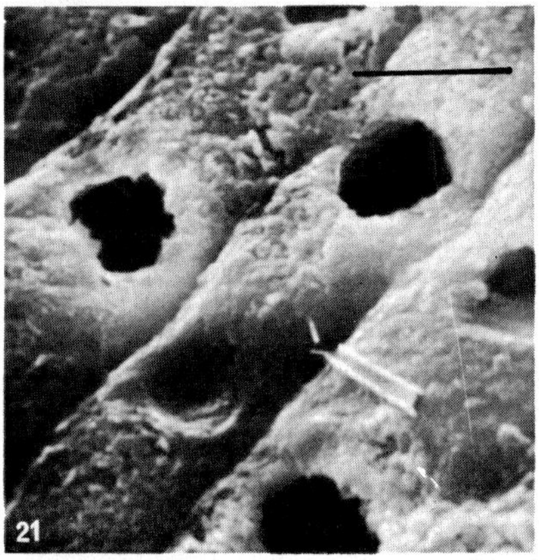
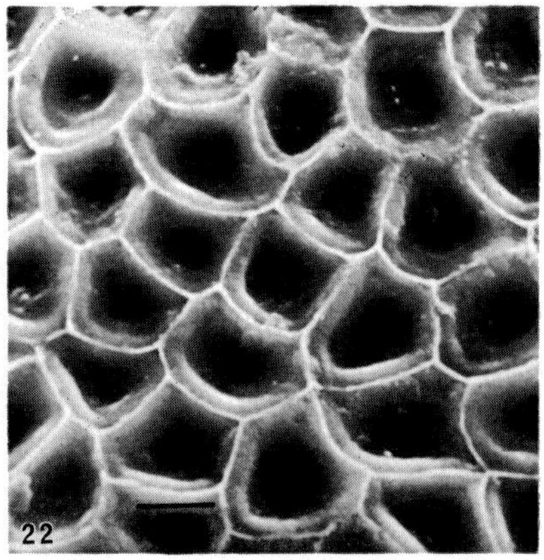
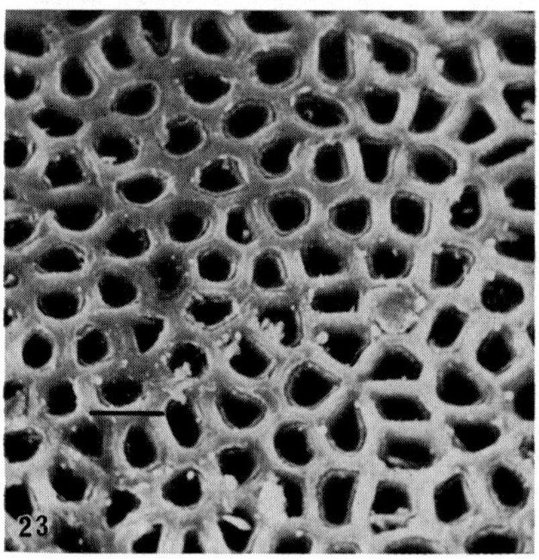
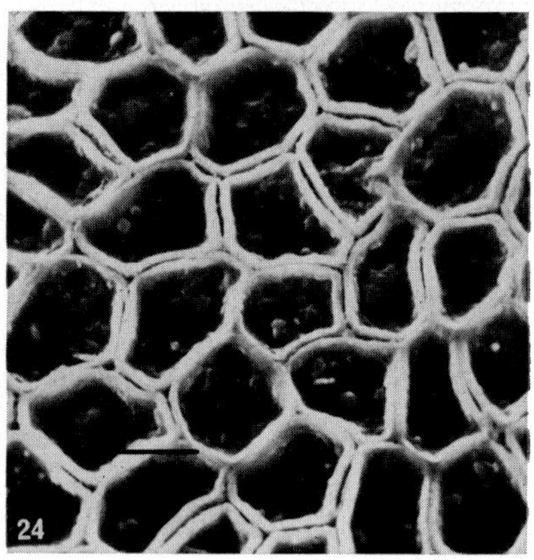

FIG. 19. Detail of surface of crust in *Fosliella minutula* with grooves separating rows of cells. Scale 5 μm.

FIG. 20. Top view of surface of conceptacle of *F. minutula* with cup-shaped cells. Scale 20 μm.

FIG. 21. Surface of *F. lejolisii*. Scale 5 μm.

FIG. 22. Surface of *Lithothamnium corallioides* with narrow ridge separating epithallial concavities. Scale 5 μm.

FIG. 23. Surface of *Lithothamnium glaciale*. Scale 10 μm.

FIG. 24. Surface of *Phymatolithon calcareum* with interconcavity matrix splitting between epithallial concavities. Scale 10 μm.

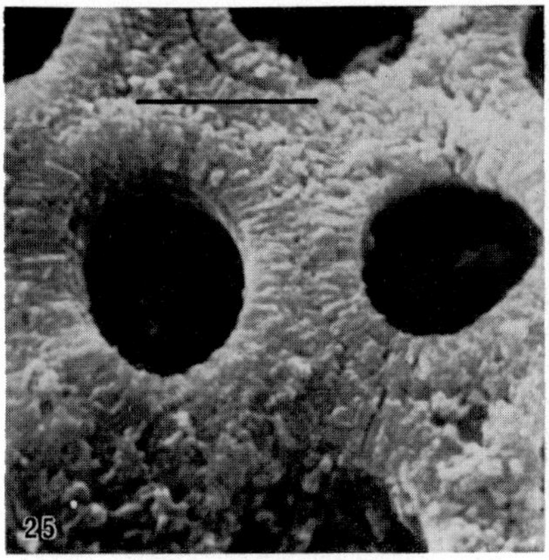

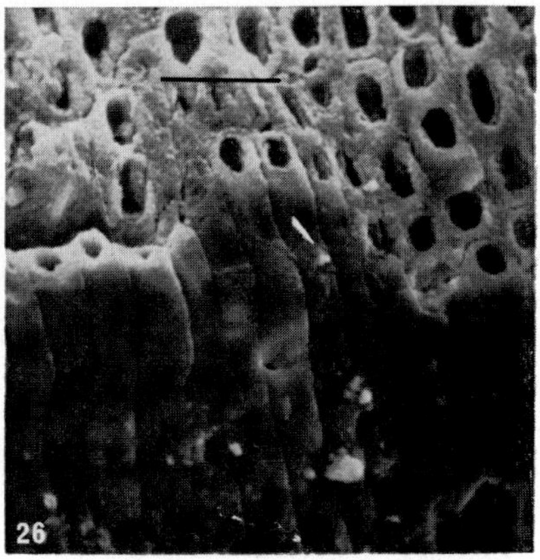

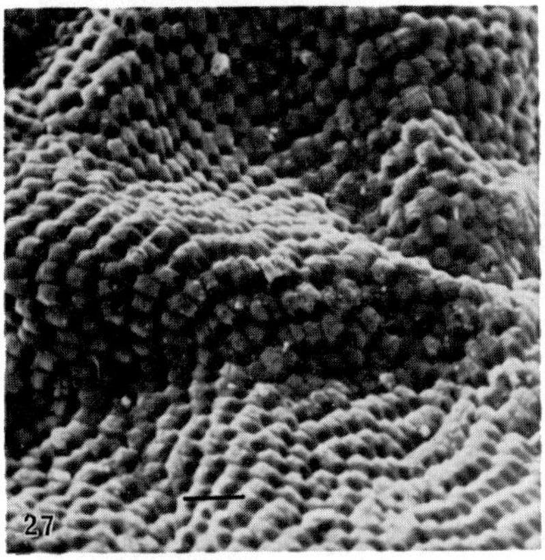

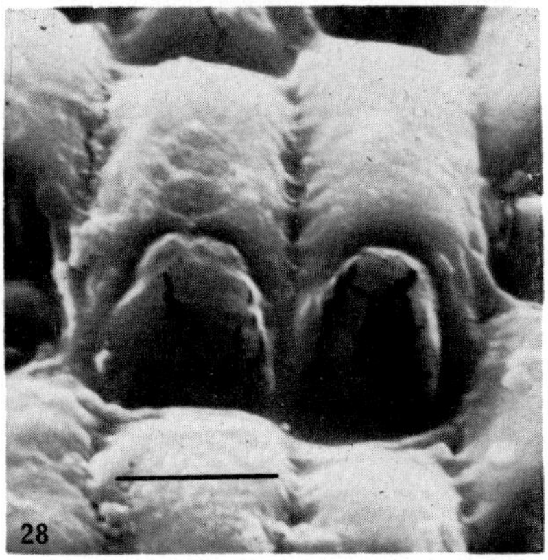

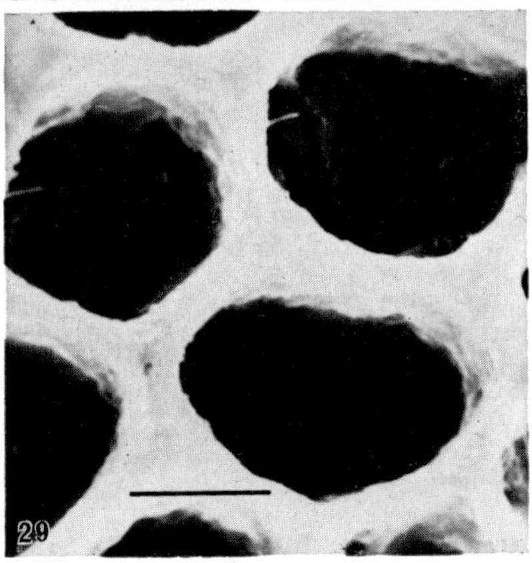

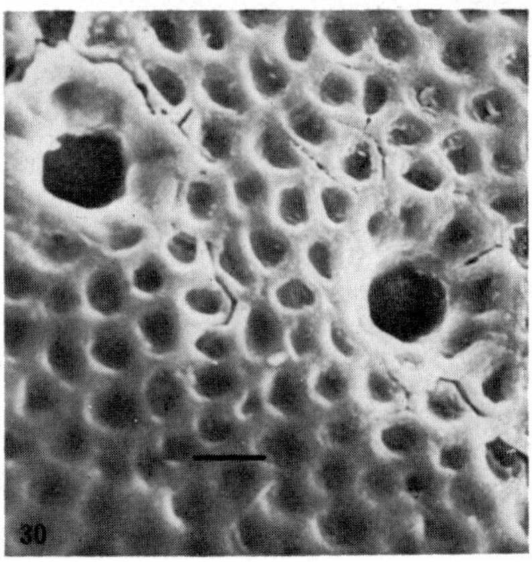

FIG. 25. *Lithophyllum incrustans*—detail of surface of crust with rosettes around epithallial concavities. Scale 5 μm.

FIG. 26. *L. incrustans*—vertical face of growing edge (darker region with grooves separating rows of cells). Scale 20 μm.

FIG. 27. *Melobesia membranacea*—general view of crust. Scale 25 μm.

FIG. 28. *M. membranacea*—detail of crust. Scale 5 μm.

FIG. 29. *Mesophyllum lichenoides*—detail of surface of crust. Scale 5 μm.

FIG. 30. *M. lichenoides*—surface of tetrasporangial conceptacle with pores. Scale 10 μm.

character appears useful in the generic delimitation of *Lithophyllum*. The fourth state is that shown by *Phymatolithon calcareum*, characterized by the splitting of the cell wall around the epithallial concavities to form two narrow ridges between concavities (Fig. 24).

2. *Surface Structure of Conceptacles*

To date eight species, in which both vegetative and conceptacle surfaces have been observed, have been examined with the SEM. In *Mesophyllum lichenoides*, *Fosliella limitata*, *Lithophyllum incrustans*, *Dermatolithon pustulatum* and *D. corallinae* there was little or no differentiation between the two surfaces. In the others examined (*Fosliella minutula*, *F. farinosa* and *D. litorale*), however, considerable differences were observed. Rather than being in the plane of the conceptacle surface, epithallial concavities and cell walls were orientated in the plane of the vegetative surface, with greater calcification on the outer wall. Thus the conceptacle surface appeared as a series of steps rather than an inclined plane (Figs 7, 20). In *Fosliella farinosa* (Figs 15–18) the conceptacle surface was less calcified than the vegetative cells (Fig. 15) and possessed epithallial concavities with irregular outlines. A second conceptacle type, also present in *F. farinosa*, showed the conceptacle surface as undifferentiated and similar to that of conceptacles of *F. limitata* (Fig. 14).

Until more species have been examined, the taxonomic usefulness of these patterns of differentiation is limited. It is possible, however, that they will be an aid in phylogenetic discussions, since presumably the conceptacle surface retains the primitive condition.

3. *Quantitative Characters*

Epithallial concavity diameter ranged from 3·9 to 15·4 μm (Table I). Although the values for all species taken together were roughly continuous, individual genera occupied only parts of this range. *Jania* spp. had the largest concavities with values ranging from 10 to 15 μm. *Fosliella* spp. clustered at the smaller end of the range with concavity sizes from 3·9 μm (*F. lejolisii*) to 6·1 μm (*F. farinosa*). In addition, it was possible to separate *Lithophyllum incrustans* and *Mesophyllum lichenoides* on the basis of this character.

There was less variation in the thickness of calcification separating the epithallial concavities. Values ranged from 1·7 to 6·0 μm. Two main groups were discernible; those with an interconcavity distance of less than 2·5 μm (*Jania* spp., *Corallina officinalis*, *Lithothamnium* spp, *Mesophyllum lichenoides*, *Phymatolithon calcareum*) and those greater than 3 μm (*Fosliella* spp., *Dermatolithon* spp., *Melobesia membranacea*, *Lithophyllum incrustans*).

Despite a correlation between epithallial concavity size and interconcavity distance (r = −0·62), there were two distinct clusters of points (Fig. 31). One group contained plants with medium to high values of interconcavity distance associated with low to medium values for diameter of epithallial concavities, whereas a second group contained species with low values for interconcavity distance and medium to high values for epithallial concavity diameter.

The proportion of the surface area occupied by epithallial concavities varied from 12 to 66%. Again two groups were apparent, those with values less than

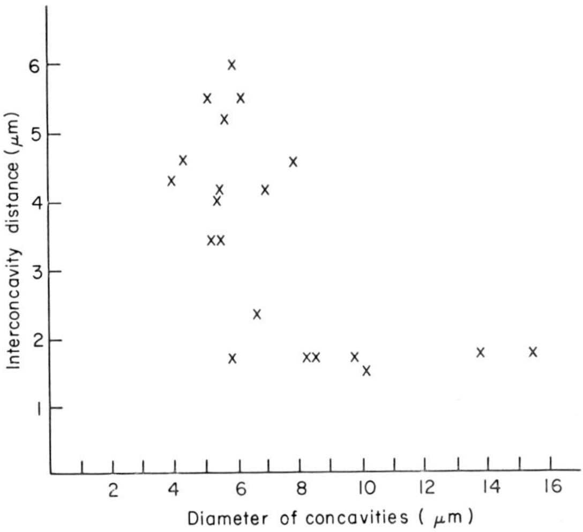

FIG. 31. The relationship between epithallial concavity size and distance between concavities for sixteen species of Corallinaceae. Data from Table I.

35% and those with values greater than 45%. When these values were plotted against corresponding measurements for interconcavity distance, similar clusters were formed (as in Fig. 31) except that the positions of *Lithothamnium glaciale* and *Lithophyllum incrustans* were exchanged.

When values for epithallial concavity size were plotted against the proportion of the surface area occupied by epithallial concavities, again two groups appeared (Fig. 32) with *Corallina*, *Jania* spp., *Mesophyllum lichenoides* and *Phymatolithon calcareum* clustered together and separated from the remaining species.

Thus these quantitative characters, especially in combination, appear to divide the Corallinaceae into two groups, with *Lithothamnium* spp. and *Lithophyllum incrustans* as possible intermediates. The lack of congruence, however, between a possible classification based on SEM characters and that based on traditional characters (see Johansen, 1976) makes this relatively simple division unlikely.

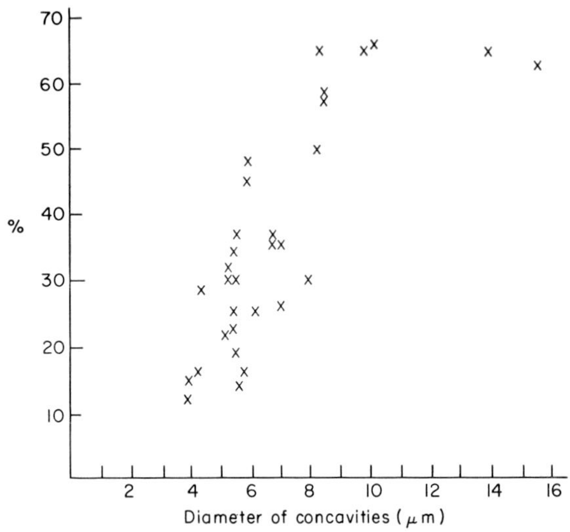

FIG. 32. The relationship between epithallial concavity size and percentage of surface area occupied by epithallial concavities for species of Corallinaceae. Data from Table I.

4. *Key to some Species of Corallinaceae*

1. Epithallial concavities in definite rows (at least near the edge) usually separated by a groove or ridge 2
1. Epithallial concavities not in rows nor with groove on upper surface . . 7
   2. Calcification heavier on one side of vegetative epithallial concavities . . . *Melobesia membrancea* (Figs 27, 28)
   2. Calcification of vegetative epithallial concavities symmetrical . . 3
3. Cells on conceptacle wall cup-shaped . . . *Dermatolithon litorale*, *Fosliella minutula* (Figs 7, 8, 19, 20)
3. Cells on conceptacle wall not cup-shaped. 4
   4. Groove (or ridge) absent in older parts of thallus and cell rows obscured at high magnifications 5
   4. Groove definitely present and cells in distinct rows 6
5. Groove present at edge of crust . . . *Dermatolithon pustulatum* (Figs 9–12)
5. Ridge present at edge of crust . . . *D. corallinae* (Figs 5, 6)
   6. Epithallial concavities 19–34% of surface area and greater than 5 μm diameter . . . *Fosliella farinosa* (Figs 15–18)
   6. Epithallial concavities 12–15% of surface area and less than 4 μm diameter . . . *F. lejolisii* (Fig. 21)
7. Epithallial concavities exposed, plant surface appearing pored . . . 8

7. Epithallial concavities covered, plant surface appearing ridged . . . 12
   8. Calcium carbonate differentiated to form rosettes around epithallial concavities or narrowing to a fine ridge 9
   8. Calcium carbonate walls not as above 10
9. Cell wall deposits in rosettes around concavities . . . *Lithophyllum incrustans* (Figs 25, 26)
9. Cell wall narrowing to a fine ridge . . . *Lithothamnium corallioides* (Fig. 22)
   10. Interconcavity matrix split, forming two ridges between concavities . . . *Phymatolithon calcareum* (Fig. 24)
   10. Interconcavity matrix not split 11
11. Epithallial concavities ≈ 8 μm diameter and greater than 50% of surface composed of concavities . . . *Mesophyllum lichenoides* (Figs 29 30)
11. Epithallial concavities ≈ 6 μm diameter and less than 50% of surface composed of concavities . . . *Lithothamnium glaciale* (Fig. 23)
   12. Epithallial concavities 10–15 μm diameter and trichocytes (?) often present . . . *Jania corniculata, J. rubens* (Figs 2–4)
   12. Epithallial concavities smaller than above and trichocytes absent . . . *Corallina officinalis* (Fig. 1).

The usefulness of the SEM to the taxonomy of the Corallinaceae is demonstrated by the capacity of the observed characters to distinguish the species under investigation. It is, however, difficult to distinguish between certain pairs of species (see above). This can be partly attributed to the accumulation of epiphytes and debris on the surface (Fig. 8), such that epithallial concavities and the cell walls then lack definition. The more species that are investigated, however, the greater the likelihood of intermediates that will "cloud over" currently obvious discontinuities. While it is possible to characterize all or most of the species in the present investigation, this may become increasingly difficult.

ACROCHAETIACEAE

Since they lack a hard cell wall, one would not suspect that the bulk of the Florideae presents suitable material for scanning electron microscopy. This might explain the paucity of published accounts. The technique has, however, produced a wealth of information on the ultrastructure of other plant groups, much of it dealing with delicate structures (see references in Cole and Behnke, 1975). As part of a general study of the Acrochaetiaceae, certain cultured isolates were subjected to scanning electron microscopy. This group is probably the most confused of all the British Rhodophyta and is in need of new characters against which proposed generic delimitations in the family (see Woelkerling, 1971; Dixon and Irvine, 1977, for reviews) can be tested. In addition, surface characters may be of use in separating closely related species and therefore in species characterization.

Plants were fixed in standard formalin—ethylene glycol–sea-water preservative for 16–24 hours. After fixation, plants were transferred to a solution containing sea-water, acetone and distilled water (1 : 1 : 2) for two hours and then dehydrated in an acetone series. Plants were dried in a critical point apparatus, coated with gold–palladium and examined in the SEM.

Surface structures in the Acrochaetiaceae were not as varied as those in the Corallinaceae. This can be attributed to the lack of hard surfaces and to the relative morphological simplicity limiting the variation. Despite these inherent limitations, certain surface characters were recognized.

Vegetative cells of *Audouinella secundata* were heavily striated (Fig. 33). In these cells, individual wrinkles in the cell surface ran 1–33% of the cell length, although there were no well-defined starting or terminating points. This condition was consistent in four isolates of the species. It should be emphasized that this surface condition was not related to collapse of the cell, since, as in views through the light microscope, cells were slightly contracted at the nodes. In addition, the wrinkles often ran from one cell to another without indication of distortion. Transmission electron microscopy (TEM) should be utilized to confirm these striations; this has been done with *Sphacelaria tribuloides* (Phaeophyta). Transverse sections of thalli in the TEM (Galatis *et al.*, 1977) confirm wrinkles seen by Newell (personal communication) with the SEM. Monosporangia of *A. secundata*, however, showed no sign of these striations and were smooth (Fig. 34).

A second type of cell wall surface was exhibited by *A. daviesii*, *A. thuretii*, *A. dasyae* (Collins) Woelkerling, *Acrochaetium densum* (Drew) Papenfuss (= *Audouinella*) and *Audouinella* sp. (a) (Figs 35, 36). In these species, longitudinal ridges were present, although they were less prominent than in *A. secundata*. In addition, there was considerable variation in the extent to which this character was exhibited. In some samples the striations were present on most of the cells, whereas in other samples (*A. daviesii*) most cells were smooth with only a few cells near the apices displaying striations. Monosporangia of *A. thuretii* showed no surface detail (Fig. 38), whereas those of *Audouinella* sp. (a) were heavily wrinkled (Fig. 37). Monosporangia of *A. daviesii* were similar to those of *Audouinella* sp. (a) except that the wrinkles were less prominent.

Vegetative cells of *A. floridula* and *Audouinella* sp. (b) had no surface structures and were smooth (Fig. 39). This is contrasted with the surfaces of the tetrasporangia in *Audouinella* sp. (b) which were slightly wrinkled (Figs 40–41). It is of note that both of these species produce only tetrasporangia and that the tetrasporophyte of *Nemalion helminthoides* also appears smooth in the SEM.

Audouinella parvula was the only species in which vegetative and reproductive

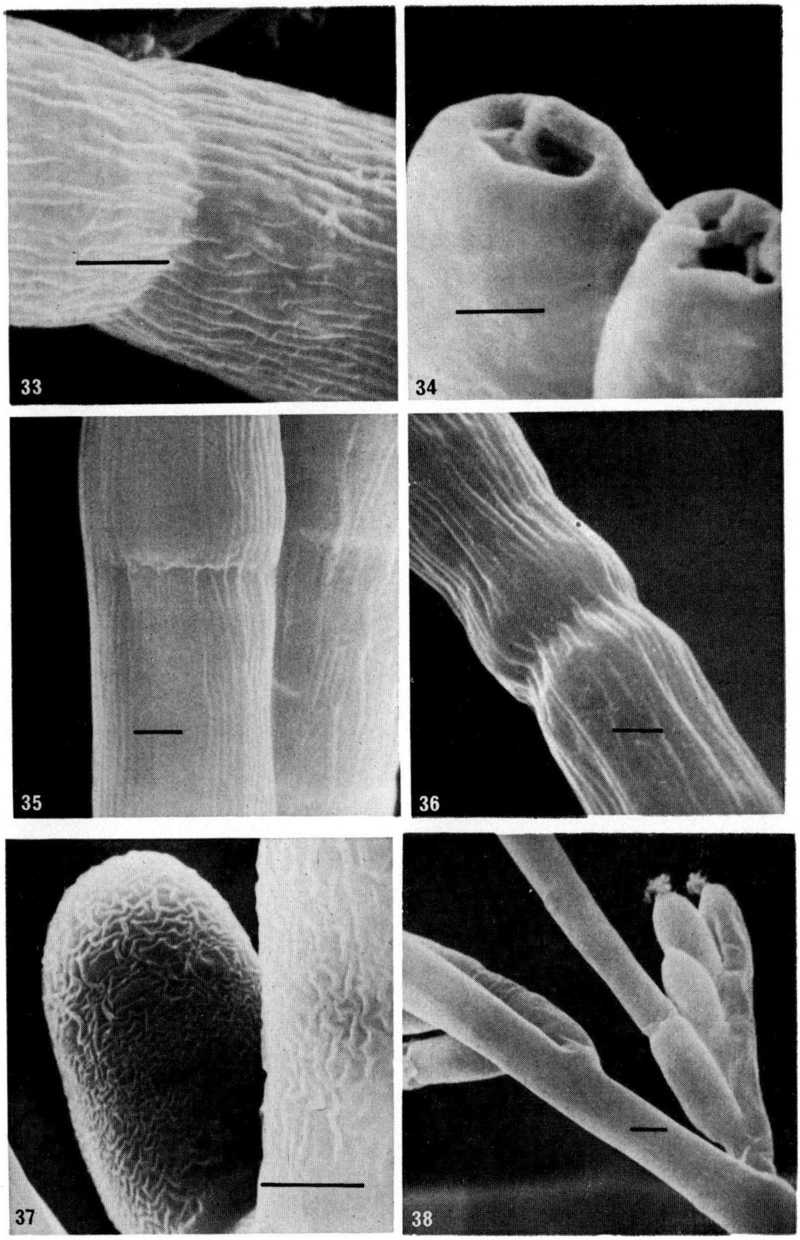

FIG. 33. *Audouinella secundata*—surface of vegetative cells with longitudinal wrinkles. Scale 2 μm.

FIG. 34. *A. secundata*—surface of monosporangia with sporangial pores. Scale 2 μm.

FIG. 35. *A. thuretii*—surface of vegetative cells with longitudinal striations. Scale 2 μm.

FIG. 36. *A. dasyae*—surface of vegetative cells with striations. Scale 2 μm.

FIG. 37. *Audouinella* sp. (a)—surface of monosporangia. Scale 5 μm.

FIG. 38. *A. thuretii*—general view of plant with vegetative cells and monosporangia. Scale 5 μm.

(monosporangial) surfaces were observed to be similar (Fig. 42). In this species, the surface was slightly wrinkled, although the wrinkles were of a nature different to those of either *A. secundata* or *A. thuretii*. The wrinkles in *A. parvula*

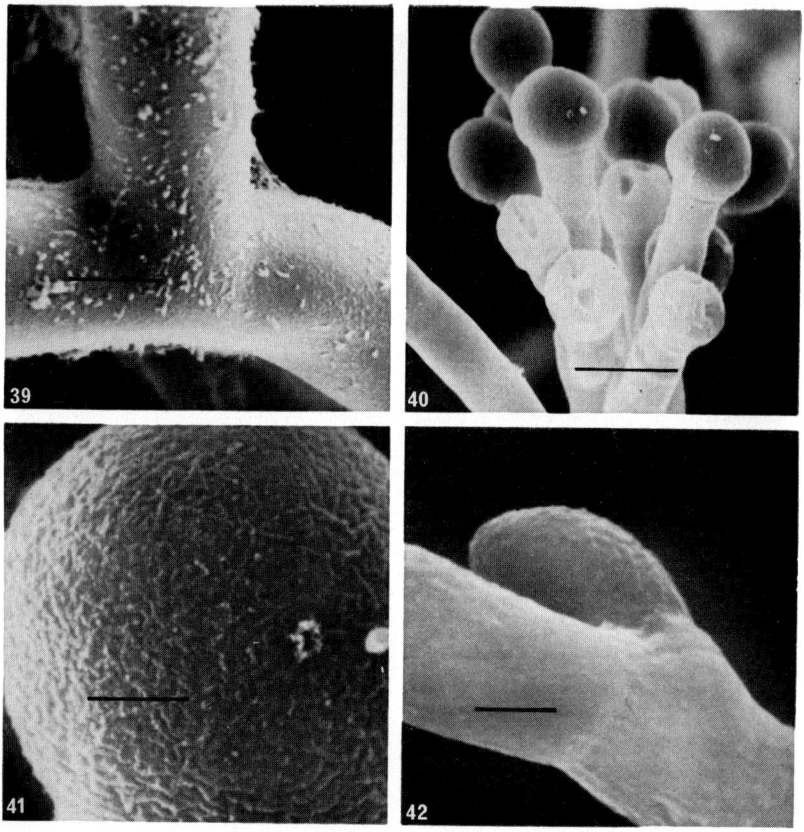

FIG. 39. *A. floridula*—surface of vegetative cells. Scale 10 μm.
FIG. 40. *Audouinella* sp. (b)—tetrasporangial cluster. Scale 10 μm.
FIG. 41. *Audouinella* sp. (b)—detail of tetrasporangial surface with wrinkles. Scale 2 μm.
FIG. 42. *A. parvula*—vegetative cells with monosporangium. Scale 2 μm.

were poorly defined, short and not necessarily running parallel to the long axis of the cell.

Although too few species have been examined in the Acrochaetiaceae even to suggest tentative relationships on the basis of surface structure, variation in surface structure does occur within the family and for a given species the structure is consistent from one sample to the next. As more species are investigated and patterns of variation are understood, a basis for assessing relationships may be possible.

POTENTIAL FOR FURTHER RESEARCH

Although at present it would be premature to suggest scanning electron microscopy as a panacea for problems in red algal taxonomy, the technique should provide insight into the structure of these organisms and the relationships between them. Applied to the Corallinaceae, scanning electron microscopy avoids problems resulting from the decalcification and sectioning of plants. In addition, the successful use of the SEM in the Acrochaetiaceae indicates application in other filamentous groups, notably the Ceramiaceae.

A major drawback of the technique as used in this study for the Corallinaceae is the impossibility of observing the structure and relationships of uncalcified cells, i.e. reproductive cells within conceptacles. Use of properly fixed living material, however, should overcome this problem and permit anatomical studies of the larger multiaxial Florideae. This possibility is currently being explored with particular reference to members of the Gigartinales.

The Corallinaceae has an advantage over most other groups of algae with regard to the taxonomic application of SEM techniques. In relation to surface structure, there is little or no deterioration of cell walls in well-kept herbarium specimens. This means that where type specimens are available, they can be realistically compared with recent collections.

The diversity and complexity of structure so far found in red algae does not approach that found in the seeds or pollen of flowering plants. Because of this, and the general absence of truly comparable surfaces among the orders of the Florideae, scanning electron microscopy will most likely not be valuable as a taxonomic aid at the familial level or above. On the other hand, the range of structures seen in this preliminary study of the Corallinaceae and Acrochaetiaceae suggests that scanning electron microscopy has potential as a tool for specific and generic delimitation.

ACKNOWLEDGEMENTS

I wish to thank the following people who contributed to this work: Mr C. J. Veltkamp for numerous hours on the SEM and advice on preparation techniques; Ms Y. Chamberlain for the use of specimens and discussions on interpretation of micrographs of the Corallinaceae; Mr W. F. Farnham for the use of specimens; Dr J. Rueness for providing a culture of *Audouinella* sp. (b); Mr H. Stegenga for providing cultures of *A. dasyae* and *A. densa*; and curators of the herbaria at the British Museum (Natural History) and Liverpool Museum for the use of material. I am also grateful to the Royal Commission for the Exhibition of 1851 for financial support during the course of this work.

REFERENCES

Bailey, A. and Bisalputra, T. (1970). A preliminary account of the application of thin-sectioning, freeze-etching, and scanning electron microscopy to the study of coralline algae. *Phycologia* **10**, 83–101.

Bailey, G. P., Rezak, R. and Cox, E. R. (1976a). A revision of generic concepts of living members in the subfamily Acetabularieae (Dasycladaceae, Dasycladales) based on scanning electron microscopy. *Phycologia* **15**, 7–15.

Bailey, G. P., Cox, E. R. and Rezak, R. (1976b). Morphological variability in the marine green alga *Acetabularia crenulata* Lamouroux (Dasycladaceae, Dasycladales). *Phycologia* **15**, 19–23.

Berger, S., Sandakhchiev, L. and Schweiger, H. G. (1974). Fine structural and biochemical markers of Dasycladaceae. *J. Microscopie* **19**, 89–104.

Borowitzka, M. A., Larkum, A. W. D. and Nockolds, C. E. (1974). A scanning electron microscope study of the structure and organization of the calcium carbonate deposits of algae. *Phycologia* **13** 195–203.

Cole, G. T. and Behnke, H.-D. (1975). Electron microscopy and plant systematics. *Taxon* **24**, 3–15.

Conway, E. and Cole, K. (1977). Studies in the Bangiaceae: structure and reproduction of the conchocelis of *Porphyra* and *Bangia* in culture (Bangiales Rhodophyceae). *Phycologia* **16**, 205–216.

Dixon, P. S. and Irvine, L. M. (1977). "Seaweeds of the British Isles", Vol. 1, Rhodophyta. Part 1, Introduction, Nemaliales, Gigartinales. British Museum (Natural History), London.

Galatis, B., Katsaros and Mitrakos, K. (1977). Fine structure of vegetative cells of *Sphacelaria tribuloides* Menegh. (Phaeophyceae, Sphacelariales) with special reference to some unusual proliferations of the plasmalemma. *Phycologia* **16**, 139–151.

Johansen, H. W. (1976). Current status of generic concepts in coralline algae (Rhodophyta). *Phycologia* **15**, 221–244.

Moss, B. (1974). Attachment and germination of the zygotes of *Pelvetia canaliculata* (L.) Dcne. et Thur. (Phaeophyceae, Fucales). *Phycologia* **13**, 317–322.

Moss, B. (1975). Attachment of zygotes and germlings of *Ascophyllum nodosum* (L.) Le Jol. (Phaeophyceae, Fucales). *Phycologia* **14**, 75–80.

Moss, B. (1977). Attaching mechanisms of some brown seaweeds. *Br. phycol. J.* **12**, 121.

Parke, M. and Dixon P. S. (1968). Check-list of British marine algae—second revision *J. mar. biol. Ass. U.K.* **48**, 783–832.

Parke, M. and Dixon P. S. (1976). Check-list of British marine algae—third revision. *J. mar. biol. Ass. U.K.* **56**, 527–594.

Schweiger, H. G. and Berger, S. (1975). *Acetabularia haemmerlingi*, a new species. *Nova Hedwigia* **26**, 33–43.

Schweiger, H. G., Berger, S., Kloppstech, K., Apel, K. and Schweiger, M. (1974). Some fine structural and biochemical features of *Acetabularia major* (Chlorophyta, Dasycladaceae) grown in the laboratory. *Phycologia* **13**, 11–20.

Woelkerling, W. J. (1971). Morphology and taxonomy of the *Audouinella* complex (Rhodophyta) in Southern Australia. *Austr. J. Bot.* (Suppl. **1**), 1–91.

10 | Investigation of Taxonomic Relationships amongst Epiphytic, Crustose Corallinaceae

Y. M. CHAMBERLAIN

Department of Biological Sciences, Portsmouth Polytechnic, The Marine Laboratory, Hayling Island, England

Abstract: Groups of epiphytes occur in both the complex and simple crustose Corallinaceae; those being investigated are of the simple crustose type, belonging to the genera *Melobesia*, *Fosliella* and *Dermatolithon*. The species concerned grow on other algae and on marine angiosperms in littoral and sublittoral habitats. The small pink crusts bear a superficial resemblance to one another and all have a rather simple vegetative structure; the thallus may be as little as two cells deep and is rarely more than twenty cells deep. Although they are vegetatively simple the reproductive structures show no corresponding simplification. It is possible that these species represent neotenic or regressive forms.

Cabioch has classified the Corallinaceae into five subfamilies on the basis of the type of secondary lateral anastomosis present in the vegetative thallus and on the type of roof structure found in the sporangial conceptacles. *Melobesia*, *Fosliella* and *Dermatolithon* belong to the Lithothamnioideae, Corallinoideae and Lithophylloideae respectively so they are thus, within the Corallinaceae, taxonomically diverse.

Techniques currently being used to investigate coralline algae generally, promise to be of use in elucidating taxonomic problems in this group; they include culture studies, regular shore collecting and scanning electron microscopy.

INTRODUCTION

The Corallinaceae is a family in the order Cryptonemiales characterized by the calcification of the cell walls with calcium carbonate in the form of calcite, which gives the plants a hard texture. Littler (1972) and Adey and Macintyre (1973) have usefully summarized historical and current knowledge of the crustose Corallinaceae; Johansen (1974) has similarly reviewed the articulated genera. Littler was mainly concerned with the epilithic crustose species and

Systematics Association Special Volume No. 10, "Modern Approaches to the Taxonomy of Red and Brown Algae", edited by D. E. G. Irvine and J. H. Price, 1978, pp. 223–246, Academic Press, London and New York.

particularly with new methods of investigating aspects such as their distribution and productivity; he did not discuss modern knowledge about the epiphytic species. A number of excellent works of a descriptive taxonomic nature which deal wholly or in part with the epiphytic crustose corallines are available and these will be listed later. Johansen includes the epiphytic species in his review of the genera of the Corallinaceae (1976). In the present paper it is proposed first to describe the main vegetative and reproductive features of the crustose corallines with special reference to one group of epiphytic genera. The suprageneric classification of the family will then be discussed in order to relate the genera concerned both to each other and to the Corallinaceae as a whole. Finally an arbitary selection of topics will be examined which have been chosen to illustrate some of the modern methods of investigating the Corallinaceae and, where possible, these methods and their results will be related to work currently in progress on British epiphytic species.

THE VEGETATIVE AND REPRODUCTIVE STRUCTURES OF *MELOBESIA*, *FOSLIELLA* AND *DERMATOLITHON*

The thallus of the Corallinaceae shows three levels of morphological complexity; briefly, these are the simple crust, the complex crust and the erect articulated form. The genera which will be discussed are *Melobesia* Lamour., *Fosliella* Howe in Britton and Millspaugh and *Dermatolithon* Fosl. (Figs 1–4). The epiphytic species of these genera comprise simple crusts which grow on other algae and on marine phanerogams. Epiphytic species also occur in complex crustose genera such as *Mesophyllum* Lemione, *Clathromorphum* Foslie and *Neopolyporolithon* Adey and Johansen, but as Mason (1953) and Adey and Johansen (1972) have investigated this group in great detail, and as only one species—*Mesophyllum lichenoides* (L.) Lemoine—occurs in Britain the group will not be considered further in this paper.

In common with all Corallinaceae, the thallus of the epiphytic crustose genera is fundamentally a polysiphonous, filamentous structure with the filaments organized into three cell layers. When a spore of these genera germinates it produces a crust composed of radiating filaments (Fig. 25) the successive cells of which are joined by pit-connections. The terminal cells of the filaments are meristematic and form a peripheral apical meristem which increases the area of

---

FIG. 1. *Dermatolithon litorale* on *Gelidium*. ×12.

FIG. 2. *Fosliella farinosa* on *Palmaria palmata*. ×2.

FIG. 3. *Fosliella limitata* and *F. lejolisii* on *Zostera marina*. ×8.

FIG. 4. *Melobesia membranacea* on *Chondrus crispus*. ×19.

Figs 1–4. Captions on opposite page.

the crust throughout its life. Immediately after differentiation, a filament cell cuts off a small cell on the upper surface of its distal end (Figs 5–8, 25). This is the epithallial cell and it does not divide again. In some species the vegetative crust remains two cells thick; in most species, however, growth in depth occurs. Since the epithallial cell does not divide, its subtending cell becomes meris-

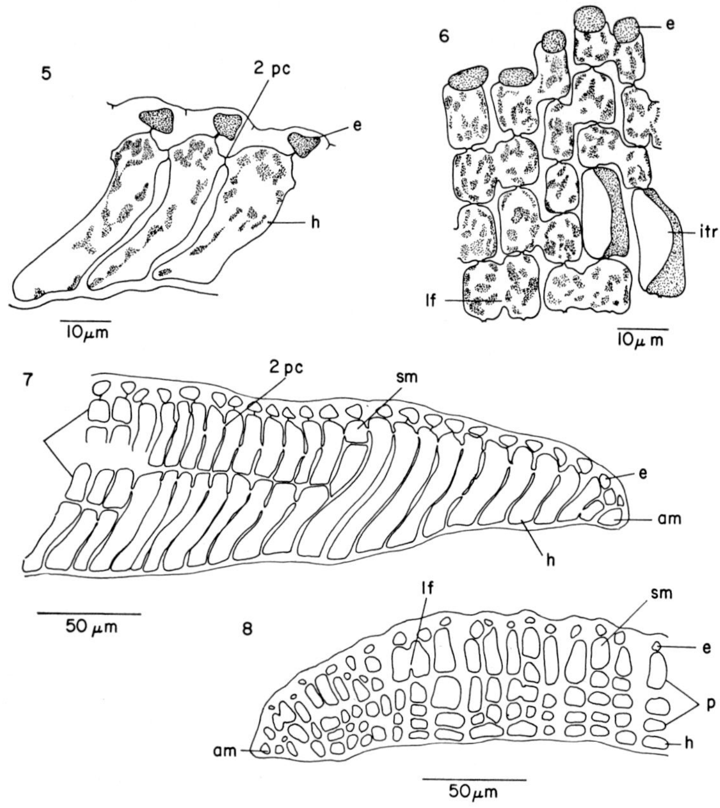

FIG. 5. V.S. *Dermatolithon litorale* crust to show secondary pit-connections.
FIG. 6. Surface view of decalcified crust of *Fosliella limitata* to show secondary cell fusions.
FIG. 7. Vertical section of crust of *Dermatolithon pustulatum*.
FIG. 8. Vertical section of crust of *Fosliella* sp.

Key—all Figs except for Fig. 16. e = epithallium, h = hypothallium, p = perithallium, lf = lateral fusion, 2pc = secondary pit-connection, itr = intercalary trichocyte, ttr = terminal trichocyte, am = apical meristem, sm = sub-apical meristem, pl = sporangial plug, t = trichogyne, sc = spermatangial conceptacle, c = carposporangium, cg = carpogonium, hy = hypogynous cell, sp = supporting cell, s = sterile cell, p = procarp, gf = gonimoblast filament, fc = fusion cell.

tematic and forms a subapical meristem, a unique feature of the Corallinaceae. This subapical meristem produces vertical filaments of cells (Figs 7, 8) which form the middle layer, or perithallium, of the thallus, while the monostromatic basal layer is called the hypothallium. In *Fosliella* and *Melobesia* (Fig. 8) the perithallial and hypothallial cells are irregular or more or less square in vertical section while in *Dermatolithon* (Fig. 7) the perithallial cells are often vertically elongated and the hypothallium has a very characteristic, sinuous appearance. When present, the perithallium becomes from one to about twenty cells deep depending on the species. The thallus of complex crusts is built up by continued growth of the hypothallium and perithallium; in most species the epithallium remains monostromatic and is composed of small cells but in some species the epithallial cells are similar in size to the perithallial cells, while other species have a polystromatic epithallium of cells formed successively towards the outer surface by the subapical meristem. Soon after they have been initiated, many of the crust cells become joined to cells of adjacent filaments by means of secondary, lateral anastomoses. In *Melobesia* and *Fosliella* these take the form of cell fusions (Figs 6, 8) whilst in *Dermatolithon* secondary pit-connections are formed (Figs 5, 7).

A feature of many members of the Corallinaceae is the presence of trichocytes or hair-bearing cells in the thallus. *Dermatolithon* has occasional trichocytes, they are absent in *Melobesia* but are a characteristic feature of *Fosliella*. Two types of trichocyte are present in *Fosliella*. The terminal trichocyte (Fig. 24) is initiated at the peripheral meristem and thus terminates a filament; it becomes buried in the thallus as adjacent filaments continue to grow round it. The intercalary trichocyte (Fig. 6) is formed by the radial division of an already differentiated crust cell. Neither trichocyte bears an epithallial cell. Terminal trichocytes are found in *Fosliella farinosa* (Lamour.) Howe, intercalary trichocytes are found in several species including *F. limitata* (Fosl.) Ganesan and *F. lejolisii* (Rosanoff) Howe. Trichocytes bear long, thin-walled hairs (Fig. 24), and although these seem to be shed quite quickly, the trichocyte itself remains apparent in the thallus. The presence or absence of trichocytes in *Fosliella* has long been a matter for contention among phycologists. Foslie, whose work on coralline algae in the last part of the nineteenth century and until his death in 1909 formed the foundation for modern studies of the group, changed his mind many times on this subject (summarized in Mason 1953, p. 334) but in his last publication (1909, p. 56) he decided that two genera were involved; he called them *Melobesia* (with trichocytes) and *Heteroderma* (without trichocytes). Since then nomenclatural research has established that the name *Melobesia* must be used for the lithothamnioid genus having *M. membranacea* (Esper) Lamour. as its type species, and because of this Howe (1920) erected the genus *Fosliella* for the trichocyte-bearing species. However, it has been found that the presence of trichocytes is very variable, even in species which unquestionably do bear them, and seems to depend on such factors as season of the year and geographical location. It seems that extensive observations of any one species throughout the seasons and throughout its geographical range are needed before it can be proved never to bear trichocytes. In Britain it has been the custom to include all the species under *Fosliella* (Parke and Dixon, 1976) whether trichocytes have been observed in them or not; this practice will be followed in this paper unless authors who use both *Fosliella* and *Heteroderma* are being quoted directly.

In the Corallinaceae the reproductive bodies are borne in conceptacles which are roofed chambers formed by the activity of vegetative cells. The various reproductive cells generally develop on the floor of the conceptacle, but in some species spermatia grow on all the surfaces. Four types of conceptacle are found in the genera under discussion (Figs 1–4, 9–14). These contain tetrasporangia, bisporangia, spermatia or carpogonia, with carposporangia developing in the carpogonial conceptacle after fertilization. The male conceptacles are usually small (not often exceeding 100 μm diameter) and do not normally protrude above the thallus surface. The spermatia are released through a central ostiole in the roof, which may be elongated into a spout (Fig. 13) in *Fosliella* and *Dermatolithon*. The other three kinds of conceptacle are domed structures which are to varying degrees prominent on the thallus surface, the diameter ranging up to 500 μm. In the female conceptacle, procarps (Fig. 13) comprising a supporting cell, one or two carpogonia and sometimes a sterile cell, are borne in the centre of the floor. The basic procarp structure is shown diagrammatically in Fig. 15 Suneson (1937) showed that the basic procarp becomes modified in different genera; he regarded the most complex structure as primitive and the simpler ones as regressive. *Melobesia* shows a very simplified structure and *Dermatolithon* a partially modified one, while in *Fosliella* it is very variable. Long trichogynes from the carpogonia protrude through the central ostiole (Fig. 13) and the spermatial contents enter through these trichogynes to fertilize the carpogonium. When at least one carpogonium has been fertilized the zygotic nucleus passes into the supporting cell and a number of these cells then fuse to form a disc-like fusion cell (Fig. 14). From this cell, usually from the periphery only but from the entire upper surface in some species, gonimoblast filaments develop, the terminal cells of which form a succession of single-celled carposporangia. The carpospores from these are released through the central ostiole.

In the Corallinaceae almost all tetrasporangia are zonately divided and contain four haploid tetraspores. The most usual type of bisporangium, and the

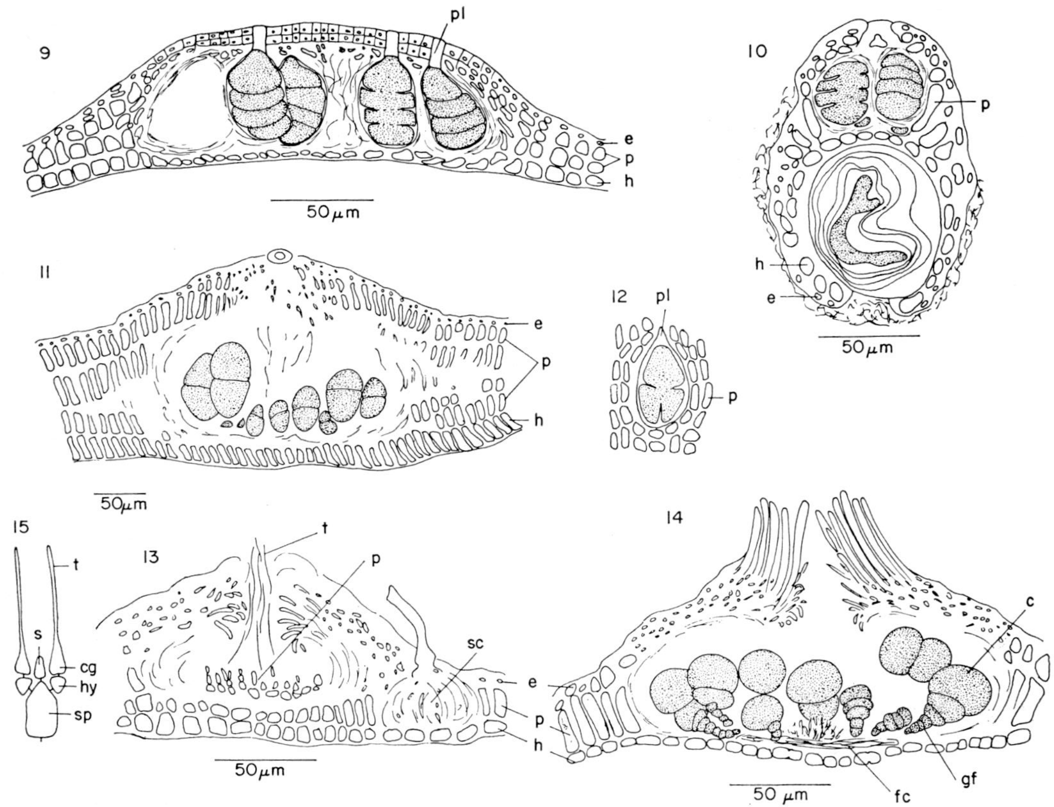

FIG. 9. Vertical section of tetrasporangial conceptacle of *Melobesia membranacea*.

FIG. 10. Vertical section of tetrasporangial conceptacle of *Fosliella minutula* on *Cladophora rupestris*.

FIG. 11. V.S. of bisporangial conceptacle of *Dermatolithon pustulatum*.

FIG. 12. Diagrammatic v.s. of cruciately divided tetrasporangium of *Sporolithon*.

FIG. 13. V.S. of monoecious gametangial crust of *Fosliella limitata*.

FIG. 14. V.S. of carposporangial conceptacle of *Fosliella limitata*.

FIG. 15. Diagrammatic representation of basic procarp structure (redrawn after Suneson).

only one found so far in British species, contains two bispores each with a diploid nucleus. Two types of sporangial conceptacle occur. In *Melobesia* (Fig. 9) the roof of the tetrasporangial conceptacle is multiperforate, but in *Fosliella* and *Dermatolithon* (Figs 10, 11) both bisporangial and tetrasporangial conceptacles have a single ostiole.

SUPRAGENERIC TAXONOMY OF THE CORALLINACEAE

Traditionally the family has been divided into the crustose and articulated genera—the Melobesioideae and Corallinoideae respectively. Although these terms used in this way have persisted until recently (e.g. Masaki, 1968), it was in fact appreciated at quite an early stage that characters other than the presence or absence of uncalcified joints (genicula) which distinguish the articulated genera were probably of more fundamental significance as far as the systematics of the family was concerned (Littler, 1972, gives a comprehensive summary of the various ways in which the family has been classified). Cabioch (1970 and 1972) and Adey and Johansen (1972, and summarized in Johansen, 1976) have recently drawn up schemes of classification which, although differing in several respects as will be explained, accept two basic characters as being of prime importance. These characters are the nature of the secondary anastomoses in the vegetative thallus and the structure of the tetra- or bisporangial conceptales. Both have been mentioned in the previous section, but in view of the significance in the present context the description needs to be somewhat expanded.

1. *Secondary Anastomoses*

(*a*) *Lateral fusions.* The walls between two adjacent cells disintegrate and the cell contents fuse (Fig. 6) to form a coenocyte. According to Cabioch (1970, 1971) this type of fusion is always accompanied by nuclear activity, but the ultimate fate of the nuclei depends on the species since sometimes they remain separate, sometimes come to lie close together and sometimes actually fuse (Fig. 16A, C, D). This type of anastomosis is seen in subfamilies Sporolithoideae (Johansen) Cabioch, Lithothamnioideae Foslie and Corallinoideae (Aresch.) Foslie emend Cabioch.

(*b*) *Secondary pit-connections.* In this case the cells anastomose by means of a narrow connection in which a plate or plug develops; this is similar to that in the primary pit-connection found between successive filament cells as is normal in the Rhodophyta (Figs 5, 7). Cabioch found that this type of anastomosing is not accompanied by nuclear activity (Fig. 16A, B); it is seen in subfamilies Sporolithoideae and Lithophylloideae Setchell emend Cabioch.

2. The Nature of the Sporangial Structures

(*a*) The sporangium is born singly in the thallus; it develops a plug at its apex which penetrates the thallus surface and the ejection of this plug allows the discharge of the spore (Fig. 12). This single sporangium is found in the subfamily Sporolithoideae.

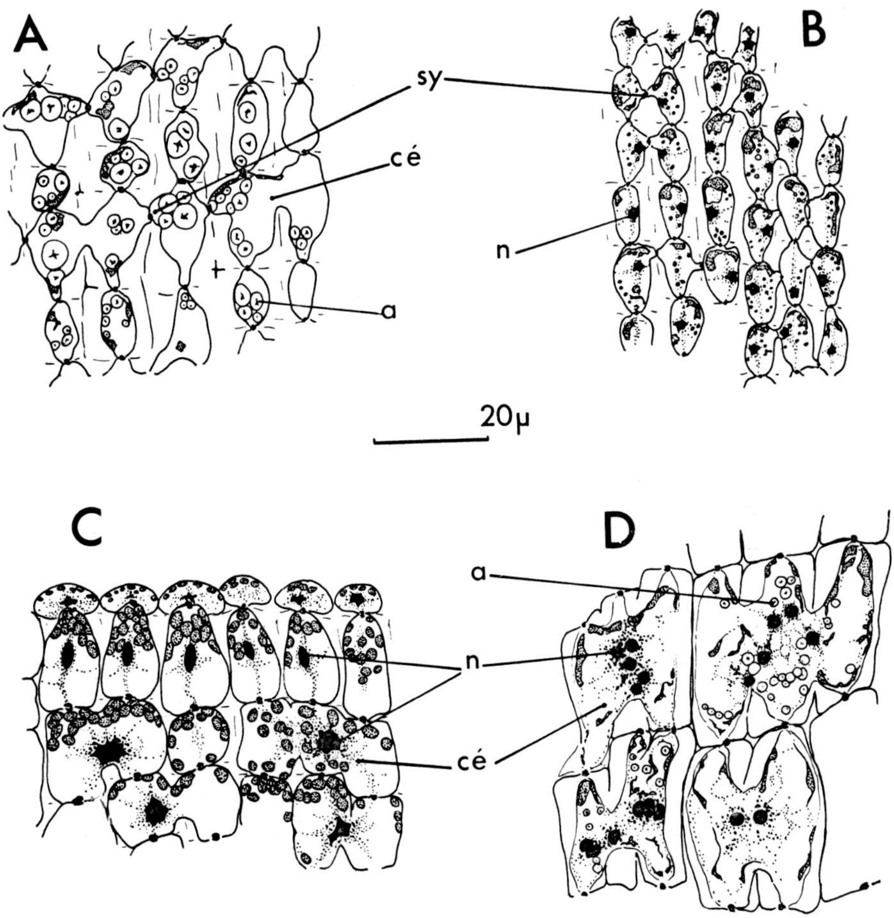

FIG. 16. Drawings of decalcified coralline crusts in surface view.
Sy = secondary pit connection; cé = lateral fusion; a = floridean starch; n = nucleus.
From Cabioch 1970 Fig. A–D. A = *Archaeolithothamnion erythraeum*; B = *Pseudolithophyllum orbiculatum*; C = *Neogoniolithon notarisii*; D = *Mesophyllum lichenoides*.

(*b*) Each sporangium has its own plug as in (*a*) but the sporangia are aggregated into a conceptacle, as in subfamily Lithothamnioideae (Fig. 9).
(*c*) Sporangia are borne in a conceptacle with a single, central ostiole, as in subfamilies Corallinoideae and Lithophylloideae (Figs 10, 11).

On the basis of a combination of these characters, Cabioch classifies the Corallinaceae into five subfamilies, four of which are shown in Fig. 17. The fifth subfamily, the Schmitzielloideae, contains a single genus, *Schmitziella*, which is endophytic and comprises uncalcified filaments, with no epithallial cells, and which grows inside algae such as *Cladophora*. Here the sporangia are not carried in conceptacles but form a sorus, the outer casing of which is formed by the host's outer membranes. As this subfamily is so different from the other four and may not even belong in the Corallinaceae it will not be discussed further.

It can be seen from Fig. 17 that Cabioch includes in her subfamilies thalli of three different levels of complexity. Central to the family is the complex crustose form which includes nearly all the principal epilithic crustose genera, either morphologically simple or massive and elaborately branched, but never bearing uncalcified genicula. These crustose genera are often a conspicuous element of the eulittoral flora; they are important contributors to the building of coral reefs and in many parts of the world dominate the sublittoral flora. Also among these forms are the free-living species which constitute maerl beds. At the top level of complexity Cabioch places the articulated genera, which she considers to have evolved independently from the crustose form within four separate tribes of the two subfamilies Corallinoideae and Lithophylloideae. The articulated forms range from the simple *Yamadaea* with one geniculum and one intergeniculum to large, much-branched species. Three genera are represented in the British Isles, all in the Corallinoideae: *Corallina, Jania* and *Haliptylon*. Finally, at the lowest level are the crusts of simple construction which rarely attain any great size except in the epilithic forms of *Pseudolithophyllum* (*sensu* Lemoine) and *Dermatolithon*. As explained previously, it is within the genera at this level that many species of epiphytic crustose coralline algae occur, together with various endophytic and parasitic genera which are indicated by downward pointing arrows. Cabioch (1971, 1972) points out that many of the simple crusts resemble early developmental stages of the more complex species and may therefore be neotenic forms. Others may be genuinely primitive or again they could be regressive. It is worth noting that although these forms are vegetatively simple, their reproductive structures do not show reduced tendencies but have the same sort of organization as the more complex members of the same subfamily. It is clear from this scheme that the three genera of epiphytic, crustose corallines being considered differ fundamentally from each other, belonging as they do

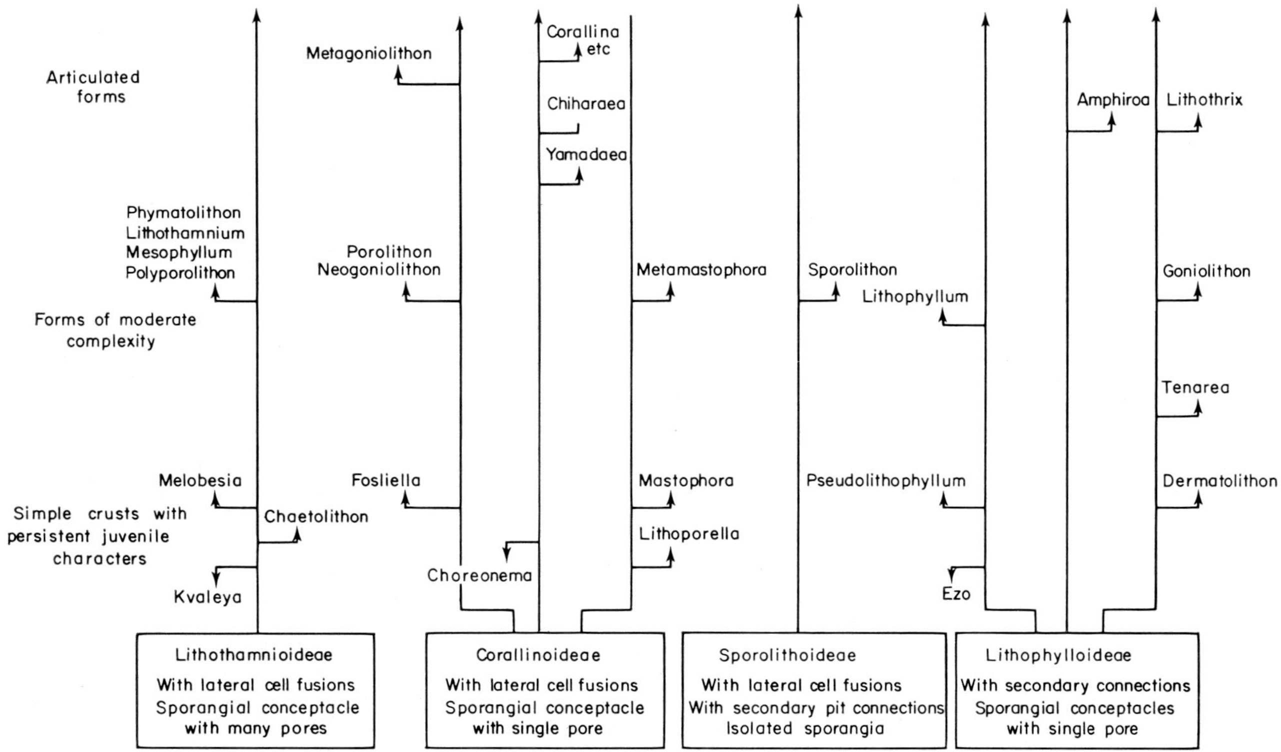

FIG. 17. Classification of the Corallinaceae according to Cabioch (1972 p. 261) modified by Y. M. Chamberlain.

to three separate subfamilies, and that the superficial resemblances among them must presumably be due to their similar life style and not to their taxonomic position.

As can be seen from Fig. 17, *Sporolithon* (*Archaeolithothamnium*), with its individually borne tetrasporangia which are sometimes cruciately and sometimes zonately divided and its possession of both lateral fusions and secondary pit-connections, occupies an isolated position. This is generally accepted as being the earliest known fossil genus which still exists; Cabioch (1970) suggests that the distinguishing features are an indication of its antiquity and that by evolution from this genus in two different directions the lithothamnioid line with cell fusions and lithophylloid line with secondary pit-connections developed. Adey and Johansen (1972) however, on the basis of the examination by Adey of a considerable range of tropical species of *Sporolithon*, consider that the resemblances between this genus and *Lithothamnium* are more significant than the differences and include it in their subfamily Melobesioideae which otherwise corresponds to Cabioch's Lithothamnioideae.

For Cabioch therefore, the subfamily represents both a taxonomic group and an evolutionary line. Adey and Johansen (Table I), while in agreement with Cabioch regarding the basic significance of lateral anastomoses and sporangial arrangement, also consider the presence or absence of the geniculum as being of fundamental importance. Although, like Cabioch, they accept that the articulated form has evolved from the crustose form more than once (three times in their opinion) they nevertheless separate off the articulated genera into three subfamilies.

THE EPIPHYTIC CRUSTOSE CORALLINACEAE—METHODS CURRENTLY BEING USED TO INVESTIGATE THEIR TAXONOMY

The general structure of these plants has been described earlier in this chapter. A considerable body of information is now available concerning the taxonomy of these species in various parts of the world. The most important authors are: Rosenvinge (1917, Denmark); Suneson (1937 and 1943, Europe—especially Sweden); Foslie (many publications listed in Foslie and Printz 1929, p. 11, cosmopolitan); Hamel and Lemoine (1953, France and North Africa); Bressan (1974, Italy); Lewalle (1961, Italy); Balakrishnan (1947, India); Masaki (1968, Japan—including a list of earlier papers by himself and Tokida); Ganesan (1971, Caribbean); Dawson (1960, Mexico, and many other papers); Lee (1970), Mason (1953) and Adey and Johansen (1972) (all Pacific North America). Until recently the group has been little studied in the British Isles but the British species are currently being investigated (Chamberlain 1977a, b) and in

Table I. Suprageneric classification of the Corallinaceae according to Adey and Johansen[a]

Main characteristics	Genera	Main characteristics	Genera
(1) Melobesioideae (Areschoug) Mason Non-articulated sporangial plugs present secondary pit-connections absent	*Clathromorphum* *Kvaleya* *Leptophytum* *Lithothamnium* *Melobesia* *Mesophyllum* *Neopolyporolithon* *Phymatolithon* *Sporolithon*	(4) Metagoniolithoideae Johansen Articulated secondary pit-connections absent genicula meristematic	*Metagoniolithon*
(2) Lithophylloideae Setchell Non-articulated sporangial plugs absent secondary pit-connections present	*Ezo* *Lithophyllum* *Tenarea*	(5) Amphiroideae Johansen Articulated secondary pit-connections present one or several genicular tiers	*Amphiroa* *Lithothrix*
(3) Mastophoroideae (Svedelius) Setchell Non-articulated sporangial plugs absent secondary pit-connections absent	*Choreonema* *Fosliella* *Heteroderma* *Hydrolithon* *Lithoporella* *Mastophora* *Neogoniolithon* *Porolithon* *Pseudolithophyllum*	(6) Corallinoideae (Areschoug) Foslie Articulated secondary pit-connections absent one genicular tier	*Alatocladia* *Arthrocardia* *Bossiella* *Calliarthron* *Cheilosporum* *Chiharaea* *Corallina* *Haliptylon* *Jania* *Marginisporum* *Serraticardia* *Yamadaea*

[a] (See Johansen 1976, p. 232)

subsequent sections of this paper the information so far gained will be related to work done elsewhere.

1. Culture Studies

A number of workers have studied various aspects of the growth of coralline algae either using cultures grown from spores or, with articulated species, using thallus fragments as the inoculum.

(*a*) *Spore size and germination patterns.* The spores of most corallines show the *Dumontia* type of germination (Cabioch 1972), that is to say that when the

18

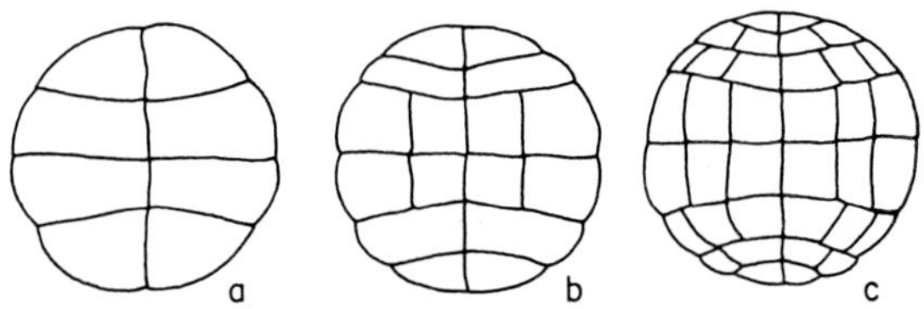

FIG. 18. a–c, stages in the germination of *Amphiroa*-type spore (diagrammatic).

19

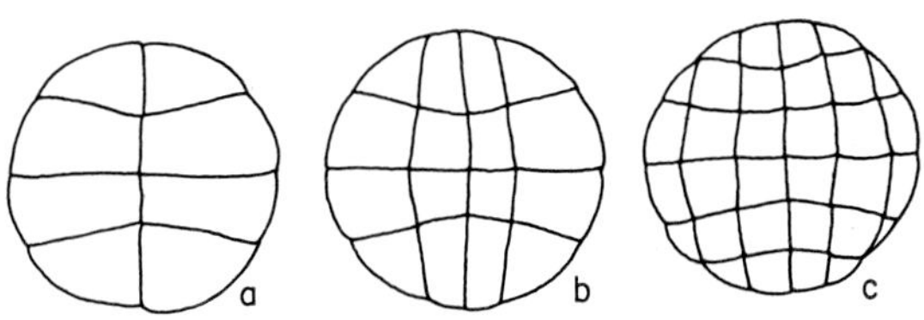

FIG. 19. a–c, stages in the germination of *Corallina*-type spore (diagrammatic).

carpospore, tetraspore or bispore has settled, the contents divide to produce up to sixteen small cells within the confines of the original spore before further cell divisions increase the area of the crust. Exceptions occur only among some parasitic and endophytic genera such as *Choreonema* (Cabioch, 1972).

Chihara (1973) germinated spores from a number of species of articulated corallines and established two basic segmentation patterns—the *Amphiroa* type and the *Corallina* type (Figs 18, 19). In a further paper (1974) Chihara described the spore segmentation patterns in a range of epiphytic and epilithic crustose corallines and while some of these correspond to the previous two types, he added four further patterns, three of them in the genus *Fosliella*. Work by Notoya (1974 and 1976a) has further confirmed the prevalence of the two main segmentation patterns, while a new pattern, for *Melobesia pacifica*, is described in the 1976a paper. In addition to investigating segmentation patterns, Chihara and Notoya measured spore diameters. Chihara's conclusion, since largely confirmed by Notoya, was that most of the corallines can be split into two groups—the *Lithophyllum*/*Amphiroa* group with *Amphiroa* type germination and relatively small spores (up to about 45 μm) and the *Lithothamnium*/*Corallina* group with *Corallina* type germination and relatively large spores (*c.* 40–70 μm

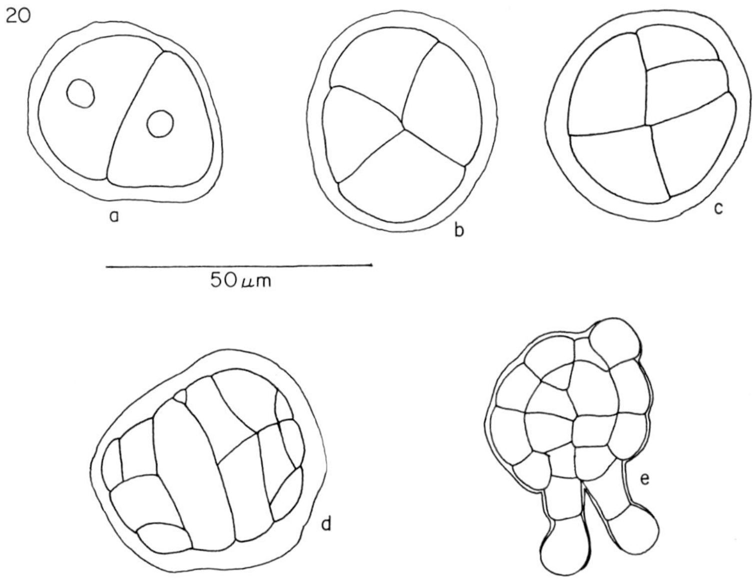

FIG. 20. a–e, germination of *Fosliella limitata* spore.

diameter). These two groups correspond to the genera with secondary pit-connections (*Lithophyllum* group) and those with lateral cell fusions (*Lithothamnium* group). A notable exception is the *Fosliella*/*Heteroderma* complex which, while it has uniformly small spores, has a multiplicity of segmentation patterns. Chihara suggested that this may be an indication of the primitive status of this complex within the Corallinaceae.

Spore segmentation patterns have been investigated in a number of British epiphytic crustose species. The pattern for *Melobesia membranacea* is quite regular (Fig. 22) and correlates clearly with Chihara's *Lithothamnium* type. Spores from a number of *Fosliella* species (Fig. 20), however, show several different patterns of segmentation and so far no clear segmentation sequence has emerged; this situation corresponds to Chihara and Notoya's results for this genus. The developmental sequence observed in the early stages of *Dermatolithon* bispore germination (Fig. 21) agrees with that obtained by Notoya.

The results of these studies correlate well with the hypothesis that the presence of secondary pit-connections is one of the most important characters differentiating one line of evolution within the Corallinaceae, since the members of the Lithophylloideae which have this type of secondary anastomosis exclusively also all have the *Amphiroa* germination pattern. So far, however, no information is available for the Sporolithoideae and no clear lines are visible in the Lithothamnioideae and Corallinoideae.

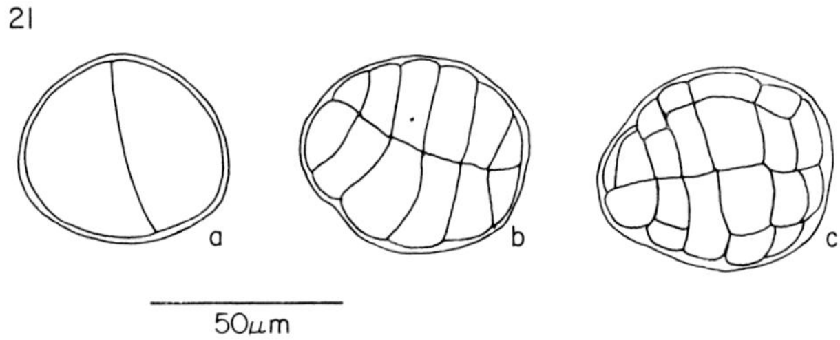

FIG. 21. a–c, germination of *Dermatolithon litorale* spore.

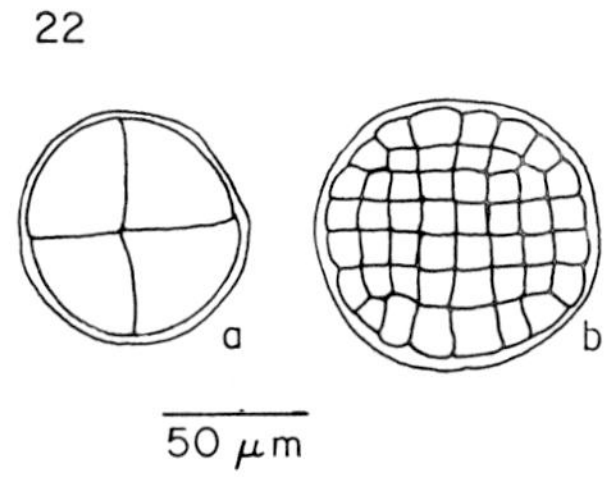

FIG. 22. a–b, germination of *Melobesia membranacea* spore.

(*a*) *Life-history*. To date, no account of a complete life-cycle established by means of cultures has been published for the crustose Corallinaceae. However, in British material, spores of various species have produced conceptacle bearing crusts in culture. Crusts grown from uninucleate bispores of *Fosliella farinosa* developed conceptacles containing further uninucleate bispores, thus proving that this is a self-perpetuating phase (Chamberlain, 1977b).

In connection with life-history studies, the observations of Suneson (1950) on the cytology of bispore formation in *Dermatolithon litorale* and *D. corallinae* are relevant since he deduced the course of the life-history on the basis of the results obtained. *D. litorale* bears conceptacles which almost exclusively contain bisporangia comprising two uninucleate bispores, although Suneson occasionally found carpogonia in the same conceptacle and regarded this as an abnormality. Suneson established that the bispores were formed as the result of an apomeiotic division of the sporangium and assumed that these diploid bispores would give rise to a further diploid, bisporangial generation. *D. corallinae* on the other hand he found to bear various types of conceptacle which contained gametangia, bisporangia composed of two uninucleate bispores, or tetrasporangia which were sometimes intermingled with bisporangia containing uni- or binucleate spores. Suneson found that the crusts bearing tetraspores or binucleate bispores were diploid and that the haploid tetra- or bispores were formed by meiotic divisions. The gametangial crusts, however, were haploid and the carpogonia and spermatia therefore produced by mitotic division. The uninucleate bispores were diploid and formed apomeiotically as in *D. litorale*. In *D. corallinae*, therefore, he supposed that there were two separate life-histories. Firstly there was a *Polysiphonia*-type life-history (cf. Dixon 1973) with a sequence of gametangial, carposporangial and tetrasporangial phases, the first and last being morphologically similar and the carposporangial phase developing on the gametangial phase. Secondly there was a self-perpetuating, diploid, bisporangial life-history as in *D. litorale*.

Since it has been found quite possible to produce conceptacle-bearing crusts of epiphytic crustose corallines in culture, it is hoped that further experimentation will enable complete life-histories to be followed in order to see whether they confirm more of the sequences deduced by Suneson.

(*c*) *Investigation of morphological variation*. It has been found in British species that there is considerable variation in morphological features such as cell size, trichocyte occurrence and the appearance of the conceptacles at different times of year. An example is *Fosliella farinosa*, which is easy to identify in summer because it bears conspicuous, terminal trichocytes on the thallus and is the only British species of the genus to do so. In winter, however, many crusts are found which, while having the general appearance of *F. farinosa* and bearing bisporangial conceptacles typical for that species (Chamberlain, 1977b), have somewhat smaller crust cells and bear no trichocytes. In an attempt to confirm the identification of this species, some bispores were cultured and the crusts grown from these spores developed terminal trichocytes within ten days of germination (Fig. 23), thus substantiating the identification. It seems possible that this technique could be used to investigate other taxonomic problems of this kind.

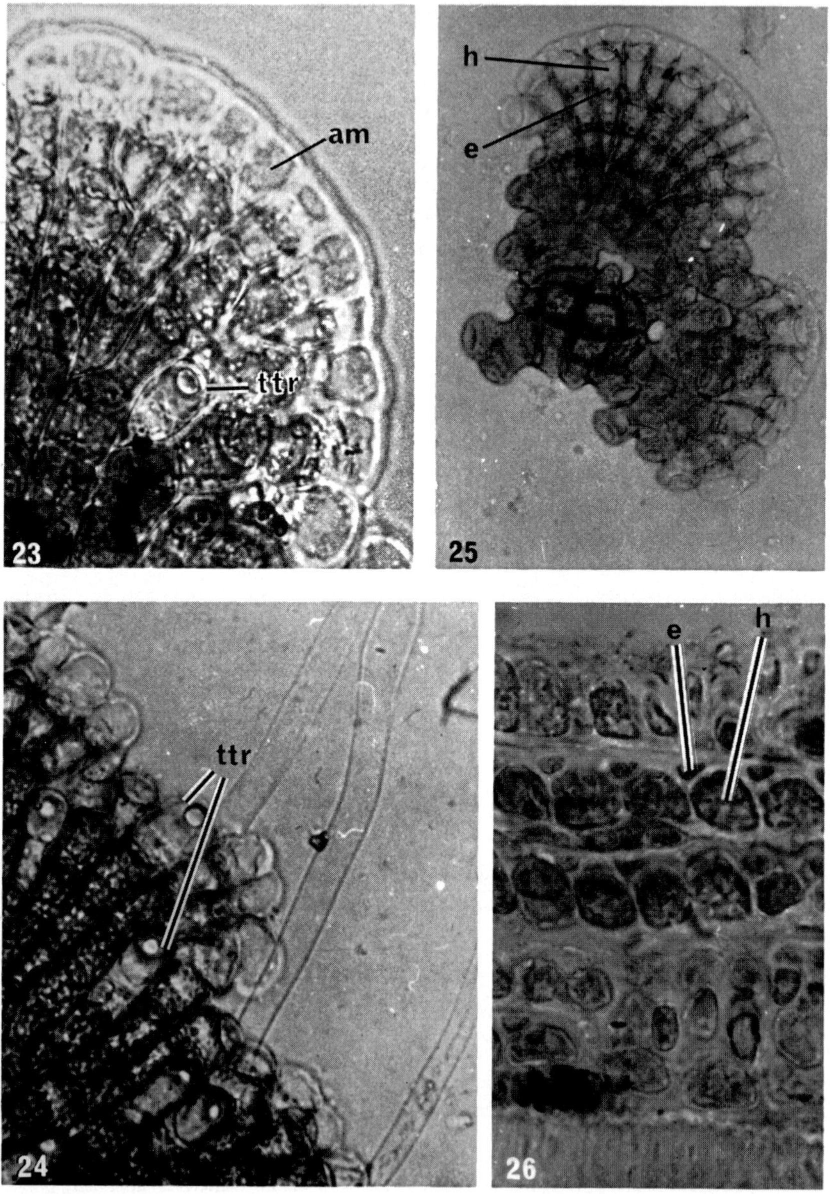

FIG. 23. *Fosliella farinosa* crust cultured from autumn bispore. ×938.
FIG. 24. *Fosliella farinosa*, cultured crust to show trichocytes and hairs. ×462.
FIG. 25. *Fosliella farinosa*, cultured crust. ×320.
FIG. 26. *Fosliella farinosa*, vertical section of superimposed thalli. ×690.

(*c*) *Effects of varying conditions on coralline cultures.* Two papers have recently appeared which report results obtained from culturing corallines under varying conditions.

Notoya (1976b) has published a preliminary report of the results of experiments to investigate the effects of varying conditions of light intensity, temperature, pH and salinity on spore germination and growth in three crustose corallines including *Melobesia pacifica,* which grows as an epiphyte on *Phyllospadix.* He found that in *M. pacifica* the optimum conditions, considered independently and not mutually varying, were pH 8·5, salinity 34% of natural sea-water, water temperature 10–20°C and light intensity about 8000 lux with growth gradually increasing to this level.

Mention should also be made of a recent paper by Brown *et al.* (1977) who reported the effect of orthophosphate concentration on the growth of three articulated coralline algae grown in culture from plant fragments. They found that these algae grow best at very much lower phosphate concentrations than most other algae in culture, presumably because the phosphate ions interfere with the formation of calcite crystals, and they suggested that coralline algae might therefore be used as sensitive indicators of high phosphate levels in the environment.

Further experiments along these lines, taken in conjunction with shore observations, should increase the understanding of variation within and between species.

1. *Light and Electron Microscopy*

(*a*) *Light microscopy.* The traditional methods of staining sections and whole mounts of coralline algae (which must be decalcified before sectioning unless ground, geological style sections are required) have enabled workers to elucidate the vegetative and reproductive structure. The most detailed studies of epiphytic species are those of Suneson, Rosenvinge, Masaki, Balakrishnan, Ganesan, Dawson and Lee. The cytology and morphogenesis of the vegetative structures are described in the works of Cabioch. Further comparative studies are now needed, for example, to increase knowledge on the development of the reproductive bodies and on post-fertilization changes leading to the establishment of the carposporophyte, since studies of this sort have already proved of use in elucidating taxonomic relationships, not only in the Corallinaceae but throughout the Rhodophyta.*

Mention should be made here of chromosome counts. The chromosomes of these algae are rather small and not easy to process for microscopy. The

*An important new interpretation of carpogonial structure and post-fertilization events appears in Lebednik (1977).

following counts for epiphytic species have, however, been published: *Dermatolithon litorale* 2n = 30 and *D. corallinae* 2n = 32 (Suneson, 1950); *Fosliella farinosa* n = ±24 (Balakrishnan, 1947).

(*b*) *Electron microscopy*. Bailey and Bisalputra (1970) made some preliminary studies by both transmission and scanning electron microscopy of three species of articulated coralline, and Garbary (this book) has extended the knowledge of the surface structure of the coralline thallus; his results indicate that differences in the surface detail promise to be of great use taxonomically. For example, the surface view of the conceptacles in *Dermatolithon litorale* and *D. pustulatum* (see Garbary, Chapter 9) show very different patterns and this has helped to confirm the separation of these two species, which are not always easy to distinguish, especially in this country where only bisporangial plants are found.

Giraud and Cabioch (1976) studied the outermost cells, both epithallial and meristematic, of various coralline species including the epiphytic *Mesophyllum lichenoides* by transmission electron microscopy. It has been known from light microscopy that the external surface of the coralline thallus is covered by a thick, stratified cuticle. Giraud and Cabioch show that the epithallial cells, previously assumed to be inactive if not dead, are in fact physiologically very active and that they contain, particularly at their outer surface, Golgi vesicles which are actively secreting polysaccharides to the exterior of the thallus where they contribute to the cuticle. This activity is shared to a lesser extent by the subapical meristematic cell. It would be interesting to extend studies of this kind to investigating cell activity at different seasons, since crusts often show differences in appearance according to the time of year.

2. Shore Studies

(*a*) *Seasonal periodicity*. One of the few reports describing the seasonal life history of an epiphytic, crustose coralline is that of Chihara (1974) who reported that *Dermatolithon tumidulum* (Fosl.) Fosl. from Japan reproduces from June to December, but most abundantly from July to November. Chihara found that the genera could be grouped according to two seasonal patterns: those, like *Dermatolithon*, which reproduce in summer and belong to his *Lithophyllum*/*Amphiroa* group (described earlier) and those which reproduce either all the year round or in any season other than summer and belong to the *Lithothamnium*/*Corallina* group. Regular observations of shores in Hampshire and Dorset over two and a half years have shown that the behaviour of the epiphytic species does not fall into a regular pattern. They settle, grow and reproduce very quickly and their occurrence seems to depend on such factors as the condition of the host plant and on local extremes of weather conditions rather than the overall seasonal cycle. For example the drought conditions of summer 1976 and a series of heavy frosts in January 1977 drastically diminished the littoral epiphytic species. They are also liable to sudden periods of great abundance, again unconnected with a seasonal pattern. Further observations are needed before firm conclusions can be reached, but it does seem that they could best be described as opportunist species.

(*b*) *Capacity to overgrow other species.* Preliminary observations suggest that whether or not the crusts of an epiphytic coralline can superimpose those either of the same or other species of epiphytic coralline and whether they can in turn be overgrown provides quite a useful taxonomic character. *Melobesia membranacea* is unable to overgrow any other species, for example, while almost every other species is able to grow on top of it. Species of *Dermatolithon* are quite often found on top of *Fosliella* spp. and *Melobesia* but they rarely overgrow either the same or other *Dermatolithon* species. *Fosliella* is variable; most species do not overgrow but *F. farinosa* (Fig. 26) can grow over many other species and is also found in compound crusts formed of many layers of individual thalli (Chamberlain, 1977b).

(*c*) *Host specificity*. A few epiphytes have a very restricted host range, *Fosliella lejolisii* being the clearest example, growing as it does only on marine phanerogams. Most other epiphytes grow on a range of algae although some are characteristic of particular hosts: *Dermatolithon corallinae* on *Corallina officinalis* and *Melobesia membranacea* on *Furcellaria lumbricalis* are two notable examples. However, it is noticeable that while some algae regularly bear numerous epiphytes others bear few or none. It is probable that culture techniques and electron microscopy could be used to examine the physical and chemical characteristics of the host/epiphyte relationship in order to establish what features make one alga a better host plant than another.

3. *Tabular Key*

In 1976 Johansen published two tabular keys for the genera of crustose and articulated Corallinaceae respectively. He quotes Newell, who describes tabular keys as being "devices useful in the identification of organisms as well as in information storage and retrieval." For the crustose Corallinaceae Johansen tabulates 21 genera against the following nine characters: habitat, hypothallium, perithallium, epithallium, secondary anastomoses, trichocytes, structure of sporangial conceptacle roof, male conceptacles, spore germination. The character variants are presented as cognitive symbols and each variant is illustrated. To give one example, secondary pit-connections are listed thus:

Character 5: Secondary pit-connections (2pc)
2pc = Secondary pit-connections present
x = Secondary pit-connections absent, cellular fusions present instead.

The resultant table enables many characters to be seen at a glance and provides a very useful reference point for further investigation. In a subsequent section, Johansen lists 22 further features which might, with more investigation, prove to be useful taxonomic characters. As more features are investigated enough data may become available to be subjected to numerical analysis.

CONCLUSIONS

The epiphytic crustose Corallinaceae belonging to the genera *Melobesia*, *Fosliella* and *Dermatolithon* are examples of the simplest level of thallus organization within the family. They comprise simple crusts from two to about twenty cells deep which grow on other algae or marine phanerogams. The transitory nature of their substrate necessitates that they should settle, germinate and reproduce quickly and on the basis of field observations so far it seems that in the littoral they tend to do this all the year round, their occurrence depending more on local extremes of climatic conditions than on general seasonal variation—drought and frost for example have been found to affect their occurrence considerably.

When the classification of the Corallinaceae into subfamilies (based on the two features secondary lateral anastomoses and sporangial roof structure) is considered, the three epiphytic genera being considered in this paper fall into three different subfamilies of the five present in Cabioch's scheme. They are thus a taxonomically diverse group united by their adaptation to a similar habitat. While they all show a simple vegetative organization, the reproductive structures do not show a correspondingly simplified construction. *Melobesia* and *Dermatolithon* seem to be rather stable genera, being fairly consistent in terms of features such as vegetative structure and germination pattern; *Fosliella* shows much wider variation, which Chihara suggests is typical of rather primitive organisms.

Advances in the knowledge of the taxonomy of these algae seem most likely to come from further investigations into the structure and development of the gametangial structures and post-fertilization changes, culture studies in conjunction with field observations and transmission and scanning electron microscopy.

REFERENCES

Adey, W. H. and Johansen, H. W. (1972). Morphology and taxonomy of Corallinaceae with special reference to *Clathromorphum*, *Mesophyllum* and *Neopolyporolithon* gen. nov. (Rhodophyceae, Cryptonemiales). *Phycologia* **11**, 159–180.

Adey, W. H. and Macintyre, I. G. (1973). Crustose coralline algae: a re-evaluation in the geological sciences. *Bull. geol. Soc. Am. B.* **84**, 883–904.

Bailey, A. and Bisalputra, T. (1970). A preliminary account of the application of thin-sectioning, freeze-etching, and scanning electron microscopy to the study of coralline algae. *Phycologia* **9**, 83–101.

Balakrishnan, M. S. (1947). The morphology and cytology of *Melobesia farinosa* Lamour. *J. Indian Bot. Sci.* **1946**, 305–319.

Bressan, G. (1974). Rodoficee calcaree dei mari Italiani. *Boll. Soc. adriat. Sci. nat.* **59**, 1–132.

Brown, V., Ducker, S. C. and Rowan, K. S. (1977). The effect of orthophosphate concentration on the growth of articulated coralline algae (Rhodophyta). *Phycologia* **16**, 125–131.

Cabioch, J. (1970). Sur l'importance des phénomènes cytologiques pour la systématique et la phylogénie des Corallinacées (Rhodophycées, Cryptonémiales). *C.r. hebd. Séanc. Acad. Sci., Paris*, sér. D271, 296–299.

Cabioch, J. (1971). Etude sur les Corallinacées, I. Caractéres généraux de la cytologie. *Cah. Biol. mar.* **12**, 121–186.

Cabioch, J. (1972). Etude sur les Corallinacées, 2. La morphogenése: conséquences systématiques et phylogénétiques. *Cah. Biol. mar.* **13**, 137–288.

Chamberlain, Y. M. (1977a). The occurrence of *Fosliella limitata* (Foslie) Ganesan (a new British record) and *F. lejolisii* (Rosanoff) Howe (Rhodophyta, Corallinaceae) on the Isle of Wight. *Br. phycol. J.* **12**, 67–81.

Chamberlain, Y. M. (1977b). Observations on *Fosliella farinosa* (Lamour.) Howe (Rhodophyta, Corallinaceae) in the British Isles. *Br. phycol. J.* **12**, 343-358.

Chihara, M. (1973). The significance of reproductive and spore germination characteristics in the systematics of the Corallinaceae: articulated coralline algae. *Jap. J. Bot.* **20**, 369–379.

Chihara, M. (1974). The significance of reproductive and spore germination characteristics to the systematics of the Corallinaceae: non-articulated coralline algae *J. Phycol.* **10**, 266–274

Dawson, E. Y. (1960). Marine red algae of Pacific Mexico. Part 3, Cryptonemiales, Corallinaceae subf. Melobesioideae. *Pacif. Nat.* **2**, 1–125.

Dixon, P. S. (1973). "Biology of the Rhodophyta." Oliver and Boyd, Edinburgh

Foslie, M (1909) Algologiske notiser VI *K. nor. Vidensk. Selsk. Skr.* **1909**, 1–63.

Foslie, M. and Printz, H. (1929). Contributions to a monograph of the lithothamnia. *Det Kongelige Norske Videnskabers Selskab Museet, Trondheim.*

Ganesan, E. K. (1971). Studies on the morphology, anatomy, reproduction and taxonomy of some Caribbean crustose Corallinaceae *In* "Symposium on Investigations and Resources of the Caribbean Sea and Adjacent Regions". UNESCO, Paris, pp 411–416.

Giraud, G. and Cabioch, J. (1976). Etude ultrastructurale de l'activité des cellules superficielles du thalle des Corallinacées (Rhodophycées). *Phycologia* **15**, 405–414.

HAMEL, G. and LEMOINE, M. (1953). Corallinacées de France et d'Afrique du Nord. *Archs Mus. natn. Hist. Paris,* Sér. 7 **1**, 17–131.

HOWE, M. A. (1920). Class 2. Algae. *In* "The Bahama Flora" (Britton, N. L. and Millspaugh, C. F. eds), pp. 553–618. Britton and Millspaugh, New York.

JOHANSEN, H. W. (1974). Articulated coralline algae. *Oceanogr. mar. Biol. ann. Rev.* **12**, 77–127.

JOHANSEN, H. W. (1976). Current status of generic concepts in coralline algae (Rhodophyta). *Phycologia* **15**, 221–244.

LEBEDNIK, P. A. (1977). Postfertilization development in *Clathromorphum, Melobesia* and *Mesophyllum* with comments on the evolution of the Corallinaceae and the Cryptonemiales (Rhodophyta). *Phycologia*, **16**, 379–406.

LEE, R. K. S. (1970). Developmental morphology of the crustaceous alga *Melobesia mediocris. Can. J. Bot.* **48**, 437–448.

LEWALLE, J. (1961). Détermination macroscopique des algues rouges calcaires (Corallinaceae et Squamariaceae partim) du Golfe de Naples. *Pubbl. Staz. zool. Napoli* **32**, 241–271.

LITTLER, M. (1972). The crustose Corallinaceae. *Oceanogr. mar. Biol. ann. Rev.* **10**, 311–347.

MASAKI, T. (1968). Studies on the Melobesioideae of Japan. *Mem. Fac. Fish. Hokkaido Univ.* **16**, 1–80.

MASON, L. R. (1953). The crustaceous coralline algae of the Pacific coast of the United States, Canada and Alaska. *Univ. Calif. Publs Bot.* **26**, 313–390.

NOTOYA, M. (1974). Spore germination in crustose corallines *Tenarea corallinae, T. dispar* and *T. tumidula. Bull. Jap. Soc. Phycol.* **22**, 47–51.

NOTOYA, M. (1976a). Spore germination in several species of crustose corallines (Corallinaceae, Rhodophyta). *Bull. Fac. Fish. Hokkaido Univ.* **26**, 314–320.

NOTOYA, M. (1976b). On the influence of various culture conditions on the early development of spore germination in three species of the crustose corallines (Rhodophyta) (preliminary report). *Bull. Jap. Soc. Phycol.* **24**, 137–142.

PARKE, M. and DIXON, P. S. (1976). Check-list of British marine algae—third revision. *J. mar. biol. Ass. U.K.* **56**, 527–594.

ROSENVINGE, L. K. (1917). The marine algae of Denmark. Part 2. Rhodophyceae 2 (Cryptonemiales). *K. danske Vidensk. Selsk. Skr.*, 7. Raekke, **7**, 155–288.

SUNESON, S. (1937). Studien über die Entwicklungsgeschichte der Corallinaceen. *Acta Univ. lund.*, N.F. Avd. 2. **33**, 1–101.

SUNESON, S. (1943). The structure, life-history and taxonomy of the Swedish Corallinaceae. *Acta Univ. lund.*, N.F. Avd. 2, **39**, 1–66.

SUNESON, S. (1950). The cytology of bispore formation in two species of *Lithophyllum* and the significance of bispores in the Corallinaceae. *Bot. Notiser* **1950**, 429–450.

11 | Hybridization in Red Algae

J. RUENESS

Department of Marine Biology and Limnology, University of Oslo, Norway

Abstract: The completion of red algal life-histories in culture makes it possible to apply to the Rhodophyta methods of experimental taxonomy, including artificial hybridization.

The significance of hybridization and reproductive isolation in delimitation of species of red algae are discussed. The frequency and degree of hybridization are considered in relation to various isolating mechanisms, among others the widespread occurrence of asexual means of reproduction in many populations.

Experimental studies and crosses between various isolates belonging to the *Ceramium diaphanum–strictum* species complex originating from Scandinavia and the American east coast show that the American isolate, referred to as *C. strictum*, and the morphologically inseparable isolate of *C. strictum* from Norway are intersterile. Of the Scandinavian entities, *C. tenuissimum* is clearly distinct and sexually isolated, while the Baltic strain of *C. tenuicorne* and *C. strictum* from Norway are interfertile. The Baltic strain, which differs genetically in various morphological and physiological features from *C. strictum*, is considered as an ecotype and taxonomically treated as a subspecies.

Infraspecific crosses between four varieties of *Antithamnion plumula* are reported. Hybrids resulting from crosses between some isolates of var. *plumula* and var. *bebbii* retained the characteristics of the parent strains through generations, while crosses using other isolates resulted in a subtle variation pattern which is not yet fully understood.

Biosystematic approaches to the taxonomy of red algae are in their infancy and there is an almost complete lack of critical data on such topics as environmental and genetic variation in populations, chromosome characteristics etc. At this stage crossability tests could only be a guide in assessing taxonomic status.

INTRODUCTION

Hybridization is a term used to designate different processes in modern biology and has to be defined more precisely. Traditionally, the words hybrid and hybridization are applied in taxonomy respectively to the organism resulting from a cross and to the process of cross-fertilization, between individuals of

Systematics Association Special Volume No. 10, "Modern Approaches to the Taxonomy of Red and Brown Algae", edited by D. E. G. Irvine and J. H. Price, 1978, pp. 247–262, Academic Press, London and New York.

different taxa. With modern techniques in cell biology, formation of cell-hybrids and DNA-hybridization are possible. Whether the latter kinds of hybrids will have taxonomic significance remains to be seen. The present paper deals only with the former aspect of sexual hybridization in red algae. Furthermore, it deals mainly with artificial hybridization effected in culture, not with naturally-occurring hybrids or hybrid analysis carried out in the field.

The development of experimental taxonomy (e.g. cytotaxonomy; cytogenetics; experimental fertilization between individuals from different populations) is only in its infancy in the red algae compared with the application of similar methods in higher plants. As was pointed out by Dixon (1963), the application of these methods to the Rhodophyta has been retarded mainly by technical difficulties.

Most difficulties have been overcome during the last fifteen years, at least in terms of laboratory cultivation techniques. Considerable advances have been achieved in knowledge of red algal reproduction, life-histories and form variation, through extensive culture studies. Amongst these culture studies, relatively few attempts at hybridization have been carried out. The first attempts at cross-fertilization between two species of red algae (*Polysiphonia*) were those of Edwards (1970); earlier Sundene (1959) was unsuccessful in crossing two varieties of *Antithamnion plumula*, although later experiments (Rueness and Rueness, 1975; Sundene, 1975) repeated the same crossings successfully, probably because of recently improved culture techniques. The only other published reports on hybridization experiments using red algae are those of Rueness (1973b), Polanshek and West (1975), Kapraun (1977) and van der Meer and Bird (1977).

SIGNIFICANCE OF HYBRIDIZATION IN DELIMITATION OF SPECIES OF RED ALGAE

The species concept and species definitions play a central role in biology (Slobodchikoff, 1976) and the primary aim in taxonomy is the delimitation and classification of species. Two major species concepts are currently used; the morphological species concept (taxonomic species) and the biological species concept. The so-called phenetic species concept is a further refinement of the morphological species, based on computer techniques in species recognition. The morphological species concept stresses similarity in morphological properties, while the biological species concept is based on reproductive isolation and is defined as the sum of interbreeding populations which evolve as a single unit. In most cases one can assume the probability that there is a high degree of coincidence between the morphologically defined species and the biologically defined species. Hybridization produces cases of non-coincidence of the morphologically defined species and the interbreeding unit.

Differences of opinion exist among higher plant taxonomists on the emphasis to be placed on crossability as a species criterion. The genecological system of categories introduced by Turesson (1922), in which the terms *ecotype*, *ecospecies* and *coenospecies* are employed as an alternative to the formal categories in taxonomy, has turned out to be of only limited practical value (Heywood, 1973).

From the point of view of red algal taxonomy the pioneer or alpha-phase still very much persists. The requirements for a step towards the "biosystematic phase" of algal taxonomy have already been indicated by Dixon (1973) and Price (1973). The problems of phenotypic plasticity and the lack of cyto-taxonomic investigations are restricting the advancement of red algal taxonomy. To quote from Dixon and Richardson (1970, p. 764):

> "The capacity for structural and reproductive variation in marine algae is greater than for any other group of plants. This degree of plasticity, particularly in members of the Florideophyceae, has resulted in numerous difficulties in taxonomic studies, ecological investigations, and the interpretation of life histories."

Since the species is a visually recognizable working concept of practical value, the study of artificial hybridization is of restricted value in its delimitation. On the other hand crossability tests, where possible, could be useful in supplementing morphological and genetic data for species recognition.

FREQUENCY AND DEGREE OF HYBRIDIZATION

Most data on red algal reproduction and life-histories are derived from culture studies. It is well-known for many species that the reproductive potential realized in culture is not necessarily realized under natural conditions in the field. In addition, considerable differences may exist in reproductive behaviour between individuals of the same species from various parts of its distribution area or in different microniche conditions on the same shore. Many deviations and irregularities from the "theoretical" scheme have been demonstrated.

There is evidence to indicate that a sexual reproductive life-cycle, comprising a regular sequence of gametophytic, carposporophytic and tetrasporophytic phases, is of restricted occurrence, particularly at the northern limits of distribution (Dixon, 1965). Vegetative propagation and asexual means of reproduction by apomixis may be of widespread occurrence. In certain areas, as for instance on the southernmost part of the Swedish west coast, at the entrance to the Baltic (Öresund), it has been shown in many red algae that reproduction by vegetative means replaces sexual reproduction (von Wachenfeldt, 1975). Much more precise information is needed before quantitative estimates of the frequency of sexual reproduction in red algae can be made.

For the same reason, even preliminary suggestions as to the frequency of hybridization between red algae are currently impossible. In contrast to the flowering plants, where numerous hybrids have been recorded (Stace, 1975), there are few records of natural hybrids in algae (e.g. *Fucus*, *Chara*), and to my knowledge no hybrids between red algae have been discovered with certainty in the wild. This apparent absence of hybrid records in red algae may be explained by one or other of two causes. The first is that hybrids may be more common than realized but have passed unnoticed due to the high degree of phenotypic plasticity in most red algae. The other is that various isolating mechanisms may be effective in preventing the formation of hybrids. With regard to the first hypothesis, there is no evidence to suggest the existence of naturally occurring interspecific hybrids, but this is probably due to lack of investigations aimed at detecting hybrids.

Amongst the possible isolating barriers, breeding behavioural isolation has already been mentioned. Other potential mechanisms include geographical, ecological and seasonal isolation. These are pre-fertilization isolation mechanisms which could be circumvented by artificial cross-fertilization under culture conditions. In culture, it will also be possible to test for the degree of genetic incompatibility which may become manifest at any stage from the beginning of the formation of the hybrid zygote. In the instance reported by Edwards (1970), complete cross-incompatibility was demonstrated between *Polysiphonia boldii* and *Polysiphonia denudata*, both species having been isolated from the Texas coast of Gulf of Mexico and originally confused. In contrast, Rueness (1973b) obtained fertilization and carposporophyte development when Edward's isolate of *Polysiphonia boldii* was crossed with the Scandinavian entity referred to as *Polysiphonia hemisphaerica*. The resultant carpospores germinated into tetrasporophytes with no indications of reduced viability. However, the products of meiosis, i.e. the tetraspores produced by F_1 tetrasporophytes, showed greatly reduced viability, indicating meiotic failure. Most of the spores degenerated after a few divisions, but a few % germinated to reach fertility and developed either spermatangia or carpogonia. These gametophytes were less vigorous than the parent gametophytes and were not used in backcrosses. The hybrid tetrasporophyte has been maintained in culture for four years. It grows vigorously and produces tetrasporangia precisely similar to those of the isolate of *P. boldii*, except that the spores are almost inviable. A cytological study of the meiotic process to reveal the possible abnormalities would be of particular interest.

Examples of hybridization between geographically widely separated individuals of *Polysiphonia* show that reproductive isolation is not complete, as evidenced by carposporophyte development; to test for hybrid viability, it is necessary to follow the development through the process of meiosis, a potentially critical stage for manifestation of inviability.

HYBRIDIZATION WITHIN THE *CERAMIUM DIAPHANUM–STRICTUM* COMPLEX

1. *Materials*

The classification and ecology of the Scandinavian *Ceramium* entities belonging to the *diaphanum–strictum* species complex have been extensively dealt with by Scandinavian phycologists (for references see Waern, 1952 and von Wachenfeldt, 1975). Cell dimensions, extent and morphology of cortical bands, and the curvature of apices are principal diagnostic features.

Ceramium tenuicorne (Kütz.) Waern (Figs 1, 2), one of the most common red algae in the Baltic, was regarded by Waern (1952), who thoroughly studied field specimens and authentic material, as a true brackish-water Baltic endemic. The principal distinguishing characteristics of this taxon are urceolate, elongated axial cells with upward-stretching cortical cells at the base of the thallus, and thin, slightly incurved apices.

Outside the Baltic, on the Norwegian south coast, three isolates of *Ceramium*, referable to the same group, i.e. with cortication limited to the nodes, have been investigated in culture. Two of these fit the description of *Ceramium strictum* Harv. *sensu* Kylin (1944) (Fig. 3), whilst the third could be identified as *Ceramium tenuissimum* (Lyngb.) J. Ag. (Figs 4, 5). One of the isolates of *C. strictum* reproduces asexually by means of paraspores (Rueness, 1973a). Paraspores give rise to new parasporangium-bearing plants and this strain could therefore not be used in crosses. The other Scandinavian isolates showed a normal *Polysiphonia*-type of life-history. One further isolate of *Ceramium strictum* Harv. (Fig. 6), from New Jersey (American east coast) and kindly provided by Dr Peter Edwards, has been grown in culture together with the Scandinavian isolates. The American alga is almost inseparable morphologically from the Norwegian isolate of *C. strictum*.

2. *Crosses*

Crosses have been carried out between the various isolates; the results are shown in Table I. In addition to crossability tests, variations in growth, reproduction and morphology were all observed under various light regimes, salinities and temperatures.

3. *Intersterility*

As seen from Table I a sterility barrier existed in all inter-isolate crosses except that between the Baltic *C. tenuicorne* and *C. strictum* from Norway, whilst the

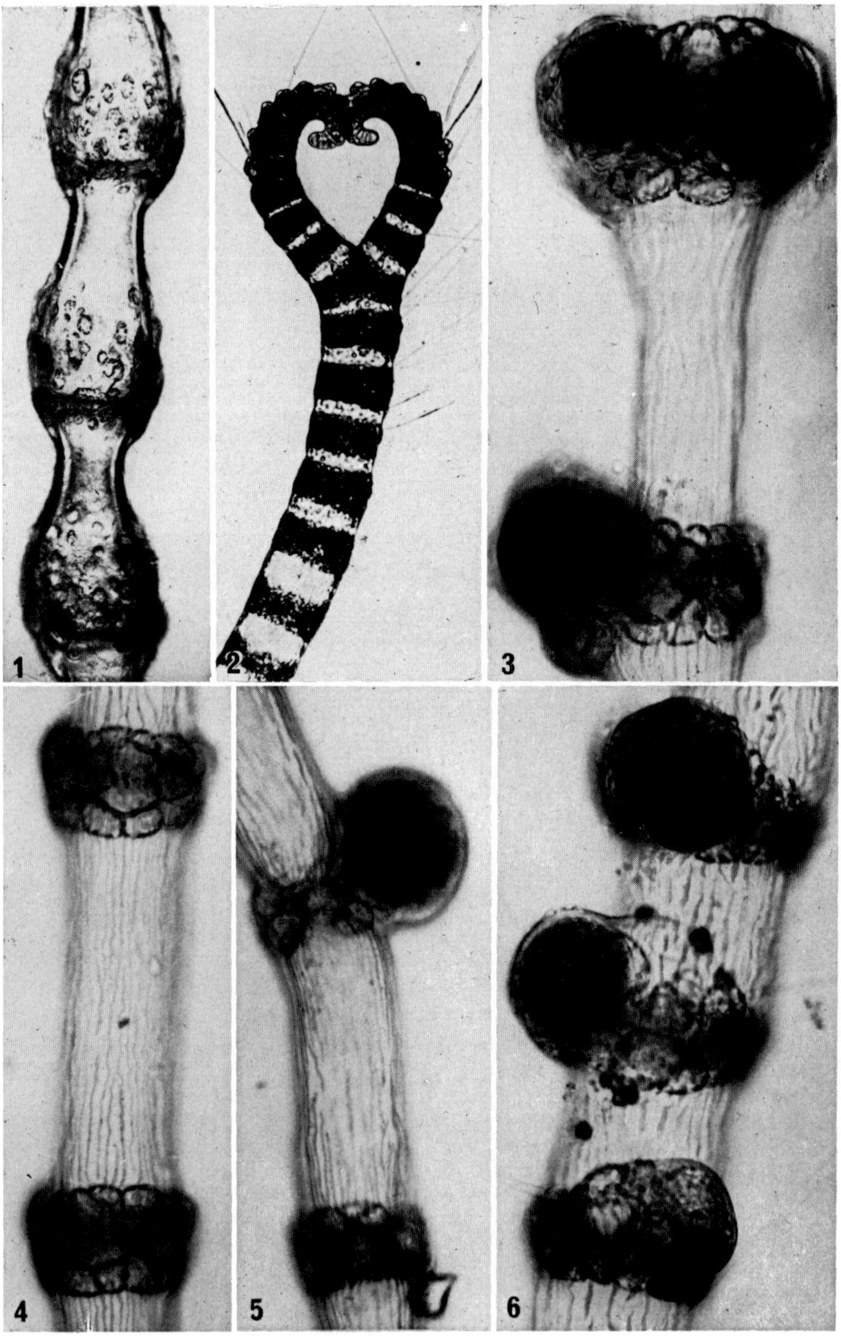

FIGS 1–6. Captions on opposite page.

controls (intra-isolate crosses) were in every case positive. *C. strictum* (Oslofjord isolate) × *C. tenuicorne*, and the offspring resulting from that cross, merit more detailed consideration later.

C. tenuissimum resembles *C. strictum* in having cortication restricted to the nodes. The patterns of cortication and cell dimensions are also almost identical when grown under standard culture conditions. However, *C. tenuissimum* is finer than *C. strictum* and has widely divergent axes. Tetrasporangia are greatly emergent and are usually formed singly on the outer face of branches (Fig. 5). The involucral branchlets developing below the carposporophyte are short and variable in number, usually 1–3, but when they are absent it is possible to misinterpret gonimoblasts as parasporangia. The morphological differences between *C. strictum* and *C. tenuissimum*, with the demonstrated breeding discontinuity, strongly suggest that two species are involved.

With regard to the allopatric pair of *C. strictum* from America and Scandinavia, complete sexual incompatibility was manifest. Even so, the two entities have more features in common than does either with *C. tenuicorne* or *C. tenuissimum*. Morphologically, the two *C. strictum* isolates are indistinguishable; the mode of growth in culture demonstrates subtle differences, the American isolate tending to grow unattached in the Petri dishes. The two isolates also differ in growth responses to varying temperatures; the American plants were stunted at about 10°C. Employing morphological features, the usual taxonomic practice, these two algae clearly belong to the same species. If specific status were to be based on cross-incompatibility, they could possibly be termed cryptic or sibling species, but delimitation of species on such grounds would be of value only as part of a biosystematic investigation of the whole complex throughout the entire North Atlantic.

FIGS 1, 2. *Ceramium tenuicorne* (Kütz.) Waern in culture. Fig. 1. Basal portion showing urceolate axial cells and upward-stretching cortical cells. × 200. Fig. 2. Female gametophyte with strongly incurved apices with a denticulate outline. × 175.

FIG. 3. *Ceramium strictum* Harv. Oslofjord strain. Tetrasporophyte in culture. The tetrasporangia are formed in whorls and are partially emergent from the cortical bands. × 450.

FIGS 4, 5. *Ceramium tenuissimum* (Lyngb.) J. Ag. in culture. Fig. 4. Axial cell and two cortical bands from the upper part of a plant. × 400. Fig. 5. Tetrasporophyte. The tetrasporangia are usually formed singly on the outer face of branches and are greatly emergent. × 400.

FIG. 6. *Ceramium strictum* Harv. New Jersey strain. Tetrasporophyte in culture. The tetrasporangia are formed in whorls in the cortical bands and are partially emergent. × 400.

4. Interfertility. C. strictum (*Oslofjord Isolate*) × C. tenuicorne
Both algae have been maintained in culture for more than two years, during which growth and reproductive responses to variations in light intensity, photoperiod, temperature and salinity have been investigated. Data on morphological features, including hair formation, shape and size of axial cells, cortex development, rhizoid formation, apical curvature and colour have been recorded. Beyond the interfertility demonstrated in Table I, no reduced viability or fertility has been observed, either in F_1 and F_2 hybrids or in the back-cross hybrids. *C. strictum* and *C. tenuicorne* differ both in growth response to

TABLE I. Number and results of crosses between different isolates of *Ceramium* from Scandinavia and USA.

Isolate	*C. strictum*[a]	*C. strictum*[b]	*C. tenuicorne*	*C. tenuissimum*	*C. strictum*[a] × *C. tenuicorne* (F_1 gametophytes)
C. strictum[a]	> 30+				
C. strictum[b]	8−	7+			
C. tenuicorne	5+	2−	10+		
C. tenuissimum	4−	4−	2−	11+	
C. strictum[a] × *C. tenuicorne* (F_1 gametophytes)	2+	2−	1+	1−	3+

[a] Oslofjord isolate. [b] New Jersey isolate.

\+ indicates carposporophyte development and release of viable carpospores.
− indicates that no carposporophyte development occurred.

various salinities (see below) and in some morphological features. The urceolate axial cells, characteristic of *C. tenuicorne* (Fig. 1), are constant at the base of the thallus as long as the enlargement of axial cells is not suppressed. A tendency to form urceolate axial cells is, however, also seen in *C. strictum* grown under conditions of physiological stress. Basal parts of *C. tenuicorne* also show an upward growth of cortex (Fig. 1), a feature not seen in *C. strictum*. This feature, emphasized as a taxonomic criterion, is thus apparently genotypically controlled. The curvature of the apices, also used as a taxonomic criterion, may vary considerably. *C. tenuicorne*, which according to the original description has thin straight apices, actually often has strongly incurved apices with a denticulate outline. As noticed also by Waern (1952), female plants in particular show this feature (Fig. 2).

The hybrids resulting from the cross *C. strictum* (Oslofjord isolate) × *C tenuicorne* were mostly of the *C. tenuicorne* type. No detailed analysis of morphological variation in the progeny from crosses has been carried out.

Influence of salinity. Tetrasporelings, 1–2 mm in length, of the Oslofjord isolate and the Baltic isolate, as well as sporelings derived from a hybrid tetrasporophyte, were grown at four salinities (5, 10, 20, 30 o/oo). Twenty sporelings of each strain were inoculated at each salinity, using five sporelings per Petri dish and four replicate dishes.

Quantitative assessment of growth is difficult in *Ceramium*. The length of each plant was measured weekly, thallus length being useful as a measure of growth in small germlings. In later stages, however, length measurements were of little value as an expression of growth due to variations in the development of adventitious branches. The dry weight of 15 plants of each strain was therefore determined for each salinity after about 30 days, by which time fertile structures had developed (Fig. 10). From the growth curves (Figs 1–9)optimum growth of *C. strictum* was obtained in the range 20–30 o/oo, and of *C. tenuicorne* in the range 5–10 o/oo. These optimal culture salinities for the two isolates correspond well with the salinity regimes in their original habitats. Attempts at growing *C. tenuicorne* at a salinity of 30 o/oo were unsuccessful. Even in plants which had been grown for six months at 20 o/oo, the response to 30 o/oo was not modified. This suggests that the salinity dependence is stable and genetically determined. Growth responses recorded for the hybrid *C. strictum* × *C. tenuicorne* provided support in that they showed an intermediate pattern relative to that of the parents. In the hybrid, growth was best in the range 10–20 o/oo (Fig. 9). Standard deviations reproduced in Fig. 9 indicate greater length variation in sporelings derived from the hybrid than in sporelings of the parent plants. This probably reflects material genotypically variable with respect to salinity tolerance. Fertile structures developed at all salinities tested in *C. strictum* and in the hybrid, with only small differences in periodicity, despite differences in vegetative growth. In *C. tenuicorne*, growth and sporulation were inhibited at 30 o/oo salinity. In Fig. 10, growth of *C. strictum*, *C. tenuicorne* and the hybrid *C. strictum* × *C. tenuicorne*, at various salinities, is expressed as an increase in dry weight. The general picture is the same as in Figs 7–9 where growth is expressed as an increase in thallus length. However, the dry weight of the hybrid exceeded that of *C. strictum* by more than would be expected from the length measurements. This could be accounted for by the greater number of adventitious branches in the hybrid.

Adventitious branches occur frequently in all the isolates studied. Their number varies considerably and it is difficult to specify the conditions which could account for their occurrence. During the salinity experiments, the number of adventitious branches was determined after 12 days in all of the twenty

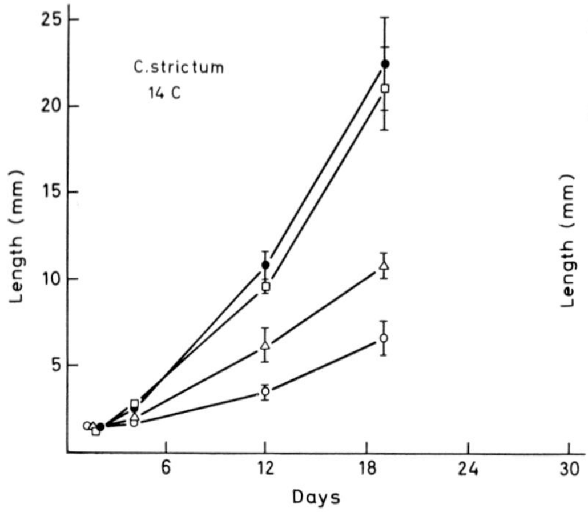

FIG. 7

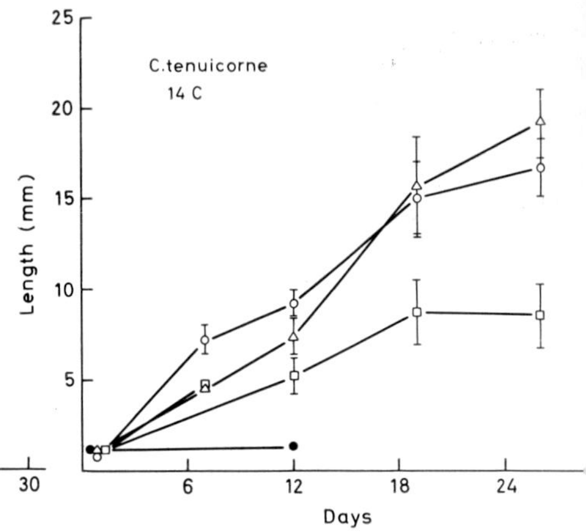

FIG. 8

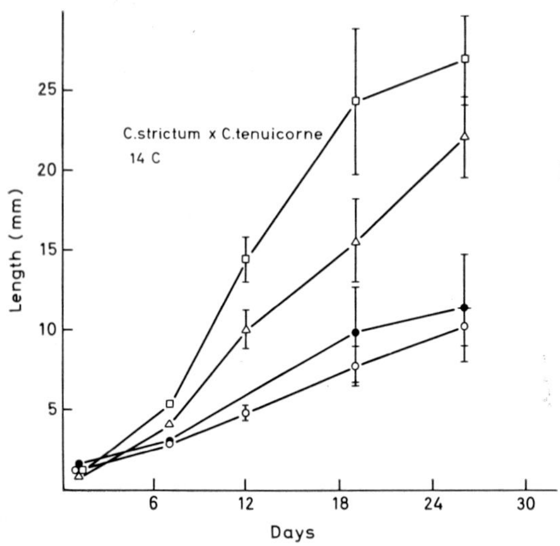

FIG. 9

FIGS 7–9. The influence of salinity on the growth of the Oslofjord strain *C. strictum* (Fig. 7), the Baltic strain *C. tenuicorne* (Fig.8) and the hybrid *C. strictum* × *C. tenuicorne* (Fig. 9). Mean length of 20 germlings at each salinity and SD indicated. Filled circles (●): 30 o/oo; rectangles (□): 20 o/oo; triangles (△): 10 o/oo; open circles (○): 5 o/oo.

plantlets of each strain at each salinity. The total number of specimens was 220. The mean lengths and the mean number of adventitious branches as well as the 95% confidence limits are presented in Table II. As can be seen from Table II and Fig. 11, the number of adventitious branches formed is generally higher in

C. tenuicorne and in the hybrid than in *C. strictum*. This is especially pronounced at the lower salinities (5 and 10 o/oo). There seems to be no direct correlation between the development of adventitious branches and salinity of the medium in the isolates, except in *C. tenuicorne* where branch production decreases with

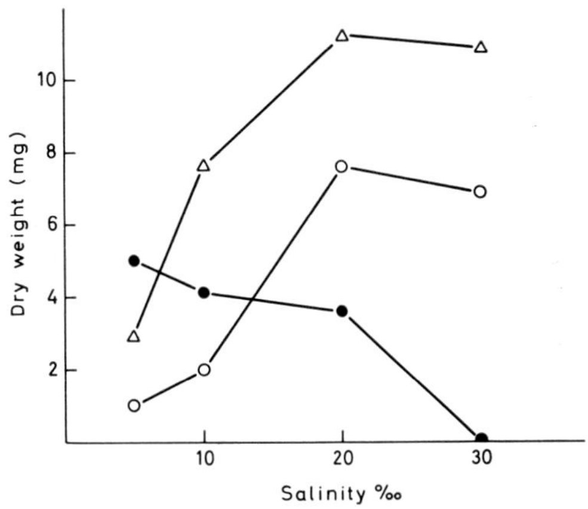

FIG. 10. Yields of the various strains in media of different salinity after 30 days. Fifteen plants of each strain were weighed. Open circles (○): *C. strictum* (Oslofjord strain); filled circles (●): *C. tenuicorne*; triangles (△): *C. strictum* × *C. tenuicorne*.

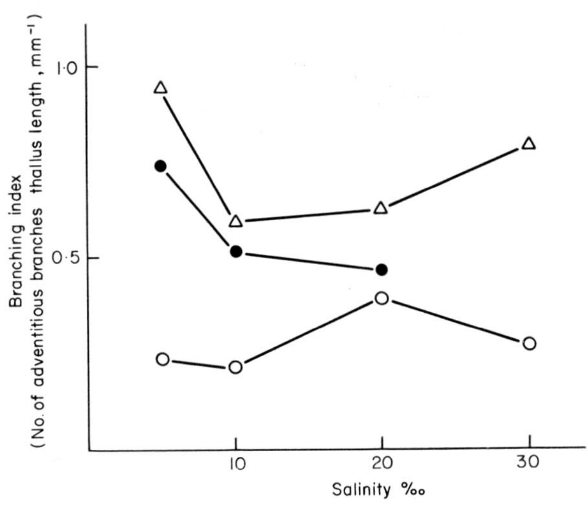

FIG. 11. Numbers of adventitious branches relative to the thallus length of the various strains in media of different salinity. Open circles (○): *C. strictum* (Oslofjord strain); filled circles (●): *C. tenuicorne*; triangles (△): *C. strictum* × *C. tenuicorne*. See also Table II.

an increase in salinity. On the other hand, conditions which favour vegetative growth also favour formation of adventitious branches in young plants. In older, fertile specimens their abundance is more variable. They occur more frequently in females than in males. This has been tested in *C. strictum* which exhibits the most pronounced differences in this respect. A ratio of about 9 : 1 in the number of adventitious branches between females and males was obtained from counts in ten plants of *C. strictum* grown to maturity, length 2–4 cm, at 20 o/oo.

The occurrence of adventitious lateral branches has been used as a diagnostic feature, e.g. in distinguishing between *C. strictum* and *C. diaphanum* (Kylin, 1924). Such branches were claimed to occur more frequently in *C. diaphanum*. In view of the great variation found within the same breeding group, this feature is of little taxonomic significance at the specific level but may possibly be of some value in distinguishing between infraspecific taxa.

TABLE II. Thallus length and number of adventitious branches in 12-day plantlets grown at different salinities. Mean values and 95% confidence limits from 20 measurements at each treatment

Salinity o/oo	*C. strictum* Thallus length mm	*C. strictum* Number of adventitious branches	*C. tenuicorne* Thallus length mm	*C. tenuicorne* Number of adventitious branches	*C. strictum* × *C. tenuicorne* Thallus length mm	*C. strictum* × *C. tenuicorne* Number of adventitious branches
30	10·8±0·7	2·9±0·4	—	—	4·8±2·34	3·8±2·6
20	9·6±0·4	3·8±0·2	5·2±0·9	2·4±1·1	14·4±1·7	8·9±1·5
10	6·2±1·0	1·3±0·4	7·4±0·9	3·8±0·4	10·0±1·1	5·9±0·4
5	3·5±0·4	0·8±0·4	9·2±0·7	6·8±1·6	4·9±0·4	4·6±0·5

5. *The Taxonomic Confusion*

The formation of fertile hybrids between the Baltic strain commonly referred to as *C. tenuicorne* and the Norwegian strain referred to as *C. strictum* is an argument for considering the two entities as a single biological species. On the other hand, culture studies have revealed that the Baltic population differs from the Norwegian population in several features of constant appearance, e.g. upward-stretching cortical cells, abundance of adventitious branches, shape of axial cells and salinity tolerance. Thus the delimitation of the morphologically defined species does not coincide with the interbreeding unit. The Baltic strain is better referred to as an ecotype, according to the genecological scheme and terminology proposed by Turesson (1922) and modified by later workers. An ecotype is "a group of individuals occupying a specific ecological habitat and forming an interbreeding population which differs genetically from other such populations" (Stace, 1975). The decision on taxonomic status will be arbitrary. The *subspecies* and *variety* are often used in nearly the same sense. According to Du Rietz (1930a), and as widely used in botanical taxonomy today (Davis and Heywood, 1963), emphasis is on the small scale, more localized range of the variety, compared with the large scale, regional basis of the subspecies.

Du Rietz (1930b) referred the whole Baltic *Ceramium* population (except *C. rubrum*) to one species, forming a part of *C. diaphanum* C. Ag., while Waern (1952) considered the Baltic population of diaphanous *Ceramium* as a proper species, *C. tenuicorne* (Kütz.) Waern, endemic to the Baltic. In view of the breeding continuity reported here, it seems reasonable to interpret the Baltic population as a subspecies, but I prefer to postpone the formal establishment of taxa until a more taxonomically oriented project can be undertaken. The differences of opinion on, and the numerous taxa described within, the *C. diaphanum–strictum* complex from the south coast of Sweden and the Öresund area are better understood if the region is considered as a hybrid zone, with a mixed population of *C. tenuicorne*, *C. strictum*, F_1 hybrids, backcross types and segregation products. Further confusion is caused by the existence of asexually reproducing clones (parasporangium-bearing plants), which evolve independently. There is evidence to indicate that the parasporangium-bearing strain from the Oslofjord has a chromosome number of about 60, while counts of gametophytes and tetrasporophytes have revealed numbers of about 20 and 40 respectively (Rueness, unpublished results). These findings are suggestive of a polyploid sequence, but more critical cytological data, in *Ceramium* as in other red algae, are required.

To decide whether *C. strictum* and *C. tenuicorne* should be reduced to synonymy with *C. diaphanum* (Lightf.) Roth requires more data, particularly on material from the British Isles and the Atlantic coasts.

INTERVARIETAL HYBRIDIZATION IN *Antithamnion plumula*

Several morphologically distinct varieties have been described in *Antithamnion plumula* (Ellis) Thur. in Le Jol.; these are only more or less geographically and ecologically separated, transitional forms having frequently been observed. In a recent paper, Sundene (1975) reported on crossings between thirteen different isolates of *A. plumula*; five isolates from the British Isles and Norway were referred to as var. *plumula*, four from Norway and Sweden as var. *bebbii* and two from the British Isles as var. *crispum*. Two Mediterranean isolates were not ascribed to any previously described taxon. The Mediterranean isolates and var. *crispum* were intersterile with all other strains tested, while var. *plumula* and var. *bebbii* were shown to be interfertile. The latter cross was further studied by

Rueness and Rueness (1975), who investigated the variation in F_1 and F_2 hybrids by tetrad-analysis and by random sampling of spore-germlings. Rather surprisingly, it was discovered that branching pattern differences between the two varieties are based on only one pair of alleles. The hybrid tetrasporophytes were always phenotypically of the var. *plumula* type, with two opposite rows of pinnae oriented in one plane. By meiosis in the tetrasporangium initials, the two varieties are segregated in equal numbers. Using material of the same varieties from the west coast of France, L'Hardy-Halos (pers. comm.) obtained similar results.

Despite the extent of morphogenetic control in *Antithamnion*, considerable phenotypic plasticity is recorded in field-collected specimens. In the Oslofjord, where both var. *plumula* and var. *bebbii* occur, it is necessary in many instances to grow the algae in culture to see whether they differ genetically from one or other of the described varieties. Vigorously growing var. *plumula* may develop a morphology approaching var. *crispum* (Rueness and Rueness, 1975). The *crispum* strain is considered sufficiently distinctive to warrant separate specific status. This is supported by the findings that sterility barriers exist between the *crispum* strain and all other *A. plumula* entities studied. The closely similar Pacific species *Platythamnion pectinatum* Kylin, considered identical by some authors (e.g. Feldmann-Mazoyer, 1941), has been compared with the *crispum* strain in a culture study (Rueness and Rueness, unpublished results) and clearly shown to be distinct.

M. Rueness (unpublished results), using eight different isolates of var. *plumula* and var. *bebbii* in fifteen different crossing combinations, obtained in some cases a rather intricate picture. The heterozygous tetrasporophytes are always of the var. *plumula* phenotype, and clear segregation into var. *plumula* and var. *bebbii* (reported previously in Rueness and Rueness, 1975) was consistently obtained through several generations. However, in some var. *plumula* × var. *bebbii* crosses, where both parents were isolated from the Oslofjord, the progeny derived from hybrid tetrasporophytes showed a subtle variation pattern, with transitional forms in addition to the var. *plumula* and var. *bebbii* phenotypes. Description of this variation pattern by morphological analysis has therefore proved exceptionally complex. Isoenzyme analysis of the different strains as a measure of genetic relationship, like that carried out on *Eucheuma* populations by Cheney and Babbel (1975), would be of considerable interest in this case.

CONCLUSIONS

Biosystematic approaches to the taxonomy of red algae are in their infancy and there is an almost complete lack of critical data on, for example, genetic and environmental variation in populations, reproductive biology, distribution and chromosome characteristics. In the present state of knowledge, artificial hybridization could serve as a useful contributory guide in arriving at taxonomic decisions. Breeding characteristics should not be overemphasized as a criterion, since sexual incompatibility may develop between individuals of different morphologically indistinguishable populations. On the other hand, at the infra-specific level, specimens differing markedly in genetically determined morphological characters may be completely interfertile.

ACKNOWLEDGEMENTS

I am very grateful to Inger Wallentinus who provided the samples from which the Baltic strain of *Ceramium* was isolated, and to Dr Peter Edwards who kindly sent a culture isolated from the American east coast. I am indebted to Professor Eystein Paasche and Mr J. H. Price who criticized a draft of this manuscript.

REFERENCES

Cheney, D. P. and Babbel, G. R. (1975). Isoenzyme variation in Florida populations of the red alga *Eucheuma*. *J. Phycol.* **11**, Suppl. 17.

Davis, P. H. and Heywood, V. H. (1963). "Principles of Angiosperm Taxonomy." Oliver and Boyd, Edinburgh and London.

Dixon, P. S. (1963). Variation and speciation in marine Rhodophyta. *In* "Speciation in the Sea" (J. P. Harding and N. Tebble, eds), pp. 51–62. Academic Press, London and New York.

Dixon, P. S. (1965). Perennation, vegetative propagation and algal life histories, with special reference to *Asparagopsis* and other Rhodophyta. *Bot. gothoburg.* **3**, 67–74.

Dixon, P. S. (1973). "Biology of the Rhodophyta." Oliver and Boyd, Edinburgh.

Dixon, P. S. and Richardson, W. N. (1970). Growth and reproduction in red algae in relation to light and dark cycles. *Ann. N.Y. Acad. Sci.* **175**, 764–777.

Du Rietz, G. E. (1930a). The fundamental units of biological taxonomy. *Svensk bot. Tidskr.* **24**, 333–428.

Du Rietz, G. E. (1930b). Studies in the taxonomy and ecology of *Ceramium diaphanum* in the Baltic. *Bot. Notiser* **1930**, 433–458.

Edwards, P. (1970). Attempted hybridization in the red algal genus *Polysiphonia*. *Nature, Lond.* **226**, 467–468.

Feldmann-Mazoyer, G. (1941). "Recherches sur les Ceramiacées de la Méditerranée occidentale." Imprimerie Minerva, Alger.

Heywood, V. H. (1973). Ecological data in practical taxonomy. *In* "Taxonomy and Ecology" (V. H. Heywood, ed.), pp. 329–347. Academic Press, London and New York.

Kapraun, D. F. (1977). Asexual propagules in the life history of *Polysiphonia ferulacea* (Rhodophyta, Ceramiales). *Phycologia* **16**, 417–426.

Kylin, H. (1924). Bemerkungen über einige *Ceramium*-Arten. *Bot. Notiser* **1924**, 443–452.

Kylin, H. (1944). Die Rhodophyceen der schwedischen Westküste. *Acta Univ. lund.* N.F. 2, **40**, 1–104.

Meer, J. P. van der and Bird, N. L. (1977). Genetics of *Gracilaria* sp. (Rhodophyceae, Gigartinales) I. Mendelian inheritance of two spontaneous green variants. *Phycologia* **16**, 159–161.

Polanshek, A. R. and West, J. A. (1975). Culture and hybridization studies on *Petrocelis* (Rhodophyta) from Alaska and California. *J. Phycol.* **11**, 434–439.

Price, J. H. (1973). Advances in the study of benthic marine algae since the time of E. M. Holmes. *Bot. J. Linn. Soc.* **67**, 47–102.

Rueness, J. (1973a). Culture and field observations on growth and reproduction of *Ceramium strictum* Harv. from the Oslofjord, Norway. *Norw. J. Bot.* **20**, 61–65

Rueness, J (1973b). Speciation in *Polysiphonia* (Rhodophyceae, Ceramiales) in view of hybridization experiments: *P. hemisphaerica* and *P. boldii*. *Phycologia* **12**, 107–109.

Rueness, J. and Rueness, M. (1975). Genetic control of morphogenesis in two varieties of *Antithamnion plumula* (Rhodophyceae, Ceramiales). *Phycologia* **14**, 81–85.

Slobodchikoff, C. N. (1976). "Concepts of Species" (C. N. Slobodchikoff, ed.). Dowden, Hutchinson and Ross, Inc., Stroudsburg, Pennsylvania.

Stace, C. A. (1975). "Hybridization and the Flora of the British Isles" (C. A. Stace, ed.). Academic Press, London and New York.

Sundene, O. (1959). Form variation in *Antithamnion plumula*. Experiments on Plymouth and Oslofjord strains in culture. *Nytt Mag. Bot.* **7**, 181–187.

Sundene, O. (1975). Experimental studies on form variation in *Antithamnion plumula* (Rhodophyceae). *Norw. J. Bot.* **22**, 35–42.

Turesson, G. (1922). The species and the variety as ecological units. *Hereditas* **3**, 100–113.

Wachenfeldt, T. von (1975). "Marine Benthic Algae and the Environment in the Öresund." I–III. (Thesis, mimeogr.). Lund.

Waern, M. (1952). Rocky-shore algae in the Öregrund Archipelago. *Acta phytogeogr. suecica* **30**, 1–297.

12 | Ecological Determination of Adult Form in *Callithamnion*: Its Taxonomic Implications

J. H. PRICE

Department of Botany, British Museum (Natural History), England

Abstract: A brief summary of the distribution and ecology of principal British species is presented. Data derived from field transplantation, from seasonal population observations and from culture have revealed that certain taxonomically-utilized characteristics vary with the environment. The robustness of growth, cortication and adventitious weft formation, axial cell form and size, apical branching pattern and form, hyaline hair formation and perennation are examined in detail. Generally, robust growth-forms or species appear in conditions of considerable wave-beat or in exposed niches on less exposed shores; delicate forms or species appear in conditions of shelter from wave-beat or in protected niches on more exposed shores. Wide-spectrum species tend closely to resemble, in gross morphology, those species generally more characteristic of the shore or niche occupied. The basis for an ecologically-orientated key to species groups is presented in diagrammatic form; it is suggested that this is the logical initial approach to keying out "species" in such a variable and so far little-known genus.

ECOLOGY

1. *Introduction*

The genus *Callithamnion* is relatively widely distributed, although certain species appear temperature-restricted within the overall generic distribution. However, few *Callithamnion* species have been recorded in the Tropics, fewer still in the Antarctic region and none in the Arctic. Reports from the Antarctic proper are doubtful because of possible misdetermination or inadequate material. In other oceanic temperature zones, particularly of the Atlantic, *Callithamnion* is both diversified and ecologically widespread.

Ecological amplitude can conveniently be described in terms of vertical levels of appearance and characteristically exploited habitat conditions. In Great

Systematics Association Special Volume No. 10, "Modern Approaches to the Taxonomy of Red and Brown Algae", edited by D. E. G. Irvine and J. H. Price, 1978, pp. 263–300. Academic Press, London and New York.

Britain and western Europe from northern Norway south to the Algarve (the principal area with which the present paper is concerned), the overall characteristics of generic and infra-generic micro-niche distribution in *Callithamnion* are consistent and interesting. Generalities of these characteristics apply equally well along the western seaboard of the north Atlantic, although several different species names are used there. The southern Atlantic also differs primarily in species names applied in the area, rather than in variation pattern. Since the bulk of present observations concern British shores, the general ecology there of *Callithamnion* is briefly considered as a background introduction.

TABLE I. The occurrence of common intertidal and shallow subtidal exposed-shore *Callithamnion*

If no:	Then no:
1. *Alaria* and steep rocks (except in E. and N. Scotland, where gentler slopes are invaded)	*C. arbuscula*
2. *Himanthalia* and shallow pools, shelved edges of deeper pools, or damp draining surfaces	*C. granulatum*
3. *Laminaria digitata* and firm rock substrata	*C. tetragonum*
4. *Himanthalia*, *Bifurcaria*, firm rock, and more sheltered flats or micro-niches than for *C. granulatum*	*C. tetricum*
5. Wide spectrum of exposure and micro-niche type	*C. hookeri*

Open rocky shores of considerable intertidal and shallow subtidal exposure to wave-action everywhere in the north-east Atlantic carry a rather specialized flora and are largely dominated by animals. Amongst the specialized flora of these conditions in Great Britain are usually at least four species of *Callithamnion*—*C. arbuscula* (Dillw.) Lyngbye, *C. granulatum* (Ducluz.) C. Ag., *C. hookeri* (Dillw.) S. F. Gray, including the form known as *C. scopulorum auct.* and *C. tetragonum* (With.) S. F. Gray. In the south there appears also *C. tetricum* (Dillw.) S. F. Gray. Only the first two of these named species tend to characterize levels on shores that lie towards the extreme of exposure. Their presence, and that of the other species, can be correlated, generally to a large extent, with the appearance of other larger algal species, either through requiring

FIG. 1. *Callithamnion arbuscula*. Close-up, Porthcurno, South Cornwall. Usual size at southern range periphery; attached to normal substratum, the inside faces of empty barnacle (*Balanus perforatus*) shells.

FIG. 2. *Callithamnion granulatum*. Close-up, Portland Bill, Dorset. Bushy growth-form in shallow pool, on *Corallina officinalis*.

Figs 1 and 2. Captions on opposite page.

them as substrata or requiring the same balance of general environmental conditions (Table 1). *C. arbuscula* occurs slightly above midlittoral in level, principally on open faces and attached to the shells of sedentary animals (Fig. 1) or to rock; *C. granulatum* is slightly lower on shore and extends to low water level, mostly epiphytic in standing water or extreme damp (Fig. 2). Blanket populations of either species, in the optimal circumstances of full vigour and distribution range centre, indiscriminately invade habitat conditions more characteristically occupied by the other.

C. hookeri is a very wide-spectrum, wide-distribution, species that has been recognized or confused under a multitude of species names, of which the commonest in Great Britain and adjacent areas include *bipinnatum* Crouan frat.; *brodiaei* Harv. in Hook.; *decompositum* J. Ag.; *dudresnayi* (Bonnem.) Crouan frat.; *polyspermum* C. Ag.; *scopulorum* and *tripinnatum* C. Ag. Even *roseum sensu* Harv. has been occasionally invoked. This is not to say that some of these taxa may not finally prove to be "good" species when adequate data on such factors as cytology, behaviour in the field and genecology are available; indeed, some are already clearly distinguishable apart from *hookeri* and related forms. At the present time, however, attempts to recognize definitive morphological discontinuities of significance and taxonomic utility within *C. hookeri sensu lato* for Great Britain inevitably produce further confusion. Misdetermination, misapplication and inconsistency have played so large a part in the nomenclatural and taxonomic tradition with which *hookeri* and the other epithets are inescapably identified that rationalization is a major problem. Here it is sufficient to state that forms presently thought worthy of inclusion in a wide concept of *C. hookeri* have habits (as individuals or as populations) that customarily embrace particular habitat niches. These niches vary widely, from the deep subtidal to the very high intertidal and supralittoral fringe, and from the calmest of conditions through to the most extreme wave-action examined.

The remaining species of consistent appearance on shores of extreme wave-action, *C. tetragonum*, is not as characterising as the first two mentioned because it is a wide-spectrum plant, rather as (but more restrictedly than) *C. hookeri*, and appears in both the shallow and the deeper subtidal. It also occurs in open, damp (but sheltered) conditions in the lower intertidal. Principally, *C. tetragonum* is epiphytic on a variety of larger algae, but will also colonize rock surfaces in favourable niches in the absence of larger forms.

The distribution of *C. tetricum* on shores and relative to wave-action is very similar to that of *C. tetragonum*, although the balance of its occurrence tends to be more often epilithic and less often epiphytic. However, that balance is not as weighted against the latter as is that of *C. tetragonum* against the former.

C. tetricum is a southern species, not recorded north of Larne and the Isle of Man on the west and of very sporadic occurrence in the east (e.g. one authentic record from Kent). It tends to be associated more in shore descriptions with the extreme exposure forms *C. arbuscula* and *C. granulatum* than does *C. tetragonum*. The primary reason for this is that *C. tetricum* is more characteristically a midlittoral species, although occasionally it invades the subtidal in Devon (Westbrook, 1930) and Cornwall (Price, unpublished), whilst *C. tetragonum* is more characteristically a shallow subtidal species that occasionally invades the intertidal in, for example, Hampshire (Price, unpublished) and Normandy (Price and Tittley, unpublished). Westbrook (1930: 193) postulated the replacement by *C. tetricum* and *C. granulatum* of the northerly *C. arbuscula* in the same extremely wave-beaten situations in the south of the country. Although *C. arbuscula* is very luxuriant in the north and less so in the south and *C. granulatum* is very well developed in the south and less so in the north, the critical habitat requirements of the two species are usually subtly different, as to both position on the shore and degree of wave-action accepted. *C. tetricum* is again slightly different in its requirements.

On many exposed shores in the south-west, all these species, in the case of *C. hookeri* in more than one growth form and niche, are often detectable on the same shore, particularly where there is varied aspect and mosaic exposure. Vellan Drang, Cornwall, provides an excellent example of this. There, an outer discontinuous firm rock "reef" takes the full force of the Atlantic fetch on its outer and western rim, being rather more sheltered as it sweeps round to east and north, partially enclosing the sides of a central boulder area and sandy lagoon that are somewhat wave-swept at fuller tides through gaps in the reef. The *Callithamnion* species of open rock shores are precisely distributed there, according exactly with the details given previously.

Shores of moderate exposure/shelter balance, but on open rocky areas, generally embrace a range of variably sheltered and deep rock pools. These, in common with the instances where similar pools provide an adequate micro-habitat in more exposed shores, introduce *Callithamnion* species that do not appear on the more open aspects typical of that particular exposure level. Mostly, *C. corymbosum* (Sm.) Lyngb. and *C. roseum* appear in these conditions although, in deeper low-level shaded pools, characteristically subtidal species such as lax forms of *C. hookeri*, *C. byssoides* Arnott ex Harv. in Hook., and the allied genera *Pleonosporium* and *Seirospora* are not uncommon. In the subtidal are known species (rarely reported but probably not really so rare of occurrence) such as *C. decompositum* (Price and Tittley, 1978) and *C. bipinnatum*, both of which may prove to be taxa recognized within the form range of the same species.

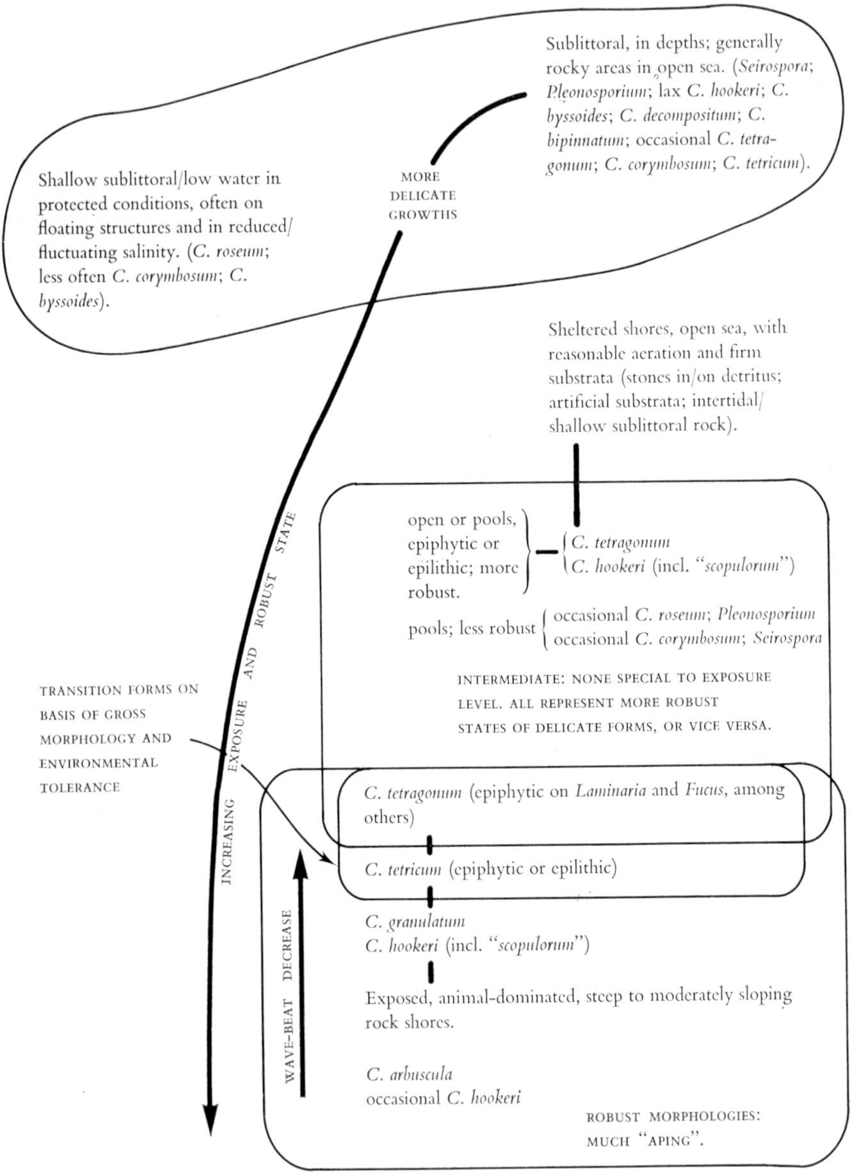

FIG. 3. *Callithamnion* and allies in Great Britain. Summary of ecological background.

In all probability, increasingly detailed subtidal surveying will (as in Mull) show such species to be more frequent and widespread.

Many of the morphologically more delicate forms in this genus are most well-developed in the extreme shelter, detrital conditions, and reduced or fluctuating salinities of river estuaries and similar areas. *C. byssoides* and *C. corymbosum*, with *C. hookeri* where the pollution and freshwater influences are not too great (Gillham, 1957a, b; Edwards, 1972a, b; Tittley and Price, 1977a, b), are frequently encountered, but the most characteristic "species" of these conditions is *C. roseum*, especially (although by no means only) where the presence of floating structures allows the plants to maintain a very shallow subtidal band without danger of exposure through tidal movement (Tittley and Price, 1977a). *C. roseum* represents a widely euryhaline morphology that also appears on open full-salinity shores.

This brief ecological survey is summarized in Fig. 3.

2. Methods

Balances of environmental factors comparable with the natural conditions (e.g. those of wave-beaten rocky shores) are often virtually impossible to recreate in the laboratory. Culture results, therefore, cannot justifiably be applied in species delimitation by morphology, unless strongly supported by carefully gathered long-term field data. Culture environments are usually highly artificial and may perpetuate phenomena that would rapidly eliminate plants or populations in nature; furthermore, responses of plants tend to drift from the field norm after much time in culture (Dixon and Richardson, 1970; Price, 1973). By contrast, culturing may reveal, especially using a gradient apparatus (Edwards and Van Baalen, 1970; Van Baalen and Edwards, 1973), that plants of the same clonal origin tolerate varied light/temperature balances and in so doing present gross morphologies with significant variations. This can elucidate otherwise mystifying field variation (Edwards, 1971).

Of other possible approaches to circumvent the taxonomic reservations on culture data, two have been found most useful and have contributed largely to the data presented here:

1. Long-term observation of wide-spectrum "species" apparently capable of growth in a wide range of environmental conditions or micro-niches, in vigorous field populations of adequate longevity. Similar observation of "species" of narrower ecological amplitude.

2. Transplantation of field material, through detachment of natural substrata with attached plants (a similar method was employed by Norton, 1969). The provision of prepared experimental artificial substrata for settle-

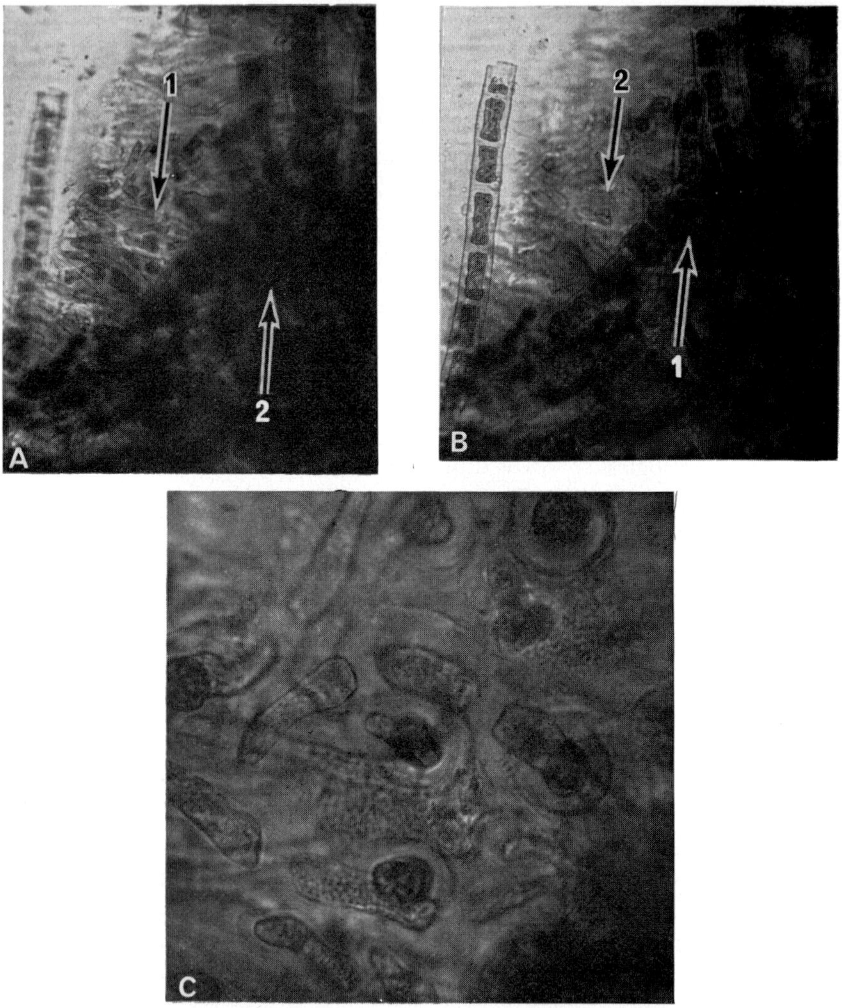

FIG. 4. *Callithamnion arbuscula*. Porthcurno, August. A. Fracture face of basal main axial break. Focused on cortication (arrow 1), with overlying adventitious weft (arrow 2) out of focus toward the viewer. × 125. B. As A, but focused on the overlying adventitious weft (arrow 1), with the underlying cortication out of focus into the picture (arrow 2). × 125. C. Magnification (× 500) of cortication at point in centre of focused cortical region of 4A. Outer parts of adventitious weft stripped off at this point, showing numbers of corticating filaments emerging extra-matrically (end-on, toward the viewer) and giving rise to adventitious lateral growths. Note frequency of emergence, compared with Fig. 5.

ment (Harlin and Lindbergh, 1977) and subsequent translocation (Sundene, 1962a, b, 1963; Neushul and Powell, 1964; Lüning, 1970) can be equally revealing and, in the case of the latter, sometimes more effective.

Although the first method has been that principally employed, the second has been utilized in attempting transplantations of species across micro-niches that support similar growth forms, and from one micro-niche type to another. Stability of criteria at both overall gross morphology level and at lower levels can be assessed in this way. The major difficulties in all such field experiments are the avoidance of more than the inevitable minimum of "house-moving trauma" and the interpretation of results in the light of that possible factor.

VARIATION PATTERNS AND THEIR RELATIONSHIP TO THE ENVIRONMENT

1. *General Gross Morphology*

(*a*) *Introduction.* The genus *Callithamnion* and immediate allies, as currently recognized, embrace forms that are:

1. Robust and heavily corticate, through both multi-layered development of down-growing corticating filaments* (Figs 4A, C; 5), at least on all but ultimate and penultimate ramification levels, and the subsequent superficial development of an often dense to impenetrable weft (sometimes referred to as a matrix) of adventitious laterals (Figs 4B, C).
2. Very delicate, with very long axial and main lateral cells that carry relatively little to no cortication of any kind, even basal to the whole plant.

These are the extremes; between them lies a whole spectrum of continuous, discontinuous or patchy clinal variation in most vegetative morphological parameters that can be recognized, even at subcellular level. It is not possible to consider all parameters here; a few of the more evident and hitherto taxonomically frequently utilized ones are examined later.

Across the spectrum, forms that are currently named as species tend to take up positions of which a central norm and a variably wide and variably identifiable "halo" of vegetative and reproductive morphological variation can be recognized; the "halos" in places overlap to a disconcerting extent with other similar "species positions" near by. Nevertheless, amongst the general overlap can be discerned groups of gross morphologies that are very much nearer together than are any of the constituents to members of other recognizable groups. These morphological types are detailed later and summarized

---

* The term "corticating filament" is retained here (cf. Dixon, 1973: 67) since, whatever their origin, such filaments *are* corticating and the term is descriptive and succinct.

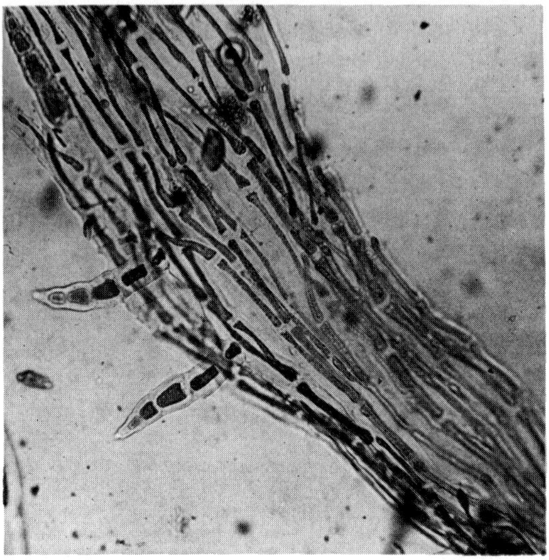

FIG. 5. *Callithamnion tetragonum*. Porthcurno, August. Multi-layered cortical strip from axis; inner layers largely out of focus, but visible just occasionally beneath outer layers. A few adventitious laterals, recognizably of this species, arise directly from outer corticating filaments. Note how many fewer per unit area than in the much smaller field of view of Fig. 4C. × 125.

in Fig. 9. Although arising from complex interrelations of many different characteristics, all of which finally depend on the relative rates of cell-growth and apical division, the vegetative gross morphology can be usefully distinguished for discussion purposes into:

1. Main axis and major lateral form: cell dimensions; extent and form of cortication.
2. Main branching pattern and sequence.
3. Apical branching form (clustered/sheathed/gross open/delicate open) and peripheral division pattern.
4. Form of individual peripheral laterals.
5. Extent and type of perennation.

The present paper takes as its main themes 1, 3 and 5 above, although occasional observations are made on the other topics.

(*b*) *Main axis and major lateral form: cells and cortication type*. The overall biological and ecological significance of axial and major lateral configuration are summarized in Fig. 6. There is very good general correlation, as would be expected, between overall delicacy of thallus form and the presence of least cortication, and between overall robustness and possession of thickest multi-layered cortication with most ramified formation of weft of adventitious branches. Although there are a few exceptions (e.g. *C. decompositum*, *C. bipinnatum*) in which paucity of corticating filament formation is accompanied by axial or major lateral cells with length/breadth ratios not exceeding 2–3 : 1 (Fig. 7), there is a strong general tendency for the presence of dense cortication

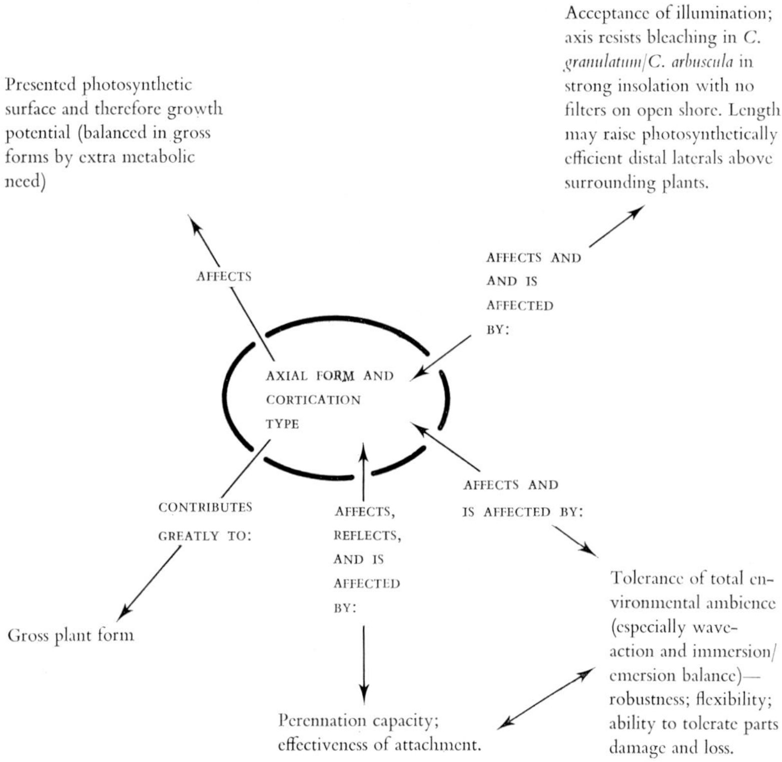

FIG. 6. The biological and ecological significance of the axial and major lateral configuration in *Callithamnion* and allies. Diagrammatic.

with adventitious weft formation to be accompanied by axial cells that are relatively short compared to their width. The reverse is also true, sparse cortication being accompanied by axial and main lateral cells that, in general, exceed 6 : 1 in length/breadth ratios and often reach extreme ratios such as 13–17 : 1 (as, for example, in *C. byssoides* and *C. corymbosum*).

There is, as yet, no good experimental evidence one way or the other but it must remain a distinct possibility that a degree of mechanical or other restriction on growth in length of axial cells is exerted by the sheer presence of intramatricular cortical growth of substantial proportions, with or without development of dense overlying adventitious weft. The interaction of all the factors would thereby cumulatively augment the robustness and axial strength. Dixon

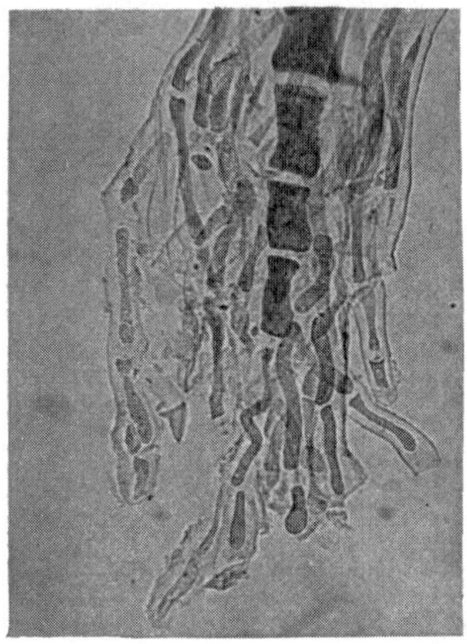

FIG. 7. *Callithamnion decompositum.* Base of plant stripped from hydroid substrate. Note rhizoidal attachment filaments at the base and fairly sparse analogous corticating filaments, from cells above, in no way obscuring the main axial base. Size of axial cells (length and breadth) increases somewhat upward from basal cell, but remains at low length/breadth ratio. × 125.

(1971) has discussed the effect of axis on lateral and lateral on axis in terms of patterns of cell enlargement in both. He demonstrated clearly:

1. The bearing axis of approximately 70 cells in a *Callithamnion corymbosum* plant contained longer axial cells in the region from five cells above the base to approximately 35 cells above the base than either above or below that region;

2. The elongation rate in lower cells of laterals diminishes after a time so that the longer cells are again about five cells above the base and distal.

Smaller cells at the bases of laterals certainly could be explained by the effect of an axis on cell enlargement, but this effect cannot explain the same phenomenon in the main axis of a plant (cf. Dixon, 1971; Fig. 1). Areas of restricted enlargement of cells at the bases of major laterals and axes are precisely those positions from which corticating and attachment filaments first develop and where the cortication is finally thickest. Considered alongside the general correlation of shorter axial cells in well-corticated, exposed shore species than in less-corticated, more sheltered shore species, this suggests that the shorter cells at bases of laterals may not be wholly a result of axial restraint, but also might relate in some mechanical and/or internal physiological manner to the cortication process. The causes of this phenomenon shown by an axial basal region in the absence of the postulated restricting agency of a dominant axis seem likely

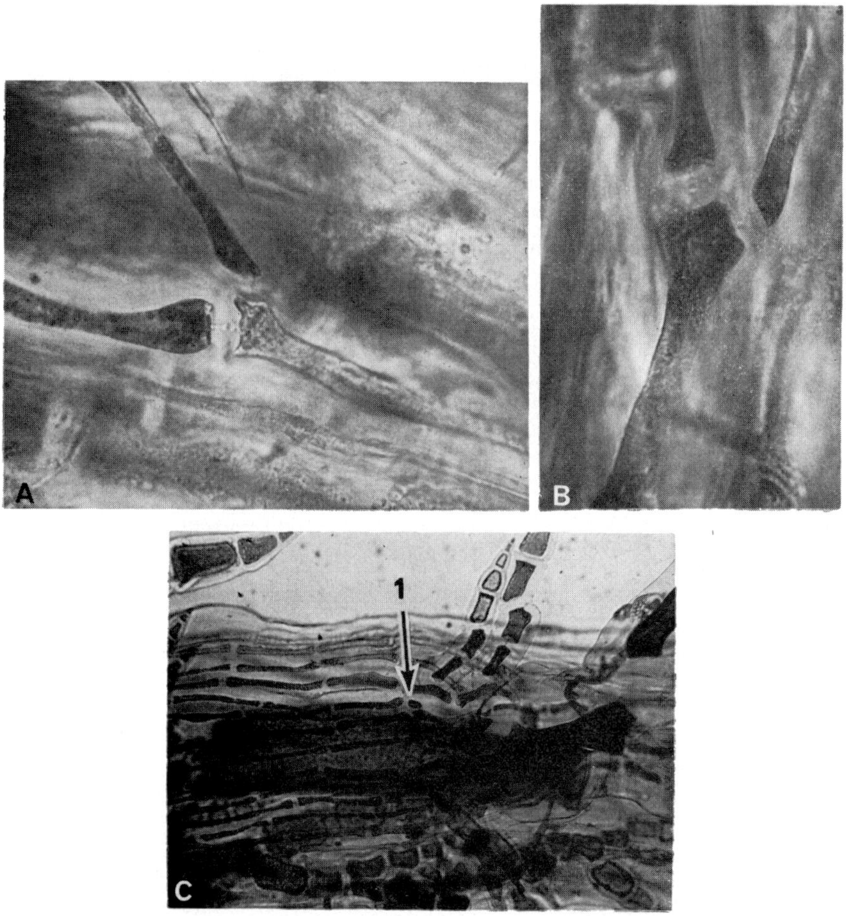

FIG. 8. A and B. *Callithamnion roseum*. Two examples of corticating filaments undergoing branching. Pit-connections can be clearly seen between branches and on-going main filaments. × 500. C. *C. tetragonum*. Cortication near base of a major lateral. A few adventitious laterals arise, top right, and a normal sequence branch can be seen centre right. Major lateral axial cell dimly visible through several cortical layers, centrally. Just above upper surface of that cell and slightly right of centre can be seen a branching corticating filament that has formed a secondary pit-connection with an adjacent filament (arrowed 1). × 125.

to be active in the similar phenomenon at the bases of laterals, even though there the bearing axis is also present to exert similar influence.

This, therefore, raises the possibility that restrictions, physiological or mechanical, on the axis by lateral development are being demonstrated by *Callithamnion*. Such a restriction by laterals themselves (represented by corticating filaments), or by their precursor internal arrangements in axial cells, would help in explaining the similar restriction (clearly indicated in Dixons' histogram, Fig. 1, 1971; see also Børgesen, 1917: 216–18; Boddeke, 1958: 592; Riouall, 1974) at the base of the main axis. Although Dixon made no comment on the occurrence in the *C. corymbosum* axis, he did indicate for *Ceramium* (p. 203) that adventitious laterals (axes) found in the older parts of the plant did have an effect, albeit reduced compared to that of laterals found apically and developed normally, on cell enlargement in the bearing axis. In any use of axial or major lateral cell dimensions, it is therefore necessary to be extremely specific as to the region within the filament that has been measured, and to make a precise statement of the morphological phenomena attendant on the measured cells.

The nature of the corticating filaments is consistent throughout the genus, despite the vast differences in the extent of development. The potential for that development exists in major laterals and axes of the whole genus and its adjacent allies (L'Hardy-Halos, 1970, 1971b); it is usually realized. The corticating filaments are best regarded as the direct homologue of the polarized attachment activity in the basal axial cell and/or adjacent cells (Fig. 7), where such activity is undertaken right from the beginning on behalf of the subsequent plant (cf. Edwards, 1977). The density of growth of corticating filaments in the multilayer-developing species leads, with the tendency for the filaments to branch in some species (e.g. *C. roseum* (Fig. 8A, B), *C. tetragonum* (Fig. 8C)), to a close appression of adjacent filaments. This juxtapositioning appears to encourage the formation of secondary pit-connections between cells in the adjacent filaments (Fig. 8C). Development of the corticating filaments can be regarded, probably correctly, as an additional aspect of the total interconvertibility of form and function between all filaments (or cells; Konrad Hawkins, 1964a, b, 1968) of the apparently simpler Floridean thalli; like other "filaments of limited growth", this filamentous form can convert to the occasional production of unlimited axes, in this case amongst the adventitious laterals.

The relationships that can be deduced, from long-term and widespread examinations of morphology and micro-niche, between cortication and wave-exposure and thus between axial cell characteristics and that same wave-beat, are summarized for the principal species in Fig. 9. This figure is self-explanatory and also shows the positive correlation between the ambient environment and

Species	Most robust morphology attained	Ecological grade penetration
arbuscula	A(i)	1
corymbosum	B/C	2/3/4
decompositum/ bipinnatum	C	3 (sublittoral)
byssoides	C	2/3/4
hookeri (all forms)	A to C (wide spectrum)	Full range (1-4)
granulatum	A(i)	1
roseum	B/C	2/3/4
tetragonum	A(ii)/B	1/2/3
tetricum	A(ii)	1/2/3
Pleonosporium	C	2/3
Seirospora	C	2/3

1

Extreme exposure. Thick cortication; adventitious matrix.

A(i): bushy peripherals giving spongy texture; *A*(ii): coarse texture.

arbuscula
granulatum
tetricum
tetragonum
aping *hookeri*.

2

Less extreme, or moderate shelter from wave-action. Shallow subtidal *tetragonum* in open exposure; intertidal *tetragonum* in more shelter; normal *hookeri* form; occasional more corticate *roseum*.

B: strong cortication; no or little matrix; branching more open. *C*: morphologies in pools (see 3 for details)—*corymbosum*; *Pleonosporium*; *Seirospora*.

3

Sheltered intertidal and shallow subtidal, or fully subtidal. Flaccid growth forms with normally some cortication basally, but often very little corticate. Peripheral branching always open, often distichous, lacking apical clumping.

C: *byssoides*; *corymbosum*; *decompositum*/*bipinnatum*; *hookeri* (full shelter form); *roseum*; *Pleonosporium*; *Seirospora*.

4

Estuarine calm; fluctuating or reduced salinity. Open growth, with cortication due to the level at which growing and the movement of water, possibly emphasized by generally adverse environment (Rosenvinge, 1924; Riouall, 1974).

C: (estuarine): *roseum*; less often *corymbosum*, *hookeri*.

FIG. 9. Relationships between certain morphological aspects and environmental conditions, in the genus *Callithamnion*. Diagrammatic.

FIG. 10. Caption on opposite page.

form and extent in another morphological phenomenon, the apical and peripheral branching form and pattern. This, together with another principally temporal phenomenon, perennation, is considered below.

(*c*) *Apical branching form and peripheral division pattern.* Figure 9 indicates that *Callithamnion* spp. of exposed shores have one of two principal peripheral patterns:

A (i), as in *C. granulatum* and *C. arbuscula*, where associated tightly-appressed bushy clusters (the "bouquets" of Debray, 1899) of peripheral laterals are not themselves especially robust but serve to retain considerable moisture against desiccation, thus giving the plant a spongy texture when wet (Figs 10, 11);

A (ii), as in *C. tetricum* and *C. tetragonum*, which are much more coarse-textured (Figs 12, 13), but still very heavily branched (*C. tetricum*) so that considerable moisture-retention is still possible in intertidal conditions, without the plants ever giving the spongy impression of group A (i). The more frequently subtidal *C. tetragonum* has a more generally open and robust peripheral branching system, except for the parts adjacent to the unlimited lateral apices, where sheathing clusters of the very robust, limited-growth peripheral laterals both protect the active apices and retain moisture to some extent (Fig. 14A). Intertidal growths of *C. tetragonum* tend to be more tightly branched than subtidal plants, whereas in *C. tetricum* the usually much-branched intertidal appearance becomes more open in the relatively uncommon subtidal occurrences. Apart from the similar sheathing protection of unlimited apices in some forms, habitats of progressively less mechanical wave/water action generally carry *Callithamnion* growth forms or species with parallel progressive opening of the peripheral branching pattern. This is achieved through longer cells (and therefore greater distances between branchings), wider branching angles, less or no lateral appression of laterals and generally fewer branches and subsidiary peripheral laterals. Apart from *C. tetragonum*, where it occurs in less exposed conditions, the more robust of the remaining morphologies is *C. roseum*; this "species" shows some apical sheathing by peripheral laterals (Fig. 14B, C), more evident (Fig. 9) in habitats of grade 2 (moderately sheltered open shores) than in those

FIG. 10. *Callithamnion arbuscula*. Porthcurno, May. A. Apical corymbose cluster, showing usual dense appearance, with individual laterals very difficult to trace accurately. Very effective form for water-retention. ×25. B. Similar, but smaller cluster spread to reveal detail. Individual peripheral laterals and their form more clearly visible (cf. those in *C. granulatum*, Fig. 11B). C. Close-up on shore, *in situ*. Size estimate can be made from spongy appearance, the individual surface bumps each representing one apical corymbose cluster, as shown detailed in Fig. 10A.

FIG. 11. *Callithamnion granulatum*. Porthcurno, August. A. Apical corymbose cluster of peripheral laterals, with clear lateral incurving in which the strict pseudodichotomy can still be discerned in places. Note densely-tipped, somewhat clavate (but not bulbous) apical hair tips. ×125. B. More open clump with fewer hairs. Spread for detailed observation, showing comparatively strict pseudodichotomy, contrasting with the lateral form and branching pattern of *C. arbuscula* (cf. Fig. 10B). Peveril Point, Dorset, May. C. Spongy material *in situ* on shore, attached to *Corallina*. Individual surface clumps (corymbose clusters) are clearly visible in this centrally placed specimen. Photograph, taken at low water in the lower intertidal, shows the immense water-retaining capacity compared with that of the "host" *Corallina*. Size estimable from the *Himanthalia* button and thong alongside.

of grades 3 or 4 (full shelter; subtidal; estuaries), although there is still some development of the clumping on the *C. roseum* plants characterizing the latter conditions. Little or no comparable clumping of peripheral laterals, either around unlimited apices or not, is evident in the fully subtidal *Callithamnion*

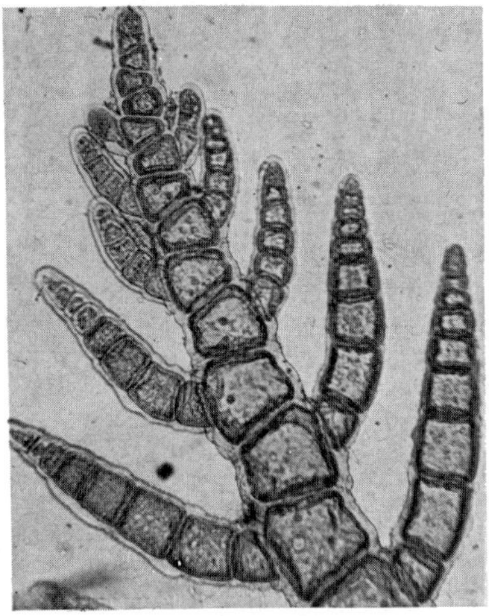

FIG. 12

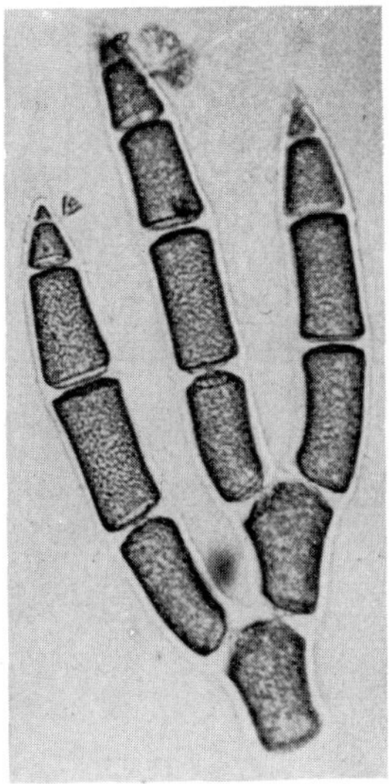

FIG. 13

FIG. 12. (Left) *Callithamnion tetricum*. Vellan Drang, Cornwall, April. Apex of unlimited growth lateral, with limited growth peripheral laterals. Note low length/breadth ratios; angled dividing walls; general overall coarse appearance and robust construction. × 125.

FIG. 13. (Right) *C. tetragonum*. Peripheral laterals, × 125. Shows comparatively large cells deriving from the small cone-shaped apical cell of the limited growth laterals. Note also the overall comparatively gross, coarse construction in this species.

growths. Species such as *C. decompositum* and *C. bipinnatum* are good examples of this extremely open condition (Fig. 15).

(*d*) *Wide-spectrum species*. This general pattern of environmental variation in cortication form and peripheral branching is not usually blurred by those "species" showing wide-spectrum habitat tolerance. *C. roseum* has been considered above, but the classic example of a wide-spectrum species is *C. hookeri* which on open exposed shores in the south of England forms populations that even the experienced eye has to examine very closely to distinguish morphologically from the widespread *C. granulatum* that characterizes such habitats.

Higher on shores (in the upper littoral and supralittoral fringe) with high and lesser degrees of wave-exposure, there is locally abundant occurrence of what

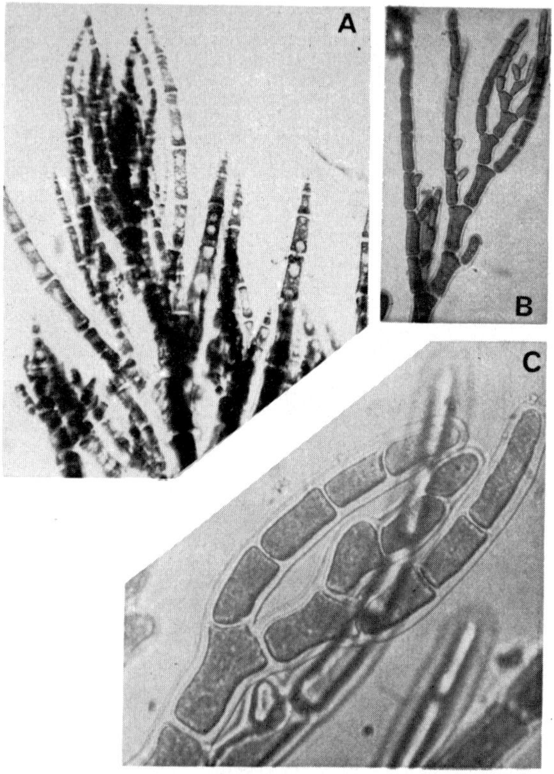

FIG. 14. *C. tetragonum*. A. Sheathing clusters of very robust limited growth laterals (in this case a of form that would be referred to as "var. *brachiatum*") around the unlimited growth apices. ×63. *C. roseum*. B. Unlimited apex showing clear sheathing by apically upsweeping limited growth laterals. ×125. C. Similar apex at higher magnification. ×500.

evidence suggests is another form of *C. hookeri*, usually known as *C. scopulorum;* this occurence presents other interesting facets. There, the gregarious outfall of spores, the presence at all levels of inter-twining "rhizoids" (Lami, 1942) and the general pattern of spreading growth develop a "turf" or "clumped" form in which the population morphology offsets the delicacy of the individual plant's morphology (Fig. 16), leading to persisting invasion of suitable shady, moisture-retaining micro-niches that would not be open in any long-term way to the individual plants.

The wide-spectrum pattern of *C. hookeri* is summarized in Fig. 17.

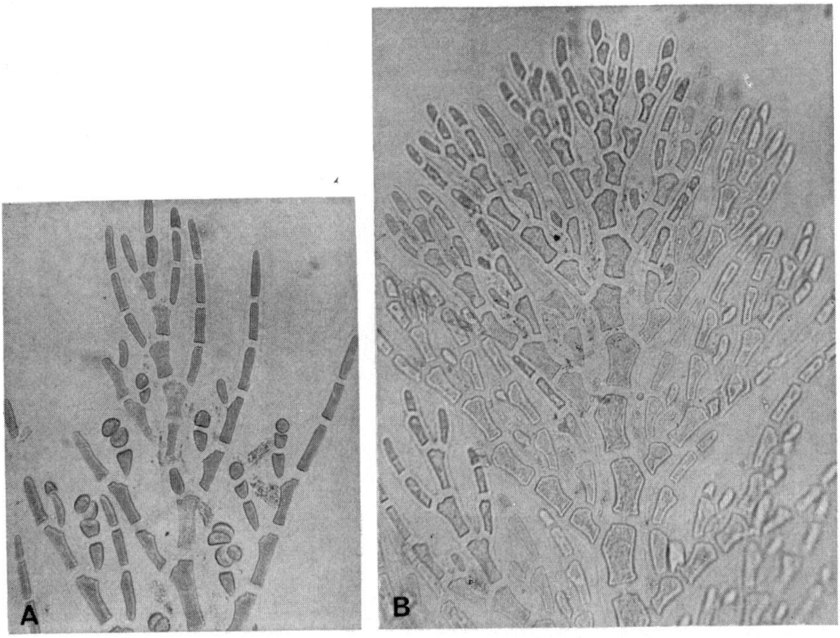

FIG. 15. *C. decompositum–bipinnatum group*. A. Parasporangial material. Shows clearly the very open distichous apical construction with few peripheral laterals, comparatively unclustered. Shetland, subtidal. × 125. B. Vegetative material, still open and distichous but rather more closely bunched than the sporangial material of 15A. × 125.

(*e*) *Extent and manner of perennation*. The ability to perennate in the erect plant of *Callithamnion* appears to be affected strongly by the presence of substantial cortication and dense formation of adventitious weft. There are as yet few field data for the more delicate forms in the morphological spectrum but such observations as are available by past authors and the present author tend to support the usual literature statements (e.g. as in Garbary, 1976) regarding the annual cycle of most such species. Edwards (1969) has convincingly demonstrated how speedily (in culture at least) the life-history can be completed in the delicate subtidal growths of *Callithamnion*; he postulated completion more than once during the growing season in nature.

FIG. 16. Caption on opposite page.

Statements in the literature to the effect that species of exposed shores (*C. arbuscula*, *C. granulatum*, *C. tetricum*, *C. tetragonum*) are ephemerophytes (present throughout the year but as overlapping generations; therefore annual) must be viewed with extreme caution. From long-term observations of southern and

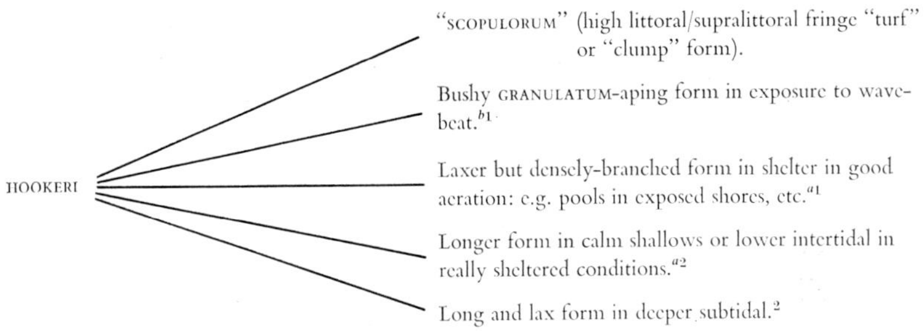

a = often called by various other names, chiefly *polyspermum* in G.B.
b = often called by various other names, chiefly *brodiaei* in G.B.
[1] = many specimens from such conditions would equate to the f. *typica sensu* Kylin (1907) and Levring (1935), and to the f. *areschougii sensu* Rosenvinge (1924) and Kylin (1944).
[2] = many specimens from such conditions would equate to the f. *elongata sensu* Kylin (1907) and Levring (1953).

FIG. 17. Morphology and habitat in *Callithamnion hookeri*: a summary.

south-western coast populations, it is clear that some plants, at least, persist for two or more years, the more susceptible apical regions of the branching systems falling away to leave plants of *C. arbuscula* and *C. granulatum* looking rather straggly and lacking in the delicately bushy appeal remarked on by many previous workers. These observations tally with comments by Børgesen (1902) on Faeroes populations. There is strong positive correlation between the extent of cortication and adventitious weft formation and the amount of the plant that

---

FIG. 16. *Callithamnion* "*scopulorum*". A. St Margaret's Bay, Kent. July. High intertidal close-up of *scopulorum* (*hookeri*) clumps; that in the centre of the field especially demonstrates typical habit and usual population micro-niche. Rock is chalk; shore comparatively sheltered from wave-action at this level and position. Scale can be deduced for the population by relation to the *Patella* (diameter *c.* 5 cm) to the left. Compare Fig. 16B. B. Portland Bill, Dorset, May. High intertidal/supralittoral fringe population in comparative shade and moisture-retaining niche on a shore providing very much more exposed overall environment and a comparatively more exposed micro-niche. Scale can be assessed from the many barnacles and, more accurately, from the penny piece at the top right.

overwinters. In the spongy-textured species, the bushy lateral clumps regenerate during the early part of the next season of active growth.

C. tetricum and *C. tetragonum* act rather differently, on the evidence so far available. *C. tetricum* loses less of its peripheral branching systems than do the bushy species; the harsh robustness of its texture and form relative to others in the genus (Fig. 12) possibly are manifestations that bear on the reasons for its more complete persistence. The winter growth form more strongly resembles the full spring/early summer form than does that in the perennated thalli of *C. arbuscula* and *C. granulatum*, although all of these are still quite easily recognized *C. tetragonum* seems to perennate less frequently as substantial portions of adult thalli, although the same light/dark colour contrasts between older axes and young growth occur on the plants as in *C. arbuscula* and *C. granulatum* in the south. In al! these forms, populations at different levels on shore and in different conditions sometimes show different perennating activity. At Portland Bill, for example, *C. granulatum* has been observed to manifest different extents of persistent (older plant) perennating material according to the plant, the year, the level on the shore and the depth of the pool. These may all be facets of a not yet fully understood sequence; this is being studied further.

2. Hyaline Terminal and Lateral Hairs

The production of terminal, often subsequently to become lateral, hyaline hairs is a common phenomenon in some species of the genus (Westbrook, 1927a), although it has never been noted for others (Table II). The hairs are both inconsistent and ephemeral in their occurrence, as well as being sometimes totally dehiscent if specimens are subjected to rough handling involving, for example, high concentrations of preservative. In my experience, the presence of hairs generally relates to high light intensities, although even then there is considerable irregularity that probably reflects the physiological state of the individual plant as a prime determinant, rather than being wholly a reaction to external environment. Chamberlain (unpublished) carried out experiments on *Ceramium rubrum* that indicated light intensity to be a principal causal factor in the production of hairs in a population on rafts in Langstone Harbour (Hants.). He also concluded, for another population on Southsea beach, that the action of waves (surf) could have knocked off all hairs. So far as *Callithamnion* is concerned, rough water seems of little or no significance in preventing development of or, indeed, removing hair points from a healthy plant, under normal circumstances. Populations of *C. arbuscula* at Porthcurno (Fig. 18) and of *C. granulatum* there (Fig. 19) and at Portland Bill regularly carry dense growths of long hairs at certain stages of the year, principally during periods of high light

TABLE II. Hyaline hair-point production in *Callithamnion* and near allies[jk]

I: Species[a]	II: Data source[b]	III: Consistent?	IV: Occurrence	V: Abundant?	VI: Position terminal	VII: Position lateral	VIII: Best developed on:	IX: Other bearing states:	X: Field or Culture data	XI: Detail of hairs[p]	XII: Number per cell[q]
arbuscula	1	No	Spasmodic	Yes	Yes	Yes	♀	All	F	Bulbous-tipped, at least the bulb lacking contents. Straight, bent, or convoluted. 8·75–12·5 μm wide × 140–200 μm long (1)	Occasionally two; then one lateral, one terminal
"*brodiaei*"[f]	3, 4, 5, 8, 9, 12, 25, 27, 34	No	Frequent in some areas (customary in S. Sweden and Denmark)	Yes, in some areas (S.E. Norway; Denmark; N. Spain; Baja California—(12), hairs abundant but species rare —(34))	Yes	Yes	N/D	⊕, ♂, ♀, ⚲, sterile (25)	F	Long; 1–2 μm wide (3). Rather short (7–9 μm wide; 160–290 μm long) (5, 9)	
byssoides	4, 5, 7, 18 [5 and 7 as *furcellariae*]	No	Rare (4, 5[n]) to occasional (7)	No	Yes	Apparently not developed (not observed) in some countries	♀ (18)	θ (5); θ/♀ (5)	F (subtidal, 5) C (18)	Much less numerous and more feeble than in, e.g. *corymbosum* (3·5 μm wide × 12–140 μm long—(5)); otherwise typical and well-developed (4)	
cordatum Børg.	20, 21, 22, 35, 36	No	Not common; "... found only a few times ... as a rule absent ..." (20); rarely (35); sometimes (36)	Probably not	Yes	N/D	Young apices in active growth (20). Only present in very young plants (22). Laterals in young parts of thallus (36)	Seldom, in older parts of thalli	F	*c.* 150 μm long × *c.* 3 μm broad; short, of no taxonomic value (35); slender, fall while still near axial apex, *c.* 45 μm long × 3 μm diam. near base (36)	
corymbosum	1, 4, 5, 6, 7, 8, 9, 10, 11, 12, 13, 14, 15, 26, 28, 29, 33	No	Customary (absent Dec./Jan.—(5)); often lacking in deeper subtidal; in adverse conditions; or with season in intertidal	Yes (April to September—(4); few in Oct./Nov.—(4))	Yes	Rarer than in robust species, but occurs (1 : 5 [Denmark])	Shoots retaining only feeble growth or ceasing growth: absent from peripheral laterals in full growth phase	⊕	F	*c.* 3·5 μm wide × 138 μm long (28)[l]. To 300 μm or > long × 4–5 μm diameter (3). 3–8 μm diameter × (40–) 160 (–180) μm long (6). 2–3 μm diameter × 114 μm long (33). *c.* 4 μm wide × 230 μm long (4, 5)[e]	Sometimes two; then one terminal, one lateral (4, 5, 11). 1–2 (–3) terminal hairs (33)
decompositum sensu Halos	2, 31, 32	No	Rare (2); "... parfois ..." (31, 32)	No	Yes	N/D	Young ♀ only (2)	⊕; ⊕/♀/♂ (31, 32)	F	diam. *c.* 3 μm; length to 60 μm (31, 32)	

TABLE II—*continued*

I: Species[a]	II: Data source[b]	III: Consistent?	IV: Occurrence	V: Abundant?	VI: Position terminal	VII: Position lateral	VIII: Best developed on:	IX: Other bearing states:	X: Field or Culture data	XI: Detail of hairs[p]	XII: Nu ber per cell[q]m
"gallicum"	2	No	Occasional	No	Yes	N/D	♀	No other states	F	N/D	
granulatum	1, 3, 24, 30	No	Spasmodic (1); "Almost all the ramuli" (30)	Yes	Yes	Yes	♀	All	F	*c.* 3 μm wide × 90 μm long (24).[e] *c.* 6 μm wide × 50 μm long (3). Slightly clavate (but not bulbous) tip with dense appearance; 8·75 μm wide × 137–140 μm long (1)	
hookeri	5	No	Rare reports (said to be absent from Denmark—(5))	No	N/D	N/D	N/D	N/D	F	N/D	
neglectum	36	No	often (36)	No	Yes	N/D	N/D	N/D	F	Long, fall while still near apex (36)	
roseum	4, 5, 29	No	Rare, very rare (29)	No	Yes	Yes	N/D	N/D	F	Little thinner than bearing filament (*c.* 7 μm); rather short, *c.* 340 μm [!][d]	
sarcodiae Børg.	12, 19	No	Often (12). On some apical cells (19)	Probably not (19)	Yes	N/D	N/D	N/D	F	3–4 μm wide × *c.* 40–50 μm long (19)	
"scopulorum"	2, 16	No	Rare reports: (spring—(16)[m] "... in some stages ...") (sometimes—(2))	No	Yes	Sometimes (16)	♀ (fertile) only	All (16), no others (2)	F	Very like trichogynes; average thickness *c.* 7 μm; length 80–160 μm, exceptionally to 530 μm	
Aristothamnion purpuriferum [*collabens*]	17, 23	No	Not common	N/D	Yes	Yes (17)	♀ only (17, 23)	No other	F	Unbranched hairs only; morphologically similar to trichogynes, but longer; average length 200 μm, occasionally to 250 μm (17)	
Aristothamnion ramelliferum Baardseth	23	No	Not common	N/D	Yes	Yes	♀ only	No other		Usually much-branched hairs, less often unbranched	cf. entry under column XI

Seirospora occidentalis Børg	20, 21 35, 36	No	"Often, but not always" (20, 21); sometimes (36)	N/D	Yes	N/D	More and larger on plants with seirosporous cystocarps[g]; on laterals, deciduous (35)	Fewer on ♂ plants[l]	F	Lengths variable (see columns VIII and IX); breadth narrow, *c.* 3 μm; falls when close to apex, 1·5 or 2·0 μm diam. × 180 μm long (36)	Single hair per cell (36)
Seirospora seirosperma	2, 4, 5, 9	No	Rare (specimens from harbour not so rarely so—(5))	No	Yes	N/D	Only on fertile ♀ (2)	N/D of any such	F	Small (4–5 × cell length and basally approximately half width of bearing cell)	

Notes on Table II:

N/D signifies no data available or traced, wherever used in the table.

[a] No reports of hairs available for species not listed; authorities are given only for species neither mentioned elsewhere in the text nor covered by Parke and Dixon (1976).

[b] Data sources are as follows:

1. Own observations.
2. Ardré (1970).
3. Feldmann-Mazoyer (1941).
4. Rosenvinge (1911).
5. Rosenvinge (1924).
6. Riouall (1974).
7. Levring (1940).
8. Kylin (1907).
9. Kylin (1944).
10. Zinova (1935).
11. Thuret and Bornet (1878).
12. Dawson (1962).
13. Harris (1962).
14. Taylor (1957).
15. Farlow (1881).
16. Westbrook (1927a).
17. Westbrook (1927b).
18. Rueness (unpublished).
19. Børgesen (1952).
20. Børgesen (1909).
21. Børgesen (1917).
22. Chapman (1963).
23. Baardseth (1941).
24. L'Hardy-Halos (1971a).
25. Sundene (1953).
26. Boddeke (1958).
27. Kützing (1861).
28. Lakowitz (1929).
29. Harris (1966).
30. De Valéra (1939).
31. Halos (1965a).
32. Halos (1965b).
33. Oliveira filho (1969).
34. Miranda (1931).
35. Itono (1971).
36. Itono (1977).

N.B. This list of source references is not exhaustive of all those authors by whom hair structures in *Callithamnion* have been mentioned. Authors whose observations were clearly not original have been largely omitted.

[c] Measured from L'Hardy-Halos (1971a, Fig. 1F).

[d] Measured from Rosenvinge (1911, Fig. 5D, as *C. byssoides*; 1924, Fig. 251D).

[e] Measured from Rosenvinge (1911, Fig. 10A; 1924, Fig. 243A).

[f] "*brodiaei*" may conceal a good taxon; at present seen or illustrated British material named as *C. brodiaei* is erroneously or doubtfully identified, chiefly representing *C. hookeri* or robustly-developed *C. roseum*. Drawings of Feldmann's (1932, Banyuls) material (in 3) are remarkably like the form and division pattern of apices of *C. granulatum* clusters, but the ecological data given (collected in January, in 12–15 m depths) are not typical of *C. granulatum*, an intertidal and very shallowly subtidal plant.

[g] Deduced from Børgesen (1909, Fig. 9A; 1917, Fig. 211A).

[i] Deduced from Børgesen (1909, Fig. 8D; 1917, Fig. 210D).

[j] Many of the taxa cited in this table are treated by all French and some other authors as belonging to the genus *Aglaothamnion* Feldmann-Mazoyer. This segregate genus is here regarded as insufficiently well-founded to permit retention, at least on currently accepted bases.

[k] Reported contrasting data under heads for a single "species" suggest (but do not prove, since there may be area variations) misapplication of names or faults in concept by one or both of the authors concerned.

[l] Measured from Lakowitz (1929, Fig. 474e).

[m] Westbrook (1927a: 133) indicated that hairs in *scopulorum* were not developed to such a conspicuous extent as in *brodiaei* or *corymbosum*.

[n] Sufficiently rare occurrence in *C. byssoides* for Rosenvinge (1924) to postulate their use as distinguishing factor between that species and the more consistently hair-bearing *C. corymbosum*.

[p] Dimensions given in the column XI relate to stated figures or figures measured (see specific notes) from drawings published. In either case, it is probable that the widths stated were not measured over the external covering of the hair but relate to the width of the contents. Comparison of the original data presented here for *C. granulatum* hair widths with past published dimensions suggests some variation in approach. The present dimensions were measured *over* the wall at the widest point (usually at the point of origin). It is assumed that the dimensions quoted by earlier authors are the maximum dimensions noted; if the widths were measured at a standard point (never specified), rather than at the widest point, that assumption may occasionally be in error.

[q] Where no entry appears for number of hairs per cell, *one* is assumed the only state of occurrence because of absence of data to the contrary.

conditions or increasing light conditions and rapid active growth. Similarly it is not possible to support observations by Harris (1962) that the reverse situation applies in *C. corymbosum*, another species that often carries dense growths of terminal hairs. Harris suggested that populations in sheltered bays were devoid of hairs, whereas adjacent populations at the exposed mouth of the bay had hairs on virtually every terminal cell. Sheltered-water populations at the entry of the

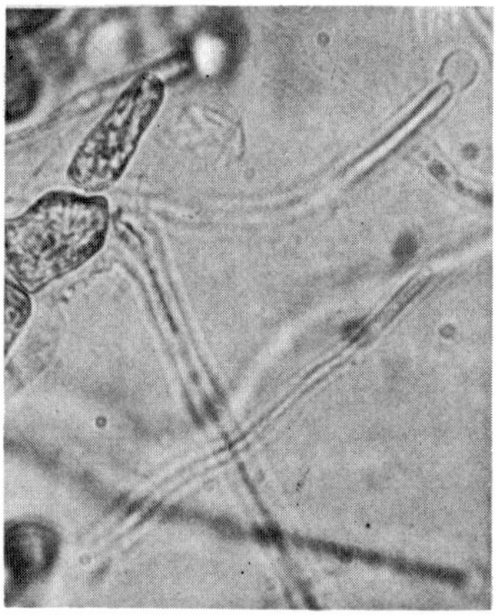

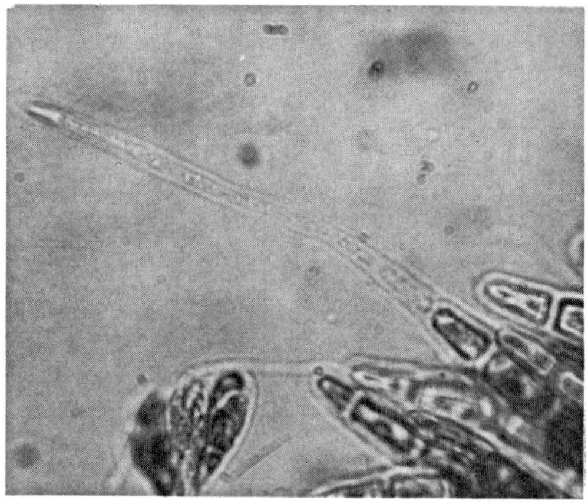

FIG. 18. (Left) *Callithamnion arbuscula*. Porthcurno, August. Group of hyaline, terminal or already by-passed, apical hairs. Note typical form of bulbous-tipped hair-point in focused hair on upper right and the extant pit-connection at the base of the by-passed hair bent to the lower right from the focused peripheral lateral. Compare *C. granulatum* hairs in Fig. 19.

FIG. 19. (Right) *Callithamnion granulatum*. Porthcurno, August. One hair focused from a local group of hairs (see Fig. 11A for such a group), showing the non-bulbous, slightly clavate, densely refractive tip of the hair. The rather more commonly persistent apical (terminal) position than for *C. arbuscula* is also shown. See Fig. 18 for comparative hairs of *C. arbuscula*. × 500.

East Fleet, Dorset, customarily carry immense and dense growths of both terminal and by-passed lateral hairs. Riouall (1974) found considerable variation in the presence and extent of hyaline hair development, in this species, amongst populations in the saline lakes of Berre and Vaïne, France.

The by-passing of terminal hairs by lateral development from the hitherto active apical cell below is a comparatively common phenomenon in some hair-bearing species of *Callithamnion* (Table II). This turns the hair into a lateral projection (Figs 18, 20); it is not clear how, whatever the function of the hair, its future utility is affected. By-passing could represent one method by which the plant dispenses with the services of a no longer useful morphological element (Fig. 20); another such method is dehiscence, so far as hair structures are concerned. That by-passing does not exclude subsequent dehiscence is clear from the occasional unexplained kinks that occur along peripheral laterals in hairy and hairless plants of *C. arbuscula*; these are probably points at which terminal hairs were by-passed.

It has been fairly generally observed (see Table II) that hyaline hairs appear most abundantly, predominantly, or only, on the female gametophytic plant, although there is variation according to species, season and geographical location. This suggests that whatever hormonal system is involved in the

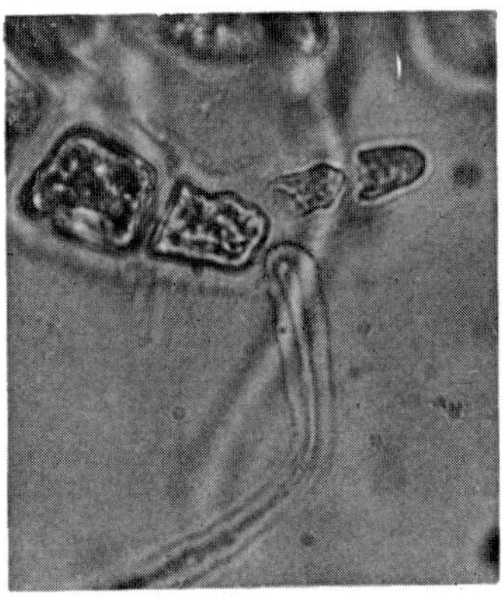

FIG. 20. *Callithamnion arbuscula*. Porthcurno, August. By-passed hair. Note signs of dehiscence but still extant pit-connection at hair-base. × 500.

production of trichogynes, essentially similar in form and position, may also lead to more ready production of vegetative terminal hyaline hairs. This possibility and the comparative ultrastructure of trichogynes and hairs in relation to function (known or conjectural), are being investigated in various Ceramiaceae (Pettitt, Price and Chamberlain in prep.).

DISCUSSION AND CONCLUSIONS

The genetic background to variations in aspects of gross morphology in the genus *Callithamnion* is little understood; even chromosome counts are few and inaccurate (Table III). Similarly, reproductive potentials or states show morphological characteristics sometimes at variance with those of the vegetative plant (Harvey, 1847; Price, unpublished; Westbrook, 1927, 1930; Børgesen, 1945, 1952; Boddeke, 1958; Harris, 1966), although usually recognizable by association with vegetative individuals in ecological groupings. The largest part of the general variation in many gross morphological aspects appears, nevertheless, primarily to result from the action of external and internal environment on

TABLE III. Chromosome counts in *Callithamnion*. (Updated after Dixon, 1966)

Species	Count (n)	Source[a]
arbuscula or *granulatum* (as *purpurascens*)	28–33	1
brodiaei	28–33	1
	29 or 30	2
byssoides	28–33	1
corymbosum	30 ($2n = 60$)[d]	3
	28–33	1
	c. 28	2
	$2n = 57–65$	7
hookeri	28–33	1
	33	2
roseum[c]	39	1
	39	2
tetragonum	9–10 ($2n = 18–20$)	4, 5[b]
	28–33	1
tetricum	*c.* 25 ($2n =$ *c.* 50)	6
	?90–100	1

[a] 1 = Harris, 1962. 2 = Harris, 1966. 3 = Hassinger-Huizinga, 1952. 4 = Mathias, 1927. 5 = Mathias, 1928. 6 = Westbrook, 1930. 7 = Whittick, 1973.
[b] According to Westbrook (1930), Mathias (1928) incorrectly identified the nucleolus as the nucleus; Mathias's further work (1932) was also criticized by Westbrook (1935). Whittick (1973) was unable to obtain preparations of such a quality that counts were possible for *C. tetragonum*.
[c] Whittick (1973) was able to obtain a diakinesis count of approximately 30 bivalents in the tetrasporangium initial; no other counts were possible for this species. Specimens utilized by Harris (1962, 1966) were multinucleate, whilst those of Whittick's studies were uninucleate. This suggests mis-determination by one or both authors, complicating further an already complex situation.
[d] This is the base-number; plants with ⊕ and ♂ or ⊕ and ♀ were reported as between $2n + 2 = 62$ and $2n + 4 = 64$.

all development stages of highly plastic morphogenetic patterns and resultant phenotypes. Therefore, although the effective factors in the environment are not yet always recognized, the question as to whether a "species" has the gross appearance which is commonly associated with it and by which it is recognized because, to some degree or other, of where and when it exists, has continually to be borne in mind in both taxonomic practice and even routine determinations.

Comparisons between morphological variation pattern throughout one or a range of "species-positions" across the generic spectrum and the conditions of the littoral or subtidal micro-niche(s) in which representative plants concerned are usually detected reveal:

1. Most of the "species-positions" are rather circumscribed as to micro-niche conditions in which recognizable morphological forms of the plants concerned naturally occur.
2. The range of morphologies which those plants there adopt is similarly rather circumscribed (Fig. 9).

Similar morphologies tend to appear in similar niches, mainly on similar shore types; that is, whatever the *Callithamnion* "species", gross morphologies adopted in similar environments manifest structural elements developed in a similar manner and to a similar degree.

More specifically, it has been indicated that the effect of ambient environment on certain morphological characteristics hitherto used taxonomically in various ways in *Callithamnion* and allies is very great. The environmental effect comes from an immediately appreciable factor (wave-action and rough water-movement) and appears consistent in the cases of cortication/adventitious weft development and of the peripheral pattern of branching. The development of hyaline hair points is less consistent to or within species (Table II), although some seem never to develop hairs; the causal factors are less immediately apparent and are not yet firmly established. Currently the general use of terminal or lateral hairs as a phenomenon by which to distinguish between taxa (or between groups, cf. Farlow, 1881) seems undesirable. It is not yet possible to judge firmly the situation in other countries, but the utilization of presence of such hairs ("*brodiaei*") as opposed to absence of hairs (*roseum*; "*furcellariae*", = *byssoides*) in distinguishing "World" forms (Dawson, 1962), Danish forms (Rosenvinge, 1924), Swedish forms (Kylin, 1944) and Norwegian forms (Rueness, 1977) is not generally applicable and may represent the acceptance of taxa partly or wholly on an inconsistent basis.

There have been two principal tendencies amongst previous authors on the genus in terms of cortication and associated axial/major lateral phenomena.

1. To utilize the matter of cortication/adventitious lateral weft directly as a criterion to key out particular species in the total absence of a demand for parallel ecological data to authenticate the cortication state (e.g. as in J. G. Agardh, 1851; Debray, 1899; Rosenvinge, 1924; Taylor, W. R., 1957; Boddeke, 1958; Dawson, 1962; Abbott and Hollenberg, 1976; Rueness, 1977).

2. To dismiss the use of an environmentally-affected characteristic of this kind for any taxonomic purposes (e.g. Harris, 1962, 1966).

Neither of these is the appropriate course of action for any treatment on other than a very local basis. Certainly, there is no one species in which cortication form is sufficiently distinct from all others and consistent enough to be utilized as an absolute in species delimitation. However, this easily-observed and readily recognizable phenomenon should not be wholly neglected simply because the form and extent of its manifestation are environmentally influenced. In its various configurations, cortication shows three major groups that can usefully form a basis for an ecologically-orientated key to the genus; one such key is in an advanced stage of preparation. Cortication, in the long-term, may also prove to be an interesting and potentially valuable phenomenon in evolutionary studies. The division of the genus into three cortication groups (multi-layered cortication with dense adventitious weft; multi-layered, but less thick, cortication with only occasional adventitious laterals; little or no cortication other than in basal regions, no weft or adventitious laterals) produces a result also entirely consistent with characteristics of the limited-growth branching patterns and forms in the peripheral laterals. The resultant overall gross morphologies are not confused in their otherwise ecologically restricted distributions by wide-spectrum forms like *C. hookeri*, which has an excellent capacity for cryptic "mimicry" of gross morphology, when in the appropriate micro-niche, of *Callithamnion* species that characterize that micro-niche or shore level.

This gross morphological "mimicry" has long been a source of confusion in both routine determination and taxonomic practice. Since *C. hookeri* adopts a very similar morphology to *C. granulatum*, when growing in similar habitats, herbarium collections of the latter frequently include specimens of both species, *C. hookeri* having been mistaken for *C. granulatum*. Similar confusion over close gross morphology has resulted in the overlooking of *C. granulatum* amongst *C. arbuscula* at northern range peripheries and of *C. arbus cula* amongst *C. granulatum* in the south; as well as in the additional confusion of different *C. hookeri* growth forms with almost every taxon in the genus!

Gross morphological "mimicry" of the *C. arbuscula*/*C. granulatum* growth form by *C. hookeri*, in exposed wave-beaten conditions, is part of a general phenomenon amongst filamentous, branched, red seaweeds in such conditions. Growth forms adopted by these algae in the conditions specified are almost always bushy or clumped in the intertidal. The legion possible examples include *Callithamnion arbuscula, C. granulatum, C. hookeri, Ceramium shuttleworthianum, Dasya hutchinsiae* and *Halurus equisetifolius*. Many of the species concerned have been confused with each other at times by past workers. Although the intertidal in strongly wave-beaten situations is subject to spray and damp or even to direct water-impact, for a large proportion of emerged time, bushiness has the advantage of retaining moisture in rapidly draining conditions and, in the south especially, in frequent periods of strong insolation. Mechanical shock of wave-beat is better resisted by the greater general robustness that also contributes to the bushiness referred to in these species. Lastly, dense peripheral laterals

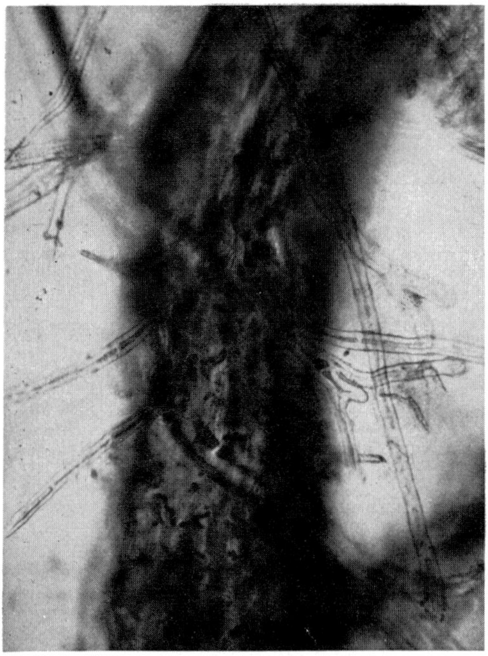

FIG. 21. *Callithamnion roseum*. Rhizoidal-type escape of corticating filaments, becoming extramatrical, along axis of the erect plant. × 125.

permit loss of a higher proportion of distal branching under wave-beat or in adverse seasons without dangerously reducing available photosynthetic capacity.

In view of the identical corticating process in all the *Callithamnion* spp. that tolerate the wave-beaten environments, it is tempting to suggest that the mechanism is evolutionarily derived from the retained tendency of basal poles to produce attachment filaments. Further, since there is a tendency, fully realized in the species producing an adventitious weft over the cortex, for *Callithamnion* to develop adventitious laterals from cortical filaments, the downgrowing corticating filaments (which frequently emerge to become extra-matrical; Fig. 21) are arguably the secondary reflection of an earlier primary heterotrichous habit. Adaptation to an exposed environment has perhaps resulted in re-employment of this existing capacity, potential in all cells, according to the position of the cell in the thallus and to the influence of determinant axial and lateral apices (and therefore to overall whole-plant requirement).

Until more is known about the fundamental genetic background to the general patterns of ecological tolerance and response in this genus, it is advisable to construct taxonomic keys to it, commencing not, as has been customary, on chiefly unrelated morphological grounds, but by group segregation on tolerance of overall environment and habitat niche;* only thereafter should separation be achieved through resultant or stable aspects of the developed morphology. The gross morphological characteristics discussed here are being used in the construction of keys for the recognition of species groups associated with particular habitat niches,* but not for delimitation of species within those groups. Apparent morphological simplicity in this genus certainly conceals morphogenetic flexibility in response of some characteristics and this results in taxonomic complexity. In their turn, these latter may well conceal currently unrecognized processes of active speciation.

ACKNOWLEDGEMENTS

At various stages in the work, many people have contributed by discussion, provision of data and particularly by collection of material from locations to which I could not travel at appropriate times. Although space is not available to list all those concerned, I would like them to know that their efforts were much appreciated. Dr J. Rueness, University of Oslo, shared valuable cultural data with me; Dr A. R. Chamberlain, University of Surrey, permitted me to make use of his unpublished observations on hair-points in *Ceramium*, as comparative background to the situation in *Callithamnion*. To all those who have helped in these and other ways, my thanks.

REFERENCES

Abbott, I. A. and Hollenberg, G. J. (1976). "Marine Algae of California." Stanford University Press, Stanford, California.

Agardh, J. G. (1851). "Species, genera, et ordines algarum, . . .", Vol. 2, "Algas Florideas complectens", pars prior. C. W. K. Gleerup, Lundae.

Ardré, F. (1970). Contribution a l'étude des Algues marines du Portugal I—La flore. *Port. Acta biol.*, ser. B **10**, 1–423.

Baardseth, E. (1941). The marine algae of Tristan da Cunha. *Results Norw. scient. Exped. Tristan da Cunha*, 1937–8, **9**, 1–173.

Boddeke, R. (1958). The genus *Callithamnion* Lyngb. in the Netherlands. A taxonomic and oecological study. *Acta Bot. neerl.* **7**, 589–604.

* In essence, the basis of such a key is provided by Fig. 9 (*q.v.*).

Børgesen, F. (1902). Marine algae. *In* "Botany of the Faeröes based upon Danish investigations" (E. Warming, ed.), Part II, pp. 339–532. Nordiske Forlag Ernst Bojesen, Copenhagen. *Note.* First published as a separate, 15 Nov. 1902, this work was also included in the overall volume (1903).

Børgesen, F. (1909). Some new or little known West Indian Florideae. *Bot. Tidsskr.* **30**(1), 1–19.

Børgesen, F. (1917). The marine algae of the Danish West Indies. Vol. 3 [error for 2]. Rhodophyceae (pp. 145–240) [part 3]. *Danske bot. Arkiv* **3**(1e), 145–240. *Note.* Also as a separate with same pagination.

Børgesen, F. (1945). Some marine algae from Mauritius III. Rhodophyceae. *K. danske Vidensk. Selsk. Skr., Biol. Medd.* **19**(10), 1–68.

Børgesen, F. (1952). Some marine algae from Mauritius. Additions to the parts previously published, IV. *K. danske Vidensk. Selsk. Skr., Biol. Medd.* **18**(19), 1–72.

Chapman, V. J. (1963). The marine algae of Jamaica. Part 2. Phaeophyceae and Rhodophyceae. *Bull. Inst. Jamaica, Science series* **12**(2), 1–201.

Dawson, E. Y. (1962). Marine red algae of Pacific Mexico. Part 7. Ceramiales: Ceramiaceae, Delesseriaceae. *Rep. Allan Hancock Pacif. Exped.* **26**(1), 1–207.

Debray, F. (1899). Florule des Algues marines du nord de la France. *Bull. scient. Dép. Nord* **32** (sér. 5, Vol. 1), 1–193.

De Valéra, M. (1939). Some new or critical algae from Galway Bay, Ireland. *K. fysiogr. Sällsk. Lund. Förh.* **9**, 91–104.

Dixon, P. S. (1966). The Rhodophyceae. *In* "The Chromosomes of the Algae" (M. B. E. Godward, ed.), pp. 168–204. Edward Arnold, London.

Dixon, P. S. (1971). Cell enlargement in relation to the development of thallus form in Florideophyceae. *Br. phycol. J.* **6**, 195–205.

Dixon, P. S. (1973). "Biology of the Rhodophyta." Oliver and Boyd, Edinburgh.

Dixon, P. S. and Richardson, W. N. (1970). Growth and reproduction in red algae in relation to light and dark cycles. *Ann. N.Y. Acad. Sci.* **175**, 764–777.

Edwards, P. (1969). The life history of *Callithamnion byssoides* in culture. *J. Phycol.* **5**, 266–268.

Edwards, P. (1971). Effects of light intensity, daylength, and temperature on growth and reproduction of *Callithamnion byssoides*. *In* "Contributions in Phycology . . ." (B. C. Parker and R. M. Brown Jr, eds), pp. 163–173. Students of Harold C. Bold, Lawrence, Kansas.

Edwards, P. (1972a). Benthic algae in polluted estuaries. *Mar. Poll. Bull.* **3**, 55–60.

Edwards, P. (1972b). Cultured red alga to measure pollution. *Mar. Poll. Bull.* **3**, 184–188.

Edwards, P. (1977). An analysis of the pattern and rate of cell division, and morphogenesis of sporelings of *Callithamnion hookeri* (Dillw.) S. F. Gray (Rhodophyta, Ceramiales). *Phycologia* **16**, 189–196.

Edwards, P. and Van Baalen, C. (1970). An apparatus for the culture of benthic marine algae under varying regimes of temperature and light-intensity. *Botanica mar.* **13**, 42–43.

Farlow, W. G. (1881). Marine algae of New England and adjacent coast. *Rep. U.S. Commnr Fish.* **1879**, 1–210.

Feldmann-Mazoyer, G. (1941). "Recherches sur les Céramiacées de la Méditerranée occidentale." Imprimerie Minerva, Alger.

GARBARY, D. (1976). Life-forms of algae and their distribution. *Botanica mar.* **19**, 97–106.

GILLHAM, M. E. (1957a). Part two—cryptogams. 1. Algae of the Exe estuary pp. 232–236, *In* Martin, W. K. and Wallace, T. J. Forty-eighth report on the botany of Devon. *Rep. Trans. Devon. Ass. Advmt Sci.* **88**, 225–245.

GILLHAM, M. E. (1957b). Vegetation of the Exe estuary in relation to water salinity. *J. Ecol.* **45**, 735–756.

HALOS, M.-TH. (1965a). Sur trois Callithamniées des environs de Roscoff. *Cah. Biol. mar.* **6**, 117–134.

HALOS, M.-TH. (1965b). Sur l'*Aglaothamnion decompositum* (Grateloup ex. J. Ag.) comb. nov. et sa position systématique. *Bull. Soc. phycol. Fr.* **10**, 18–19, 1964.

L'HARDY-HALOS, M.-TH. (1970). Recherches sur les Ceramiacées (Rhodophycées–Céramiales) et leur morphogénèse. I.—Structure del'appareil végétatif et des organes reproducteurs. *Rev. gén. Bot.* **77**, 211–287.

L'HARDY-HALOS, M -TH. (1971a). Recherches sur les Ceramiacées (Rhodophycées–Céramiales) et leur morphogénèse. II.—Les modalités de la croissance et les remaniements cellulaires. *Rev. gén. Bot.* **78**, 201–256.

L'HARDY-HALOS, M -TH. (1971b). Recherches sur les Ceramiacées (Rhodophycées–Céramiales) et leur morphogénèse. III.—Observations et recherches expérimentales sur la polarité cellulaire et pour la hiérarchisation des éléments de la fronde. *Rev. gén. Bot.* **78**, 407–491.

HARLIN, M. M. and LINDBERGH, J M. (1977). Selection of substrata by seaweeds: optimal surface relief. *Mar. Biol. Wash.* **40**, 33–40.

HARRIS, R. E. (1962). Contribution to the taxonomy of *Callithamnion* Lyngbye emend. Naegeli. *Bot. Notiser* **115**, 18–28.

HARRIS, R. E. (1966). Contributions to the genus *Callithamnion* Lyngbye emend. Naegeli: taxonomy of the species indigenous to the British Isles. *Advg Front. Pl. Sci.* **14**, 109–131.

HARVEY, W. H. (1847). "Phycologia britannica: . . .", Vol. 2. (Pl. 129). 1849. Reeve and Benham, London.

HASSINGER-HUIZINGA, H. (1952). Generationswechsel und Geschlectsbestimmung bei *Callithamnion corymbosum* (Sm.) Lyngb. *Arch. Protistenk.* **98**, 91–124.

ITONO, H. (1971). The genera *Callithamnion*, *Aglaothamnion*, *Seirospora*, *Pleonosporium* and *Mesothamnion* (Ceramiaceae, Rhodophyta) in Southern Japan. *Mem. Fac. Fish. Kagoshima* **20**, 217–237.

ITONO, H. (1977). Studies on the Ceramiaceous alga (Rhodophyta) from southern parts of Japan. *Bibliotheca phycol.* **35**, 1–499.

KONRAD HAWKINS, E. (1964a). Developmental studies on regenerates of *Callithamnion roseum* Harvey, Part I. The development of a typical regenerate. *Protoplasma* **58**, 42–59.

KONRAD HAWKINS, E. (1964b). Developmental studies on regenerates of *Callithamnion roseum* Harvey, Part II. Analysis of apical growth. Regulation of growth and form. *Protoplasma* **58**, 60–74.

KONRAD HAWKINS, E. (1968). Induction of cell differentiation in dissociated cells and fragments of *Callithamnion roseum*. *Am. J. Bot.* **55**, 255–264.

KÜTZING, F. T. (1861). "Tabulae phycologicae oder Abbildungen der Tange." XI. Band. (Tab. 100_{II}). Kützing, Nordhausen.

Kylin, H. (1907). "Studien über die Algenflora der schwedischen Westküste." Lund, Akademische Abhandlung.

Kylin, H. (1944). Die Rhodophyceen der schwedischen Westküste. *Lunds Univ. Årsskr.* N.F. Avd. 2, **40**(2), 1–104.

Lakowitz, K. (1929). "Die Algenflora der gesamten Ostsee (Ausschl. Diatomeen)." R. Friedländer und Sohn (for Westpreussischen Botanisch-Zoologischen Verein), Danzig.

Lami, R. (1942). La végétation algale du "Trou-du-Chat" en Saint-Lunaire (Ille-et-Vilaine). *Bull. Lab. marit. Dinard* **24**, 80–88.

Levring, T. (1935). Zur Kenntnis der Algenflora von Kullen an der schwedischen Westküste. *Lunds Univ. Årsskr.*, N.F. Avd. 2, **31**(4), 1–64.

Levring, T. (1940). "Studien über die Algenvegetation von Blekinge, Südschweden." Håkan Ohlssons, Lund.

Lüning, K. (1970). Cultivation of *Laminaria hyperborea* in situ and in continuous darkness under laboratory conditions. *Helgoländer wiss. Meeresunters.* **20**, 79–88.

Mathias, W. T. (1927). The cytology of *Callithamnion*. *Rep. Br. Ass. Advmt Sci.* **1927**(95), 380.

Mathias, W. T. (1928). The cytology of *Callithamnion brachiatum* Bonnem. *Publs Hartley bot. Lpool Univ.* **5**, 1–27.

Mathias, W. T. (1932). The cytology of *Callithamnion brachiatum*, Bonnem. *Ann. Bot.* **46**, 185–187.

Miranda, F. (1931). Sobre las algas y Cianofíceas del Cantábrico especialmente de Gijón. *Trab. Mus. nac. Cienc. nat. Madr., Ser. Bot.* **25**, 1–106.

Neushul, M. and Powell, J. H. (1964). An apparatus for experimental cultivation of benthic marine algae. *Ecology, Brooklyn* **45**, 893–894.

Norton, T. A. (1969). Growth form and environment in *Saccorhiza polyschides*. *J. mar. biol. Ass. U.K.* **49**, 1025–1045.

de Oliveira filho, E. C. (1969). Algas marinhas do sul do estado do Espirito Santo (Brasil). I—Ceramiales. *Bolm Fac. Filos. Ciênc. Univ. S. Paulo* **243**, *Bôtanica* **26**, 1–277.

Parke, M. [W.] and Dixon, P. S. (1976). Check-list of British marine algae—third revision. *J. mar. biol. Ass. U.K.* **56**, 527–594.

Price, J. H. (1973). Advances in the study of benthic marine algae since the time of E. M. Holmes. *Bot. J. Linn. Soc.* **67**, 47–102.

Price, J. H. and Tittley, I. (1978). Marine algae. *In* "The Flora of Mull and Adjacent Islands" (A. C. Jermy and J. A. Crabbe, eds), BMNH, London.

Riouall, R. (1974). Particularités de quelques spécimens de la flore algale des étangs de Berre et de Vaïne (Bouches-du-Rhône). *Bull. Soc. phycol. Fr.* **19**, 3–16.

Rosenvinge, L. K. (1911). Remarks on the hyaline unicellular hairs of the Florideae. *In* "Biologiske Arbejder tilegnede Eug. Warming paa hans 70 aars Fødselsdag den 3. November 1911" (L. K. Rosenvinge, ed.), pp. 203–216. H. Hagerups Forlag, Köbenhavn.

Rosenvinge, L. K. (1924). The marine algae of Denmark. Contributions to their natural history. Part III. Rhodophyceae III. (Ceramiales). K. danske Vidensk. Selsk. Skr., 7. Raekke, Nat. og Mat. Afd. **7**(3), 285–488.

Rueness, J. (1977). "Norsk Algeflora". Universitetsforlaget, Oslo, Bergen, Tromsø.

Note. A few copies of the original 1976 issue were released for sale before the withdrawal of the edition because of inadequacies in printing of plates and figures.

SUNDENE, O. (1953). The algal vegetation of Oslofjord. *Skr. norske Vidensk-Akad. mat.-nat. Kl.* **1953**(2), 1–245.

SUNDENE, O. (1962a). Growth in the sea of *Laminaria digitata* sporophytes from culture. *Nytt Mag. Bot.* **9**, 5–24.

SUNDENE, O. (1962b). The implications of transplant and culture experiments on the growth and distribution of *Alaria esculenta. Nytt Mag. Bot.* **9**, 155–174.

SUNDENE, O. (1963). Reproduction and ecology of *Chorda tomentosa. Nytt Mag. Bot.* **10**, 159–167.

TAYLOR, W. R. (1957). "Marine Algae of the Northeastern Coast of North America", Revised 2nd edition, University of Michigan Press, Ann Arbor (Univ. of Michigan Science Studies, Vol. 13).

THURET, G. and BORNET, É. (1878). "Études phycologiques. Analyses d'algues marines." Libraire de l'Académie de Médecine, Paris.

TITTLEY, I. and PRICE, J. H. (1977a). An atlas of the seaweeds of Kent. *Trans. Kent Fld Club* **7**, 1–80.

TITTLEY, I. and PRICE, J. H. (1977b). The marine algae of the tidal Thames. *Lond. Nat.* **56**, 10–17.

VAN BAALEN, C. and EDWARDS, P. (1973). Light-temperature gradient plate. *In* "Handbook of phycological methods. Culture methods and growth measurements", (J. R. Stein, ed.). University Press, Cambridge.

WESTBROOK, M. A. (1972a). *Callithamnion scopulorum* C. Ag. *J. Bot., Lond.*, **65**, 129–138.

WESTBROOK, M. A. (1972b). *Callithamnion purpuriferum* J. G. Ag. *J. Bot., Lond.* **65**, 161–167.

WESTBROOK, M. A. (1930). *Callithamnion tetricum* (Dillw.) Ag. *J. Bot., Lond.* **68**, 193–203.

WESTBROOK, M. A. (1935). Observations on nuclear structure in the Florideae. *Beih. bot. Zbl.* **53** (Abt. A), 564–585.

WHITTICK, A. (1973). "The taxonomy, life history and ecology of some species of the Ceramiaceae (Rhodophyta) in the north-west Atlantic." Thesis for Ph.D., Department of Biology, Memorial University of Newfoundland.

ZINOVA, E. S. (1935). Algues de la Baie Novorossijsk dans la Mer Noire et leur utilisation. *Trudȳ Sevastopol. biol. Sta.* **4**, 1–136.

13 | Criteria Used in Systematic Studies in the Sphacelariales

W. F. PRUD'HOMME VAN REINE

Rijksherbarium, Leiden, The Netherlands

Abstract: The Sphacelariales form a natural group within the Phaeophyceae. Their main characteristics are apical growth, marked heterotrichy and blackening of the cell-walls when treated with bleaching liquor (eau de Javel). The six or seven European genera (and two genera of dubious affinity) can be classed into four or five families. Several subdivisions of the genus *Sphacelaria* (subgenus, series), some infraspecific taxa (formae) in *Sphacelaria* and some other genera can be distinguished.

In systematic studies on the Sphacelariales morphological characters (branching, hair-formation, growth and division of segments, cortex-formation) can be used, as well as characters related to reproduction, ecology and biogeography. A proposed systematic classification, based on intuitive weighting of characters, is given.

Numerical taxonomic methods have been used by Russell and Fletcher (1975) in their study of the British Phaeophyta. Similar methods of analysis are used in the present study for four sets of data. One of these sets consists of the original computer cards for the Sphacelariales used by Russell and Fletcher. In the other sets corrections and additional data are incorporated.

The results of these analyses are discussed and are compared with the proposed systematic classification.

INTRODUCTION

Kützing (1843) and J. G. Agardh (1848) were among the first authors who separated *Sphacelaria* and *Cladostephus* from the other filamentous brown algae and placed them together in the family Sphacelariaceae. Both agreed that this family was well defined. Later Migula (1909) created the order Sphacelariales, and Fritsch (1945) stated: "the Sphacelariales form a clearly defined group which is almost certainly a natural one . . .". The present author (Prud'homme van Reine, 1974) and Russell and Fletcher (1975) expressed a similar opinion.

Systematics Association Special Volume No. 10, "Modern Approaches to the Taxonomy of Red and Brown Algae", edited by D. E. G. Irvine and J. H. Price, 1978, pp. 301–323, Academic Press, London and New York.

The systematic classification of the Sphacelariales is mainly based on taxonomic concepts of phycologists from the nineteenth and early twentieth century. The scientific works of Geyler (1866), Magnus (1873), Pringsheim (1873), Reinke (1891) and Sauvageau (1900–1914) are especially important.

Kylin (1917) suggested that the Choristocarpaceae should be included in the Sphacelariales, a suggestion which has been discussed several times since.

The results of recent research on the Sphacelariales will be added to the many facts previously known and a modern systematic classification based on intuitive weighting of characters will be proposed.

Numerical taxonomic methods have been used by Russell and Fletcher (1975) to classify all the Phaeophyta reported to occur in Britain. Their first purpose was to record the incidence of taxonomically important features throughout this group of algae. They also used this record of features in a numerical taxonomic study, in which they used the facts in an objective and unweighted manner. In this way they found that the British Phaeophyta fall into a number of discrete groups, corresponding with seven major taxa (orders). Their data on British Sphacelariales have been used in the present study as primary information but several corrections and additions have been incorporated.

Apart from listing most of the taxonomically important features among all known taxa of European Sphacelariales, the purpose of this study is also to compare the clustering of taxa within the groups in the proposed intuitive classification. It is also my purpose to incorporate the information collected into a possible new classification of the European Sphacelariales.

METHODS

The numerical taxonomic methods used by Russell and Fletcher (1975) were also used in the present study, but with some differences.

The data were computed at the Central Computing Centre of the State University at Leiden, using the program system for biological pattern analysis (BIOPAT) developed by P. Hogeweg and B. Hesper. The basic calculation of my analyses was the Jaccard coefficient, a coefficient similar to the one used by Russell and Fletcher and also modified into "percentage similarities". Negative matched characters were not taken into consideration.

Four different sets of data have been used, as summarized in Table III.

The starting points for the list of taxa were the numbers 156–174 of the list given in Table 1 of Russell and Fletcher's paper (my Table IA). In set 3, five taxa of Sphacelariales were added. These taxa are also known from British coasts (my Table IB). In set 4, ten other European taxa were added (Table IC). Only 58 of the 132 taxonomic characters used by Russell and Fletcher have been

TABLE I. Key to the code used for the taxa of European Sphacelariales

A.	
C. TENELL	— *Choristocarpus tenellus* (Kütz.) Zanard.
B. MIRABI	— *Battersia mirabilis* Batt. ex. Reinke
S. BIPINN	— *Sphacelaria bipinnata* (Kütz.) Sauv.
S. BRITAN	— *S. britannica* Sauv.
S. CAESPI	— *S. caespitula* Lyngb.
S. CIRROS	— *S. cirrosa* (Roth) C. Ag.
S. FUSCAA	— *S. fusca* (Huds.) S. F. Gray (*sensu* Harvey)
S. HYSTRI	— *S. hystrix* Suhr ex Reinke
S. PLUMIG	— *S. plumigera* Holmes
S. PLUMOS	— *S. plumosa* Lyngb.
S. PLUMUL	— *S. plumula* Zanard.
S. RACEMO	— *S. racemosa* Grev.
S. RADICA	— *S. radicans* (Dillw.) C. Ag.
S. TRIBUL	— *S. tribuloides* Menegh.
S. SUBTIL	— *Sphacella subtilissima* Reinke[a]
H. FILICI	— *Halopteris filicina* (Grat.) Kütz.
H. SCOPAR	— *H. scoparia* (L.) Sauv. or *Stypocaulon scoparium* (L.) Kütz.
C. SPONGI	— *Cladostephus spongiosus* (Huds.) C. Ag.
C. VERTIC	— *C. spongiosus* forma *verticillatus* (Lightf.) P.v.R.
B.	
S. FURCIG	— *Sphacelaria furcigera* Kütz.
S. PATENT	— *S. cirrosa* forma *patentissima* Grev.
S. SAXATI	— *S. saxatilis* (Kuck.) Kuck. ex Sauv.
H. RECURV	— *Halopteris filicina* forma **recurva** (Mont.) P.V.R., **nov. comb.**[b]
H. SCOSCO	— *Halopteris* (or *Stypocaulon*) *scoparia* forma **scoparioides** (Lyngb.) P.V.R., **nov. comb.**[b]
C.	
S. ARCTIC	— *Sphacelaria arctica* Harv.
S. BRACHY	— *S. brachygonia* Mont.
S. PINNAT	— *S. plumigera* forma *pinnata* (Reinke) Lund
S. SYMPOD	— *S. sympodiocarpa* Sauv.
H. SPINUL	— *Halopteris* (or *Stypocaulon*) *spinulosa* (Lyngb.) Sauv.
H. PILULI	— **H. pilulifera** P.V.R,. **nov. sp.**[b]
C. LACUSA	— *Cladostephus spongiosus* forma *laxus* (C. Ag.) Aresch.
C. HEDWIG	— *C. spongiosus* forma *hedwigioides* (Bory) P.v.R.
D. RETICU	— *Disphacella reticulata* (Lyngb.) Sauv.
D. MESART	— *Discosporangium mesarthrocarpum* (Menegh.) Hauck

[a] *Sphacella subtilissima* of set 1 is an ectocarpoid alga. Sets 2-4 include the characteristics of Mediterranean *S. subtilissima*. [b] Effected in full elsewhere.

employed by them for members of the Sphacelariales (compare Table 2 of their paper with Table II in the present publication). These 58 characters have been used in my set 1. In order to use the original computer cards of Russell and Fletcher, I used exactly the same code for taxonomic characters. I do not list those of the original characters which are not relevant for the present study; consequently many numbers have been omitted.

TABLE II. Key to the taxonomic characters used for Sphacelariales

1. Erect vegetative system of any sort[1]
3. Free, erect, branched assimilatory filaments[1]
4. Filament branching irregular
5. Filament branching opposite
6. Filament branching pseudodichotomous
7. Filament branching secund or subsecund
8. True hairs
11. Fused assimilatory filaments[1]
14. Periclinal cell divisions
31. Meristoderm
33. Cortex
34. Medulla
39. Growth apical[1]
40. Chloroplasts discoid[1]
44. Chloroplasts more than one per cell[1]
47. Eau de Javel
48. Prostrate vegetative system of any sort
49. Prostrate system of free filaments
51. Disc
53. Disc uniaxial
54. Periclinal cell divisions in prostrate system
57. Rhizoids
63. Propagules
64. Unisporangia
65. Plurisporangia
66. Meiosis
67. Isogamy
68. Anisogamy
69. Oogamy[1]
77. Sporangia exserted
79. Sporangia terminal
81. Sporangia scattered
82. Sporangia in sori
84. Gametangia exserted
86. Gametangia terminal
88. Gametangia scattered
89. Gametangia in sori
93. Monoecious
94. Dioecious
95. Chromosomes, x = 0–12
96. Chromosomes, x = 13–24
97. Chromosomes, x = 25–inf.[2]
102. Life history monomorphic
103. Life history dimorphic
106. Life history diplontic
107. Life history diplohaplontic
109. Germ tube
111. Germ cell evacuated
112. Plethysmothallus
113. Protonema[2]
115. Annual
116. Phanerophyte
117. Hemiphanerophyte or hemicryptophyte[2]
118. Chamaephyte
119. Epiphyte
120. Endophyte[2]
121. Epizooite[1]
123. Littoral fringe
124. Eulittoral zone
125. Sublittoral zone or sublittoral fringe
126. Rock pools
127. Estuarine[2]
128. Caves
129. Aestival[2]

TABLE II—*continued*

New characters, only used in set 3 and set 4

133. Filament branching distichous
134. Filament branching alternate
135. Filament branching verticillate
136. Determinate and indeterminate laterals
137. Only indeterminate laterals
138. Disc monostromatic
139. Disc polystromatic
140. Propagules tribuliform
141. Propagules with long arms
142. Propagules with apical hair
143. Always vegetative (never fertile)
144. Unattached
145. Hemiblastic laterals
146. Acroblastic laterals
147. Polyblastic
148. Acrohomoblastic hairs
149. Acroheteroblastic hairs
150. Hypacroblastic hairs
151. Segments with secondary growth in length
152. Segments with secondary growth in width
153. Secondary transverse walls in secondary segments
154. Pericysts
155. Rhizoids form a complete pseudo-cortex
156. Rhizoids form an incomplete pseudo-cortex
157. Rhizoids only in the plane of branching
158. Periclinal longitudinal walls
159. Radial longitudinal walls
160. Arctic seas
161. Baltic Sea
162. North Sea
163. Atlantic coasts
164. Mediterranean
165. Black Sea
166. Norwegian coast (Bergen—Tromsö)

[1] Only used in set 1. [2] Not used in set 1.

Of the above-mentioned 58 characters six were common to all British Sphacelariales (numbers 1, 3, 39, 40, 44, 48) and four characters were irrelevant, uncertain, or wrongly applied (numbers 7, 11, 69, 121). However, two of these characters (7 and 48) can be used for other taxa in the other sets of data.

In set 2, the data of set 1 have been corrected. The characters which are in common for all Sphacelariales (1, 3, 39, 40, 44) and the uncertain characters (7, 11, 69, 121) have not been used, but five characters of the original list were added (113, 117, 120, 127, 129).

In set 3, one of the characters of the original list (97) was added, together with 32 new characters (133–160, 162–164, 166). In set 4, character 7 of the original list was added, together with two new characters.

Of the characters given in Table II, numbers 1–62, 133–139 and 145–159 are morphological, 63–113 and 140–143 are concerned with reproduction or life-history, 115–129 and 144 are mainly ecological and 160–166 are phyto-geographical.

TABLE III. Sets of data for numerical taxonomic analysis

Set	Number of taxa	Total number of characters	Maximum number of characters used for a taxon	Minimum number of characters used for a taxon	Number of characters used only twice	Number of characters used only once
1	19	58	34 S. BIPINN S. FUSCAA H. FILICI	16 S. SUBTIL	7	9
2	19	54	36 S. BIPINN	13 S. SUBTIL	7	4
3	24	87	48 S. BIPINN C. SPONGI C. VERTIC	15 S. SUBTIL H. RECURV	5	4
4	34	90	49 C. VERTIC	14 D. RETICU D. MESART	4	4

MAIN CRITERIA

For the main characteristics and the description of apical growth of the Sphacelariales see Prud'homme van Reine (1974).

1. Morphological Characters

(*a*) *Branching*. Two different patterns in the formation of laterals can be distinguished within the Sphacelariales. In the Sphacelariaceae, all laterals are formed as outgrowths of secondary segments (Fig. 1B, g). In these cases the laterals have been formed below the apical cell and for this pattern of formation the term "hypacroblastic branching" can be used. If the width of the basal part of a lateral is equal to the length of the secondary segment of the filament from which it sprouts, I call this "hemiblastic branching". If the lateral sprouts from a smaller cell (not as long as the secondary segment) the hypacroblastic branching is called "meriblastic".

In the Stypocaulaceae branching is mainly initiated in the apical cell (acroblastic braching). The apical cell cuts off a small apical lenticular branch-initial

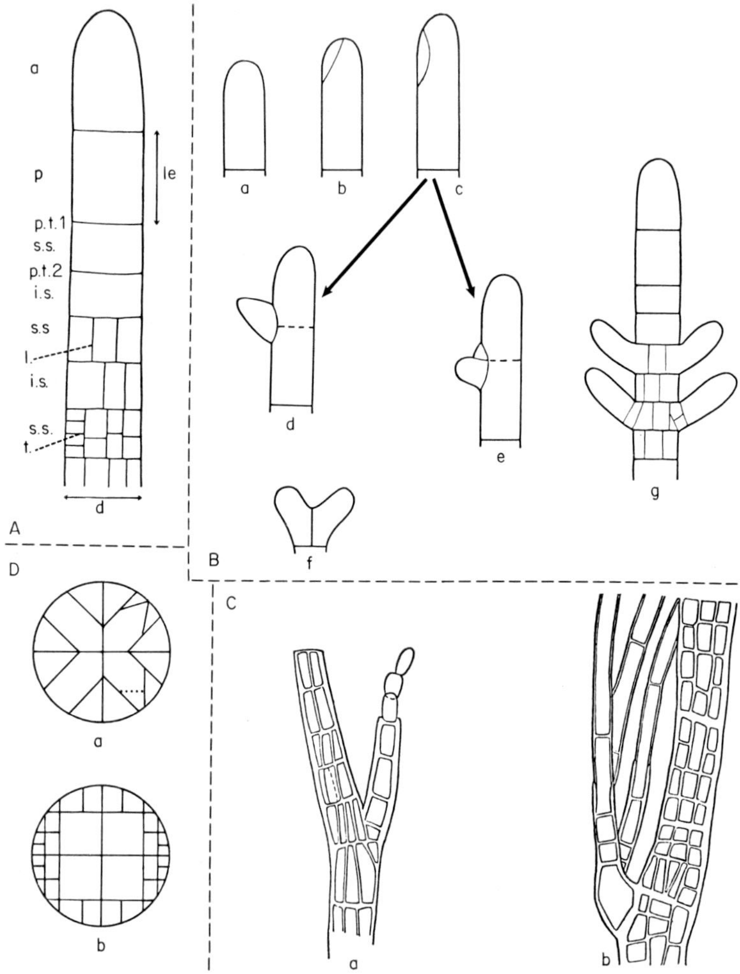

Fig. 1. Morphological characters in Sphacelariales. A. Diagram of apical growth of an unbranched filament (based on *Sphacelaria* sp.): a = apical cell; d = diameter or width; i.s. = inferior secondary segment; l = longitudinal cell-wall; le = length; p = primary segment; p.t. = primary transverse wall; s.s. = superior secondary segment; t = (secondary) transverse cell-wall. B. Diagram illustrating descriptive terminology used for branching: a–f, acroblastic branching, longitudinal sections; a–c, formation of the lenticular cell. The arrows point to further possible development of this lenticular cell; d, direct outgrowth (acrohomoblastic); or e, formation of two cells from which different structures develop (acroheteroblastic); f, dichoblastic branching; g, hypacroblastic (hemiblastic) branching. C. Hairs of *Sphacelaria plumosa*: a, solitary acrohomoblastic hair; b, bundle of four hypacroblastic hairs (only three have been drawn). D. Transverse sections through filaments of *Sphacelaria*: a, radial pattern; b, periclinal pattern.

(Fig. 1B, a–c). Occasionally this small lenticular branch-initial sprouts directly into a lateral (Fig. 1B, d)—this branching is called "acrohomoblastic branching". More often this branch-initial divides horizontally at an early stage, resulting in a small upper cell and a large lower cell (Fig. 1B, e). I use the term "acro-heteroblastic branching" for this kind of branching, where two different structures originate from one lenticular initial. The lower cell forms the lateral, while the upper cell may give rise to a bundle of hairs, to one or more sporangia or to a group of dormant cells. Usually the larger part of the original apical cell elongates and functions again as the apical cell of the main axis and the small lenticular branch-initial will be pushed into a lateral position (Fig. 1B, a–e).

A special case of acrohomoblastic branching is dichotomy occurring in the apical cell (dichoblastic branching, Fig. 1B, f). Dichoblastic branching is found only in *Disphacella*.

In the Cladostephaceae both patterns of formation of laterals can be observed. The whorls of small laterals around the main axes have been formed by hemi-blastic or meriblastic branching. The secondary laterals on these small laterals show acroheteroblastic branching. This complex branching is often called "polyblastic branching".

In the Choristocarpaceae branching has not been studied in detail, but there is no sign of acroblastic branching.

The small upper cell of the lenticular initial in Stypocaulaceae may form a small filament, one or two sporangia or 2–4 hairs in *Halopteris*, but in *Stypocaulon* dormant cells, several hairs or many sporangia occur.

The pattern of branching is an important character in several genera. In *Sphacelaria* this pattern may be distichous, helicoid or irregular; laterals, often determinate, may differ from the main axes in diameter and growth. In other species only laterals of indeterminate growth occur. In *Halopteris* and *Stypocaulon* the alternate laterals may be distichous or more or less helicoid.

(*b*) *Hair formation*. In the Sphacelariaceae hairs arise by acrohomoblastic branching (Fig. 1C, a), and occasionally also by hypacroblastic branching (Fig. 1C, b). Hairs may be solitary, geminate or in bundles.

In the Stypocaulaceae and Cladostephaceae almost all hairs arise by acro-blastic branching. In most cases this is acroheteroblastic, rarely acrohomoblastic. In the Choristocarpaceae no hairs are known.

(*c*) *Growth and mode of division of segments*. In most European Sphacelariales the segments exhibit no further enlargement after their formation. In these "leptocaulous" Sphacelariales the diameter of the segments of the filaments equals the diameter of the apical cell at the time of formation of these segments. The length of the secondary segments does not alter after formation in leptocaulous Sphacelariales (Fig. 1A).

The Cladostephaceae, however, are "auxocaulous" Sphacelariales, in which secondary segments enlarge conspicuously both in length and diameter.

In *Choristocarpus*, *Discosporangium* and *Sphacella* secondary segments are inconspicuous or of irregular occurrence and no further division of the large cells of the filaments can be observed.

The secondary segments of all other Sphacelariales divide further to produce the mature polysiphonal tissue (Fig. 1A). The first longitudinal cell-walls are mostly radial and the next two longitudinal walls are perpendicular to this primary wall. Later longitudinal walls may be of the radial type or of the periclinal type. In the radial type these walls form acute angles with those formed before (Fig. 1D, a); in the periclinal type all walls are perpendicular to each other (Fig. 1D, b). In *Sphacelaria* subgenus *Sphacelaria* all filaments are of the periclinal type, but in the subgenera *Acladus* and *Propagulifera* the pattern is always radial. Corresponding differences are found in the filaments of *Halopteris* (altered radial type) and *Stypocaulon* (strictly periclinal).

In secondary segments already divided by longitudinal walls secondary transverse walls may also occur. In *Sphacelaria* the occurrence or absence of secondary transverse walls is an important distinction at the subgenus and species levels.

(*d*) *Cortex*. The only means of secondary enlargement for leptocaulous Sphacelariales is by downward growing corticating filaments originating from the peripheral cells of the axial filaments. Many of these corticating filaments may extend down to the basal system and spread over it, thus functioning as a part of the attachment organ. These corticating filaments are specialized rhizoidal filaments, commonly called rhizoids. In *Sphacelaria* presence or absence of a cortex is a taxonomic criterion at species level.

(*e*) *Pericysts*. Some species of *Sphacelaria*, *Stypocaulon* and *Disphacella* have distinct dormant branch-initials (pericysts). Some peripheral cells of the filaments of these algae are not divided by secondary transverse walls and remain the same length as a secondary segment. In *Sphacelaria* pericysts occur in one subgenus only. The presence or absence of pericysts can also be used to separate *Halopteris* and *Stypocaulon*.

(*f*) *Rhizoids and attachment*. The formation of rhizoids is different between *Halopteris* and *Stypocaulon*. In the former genus the rhizoids sprout from laterals of the first order, but in *Stypocaulon* rhizoids sprout from peripheral cells (pericysts) of the axis.

In some species of *Sphacelaria* the rhizoids sprout very regularly from the axis, in other species irregularly or in all directions. The mode of attachment of most species is by a basal disc or by rhizoids. The basal disc may be monostromatic or polystromatic and can be used as a criterion to distinguish subgenera in *Sphacelaria*. In *Halopteris* a basal disc is formed but *Stypocaulon* is attached to the substrate by rhizoids only.

2. *Chemical Characters*

The blackening of the cell-wall when treated with bleaching liquid (eau de Javel) is characteristic for all Sphacelariaceae, Stypocaulaceae and Cladostephaceae, but not for the Choristocarpaceae.

3. *Characters Related to Reproduction*

Information on the reproduction and life-history of most species is still lacking or incomplete.

(*a*) *Oogamy and (an–) isogamy*. Anisogamy is observed or supposed to occur in *Sphacelaria furcigera*, *S. hystrix* and *Halopteris filicina*. Isogamy occurs (according to the literature) in *Sphacelaria bipinnata* and *Cladostephus spongiosus* (in forma *spongiosus* as well as in forma *verticillatus*). The supposed oogamy described by Sauvageau (1900–1914) for *Stypocaulon scoparium* is not evident. The suggestion by Higgins (1931) that Sauvageau observed abnormal unilocular sporangia is reasonable. No drawings or herbarium material are left of the only and badly preserved specimen on which Sauvageau observed unilocular oogonia and plurilocular antheridia. The information given by Moore (1951) about probable oogamy in five species of *Halopteris* from New Zealand is more convincing.

(*b*) *Propagules*. Species in the subgenus *Propagulifera* of *Sphacelaria* are capable of forming small branched laterals specialized for vegetative propagation. The form of these propagules is an important character to delimit series and species in this group of Sphacelariales. Similar but less complex structures occur in *Choristocarpus*.

4. *Characters related to Ecology*

Information about the ecology of most Sphacelariales is scanty.

(*a*) *Unattached forms*. Two species and several forms of Sphacelariales are recorded as unattached specimens amongst seagrasses (*Zostera*) or large algae.

These loose specimens have no (or very few) rhizoids and are sterile. Propagation takes place by fragmentation or very occasionally by propagules.

(*b*) *Parasitism and epiphytism*. Reinke (1891) distinguished two main groups in the genus *Sphacelaria*: parasitic and non-parasitic species. Parasitic species have long rhizoids which penetrate into the tissue of the host. Subsequent research has shown that a species may be epiphytic on one large alga, endophytic in another and epilithic on the stones in between. Epizoic growth of Sphacelariales has not been carefully studied.

(*c*) *Habitat and growth-form*. Specimens from shallow or brackish water may show a morphology very different to other specimens of the same species from deep water or from exposed localities; unattached forms are extreme examples of this variation. The attached specimens from different localities may differ in their mode of branching, their formation of prostrate structures or their reproduction.

5. *Characters related to Phytogeography*

In *Sphacelaria*, the species of the subgenus *Sphacelaria* always occur in the northern part of Europe. Most species of the subgenus *Propagulifera* are found on more southerly coasts. Different forms of one species often occur in different geographical areas.

SYSTEMATIC CLASSIFICATION

In Fig. 2 a systematic classification is proposed; this is based on intuitive weighting of the characters mentioned above. Most weight has been given to morphological characters.

The relation of the Choristocarpaceae to the Sphacelariales (and of *Discosporangium* to the Choristocarpaceae) remains dubious.

In the Sphacelariaceae, *Sphacella subtilissima* stands apart from the other genera and close to the Choristocarpaceae. Three subgenera (of which one, the subgenus *Acladus*, is provisional) and one aberrant species are mentioned, and in some subgenera several series are delimited. *Disphacella reticulata* (perhaps the only species in the aberrant family Disphacellaceae) has some possible relation to *Sphacelaria caespitula*. *Battersia mirabilis* may be related to *Sphacelaria plumigera*. The complex of *Sphacelaria britannica* and *S. saxatilis* has not been incorporated into a subgenus or a section and the relations between this complex and other species of *Sphacelaria* (*S. furcigera*, *S. radicans*, *S. arctica*) are not yet clear.

In the Stypocaulaceae the status of *Stypocaulon*? *spinulosum* is not clear. This

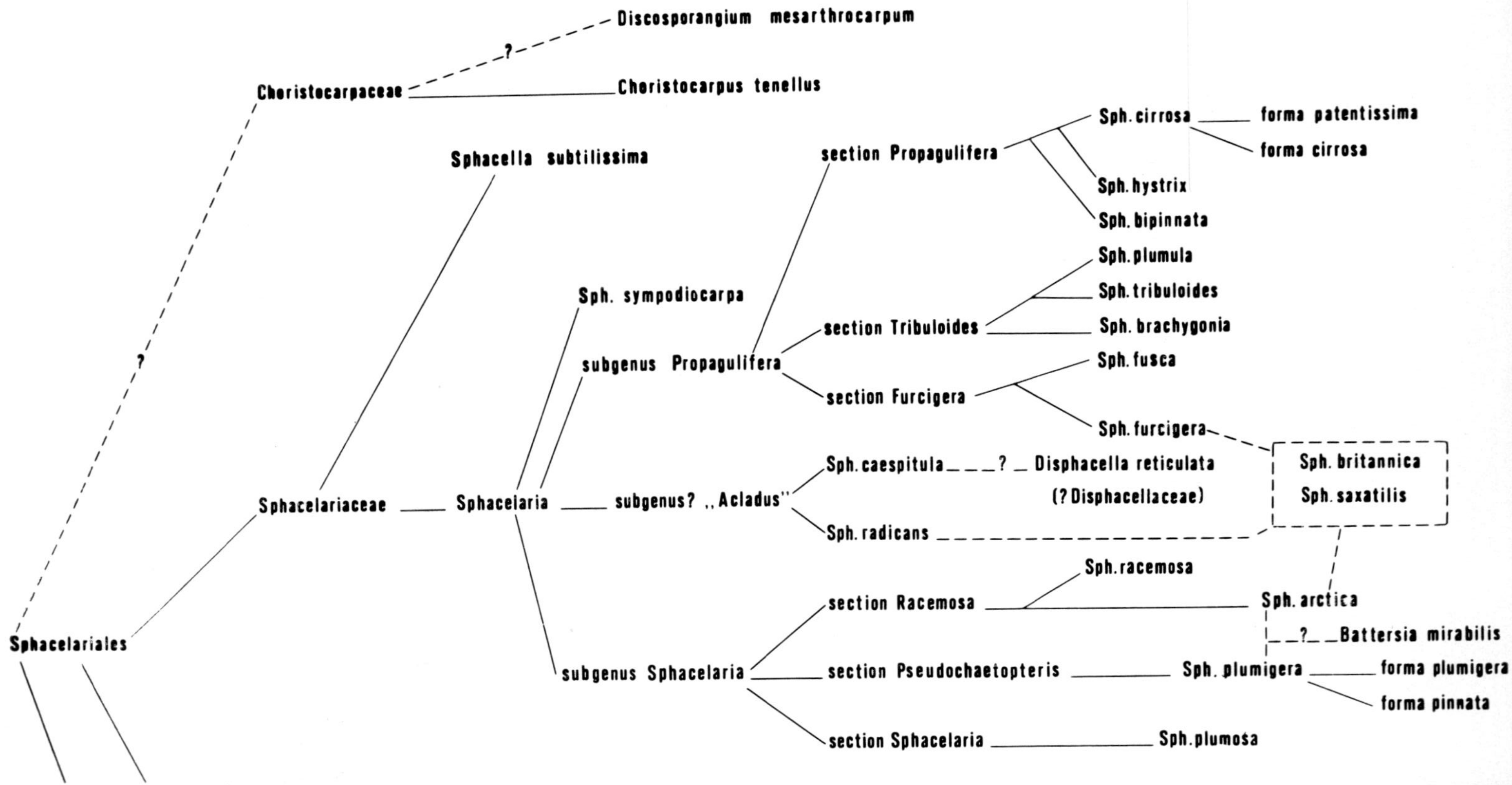

Discosporangium mesarthrocarpum
?
Choristocarpaceae
Choristocarpus tenellus
Sph. cirrosa
forma patentissima
forma cirrosa
Sphacella subtilissima
section Propagulifera
Sph. hystrix
Sph. bipinnata
Sph. plumula
Sph. tribuloides
Sph. sympodiocarpa
section Tribuloides
Sph. brachygonia
?
subgenus Propagulifera
Sph. fusca
section Furcigera
Sph. furcigera
Sph. britannica
Sph. saxatilis
Sph. caespitula
?
Disphacella reticulata
(? Disphacellaceae)
Sphacelariaceae
Sphacelaria
subgenus? „Acladus"
Sph. radicans
Sph. racemosa
section Racemosa
Sph. arctica
?
Battersia mirabilis
Sphacelariales
subgenus Sphacelaria
section Pseudochaetopteris
Sph. plumigera
forma plumigera
forma pinnata
section Sphacelaria
Sph. plumosa

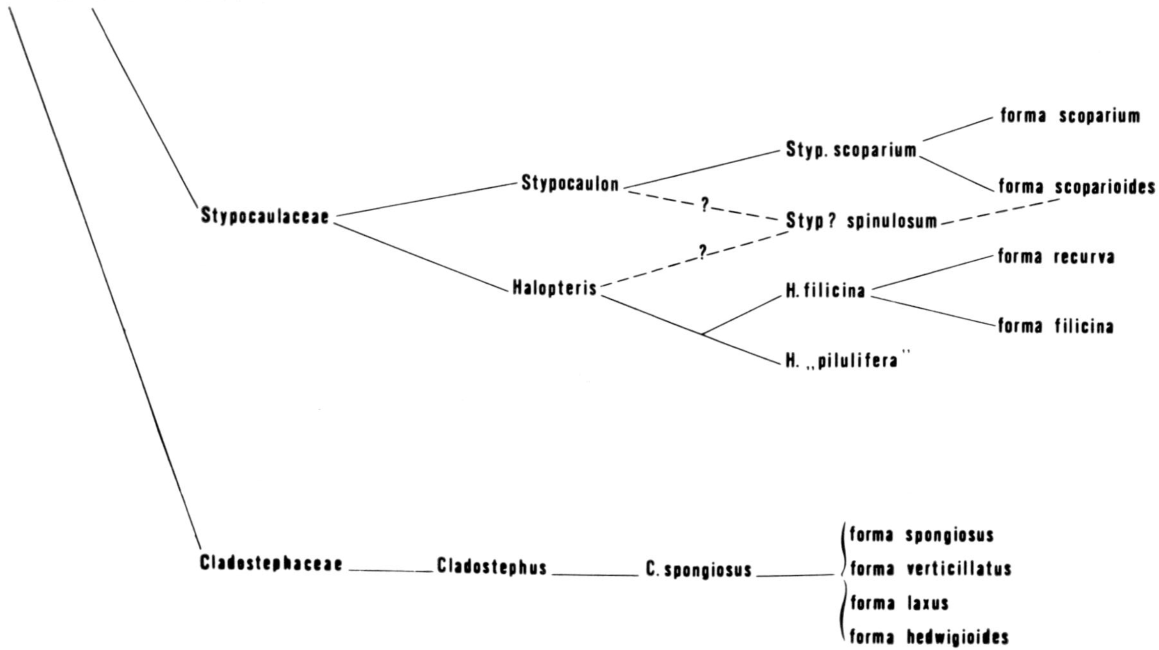

FIG. 2. Diagram of the proposed systematic classification of European Sphacelariales. The connecting lines indicate obvious relationship, the broken lines indicate affinities and a query mark indicates dubious affinity.

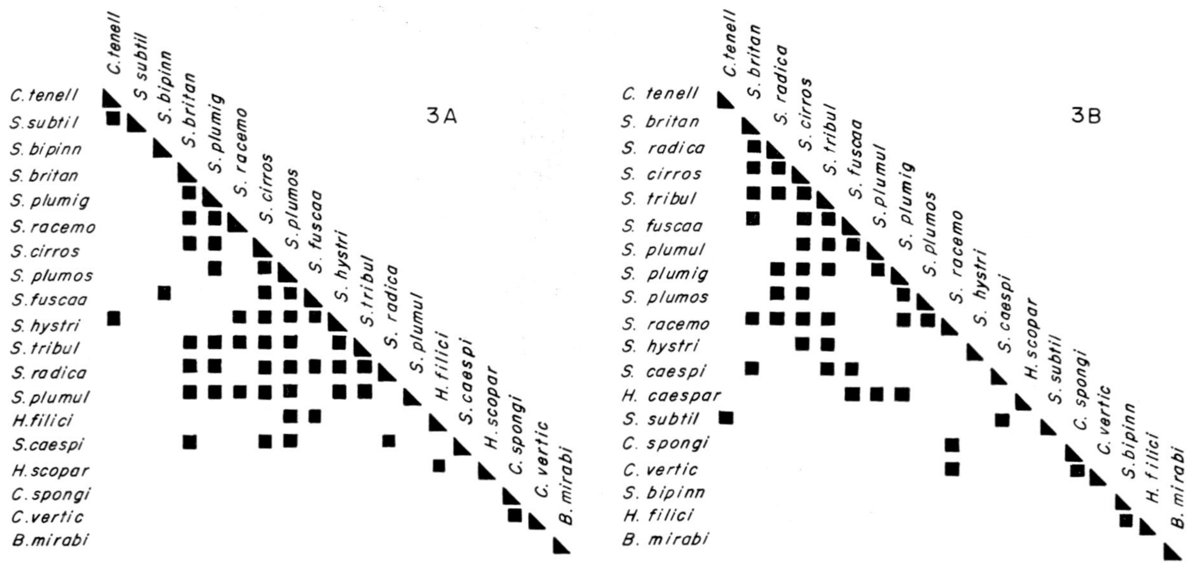

FIG. 3. Similarity matrix. Similarities above 89%. A. Set 1; B. Set 2; C. Set 3; D. Set 4.

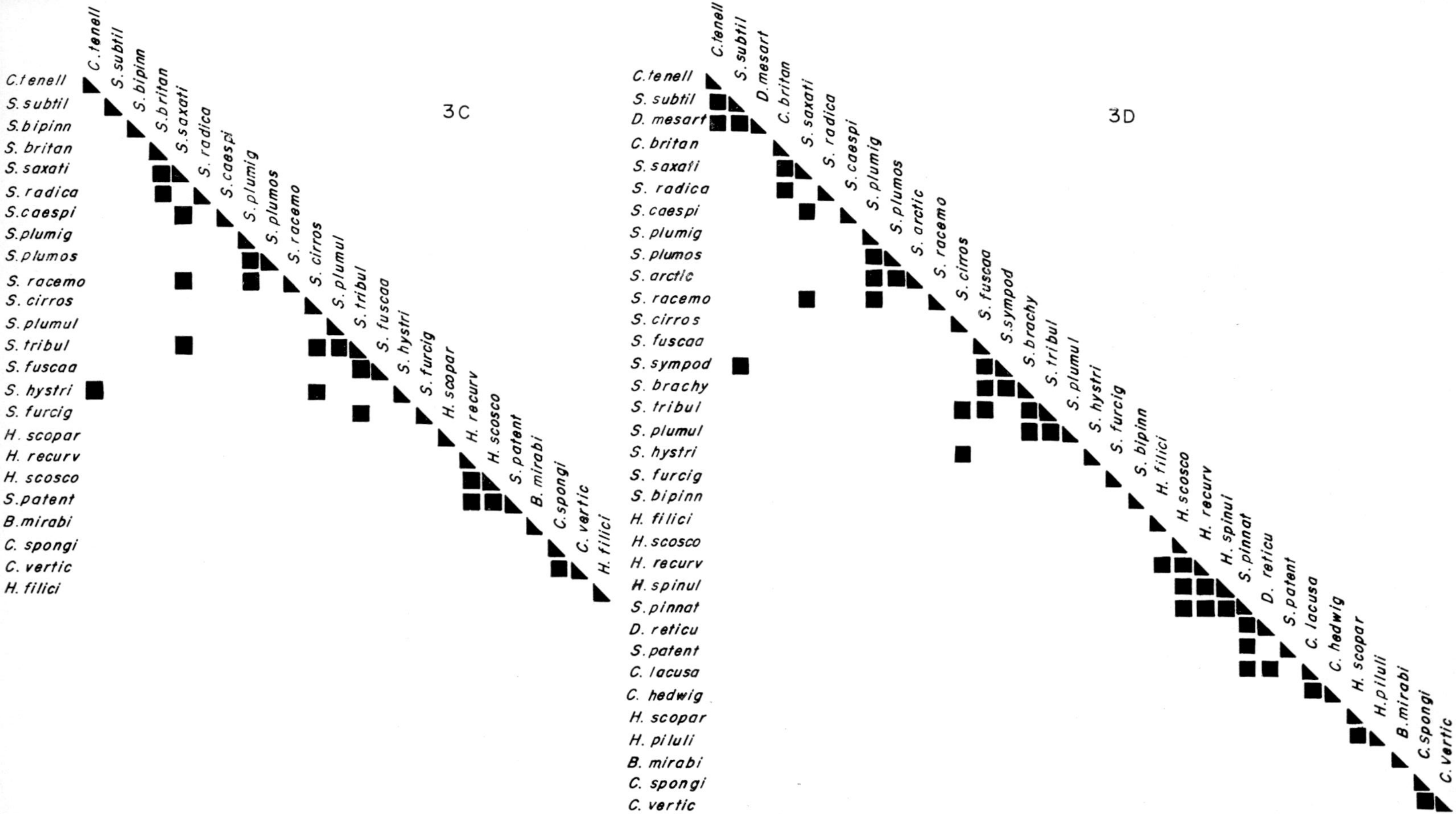
3C
C.tenell
S.subtil
S.bipinn
S.britan
S.saxati
S.radica
S.caespi
S.plumig
S.plumos
S.racemo
S.cirros
S.plumul
S.tribul
S.fuscaa
S.hystri
S.furcig
H.scopar
H.recurv
H.scosco
S.patent
B.mirabi
C.spongi
C.vertic
H.filici
3D
C.tenell
S.subtil
D.mesart
C.britan
S.saxati
S.radica
S.caespi
S.plumig
S.plumos
S.arctic
S.racemo
S.cirros
S.fuscaa
S.sympod
S.brachy
S.tribul
S.plumul
S.hystri
S.furcig
S.bipinn
H.filici
H.scosco
H.recurv
H.spinul
S.pinnat
D.reticu
S.patent
C.lacusa
C.hedwig
H.scopar
H.piluli
B.mirabi
C.spongi
C.vertic

unattached alga has some characters of *Stypocaulon* but others of *Halopteris*. Perhaps Sauvageau (1900–1914) was right in uniting *Halopteris* and *Stypocaulon* into one genus. *Halopteris "pilulifera"* is probably a new species. For a delimitation of the forms of *Cladostephus spongiosus* see Prud'homme van Reine (1972).

NUMERICAL ANALYSIS

As discussed under Methods and summarized in Table III, four sets of data have been used for numerical analysis. The results of these analyses are presented in different ways which are based on different sorting strategies. Compare also with the results in Russell and Fletcher (1975).

Set 1. British Taxa and Taxonomic Characters also used by Russell and Fletcher

(*a*) *Similarity matrix*. The similarities between the taxa used for this analysis were always above 75%. Three groups of greatest similarity are found (Fig. 3A).

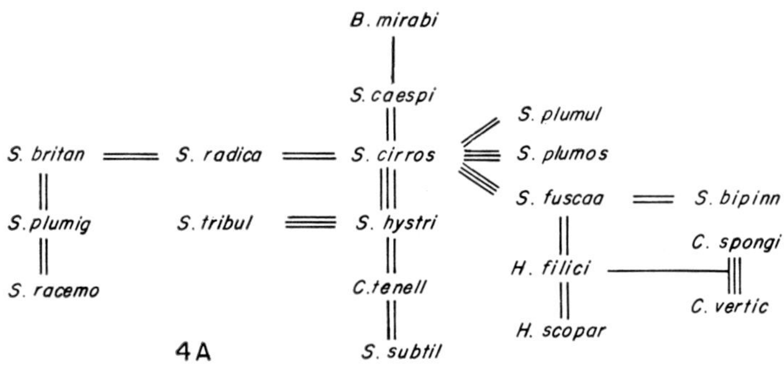

4A

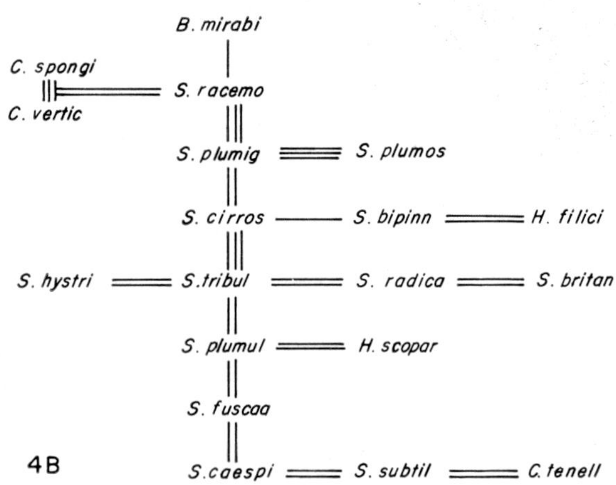

4B

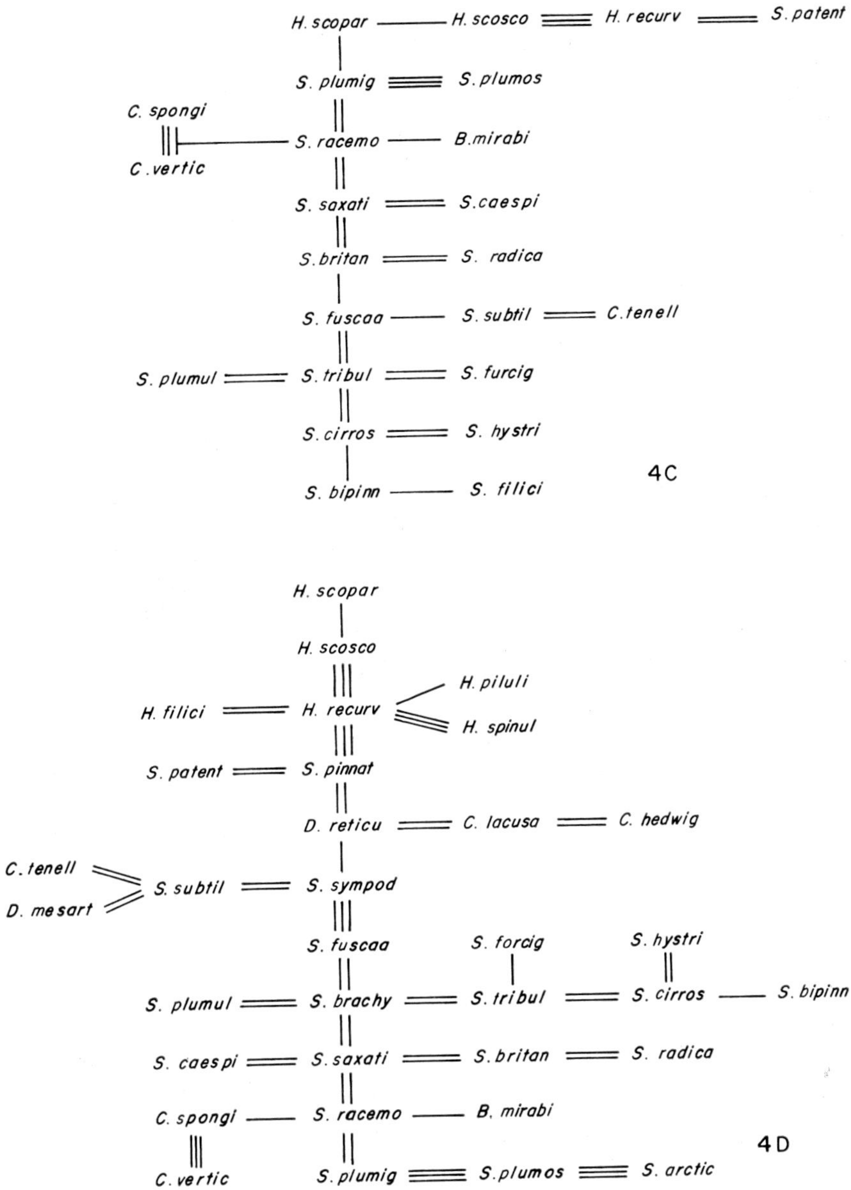

FIG. 4. Minimum spanning tree. Single line corresponds with similarities of 80–89%, a double line denotes similarities of 90–94% and a triple line 95–100%. A. Set 1; B. Set 2; C. Set 3; D. Set 4.

Sphacelaria bipinnata, *Sphacelaria caespitula*, both species of *Halopteris* and *Battersia mirabilis* are not incorporated in one of the clusters.

(*b*) *Minimum spanning tree*. This has to be prepared by hand. The lowest similarity used was 87%. This results in a figure similar to Russell and Fletcher's (compare their Fig. 2D with my Fig. 4A), with *Sphacelaria cirrosa* in a central position.

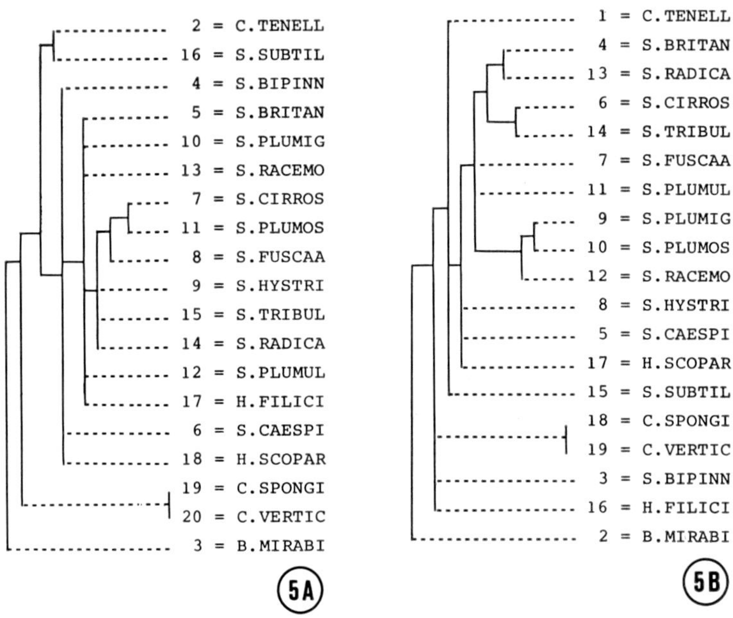

FIG. 5 Dendrogram, single linkage. A. Set 1; B. Set 2.

Sphacelaria plumigera (and *S. racemosa*) have been connected to *S. britannica* in this figure. But it is also possible to connect *S. plumigera* to *S. plumosa*. In both cases similarity is 94%.

(*c*) *Dendrogram*. When clustered by single linkage the same groups can be found as in the similarity matrix (Fig. 5A).

Set 2. British Taxa, with Corrections

(*a*) *Similarity matrix*. All similarities were above 74%. Three groups of greatest similarity are found (Fig. 3B). One large and indifferent group is formed by all *Sphacelaria* species (except *Sphacelaria bipinnata*, which can be found in a small group together with *Halopteris filicina*). *Choristocarpus tenellus*, *Halopteris scoparia*, *Sphacella subtilissima* and *Battersia mirabilis* are not incorporated into the main clusters.

(*b*) *Minimum spanning tree.* Lowest similarity 89%. Central positions for *Sphacelaria cirrosa* and *S. tribuloides* (Fig. 4B). *Choristocarpus tenellus* and *Sphacella subtilissima* are nearest neighbours in this figure.

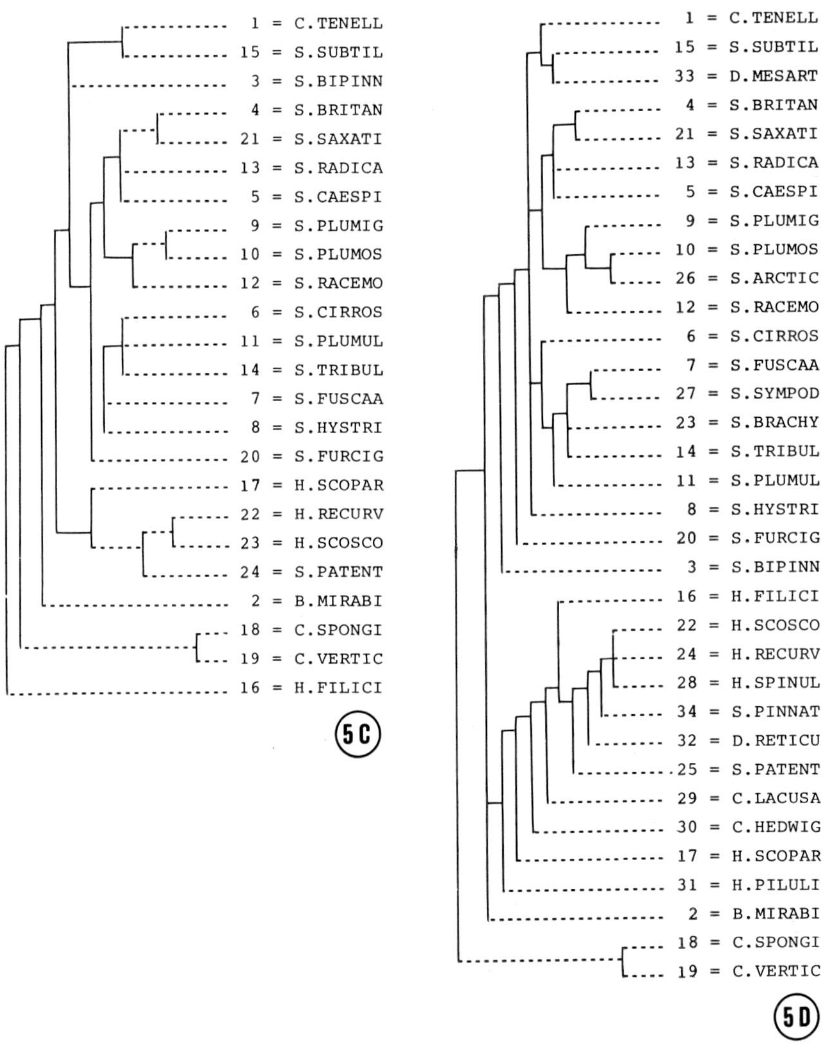

FIG. 5. Dendrogram, single linkage. C. Set 3; D. Set 4.

(*c*) *Dendrogram.* Single linkage (Fig. 5B). Apart from *Sphacelaria bipinnata* all *Sphacelaria* species are found in one cluster (together with *Halopteris scoparia*).

Set 3. British Taxa, with Corrections and Additions

(*a*) *Similarity matrix.* All similarities were above 69%. One large group (with three subgroups) and two small groups of greatest similarity are found (Fig. 3C). *Choristocarpus tenellus*, *Sphacella subtilissima*, *Sphacelaria bipinnata*, *Halopteris scoparia*, *Battersia mirabilis* and *Halopteris filicina* are not incorporated into any of the clusters.

(*b*) *Minimum spanning tree.* Lowest similarity used is 82%. There is no distinct central position (Fig. 4C). In this figure *Halopteris filicina* is separated from other taxa in *Halopteris* by the large group of *Sphacelaria* species.

(*c*) *Dendrogram.* Clustering by single linkage (Fig. 5C). All *Sphacelaria* taxa apart from *Sphacelaria bipinnata* and *Sphacelaria cirrosa* forma *patentissima* are in one cluster. *Halopteris filicina* has an outlying position.

Set 4. All European Taxa of Sphacelariales

(*a*) *Similarity matrix.* All similarities were above 68%. Two large groups (one with three subgroups) and four small groups of greatest similarity are found (Fig. 3D). *Sphacelaria furcigera*, *S. bipinnata*, and *Battersia mirabilis* are not incorporated in any of the clusters.

(*b*) *Minimum spanning tree.* Lowest similarity used is 84%. There are several central positions (*Halopteris pilulifera*, *Sphacelaria brachygonia*, *S. saxatilis* and *S. racemosa*, Fig. 4D). All taxa in *Halopteris* can be found together, but the unattached or sterile taxa in *Sphacelaria* and *Cladostephus* are separated from the attached ones.

(*c*) *Dendrogram.* Clustering by single linkage (Fig. 5D). All *Sphacelaria* species (except the unattached forms) are in one cluster. The taxa in *Halopteris* are in another cluster but the taxa in *Cladostephus* are found in two different clusters.

DISCUSSION

The results of the numerical analysis have been compared with the proposed systematic classification. The results of set 1 will not be discussed.

1. Similarity Matrix
There are no distinct taxonomic groups which can be compared with the groups of greatest similarity in Fig. 3B. In Fig. 3C (set 3) almost all *Sphacelaria* species can be found in one large group. The three subgroups can be related to the subgenera mentioned in Fig. 2. *Sphacelaria bipinnata* has an outlying position and *Sphacelaria cirrosa* forma *patentissima* can be found in a group together with the two unattached forms in *Halopteris*. *Halopteris filicina* and *Halopteris* (*Stypocaulon*) *scoraria* are not clustered.

In Fig. 3D (set 4) there is again a large *Sphacelaria*-group with three subgroups (related to the subgenera mentioned in Fig. 2). Now *S. bipinnata* can be found near the other species of the subgenus *Propagulifera*. In this group is also merged *Sphacelaria sympodiocarpa* (separated in my proposed classification). *Sphacella subtilissima* can be found in a group together with *Choristocarpus tenellus* and *Discosporangium mesarthrocarpum*. The unattached forms of several genera can be found in the second large group in Fig. 3D. There are small clusters of the sterile *Cladostephus* forms, of *Halopteris* (*Stypocaulon*) *scoparia* and *H. pilulifera* and of the fertile *Cladostephus* forms.

2. Minimum Spanning Tree
In Fig. 4B (set 2) three species of the subgenus *Sphacelaria* are nearest neighbours. *Sphacelaria racemosa* is also connected to *Battersia mirabilis* and to the two British forms of *Cladostephus spongiosus*. The *Sphacelaria* species of the subgenus *Propagulifera* are directly connected to each other but *Sphacelaria radicans* and *S. caespitula* are separated, as are the two British species of *Halopteris*.

In Fig. 4C (set 3) not much is changed, but *Sphacelaria radicans* and *S. caespitula* are no longer far removed. The similarity of *Sphacelaria cirrosa* forma *patentissima* to the unattached *Halopteris* forms is distinct in this figure.

In Fig. 4D (set 4) all *Halopteris* taxa are found together. The unattached *Sphacelaria* forms are near-by and these are separated from the other Sphacelarias by *Disphacella reticulata*. The subgenera of *Sphacelaria* form distinct branches of the tree, and *Sphacelaria sympodiocarpa* has an outlying position. The sterile and the fertile *Cladostephus* forms are still separated.

3. Dendrogram (Single Linkage)
In Fig. 5B all *Sphacelaria* species (except *S. bipinnata*) are already found in one large cluster, in which the British species of the subgenus *Sphacelaria* form a distinct small cluster. The clustering of *Sphacelaria bipinnata, Halopteris filicina* and the forms of *Cladostephus* is probably due to the many characters in the life-history which are lacking for the other species.

In Fig. 5C (set 3) the three subgenera (again excepting *Sphacelaria bipinnata*) form separate clusters but the unattached *Sphacelaria cirrosa* forma *patentissima* is clustered with the unattached taxa in *Halopteris*. The complex of *Sphacelaria britannica* and *S. saxatilis* is found in one cluster with the species of the subgenus *Acladus*. *Halopteris filicina* and *H.* (*Stypocaulon*) *scoparia* are found in different clusters. In Fig. 5D (set 4) little is changed, but all unattached Sphacelariales are here clustered; *Sphacella subtilissima* is found in one cluster with the Choristocarpaceae.

CONCLUSIONS

The proposed systematic classification, based on intuitive weighting of characters, can be supported in part by using numerical taxonomic methods. The classification in the genus *Sphacelaria* is fully endorsed by numerical analysis and the *britannica*/*saxatilis* complex can probably be incorporated into the subgenus *Acladus*. Often the number of characters related to presence or absence of attachment and/or reproduction results in clustering of morphologically similar but not taxonomically related taxa.

Perhaps *Halopteris* and *Stypocaulon* are not as closely related as proposed in my classification and probably the Choristocarpaceae are more closely related to the Sphacelariales than I supposed. The taxonomic status of the genera *Disphacella* and *Battersia* remains uncertain.

ACKNOWLEDGEMENTS

The writer wishes to express his thanks to the directors and keepers of the following Botanical Institutes for loan of material or permission to examine specimens: AAU, BG, BM, BR, BREM, C, CHE, CN, DIN, E, FI, GB, GL, GOET, H, HBG, ICEL, KIEL, L, LD, LISU, LIVU, M, MPU, NMW, O, OXF, PC, S, STA, TCD, U, UPS, W, WU, ZA. (Abbreviations according to the Index Herbariorum.)

He also acknowledges the directors of the marine biological institutes in Askö, Banyuls, Bergen (Espegrend), Concarneau, Dinard, Drøbak, Helgoland, Kristineberg, Roscoff, St Andrews and Wimereux for permission to stay and study at these institutes. Funds for the journeys have been provided by the Netherlands Organization for the Advancement of Pure Research (Z.W.O.), the Centre National de la Recherche Scientifique (C.N.R.S.) and the State University at Leiden.

Drs G. D. E. Povel (Leiden) and Drs P. Hogeweg (Utrecht) and the staff of the Centraal Reken Instituut (CRI = Central Computing Centre) of the State University at Leiden guided me through the many difficulties in running the classificatory computer programs on the data I provided.

REFERENCES

AGARDH, J. G. (1848). "Species, Genera et Ordines Algarum . . .", Vol. 1, ". . . Fucoid-earum." C. W. K. Gleerup, Lund.

FRITSCH, F. E. (1945). "The Structure and Reproduction of the Algae", Vol. II. University Press, Cambridge.

GEYLER, T. (1866). Zur Kenntnis der Sphacelarieen. *Jb. wiss. Bot.* **4**, 479–535.

HIGGINS, E. M. (1931). A cytological investigation of *Stypocaulon scoparium* (L.) Kütz. *Ann. Bot.*, **45**, 345–353.

KÜTZING, F. T. (1843). "Phycologia Generalis." F. A. Brockhaus, Leipzig.

KYLIN, H. (1917). Ueber die Entwicklungsgeschichte und die systematische Stellung der Tilopterideen. *Ber. dt. bot. Ges.* **35**, 298–310.

MAGNUS, P. (1873). "Zur Morphologie der Sphacelarieen . . .". Ferd Dümmlers Verlagsbuchhandlung, Berlin.

MIGULA, W. (1909). "Kryptogamen-Flora von Deutschland, Deutsch-Österreich und der Schweiz", Band II (2). Verlag Friedrich von Zezschwitz, Gera, R.

MOORE, L. B. (1951). Reproduction in *Halopteris* (Sphacelariales). *Ann. Bot., Lond.* **15**, 265–278.

PRINGSHEIM, N. (1873). Ueber den Gang der morphologischen Differenzirung in der Sphacelarien-Reihe. *Phys. Math. Abh. K. Akad. Wiss. Berlin.* **1873**, 137–191.

PRUD'HOMME VAN REINE, W. F. (1972). Notes on Sphacelariales (Phaeophyceae) II. *Blumea* **20**, 138–144.

PRUD'HOMME VAN REINE, W. F. (1974). Geographic distribution of European *Sphacelaria*-species in the world (Phaeophyceae, Sphacelariales). *Bull. Soc. phycol. Fr.* **19**, 171–177.

REINKE, J. (1891). Beiträge zur vergleichenden Anatomie und Morphologie der Sphacelariaceen. *Biblthca bot.* **23**, 1–40.

RUSSELL, G. and FLETCHER, R. L. (1975). A numerical taxonomic study of the British Phaeophyta. *J. mar. biol. Ass. U.K.* **55**, 763–783.

SAUVAGEAU, C. (1900–1914). "Remarques sur les Sphacélariacées." Partly published in *J. Bot., Paris*, but also published as a separate publication with altered pagination. Féret et fils, Bordeaux.

14 | The Application of Computers to the Identification of the Phaeophyta

R. J. PANKHURST and I. TITTLEY

Department of Botany, British Museum (Natural History), England

Abstract: This paper demonstrates the application of computers to the identification of the brown seaweeds. The study is based on a suite of computer programs already applied (by R. J. P.) to many other kinds of organisms. All these programs make use of the same set of data, derived from the numerical classification studies of Russell and Fletcher (1975) and other major texts on the brown algae (e.g. Hamel, 1931–1939; Rosenvinge, 1941–1950).

The main problems encountered during the preliminary survey of data involved the standardization of data and the accurate definition of characters and character states. The further problem of incomplete data is unfortunately inevitable in any broad survey. From the data have been derived:

1. A variety of diagnostic keys.
2. A punched card multi-access key (polyclave).
3. A printed description, similar to that in a published flora or monograph, of each species.

Advantages of this approach are that it is easy to store and scan large amounts of data, to make corrections to the base data and to produce alternative versions of keys.

INTRODUCTION

Computer-aided identification methods were first described in the 1960s and a considerable variety is now available. It should be clear at the outset that, as in more traditional methods, the final decision about the correct identification remains with the biologist, the computer acting only in an auxiliary rôle. With few exceptions, it is still necessary for the worker to examine specimens and assess their characters. For a detailed acount of the different methods of biological identification with the aid of a computer see Pankhurst (1975).

Systematics Association Special Volume No. 10, "Modern Approaches to the Taxonomy of Red and Brown Algae" edited by D. E. G. Irvine and J. H. Price, 1978, pp. 325–337, Academic Press, London and New York.

Taking the broad view, there are two main kinds of identification method. These are:

1. Sequential or monothetic. Diagnostic and most punched-card keys are of this type. Each character, separately or in group, is used in a step-by-step procedure which eliminates all but one taxon. Whilst the methods can be quite sparing of the number of different characters required, they are sensitive to error if even one character is wrongly observed.

2. Simultaneous or polythetic. Various tabular, comparison or matching methods belong here, as well as some aspects of on-line conversational programs for which the computer must be used. These methods work best with as complete as possible a description of the specimen but can still give a correct result even if some characters are wrongly scored.

The diagnostic key is the only method among these which is in regular use. The reason for this is probably its practicality, which is paramount despite obvious faults. The use of the computer, mostly in an indirect fashion, makes a much wider range of methods readily available; this is demonstrated below.

METHODS

Initially descriptions of all the relevant taxa are compiled in the form of a data matrix, and then the data are subjected to a series of computer programs in order to produce different identification methods for the same taxonomic group.

One method, the punched-card key (polyclave), was favoured from the outset. Although it still has some of the faults of sequential methods, it permits any choice of characters to be used in any order. It is therefore a multi-access key and is very convenient for dealing with fragmentary or incomplete material. This has always been true of punched-card keys, but now their production can be carried out by the computer from the data matrix, making them readily available in quantity at reasonable cost.

Although the main aim of the present project was to create a punched-card key, little extra effort was required to apply the various other programs to the same matrix. We therefore also show some results from programs for key construction; identification by matching; and the construction and printing of descriptions of taxa as in a flora or monograph.

The brown algae were chosen for experiment with computer-based aids to identification for two main reasons:

1. The only complete set of diagnostic keys to the group is now out of date and more recent productions cover only restricted aspects.

2. The recent numerical taxonomic study (Russell and Fletcher, 1975) collated data into a concise form which could easily be adapted.

Russell and Fletcher (1975) surveyed 132 taxonomic characters for all the British species of the Phaeophyta. Some of these are not practical for use in identification, for example, flagellar structure, chromosome number and aspects of life-history; these have not been used. In order to amplify and clarify data, other standard reference works have been consulted (Hamel, 1931–1939; Newton, 1931; Rosenvinge and Lund, 1941; Pankow, 1971). Forty-one morphological, reproductive or ecological characters have been scored for the Ectocarpaceae in the following way:

character: Erect system
score: 0 No data
1 Hairs only [first character state]
2 Assimilatory filaments only [second character state]
4 Hairs and assimilatory filaments [third character state]
8 . . . . . .

Character states are numbers by powers of two, which allows a combination of them to be expressed unambiguously. Dependent characters may, in certain cases, be scored differently. For example:

character: Erect assimilatory filament/branching
first state: Unbranched
second state: Branched
dependent character: Branching type

If the first character state (Erect assimilatory filament/Unbranched) applies, then -1 is scored against the subsequent (dependent) character. Since several sources of data have been consulted, a certain amount of standardization in the definitions of characters and states of characters is inevitable and may have led to oversimplification. A list of the characters and states used in describing the Ectocarpaceae appears in Table I.

Six weights have been used for characters in the compilation of our diagnostic keys. The lowest weighting indicates that a character should not be used, the highest that a character is reliable and should be used.

Once the data have been collated, they are stored on file in the computer, together with lists of characters and character states, conditioned characters, character weightings and species.

A polyclave finally consists of a pack of computer cards, each of which represents a character state (Fig. 1). The character state is printed at the centre top of each card, which is also numbered at the right-hand end. At the left-hand end is a series of holes. Each hole represents a species which shows the particular character state. In the production of a polyclave, variable and missing information is treated specially. If data for the states of a character are missing for a

TABLE I. List of characters and character states used in describing the Ectocarpaceae

Character	State	Description
1		Thallus/form
	1	Uniseriate filaments
	2	Pseudoparenchymatous
	3	Parenchymatous
2		Main vegetative system/type
	1	Erect
	2	Erect and prostrate
	3	Prostrate
3		Thallus/size
	1	Microscopic
	2	Macroscopic
4		Macroscopic thallus/size
	1	Less than 1 cm
	2	More than 1 cm
5		Erect system/size versus prostrate
	1	Less than or equal to prostrate
	2	Greater than prostrate
6		Erect system as/type
	1	Colourless hairs only
	2	Assimilatory filaments only
	3	Hairs and filaments
7		Erect assimilatory filaments/branching
	1	Unbranched
	2	Branched
8		Right angle branches/presence
	1	Absent
	2	Present
9		Filament branching/pseudo-dichotomous or alternate
	1	Not pseudodichotomous or alternate
	2	Pseudodichotomous or alternate
10		Filament branching/opposite
	1	Not opposite
	2	Opposite
11		Filament branching/secund
	1	Not secund
	2	Secund
12		Filament branching/abundance
	1	Sparse
	2	Abundant
13		Filament branching/regularity
	1	Irregular
	2	Regular
14		Hairs/type
	1	Pseudohairs
	2	True hairs
15		Prostrate system as/type
	1	Free filaments
	2	Pseudodisc
	3	Disc
16		Periclinal cell divisions of prostrate vegetative system/presence
	1	Absent
	2	Present
17		Rhizoids/presence
	1	Absent
	2	Present
18		Corticating filaments/presence
	1	Absent or rare
	2	Present
19		Longitudinal divisions in older vegetative cells of erect filaments/presence
	1	Rare
	2	Common
20		Filaments/interwoven
	1	Not interwoven
	2	Densely interwoven
21		Growth/type
	1	Diffuse
	2	Intercalary
	3	Apical
22		Chloroplasts/shape
	1	Discoid
	2	Linear
23		Choroplasts/number per cell
	1	One
	2	Not more than three
	3	More than three
24		Unilocular sporangia/presence
	1	Absent
	2	Present

TABLE I—*continued*

25		Plurilocular sporangia/presence
	1	Absent
	2	Present
26		Monosporangia/presence
	1	Absent
	2	Present
27		Sporangia/organization
	1	Not organized or scattered
	2	Organized in sori
28		Sporangia/position
	1	Terminal
	2	Intercalary
29		Sporangia/exsertion
	1	Sunk
	2	Exserted
30		Plurilocular sporangia/structure
	1	Uniseriate
	2	Uni- and bi-seriate
	3	Multiseriate
	4	Clustered
31		Plurilocular sporangia of/ recognizable types
	1	One type
	2	Several types (micro, meio, mega)
32		Plurilocular spornagia/terminal hair
	1	Without terminal hair
	2	With terminal hair
33		Multiseriate plurilocular sporangia/shape
	1	Elongate
	2	Conical
	3	Ovoid
	4	Irregular
34		External mucilage/presence
	1	Absent
	2	Present
35		/Attachment
	1	Pelagic
	2	Benthic
36		Substrate/type
	1	Epilithic
	2	Epibiotic
	3	Endobiotic
37		/Epibiotic
	1	Epiphytic
	2	Epizoic
38		/Endobiotic
	1	Endophytic
	2	Endozoic
39		/Shore position
	1	Supralittoral
	2	Eulittoral
	3	Infralittoral
40		/Salinity
	1	Freshwater
	2	Estuarine
	3	Marine
41		/Caves
	1	Not in caves
	2	In caves

species, holes for that species are punched on all the relevant character state cards. Similarly, if a species is known to be variable for a particular character, holes are punched in the cards for each known possible character state.

RESULTS AND UTILIZATION

To use the polyclave (Figs. 1 and 2), cards are chosen in any order to represent the character states shown by the specimen to be identified. The chosen cards are laid on top of each other and held to the light. The places where holes still show indicate the species possessing that selection of characters. The process is

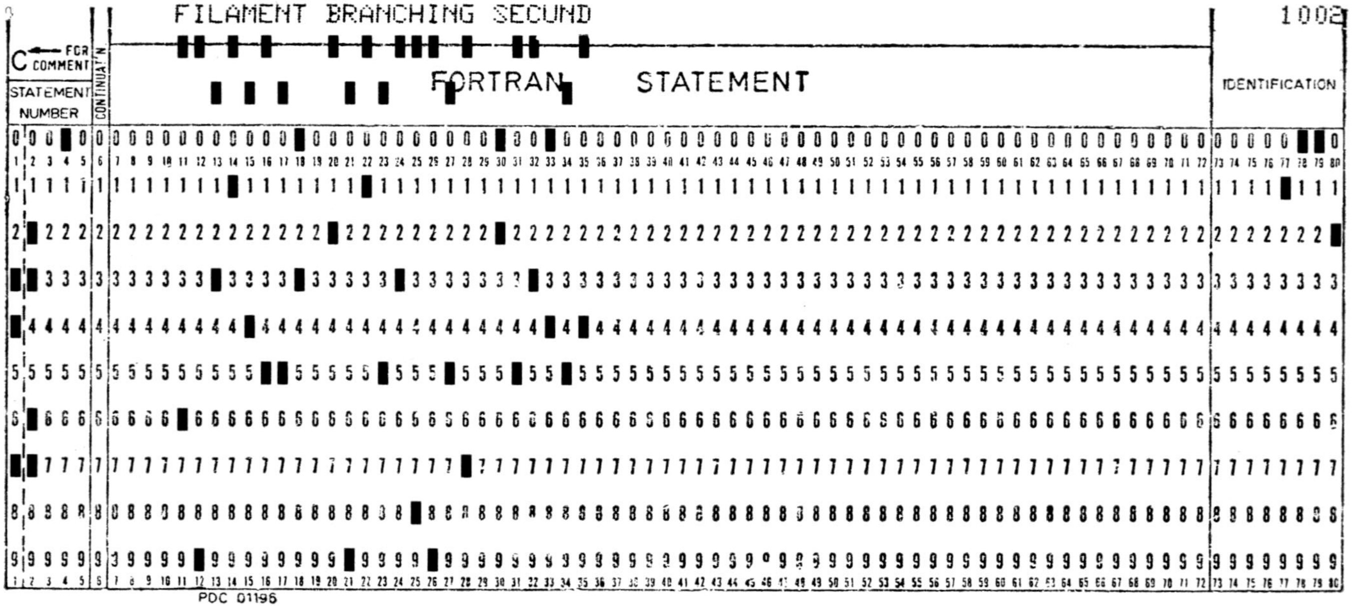
FILAMENT BRANCHING SECUND
1002
C FOR COMMENT
STATEMENT NUMBER
CONTINUAT'N
FORTRAN STATEMENT
IDENTIFICATION
PDC 01196

FIG. 1A

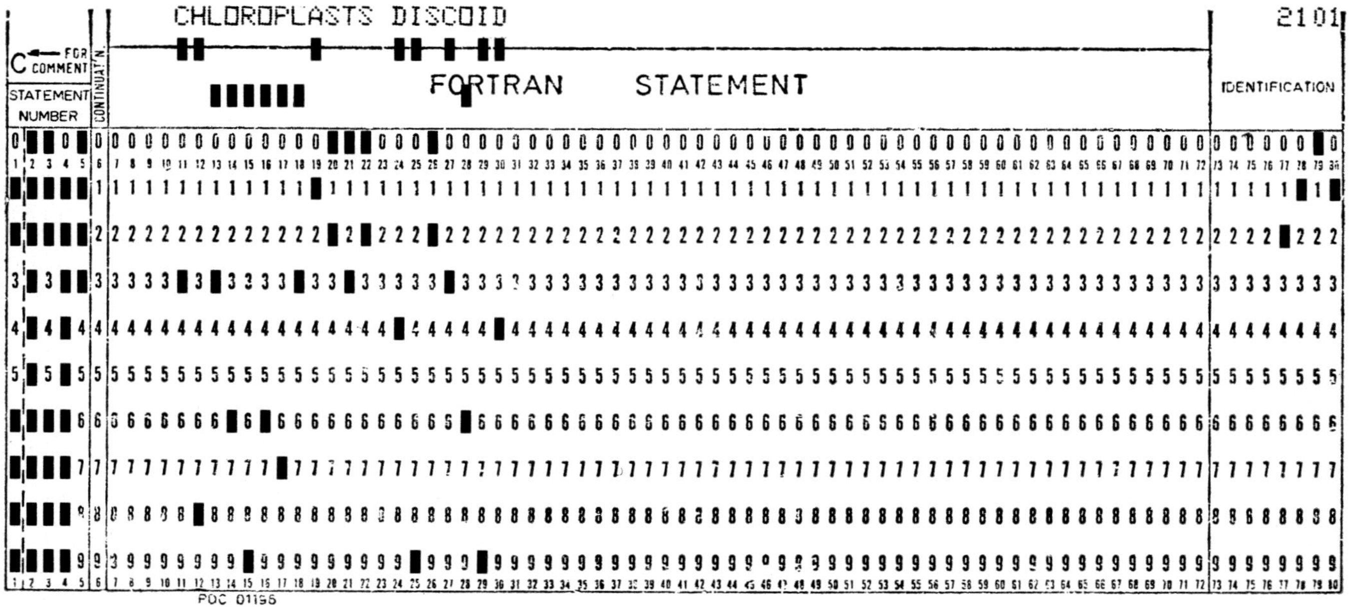

Fig. 1b

Figs 1 and 2. Demonstration of polyclave utilization. 1. States for individual characters. 2. Reduction in punched holes when state-cards are overlain.

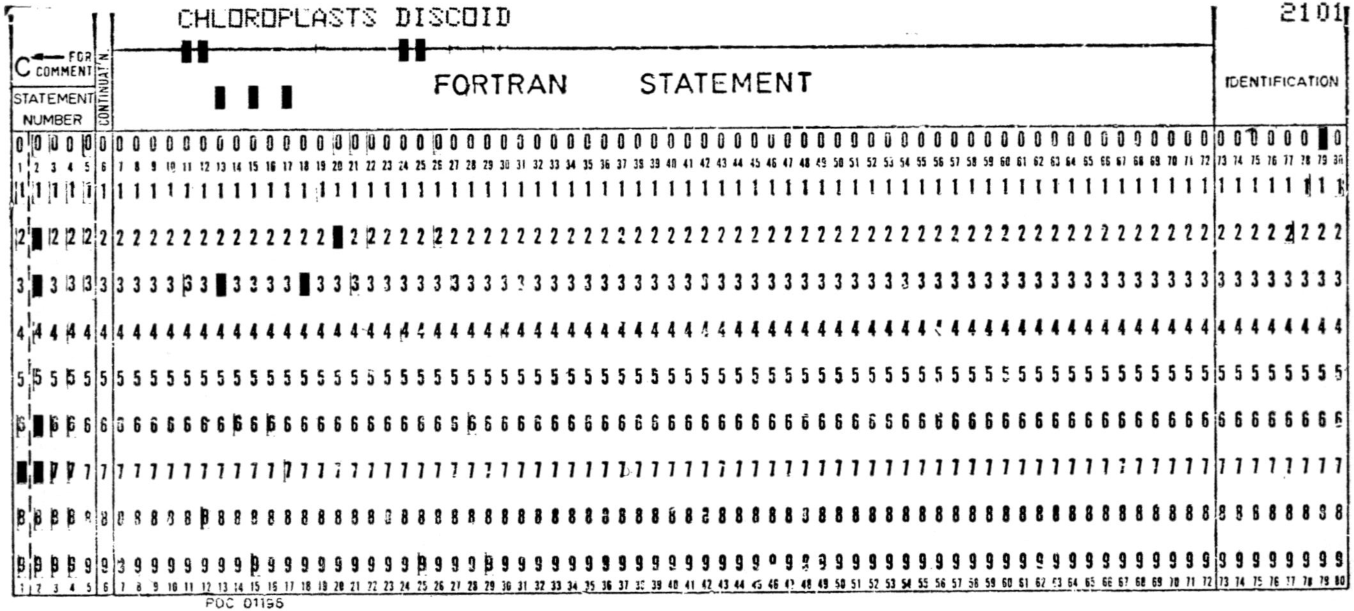

FIG. 2. Caption on previous page

continued until only one hole remains. If no holes remain the cards for states about which the user is not certain are discarded and the process repeated.

In addition to the polyclave, diagnostic keys have been derived; one of these is presented in Fig. 3. Different versions of the key are produced by applying different weights to characters. In one case we have produced a key using vegetative characters only. Missing or variable data are treated as they were for the polyclave and this often results in a species keying out in several places.

Identification by matching uses the computer directly in a batch processing mode. The program accepts a description of a specimen, computing a similarity coefficient between the specimen and each of the possible taxa and printing a list of those that most closely agree. The description fed into the computer need not be a complete species description, but simply information derived from the specimen under examination. The first line of output (Fig. 4) for each specimen is the title; a list of special characters which the user considers important for identification follows. For each taxon in the list, the following items are printed:

1. Sequence number.
2. Percentage similarity between specimen and taxon.
3. Number of comparable characters which are known for both the specimen and taxon.
4. For each character, a plus sign if the taxon agrees with the character, a blank if it does not. These appear in the same sequence as the list of special characters.
5. Taxon name.

There follows the number of the group (in this case, genus) which the specimen most resembles. If other (special) taxa have been selected for comparison, there then follows a line for each of these, giving the sequence number, percentage similarity and name for each taxon. Finally, ideally, a report may appears on the discrepancies between the description of the specimen and that of the taxon to which it is supposed to belong. If there are several taxa to which the specimen could be assigned, the report is undertaken only for the first. For each combination of character states in the specimen but absent from the taxon to which the specimen is supposed to belong, the program gives character and states numbers. It also lists those taxa which differ from the special taxon, but which would not do so if the missing character or characters were added to the data matrix. This is intended as a warning against assuming that the specimen is correctly described and identified, and an indication that the specimen shows more variation than has so far been recorded. The "correct" determination is left to the user to decide; the final computer determination

1 MAIN VEGETATIVE SYSTEM ERECT. STREBLONEMA INTESTINUM
1 MAIN VEGETATIVE SYSTEM PROSTRATE. 2
2 CHLOROPLASTS MORE THAN THREE, RHIZOIDS PRESENT, PLURILOCULAR SPORANGIA MULTISERIATE, INFRALITTORAL. ENDODICTYON INFESTANS

2 CHLOROPLASTS NOT MORE THAN THREE, RHIZOIDS ABSENT, PLURILOCULAR SPORANGIA UNISERIATE, EULITTORAL. 3
3 SPORANGIA TERMINAL. MIKROSYPHAR PORPHYRAE
3 SPORANGIA INTERCALARY. MIKROSYPHAR POLYSIPHONIAE MIKROSYPHAR PORPHYRAE

1 MAIN VEGETATIVE SYSTEM ERECT AND PROSTRATE. 4
4 ERECT SYSTEM LESS OR EQUAL PROSTRATE. 5
5 THALLUS MACROSCOPIC. 6
6 ERECT SYSTEM AS ASSIMILATORY FILAMENTS ONLY, UNILOCULAR SPORANGIA PRESENT. HERPONEMA VELUTINUM
6 ERECT SYSTEM AS HAIRS AND FILAMENTS, UNILOCULAR SPORANGIA ABSENT. 7
7 ERECT ASSIMILATORY FILAMENTS UNBRANCHED, PSEUDOHAIRS, RHIZOIDS PRESENT, PLURILOCULAR SPORANGIA MULTISERIATE. HERPONEMA VALIANTEI
7 ERECT ASSIMILATORY FILAMENTS BRANCHED, TRUE HAIRS, RHIZOIDS ABSENT, PLURILOCULAR SPORANGIA UNISERIATE. STREBLONEMA STILOPHORAE
5 THALLUS MICROSCOPIC. 8
8 PROSTRATE SYSTEM AS PSEUDODISC, PERICLINAL CELL DIVISIONS OF PROSTRATE VEGETATIVE SYSTEM PRESENT. PHAEOSTROMA PUSTULOSUM

8 PROSTRATE SYSTEM AS FREE FILAMENTS, PERICLINAL CELL DIVISIONS OF PROSTRATE VEGETATIVE SYSTEM ABSENT. 9
9 CHLOROPLASTS NOT MORE THAN THREE. 10
10 ERECT SYSTEM AS ASSIMILATORY FILAMENTS ONLY, MULTISERIATE PLURILOCULAR SPORANGIA OVOID. STREBLONEMA BREVE
10 ERECT SYSTEM AS COLOURLESS HAIRS ONLY, MULTISERIATE PLURILOCULAR SPORANGIA ELONGATE. 11
11 SPORANGIA EXSERTED. STREBLONEMA EFFUSUM
11 SPORANGIA SUNK. STREBLONEMA EFFUSUM STREBLONEMA FASCICULATUM

9 CHLOROPLASTS ONE. 12
12 ERECT SYSTEM AS ASSIMILATORY FILAMENTS ONLY, UNILOCULAR SPORANGIA PRESENT, PLURILOCULAR SPORANGIA ABSENT, SPORANGIA SUNK. STREBLONEMA HELOPHORUM
12 ERECT SYSTEM AS HAIRS AND FILAMENTS, UNILOCULAR SPORANGIA ABSENT, PLURILOCULAR SPORANGIA PRESENT, SPORANGIA EXSERTED. 13
13 ERECT ASSIMILATORY FILAMENTS UNBRANCHED, INFRALITTORAL. STREBLONEMA PARASITICUM
13 ERECT ASSIMILATORY FILAMENTS BRANCHED, EULITTORAL. STREBLONEMA TENUISSIMUM
9 CHLOROPLASTS MORE THAN THREE. 14
14 ERECT SYSTEM AS COLOURLESS HAIRS ONLY. DICHOSPORANGIUM CHORDARIAE
14 ERECT SYSTEM AS HAIRS AND FILAMENTS. 15
15 PSEUDOHAIRS, PLURILOCULAR SPORANGIA OF SEVERAL TYPES (MICRO, MEIO, MEGA). HERPONEMA SOLITARIUM
15 TRUE HAIRS, PLURILOCULAR SPORANGIA OF ONE TYPE. 16
16 RHIZOIDS PRESENT, UNILOCULAR SPORANGIA ABSENT. 17
17 SPORANGIA SUNK, PLURILOCULAR SPORANGIA UNISERIATE, OR UNI AND BI-SERIATE. STREBLONEMA ZANARDINII
17 SPORANGIA EXSERTED, PLURILOCULAR SPORANGIA MULTISERIATE. STREBLONEMA VOLUBILE
16 RHIZOIDS ABSENT, UNILOCULAR SPORANGIA PRESENT. 18
18 PLURILOCULAR SPORANGIA UNISERIATE, OR UNI AND BI-SERIATE. STREBLONEMA SPHAERICUM
18 PLURILOCULAR SPORANGIA MULTISERIATE. STREBLONEMA AEQUALE
4 ERECT SYSTEM GREATER THAN PROSTRATE. 19
19 MACROSCOPIC THALLUS LESS THAN 1CM. 20
20 ERECT SYSTEM AS ASSIMILATORY FILAMENTS ONLY, CHLOROPLASTS LINEAR, CHLOROPLASTS ONE. LAMINARIOCOLAX TOMENTOSOIDES

20 ERECT SYSTEM AS HAIRS AND FILAMENTS, CHLOROPLASTS DISCOID, CHLOROPLASTS MORE THAN THREE. 21
21 UNILOCULAR SPORANGIA ABSENT. 22
22 PLURILOCULAR SPORANGIA OF ONE TYPE. FELDMANNIA LEBELII
22 PLURILOCULAR SPORANGIA OF SEVERAL TYPES (MICRO, MEIO, MEGA). FELDMANNIA PADINAE
21 UNILOCULAR SPORANGIA PRESENT. 23
23 GROWTH INTERCALARY. 24
24 FILAMENT BRANCHING NOT OPPOSITE, BRANCHING ABUNDANT, SUBSTRATE EPIBIOTIC. FELDMANNIA GLOBIFERA

```
   24    FILAMENT BRANCHING OPPOSITE, BRANCHING SPARSE, SUBSTRATE ENDOBIOTIC.                     FELDMANNIA SIMPLEX
  23    GROWTH DIFFUSE.                                                                                            25
   25    PLURILOCULAR SPORANGIA UNISERIATE, OR UNI AND BI-SERIATE, SUBSTRATE EPILITHIC, IN CAVES.
                                                                                                   WAERNIELLA LUCIFUGA
   25    PLURILOCULAR SPORANGIA MULTISERIATE, SUBSTRATE EPIBIOTIC, NOT IN CAVES.                                   26
    26    MULTISERIATE PLURILOCULAR SPORANGIA ELONGATE, PLURILOCULAR SPORANGIA OF SEVERAL TYPES (MICRO, MEIO,
          MEGA).                                                                                FELDMANNIA IRREGULARIS
    26    MULTISERIATE PLURILOCULAR SPORANGIA OVOID, PLURILOCULAR SPORANGIA OF ONE TYPE.     KUETZINGIELLA BATTERSI
19    MACROSCOPIC THALLUS MORE THAN 1CM.                                                                           27
 27    CHLOROPLASTS LINEAR.                                                                                        28
  28    FILAMENT BRANCHING PSEUDODICHOTOMOUS OR ALTERNATE, PLURILOCULAR SPORANGIA WITH TERMINAL HAIR.
                                                                                                 ECTOCARPUS SILICULOSUS
  28    FILAMENT BRANCHING NOT PSEUDODICHOTOMOUS OR ALTERNATE, PLURILOCULAR SPORANGIA WITHOUT TERMINAL HAIR.      29
   29    RIGHT ANGLE BRANCHES ABSENT, FILAMENT BRANCHING OPPOSITE, FILAMENTS NOT INTERWOVEN, CHLOROPLASTS MORE
         THAN THREE, EXTERNAL MUCILAGE ABSENT.                                                ECTOCARPUS FASCICULATUS
   29    RIGHT ANGLE BRANCHES PRESENT, FILAMENT BRANCHING NOT OPPOSITE, FILAMENTS DENSELY INTERWOVEN,
         CHLOROPLASTS NOT MORE THAN THREE, EXTERNAL MUCILAGE PRESENT.                          SPONGONEMA TOMENTOSUM
 27    CHLOROPLASTS DISCOID.                                                                                        30
  30    ERECT SYSTEM AS ASSIMILATORY FILAMENTS ONLY, CHLOROPLASTS ONE, MONOSPORANGIA PRESENT. ACINETOSPORA CRINITA
  30    ERECT SYSTEM AS HAIRS AND FILAMENTS, CHLOROPLASTS MORE THAN THREE, MONOSPORANGIA ABSENT.                  31
   31    TRUE HAIRS.                                                                                               32
    32    RHIZOIDS ABSENT, SPORANGIA NOT ORGANISED OR SCATTERED, PLURILOCULAR SPORANGIA MULTISERIATE.
                                                                                                  POLYTRETUS REINBOLDII
    32    RHIZOIDS PRESENT, SPORANGIA ORGANISED IN SORI, PLURILOCULAR SPORANGIA CLUSTERED.   SOROCARPUS MICROMORUS
   31    PSEUDOHAIRS.                                                                                              33
    33    LONGITUDINAL DIVISIONS IN OLDER VEGETATIVE CELLS OF ERECT FILS. COMMON.              PILAYELLA LITTORALIS
    33    LONGITUDINAL DIVISIONS IN OLDER VEGETATIVE CELLS OF ERECT FILS. RARE.                                     34
     34    CORTICATING FILAMENTS ABSENT OR RARE.                                                                   35
      35    UNILOCULAR SPORANGIA PRESENT.                                                                          36
       36    SPORANGIA NOT ORGANISED OR SCATTERED, EULITTORAL.                                 KUETZINGIELLA HOLMESII
       36    SPORANGIA ORGANISED IN SORI, INFRALITTORAL.                                          GIFFORDIA SANDRIANA
      35    UNILOCULAR SPORANGIA ABSENT.                                                                           37
       37    FILAMENT BRANCHING NOT PSEUDODICHOTOMOUS OR ALTERNATE, FILAMENT BRANCHING OPPOSITE, SPORANGIA
             ORGANISED IN SORI.                                                                       GIFFORDIA OVATA
       37    FILAMENT BRANCHING PSEUDODICHOTOMOUS OR ALTERNATE, FILAMENT BRANCHING NOT OPPOSITE, SPORANGIA NOT
             ORGANISED OR SCATTERED.                                                                               38
        38    BRANCHING SPARSE, PLURILOCULAR SPORANGIA OF ONE TYPE.                            GIFFORDIA FENESTRATA
        38    BRANCHING ABUNDANT, PLURILOCULAR SPORANGIA OF SEVERAL TYPES (MICRO, MEIO, MEGA).
                                                                                                GIFFORDIA MITCHELLAE
     34    CORTICATING FILAMENTS PRESENT.                                                                          39
      39    FILAMENT BRANCHING OPPOSITE.                                                                           40
       40    FILAMENT BRANCHING NOT SECUND, UNILOCULAR SPORANGIA ABSENT, SPORANGIA ORGANISED IN SORI, MULTISERIATE
             PLURILOCULAR SPORANGIA ELONGATE.                                                         GIFFORDIA OVATA
       40    FILAMENT BRANCHING SECUND, UNILOCULAR SPORANGIA PRESENT, SPORANGIA NOT ORGANISED OR SCATTERED,
             MULTISERIATE PLURILOCULAR SPORANGIA OVOID.                                           GIFFORDIA GRANULOSA
      39    FILAMENT BRANCHING NOT OPPOSITE.                                                                       41
       41    FILAMENT BRANCHING PSEUDODICHOTOMOUS OR ALTERNATE, FILAMENT BRANCHING NOT SECUND, BRANCHING SPARSE,
             UNILOCULAR SPORANGIA ABSENT.                                                        GIFFORDIA FENESTRATA
       41    FILAMENT BRANCHING NOT PSEUDODICHOTOMOUS OR ALTERNATE, FILAMENT BRANCHING SECUND, BRANCHING
             ABUNDANT, UNILOCULAR SPORANGIA PRESENT.                                                               42
        42    ERECT FILAMENT BRANCHING REGULAR, SPORANGIA NOT ORGANISED OR SCATTERED, PLURILOCULAR SPORANGIA OF
              SEVERAL TYPES (MICRO, MEIO, MEGA), MULTISERIATE PLURILOCULAR SPORANGIA OVOID.      GIFFORDIA SECUNDA
        42    ERECT FILAMENT BRANCHING IRREGULAR, SPORANGIA ORGANISED IN SORI, PLURILOCULAR SPORANGIA OF ONE
              TYPE, MULTISERIATE PLURILOCULAR SPORANGIA ELONGATE, OR CONICAL.                                      43
         43    MULTISERIATE PLURILOCULAR SPORANGIA ELONGATE.                                      GIFFORDIA SANDRIANA
         43    MULTISERIATE PLURILOCULAR SPORANGIA CONICAL.                                       GIFFORDIA HINCKSIAE
```

FIG. 3. Computer-generated diagnostic key to the Ectocarpaceae.

```
/GIFFORDIA HINCKSIAE BMSLC11826
SPECIAL CHARACTERS ARE -
    CHLOROPLASTS/NUMBER PER CELL
    SPORANGIA/ORGANISATION
    SPORANGIA/POSITION

 SEQ     SIM.  COUNT            SPECIES

     1     91.5    19   *++        GIFFORDIA HINCKSIAE
     2     85.4    19   *++        GIFFORDIA SANDRIANA
     3     82.5    19   *+         GIFFORDIA SECUNDA
     4     76.8    19   *++        GIFFORDIA OVATA
     5     76.4    19   *+         GIFFORDIA GRANULOSA

RESEMBLES GROUP     6

SPECIAL TAXA COMPARED
   10      71.3                    ECTOCARPUS SILICULOSUS
   22      49.8                    FELDMANNIA SIMPLEX
    3      82.5 *                  GIFFORDIA SECUNDA
   21      51.8                    PHAEOSTROMA PUSTULOSUM
   39      16.5                    STREBLONEMA BREVE
   18      58.3                    KUETZINGIELLA BATTERSI


REPORT ON TAXON       4
CHARACTER VALUE
    15     2
     DISTINGUISHED TAXON    26
    21     1
     DISTINGUISHED TAXON    1    2    6    7    8    9   10   11   12   13   14   15   16   17   18   19
  20    21    22    26    27   28   29   31   32   33   34   35   37   38   40   41
  42    43
    26     2
     DISTINGUISHED TAXON    13   15   16   29
```

FIG. 4. Demonstration of matching program.

```
PILAYELLA

PILAYELLA LITTORALIS
    Thallus uniseriate filaments. Main vegetative system erect and prostrate. Thallus macroscopic more than 1cm. Erect
system greater than prostrate, hairs and filaments. Erect assimilatory filaments branched. Branching abundant,
irregular or regular. Filament branching all pseudodichotomous, not secund. Right angle branches absent. Pseudohairs.
Prostrate system as free filaments. Periclinal cell divisions of prostrate vegetative system absent. Rhizoids
Corticating filaments absent or rare. Longitudinal divisions in older vegetative cells of erect fils. common. Filaments
not interwoven. Growth diffuse. Chloroplasts discoid, more than three. External mucilage absent.
    Unilocular sporangia present. Plurilocular sporangia present. Monosporangia absent. Sporangia not organised or
scattered, terminal or intercalary, exserted. Plurilocular sporangia multiseriate, one type, without terminal hair,
elongate.
    Benthic, epilithic or epibiotic and epiphytic or epizoic. Eulittoral. Estuarine or marine. Not in caves.
```

FIG. 5. Computer-written description of *Pilayella litto-alis* (L.) Kjellm.

will be towards the top of the list, should show most of the special characters and will usually have an asterisk beside its name. The asterisk signifies taxa which belong to the most similar intermediate taxon, in this case the genus.

From the basic data matrix, we have also derived computer-written species descriptions, one of which appears as Fig. 5.

There are several advantages to using a computer as an aid in identification:

1. The base data can be stored with ease and recalled immediately, either on a visual display unit or as a print-out.
2. Alterations and corrections can easily be made.
3. The system allows considerable flexibility in the production of polyclaves and diagnostic keys, which can be tailor-made to requirements.
4. The polyclave is extremely flexible, unlike the diagnostic key, allowing the user to select any character in any sequence to attempt an identification.

The present survey has so far resulted in the compilation of a polyclave and diagnostic keys for the Ectocarpaceae. These results are preliminary and we expect to make a number of alterations to the data matrix, since we now consider some of the characters used in our original survey to be inappropriate (e.g. characters 16, 21, 31). The basic data matrix is complete for the Ectocarpales and will be extended to incorporate the remaining orders of the British Phaeophyta.

REFERENCES

HAMEL, G. (1931–39). "Phéophycées de France." Paris.

NEWTON, L. (1931). "A Handbook of British Seaweeds". B.M. (N.H.), London.

PANKHURST, R. J. (ed.) (1975). "Biological Identification with Computers." Academic Press, London.

PANKOW, H. (1971). "Algenflora der Ostee. I. Benthos . . .". VEB Gustav Fischer Verlag, Jena.

ROSENVINGE, L. K. and LUND, S. (1941). The marine algae of Denmark . . . *K. danske Vidensk Selsk. Skr.* **1**(4), 1–79.

RUSSELL, G. and FLETCHER, R. L. (1975). A numerical taxonomic study of the British Phaeophyta. *J. mar. biol. Ass. U.K.* **55**, 763–783.

15 | Environment and Form in the Discrimination of Taxa in Brown Algae

G. RUSSELL

Hartley Botanical Laboratories, University of Liverpool, England

Abstract: Taxonomic discrimination in brown algae is conducted mainly on the basis of phenotypic characters observed in the field and in culture. Two conditions of the phenotype which have caused great taxonomic difficulty are: the developmental state of the plant and its responses to habitat conditions.

The taxonomic problems raised by developmental state are most pronounced in annuals, which commonly attain reproductive maturity before full development of the thallus. The importance of thallus development in the family Ectocarpaceae is illustrated by numerical taxonomic methods. The results of this study suggest that the circumscription of ectocarpoid genera is unsatisfactory and that the species form a single complex of interrelated expressions of the heterotrichous filament.

The responses of the phenotype to the environment are often most marked in perennial species. Intraspecific variation in relation to habitat is illustrated by means of examples selected from the order Fucales. The history of taxonomic inconsistency and uncertainty in dealing with intraspecific variation is discussed.

It is concluded that where intraspecific variation involves clearly distinct phenotypes in facultative rather than obligate alternation, as for example in the family Scytosiphonaceae, a conservative approach to species names should be exercised, for the present. Less sharp intraspecific variation, whether associated with habitat conditions or stages in morphogenesis, should also be recognized, if only to stimulate enquiry into its nature and origins. It is suggested that the deme terminology is a suitable vehicle for this purpose.

INTRODUCTION

Every brown alga comes into the world equipped with a genotype that permits a wide range of phenotypic expression, and a phenotype that responds, often with considerable sensitivity, to differences in environment. These two properties have combined to create an army of taxonomic problems which have

---

Systematics Association Special Volume No. 10, "Modern Approaches to the Taxonomy of Red and Brown Algae", edited by D. E. G. Irvine and J. H. Price, 1978, pp. 339–369. Academic Press, London and New York.

harassed systematists for well over a century and sorely tried their judgment. The number of problems and their quality have varied from alga to alga but few taxa have caused little trouble, and in still fewer have the troubles been resolved to everybody's satisfaction.

There is relatively little to be said, unfortunately, about the rôle of the genotype in these problems. The genetics controlling thallus morphology in the brown algae are largely unknown; indeed our knowledge of inherited characteristics of any sort is vestigial. The most obvious examples of morphological differences resulting entirely from genetic factors are to be found in the sexual and asexual phases of polymorphic life-histories, for example, *Cutleria*, *Laminaria*. Similar but less obvious distinctions occur in some of the simpler forms such as *Sphacelaria furcigera*,* (van den Hoek and Flinterman, 1968) and *Ectocarpus fasciculatus* (Müller, 1972). However all these examples can be interpreted more correctly as responses to differences in ploidy level.

In the genus *Fucus*, interspecific crosses have been carried out successfully ($> 90\%$) and the resultant hybrids have been shown to develop morphologies intermediate in character between those of the parent species (Burrows and Lodge, 1951), although recent work by Bolwell *et al.* (1977) would seem to contradict this. Similar experiments have demonstrated an absence of sterility barriers between certain non-digitate species of *Laminaria* and have established that some of the diagnostic morphological features are not genetically based (Chapman, 1975). Conversely, it has been shown that frond bullation in *Laminaria saccharina*, another non-digitate species, is a genetically transmitted and dominant character which is expressed in both homozygous and heterozygous conditions of the sporophyte (Lüning, 1975).

Elsewhere in the literature one may find instances of different growth or reproductive patterns between isolates of a single species under controlled culture conditions; there are several good examples of this in the work of Wynne (1969). These differences in response presumably denote a degree of genetic variation within the species.

Other cases of intraspecific genetic variation take the form of differences in physiological properties. Certain cultured strains of *Ectocarpus siliculosus* and *Pilayella littoralis* have been found to possess especially marked ability to grow in waters of reduced salinity or on substrates rich in copper, properties which appear to be inherited by progeny cultured over several generations (Russell, 1963; Russell and Morris, 1973; Russell and Bolton, 1975). In Helgoland, the

* The authorities for the species names given in this text follow Parke and Dixon (1976), except where stated.

native smooth-fronded populations of *Laminaria saccharina* have higher temperature tolerances than the bullate members of the same species on the Isle of Man (Lüning, 1975).

The adaptive value of these attributes in particular habitats has been demonstrated but other cases of physiological difference are less easy to interpret. The differing proportions of unilocular and plurilocular sporangia produced under various controlled temperatures by isolates of *Ectocarpus siliculosus* from different geographical localities (Müller, 1963) do not obviously correlate with climatic–hydrographic conditions, or confer any ecological advantages. The adaptive value of responses such as these remains something of a mystery.

An essay on brown algal genetics would serve only to emphasize that our knowledge in this field is poor, scattered and rather inconclusive. The scant treatment of the brown algae in the recent review by Fjeld and Løvlie (1976) indicates this paucity. It seems safe to predict that a good understanding of such heritable aspects as branching patterns and sporangial morphology will develop slowly.

Chromosome studies have now been carried out on numerous species throughout the brown algae (Cole, 1967, 1968) but as an aid to systematics these have been generally disappointing. Most taxonomic decisions on the brown algae, in the foreseeable future, are therefore likely to be made on observations of the phenotype in the field and in culture.

Of the numerous factors which determine the phenotypes of brown algae, there are two which have played a major part in creating taxonomic difficulties. These are, first, the developmental state of the individual and, second, its responses to habitat conditions. These two considerations are in fact closely interrelated but it is probably helpful if, initially, an attempt is made to treat them separately.

DEVELOPMENTAL STATE

1. *Introduction*

Cases of taxonomic confusion resulting from failure to distinguish developmental states have arisen in various species of brown algae. Most of these species have life-forms that belong to one category or another of annual, and the taxonomic difficulties have almost always arisen from the fact that plants of this sort are very liable to reach reproductive maturity at a juvenile state in the development of the thallus. This applies not only to the microscopic phases (plethysmothalli, etc.) but also to the macrophases. A good illustration of this problem is provided by the genus *Desmotrichum*, of which four species, *D.*

undulatum (J. Ag.) Reinke, *D. balticum* Kütz., *D. scopulorum* Reinke and *D. repens* Kylin, appear to form a developmental sequence, and Rosenvinge (1941) therefore considered them to be conspecific.

A similar, but rather more complicated, problem seems to exist in the Ectocarpaceae. All members of this family have as the basic thallus form a heterotrichous filament, as defined by Fritsch (1929). In no species is there a development of true parenchyma or pseudoparenchyma, although occasional periclinal cell divisions may occur in some species, for example *Pilayella littoralis*, while others often show extensive axial cortication by rhizoidal filaments, for example *Giffordia granulosa.* The species and genera of the Ectocarpaceae are distinguished by their particular expressions of the heterotrichous filament and by their sporangial morphology. Other features, such as endophytism, endozoism and the identities of host species, though widely used in the discrimination of taxa may be considered as, at best, of secondary value. The variation in expression of the heterotrichous filament shown by ectocarpoid algae ranges from those microscopic forms with a thallus consisting of little more than the horizontal branch systems (*Mikrosyphar*, *Phaeostroma*), to macroscopic forms with numerous orders of well-developed, erect branches in characteristic pattern (*Giffordia*, *Pilayella*).

The importance of developmental state in the discrimination of these taxa may be illustrated by the techniques of numerical taxonomy. In a recent numerical study of the brown algae, Russell and Fletcher (1975) have suggested that the genera of the Ectocarpaceae are not sharply circumscribed. The purpose of that work was to consider the gross classification of the brown algae and it was based upon a choice of characters and character states appropriate to the group as a whole. Most of these were inapplicable to members of the Ectocarpaceae and to have made any taxonomic decisions on generic status could not have been justified. However, even if the investigation had been aimed specifically at the Ectocarpaceae, it might have proved impossible to draw any conclusions using the single-linkage sorting procedure adopted by the authors. This method, even with the benefit of a coefficient of similarity employing positive matches only, often results in the formation of long straggly clusters so that natural groupings within an array of similar plants might pass unrecognized (Sneath and Sokal, 1973). The work of Russell and Fletcher (1975) is perhaps also open to criticism for the use made of ecological parameters but such matters are now very intimately linked with taxonomic discrimination in the brown algae. The authors might be criticized also for embarking upon a numerical exercise with an assumption that the species were sound. This too may be defended, since the aim of the project was to study taxonomic groupings above the rank of species.

The general lack of information on brown algae at the individual plant level in any case precludes a firmer null hypothesis.

Therefore, the following evaluation of ectocarpoid genera has been based upon the species of the Ectocarpaceae in Britain, as they are currently conceived.

2. Materials and Methods

The work was carried out using characters and character states which were entirely morphological and no ecological information of any sort was taken into consideration. The details of the characters adopted are given in Table I. The characters and their various states are fairly evenly balanced between vegetative and reproductive attributes and all have been used at some time or another to distinguish species and genera in the family.

TABLE I. List of characters and their possible expressions as employed in the numerical study of the Ectocarpaceae. The grouped characters are not mutually exclusive; some species for example have both plate-like and ribbon-like chloroplasts. The characters listed above the middle line refer to vegetative morphology, those below the line to reproductive morphology

Characters	Expressions
Chloroplast morphology	Plate-like: ribbon-like
Diameter of assimilatory cells of erect system	0–10 μm: 11–20 μm: 21–30 μm: > 30 μm
Height of erect system	0–5 mm: 6–10 mm: 11 mm–5 cm: > 5 cm
Branching of erect system	Unbranched: primary branches: secondary branches: tertiary and greater orders of branches
Branch pattern	Irregular: alternate: opposite: unilateral
Branch configuration	Fasciculate: straight: recurved: cable rolling
Branch variants	Crampons: true hairs: pseudo-hairs
Unilocular sporangia	Presence
Insertion of unilocular sporangia	Terminal: intercalary: solitary: in chains: in clusters: sessile: pedicellate
Plurilocular sporangia/gametangia	Presence
Insertion of plurilocular structures	Terminal: intercalary: solitary: in chains: in clusters: sessile: pedicellate
Plurilocular morphology	Conical: ovate-lanceolate: cylindrical: irregular: linear: multiseriate: straight: curved: simple: branched: spherical
Monosporangia	Presence

The information on which the species were scored was obtained from published descriptions, examination of herbarium material and examination of plants personally collected. Positive scores were accepted for any stage in the development of a species. The 43 species of Ectocarpaceae in the current check-list of British Marine Algae—third revision by Parke and Dixon (1976)—were compared using the coefficient of similarity (2A/(2A+B+C) given in Sneath and Sokal (1973); this coefficient uses both positive and negative matches. The raw data were computed at the University of Liverpool using the CLUSTAN program and were sorted using the average linkage procedure of Sneath and Sokal (1973). The results are presented in the form of dendrograms (Figs 1 and 2).

3. *Results and Discussion*

If vegetative morphology is considered first, it can be seen (Fig. 1) that the ectocarpoid species may be divided initially (line A) into two major groups plus a single outrider. This single isolated species, *Mikrosyphar polysiphoniae*, is the most reduced member of the Ectocarpaceae, consisting of only the horizontal branch systems without hairs or erect assimilatory filaments; its solitary position is the product of this lack of morphological features.

The first of the two main groups comprises the macroscopic plants. These have erect filaments of large diameter, bearing various orders of branches, often in a characteristic pattern. The other group contains the small or microscopic species with narrow erect assimilatory filaments and little or no branching.

A more divisive treatment, as indicated by line B, raises the number of species clusters to seven and these are composed as follows:

Cluster 1: This group consists of moderately sized plants with irregular branching of the primary or, occasionally, of the second order; for example, *Acinetospora*.

Cluster 2: This group consists of large plants with broad erect filaments bearing branches of the tertiary or greater order and often in characteristic pattern; for example, *Pilayella*.

Cluster 3: This group consists of microscopic plants with narrow, unbranched, erect, assimilatory filaments; for example, *Streblonema*.

Cluster 4: This group also consists of microscopic plants but without erect assimilatory filaments, although colourless hairs may be present; for example, *Phaeostroma*.

Cluster 5: This group consists of small plants with assimilatory filaments of moderate widths. These are either simple or with a few primary branches; for example, *Herponema*.

Cluster 6: This group consists of minute plants with narrow erect filaments which are branched, primary branches only being formed; for example, *Laminariocolax*.

Cluster 7: This contains only *Mikrosyphar polysiphoniae*, which may be regarded as the most reduced member of Cluster 4.

It is a fairly simple matter to detect the developmental aspects of these groupings. Field observation and culture studies of *Ectocarpus* spp. and of *Spongonema tomentosum* have shown that members of these species pass through early phases in development which correspond with the morphologies of some small or even microscopic taxa (Kornmann in Kuckuck, 1961; Russell, 1966). Indeed, the development of all macroscopic members of the Ectocarpaceae starts with a horizontal filament which next produces narrow, unbranched, erect filaments and eventually metamorphoses into the characteristic mature form.

Other members of the Ectocarpaceae show only a limited capacity for thallus elaboration. For example, *Mikrosyphar porphyrae* in culture produces short erect assimilatory filaments and a more streblonematoid organization than is found in nature (Levring, 1945).

Another developmental problem occurs in *Endodictyon infestans*. Endozoic plants, apparently indistinguishable from those in the original description by Gran (1897) have been observed in nature to metamorphose into branched macroscopic filaments identifiable with a species of *Ectocarpus* (Irvine *et al*., 1975). On the other hand, Levring (1945) has found that *Endodictyon* reproduces only the typical form in culture. Here is perhaps an example of intraspecific variation in life-history comparable with that of *Hecatonema maculans*, which has been shown to exist both as an autonomous, self-reproducing plant by Edelstein *et al*. (1971) and as a plethysmothallus of *Myriotrichia sp*. by Loiseaux (1969).

A similar instance may be quoted in the myrionematoid variant of *Ectocarpus fasciculatus* discovered in culture by Baker and Evans (1971). Early in their culture series, the microscopic plants were observed to metamorphose into the macrophase, but by repeated subculturing the myrionematoid form was isolated and bred true. Clayton (1972) by a similar process of selective subculturing was able to remove all such microscopic forms from her cultures of *Ectocarpus* and *Giffordia* species and arrive at unmixed populations of the macroscopic plants. It follows, therefore, that the developmental state of one of these plants does not necessarily bear any relation to its age.

This does not mean that all microscopic forms are simply neotenic juvenile states of larger species. Cultures of certain *Streblonema* species and of *Laminariocolax tomentosoides* have been shown by Dangeard (1970) and Russell (1964) to be self-reproducing forms with no capacity for further vegetative development. Cultures of *Waerniella lucifuga* have now been maintained in the laboratory at Liverpool for over 18 months without showing any morphology other than the typical one, or any increase in the naturally slow rate of growth. However, it is perfectly possible that other isolates of this species under different conditions might respond differently.

The Fig. 1 dendrogram illustrates another taxonomic problem. In the absence of data on reproductive morphology some species are indistinguishable from

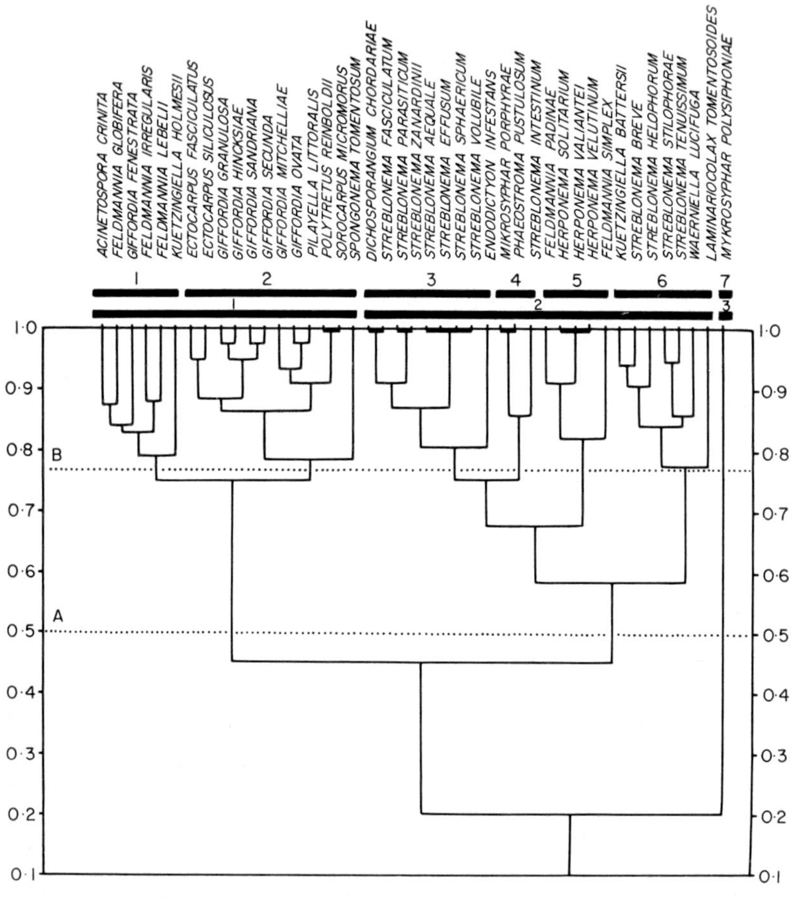

FIG. 1. Cluster analysis of the species of Ectocarpaceae in the British flora using only characters of vegetative morphology. Dendrogram based on average linkage and similarity coefficient 2A/(2A+B+C), see Sneath and Sokal (1973). The lines A and B cut off the species groups discussed in the text.

one another, a fact which applies especially to the microscopic forms (Fig. 1). However, generic circumscription in the Ectocarpaceae does not depend solely upon vegetative characteristics; the effect of including reproductive features may be seen in Fig. 2. Sporangial characteristics now make it possible, though not always easy, to distinguish every species. They also bring about a further differentiation of the species clusters delimited in Fig. 1. Thus the line A drawn at a level close to that of line B on Fig. 1 now cuts off 15 clusters. The species composition of these is given in Fig. 2 but the combined influences of vegetative

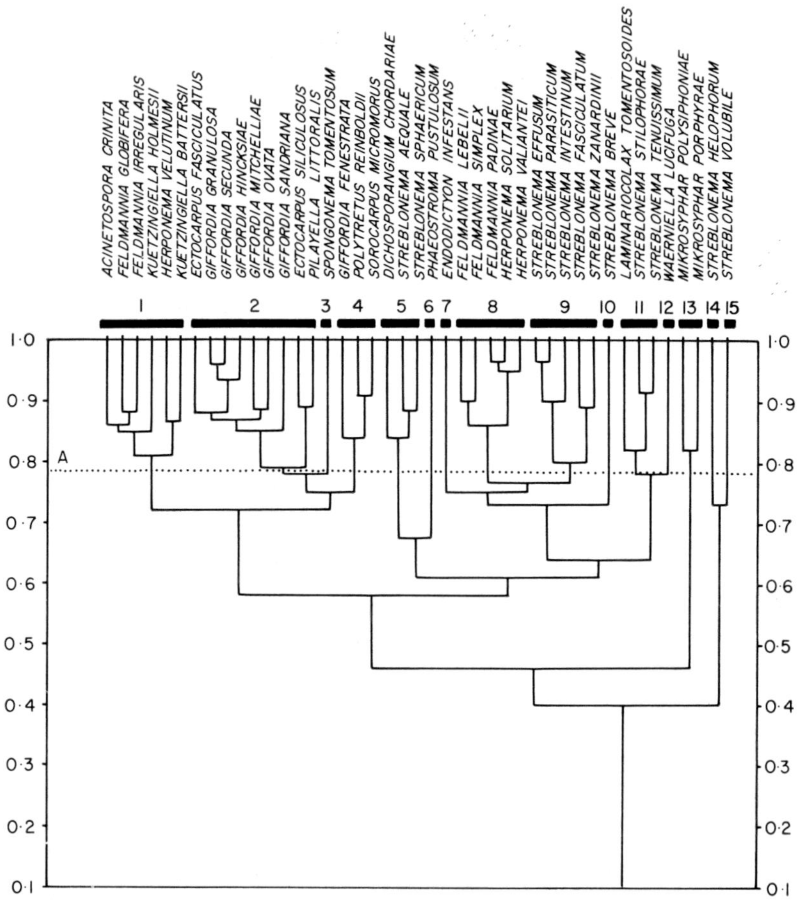

FIG. 2. Cluster analysis of the species of Ectocarpaceae in the British flora using characters of vegetative and reproductive morphology. Dendrogram based on average linkage and similarity coefficient 2A/(2A+B+C), see Sneath and Sokal (1973). The species groups cut off by line A are discussed in the text.

and reproductive morphology on their formation is perhaps more easily grasped by reference to Fig. 3. This drawing shows examples of species with the six main types of vegetative morphology described from Fig. 1, and the clusters shown in Fig. 2 may be derived from these, as follows:

Cluster 1: This group is derived from Fig. 3A and contains, for example, *Acinetospora crinita* and *Feldmannia irregularis*.

Cluster 2: This group is derived from Fig. 3B; for example, *Giffordia hinksiae*, *Pilayella littoralis*.

Cluster 3: This consists only of *Spongonema tomentosum*, which has been segregated from Fig. 3B by virtue of its exceptionally wide range of morphological features including some of the more unusual characters such as crampons.

Cluster 4: This group is another expression of Fig. 3B, separated from the other by the rather irregular alternate branching pattern and by the presence of plurilocular reproductive structures only. It is exemplified by *Polytretus reinboldii*, *Giffordia fenestrata*.

Cluster 5: This group is derived from Fig. 3C and is distinguished from other members of it by the presence of unilocular *and* plurilocular structures, for example *Dichosporangium chordariae*, *Streblonema aequale*.

Cluster 6: This consists only of *Phaeostroma pustulosum*, which has been separated from *Mikrosyphar* spp. (Fig. 3D) by its possession of unilocular as well as plurilocular sporangia.

Cluster 7: This consists of *Endodictyon infestans*, also isolated from Fig. 3C.

Cluster 8: This group is derived from Fig. 3E and may be exemplified by *Feldmannia padinae* and *Herponema valiantei*.

Cluster 9: This group is another derivative from Fig. 3C and has been formed from species with plurilocular structures only, for example *Streblonema fasciculatum*.

Cluster 10: This consists of *Streblonema breve* and it has been removed from Fig. 3F by virtue of the morphology of its plurilocular structures, the only kind of sporangia known.

Cluster 11: This group has the morphological features of Fig. 3F and is consolidated by the common presence of linear plurilocular sporangia and no unilocular sporangia. For example, *Laminariocolax tomentosoides*, *Streblonema tenuissimum*.

Cluster 12: This consists of *Waerniella lucifuga*, which has the morphology of Fig. 3F but differs from the taxa in cluster 11 through the presence of unilocular as well as plurilocular sporangia.

Cluster 13: This group is derived from Fig. 3D and consists of the two *Mikrosyphar* species.

Cluster 14: This consists of *Streblonema helophorum*. It has morphological affinities with Fig. 3F but differs from other such species in having only unilocular sporangia.

Cluster 15: This consists of *S. volubile* and has been segregated from the other Fig. 3C species by the fact that it too bears only unilocular sporangia.

The number of clusters formed in Fig. 2 is close to that of the genera in the Ectocarpaceae in Britain (18), but the species compositions of the clusters are often very different. It is equally evident that the criteria which were used to establish the genera have a somewhat arbitrary nature. The excessive importance of reproductive features among the microscopic taxa is another very obvious factor and it is the inevitable product of their lack of distinctive vegetative morphologies.

It would also seem to be the case that authors have been inconsistent in matters of generic status. Thus if *Pilayella* is to be retained as a distinct genus in Fig. 2, then the magnitude of difference required to do so must also have the logical consequence of raising the total number of clusters to 33. We are clearly far removed from the optimistic days when a particular level on a dendrogram

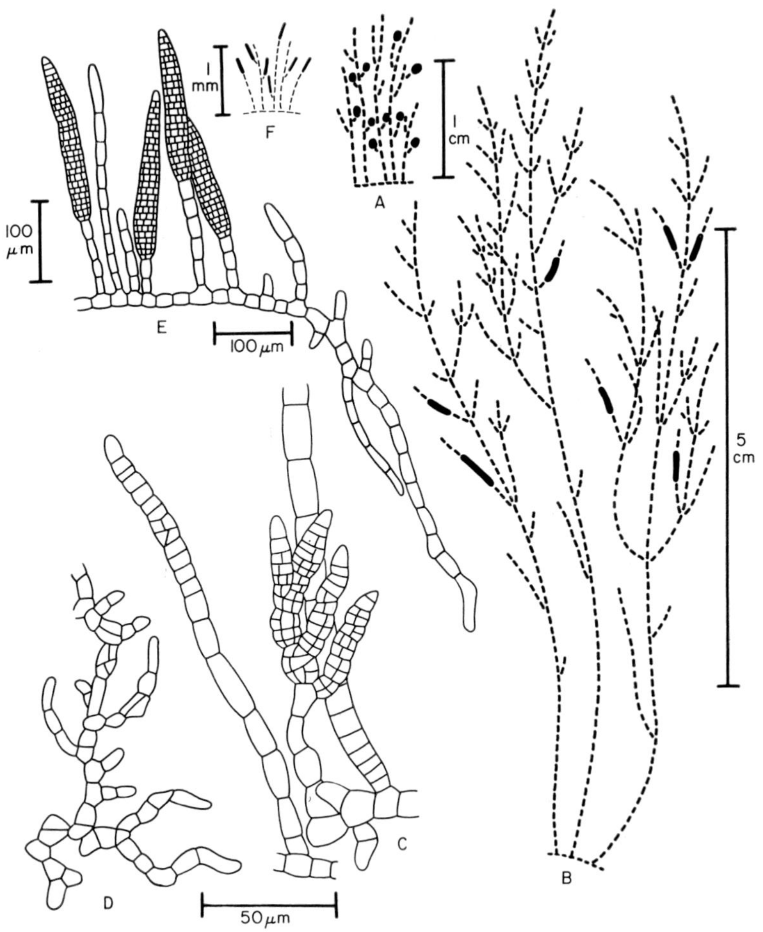

FIG. 3. Diagram showing the six ectocarpoid morphologies from which the species clusters in Fig. 2 may be derived. The examples illustrated are: A—*Feldmannia globifera*, B—*Pilayella littoralis*, C—*Streblonema fasciculatum*, D—*Mikrosyphar polysiphoniae*, E—plethysmothallus of *Ectocarpus siliculosus*, F—*Laminariocolax tomentosoides*.

can be used as a measure of generic delimitation (see Davis and Heywood, 1963, Fig. 1).

This particular exercise is open to some criticism on grounds of character weightings, sorting method and so forth. It is, however, only part of a more detailed study undertaken by Mr D. Garbary and myself. Blocking out certain characters and the use of different coefficients and alternative sorting methods do nothing to improve the suspect nature of these genera, as illustrated in this brief account.

4. *Conclusions*

The importance of recognizing patterns of morphogenesis in taxonomic studies of brown algae cannot be over-stressed. Dixon (1963) maintained that, with regard to the red algae, it was feasible to argue that the species *is* a pattern of development. The passage of time since the publication of that comment has not weakened its relevance to the simpler forms of brown algae, for most of which the species exists only in terms of morphological characteristics.

In the case of the Ectocarpaceae it would appear that we have been over-liberal in the establishment of genera. An analagous problem has existed in the family Acrochaetiaceae (Rhodophyta) which contains a similarly wide range of heterotrichous filamentous species ranging from microscopic endophytes to relatively large epilithic forms. Parke and Dixon (1976) have recently adopted the single generic name *Audouinella* for all such species in the British flora and a return to *Ectocarpus s. lat.* for their brown algal counterparts would seem, on this evidence, to be a sensible decision, in spite of the reservations of Russell and Fletcher (1975).

This over-liberality in the establishment of genera has probably also occurred in other families of the simpler brown algae. In the Myrionemataceae, for example, the distinctiveness of *Chilionema*, *Compsonema*, *Myrionema*, etc. should perhaps be more critically re-examined.

At the species level in the brown algae there is a good case for the exercise of some taxonomic conservatism. We now know that plants recognizable as *Ralfsia clavata* may have progeny which develop into the thallus form of *Petalonia fascia* (Edelstein *et al.*, 1970; Roeleveld *et al.*, 1974) and, in the same way, *Microspongium gelatinosum* may give rise to *Scytosiphon lomentaria* (Lund, 1966). Loiseaux (1970) has also described the development of a *Scytosiphon* thallus from species of *Streblonema* and *Compsonema*. It could be argued from evidence of this sort that taxa so linked should be united under a single species name (Edelstein *et al.*, 1970). However, it is by no means certain that an obligate alternation exists between these forms. Indeed, there is strong evidence to suggest that the pattern of thallus development obtained is often heavily dependent upon environmental conditions and that the crustose or filamentous state may be repeated (Wynne, 1969; Edelstein *et al.*, 1970). *Hecatonema maculans*, as already mentioned, seems to be another example of facultative rather than obligate alternation but if the names of the ralfsioid and other microscopic phases in the Scytosiphonales are to be discarded, logic also requires that *Hecatonema maculans* disappears as a taxon. Where the alternation is only facultative and the thallus morphologies are as distinctive as in these examples, it seems acceptable to retain the species names of both phases for the present as Parke and Dixon (1976) have chosen to do. Nevertheless, there is a taxonomic decision to be taken here with obviously important repercussions. It should be based, therefore, on a thorough understanding of the life-history and the relationships of the component phases.

PHENOTYPE AND DEVELOPMENT

1. *Introduction*

The sequence of developmental states through which a brown alga may pass in the course of its life span is also the partial product of its environment. Culture experiments on *Petalonia fascia*, for example, have shown that the concentration of dissolved iodide in the culture medium is of great importance in determining the pattern of morphogenesis (Hsiao, 1969). We also know that different levels of phosphate phosphorus produce similar effects in *Scytosiphon lomentaria* (Roberts and Ring, 1972). The development of the *Scytosiphon* and *Petalonia* thalli, in contrast with their ralfsioid or filamentous micro-phases, has been found to respond to combined light and temperature conditions (Wynne, 1969; Lüning and Dring, 1973). In the case of *Scytosiphon* the most important mechanism is now known to be a blue-light activated photoperiodism (Dring and Lüning, 1975), although the response is by no means uniform and certain isolates have already been shown to be unresponsive to this photoperiodic effect (Lüning, pers. comm.). The seasonality of these phases in the field has also been demonstrated in *Petalonia* by Roeleveld *et al.* (1974).

Scytosiphon and *Petalonia* provide particularly good examples of brown algal phenotypes which respond directly to their environment but it seems increasingly probable that they will not be isolated cases. Reported differences in life-history between and within species belonging to the Ectocarpales may well result from differences in culture conditions, and the observed patterns of development in many species will doubtless be found to depend on a subtle interplay of physical and chemical conditions. Such findings will inevitably generate fresh cohorts of taxonomic and nomenclatural problems. However, the problems of environment-associated variations which at present afflict us are usually linked with the more obvious habitat factors, such as wave action. Their effects are usually known from the macrophase alone.

Even among relatively short-lived annual species these responses can be very marked. Typical attached plants of *Ectocarpus fasciculatus* and *Pilayella littoralis*,

FIG. 4. *Halidrys siliquosa* from A—sublittoral, and B—intertidal habitats, photographed from herbarium material. Scale = 10 cm.

when subjected to experimentally contrived free-living conditions, have been found to develop short, divaricate lateral branches and small sporangia which, in the case of *Pilayella*, tend to be subterminal or terminal in position rather than intercalary (Russell, 1967). These morphological changes, proding thallus forms characteristic of *Ectocarpus distortus* Carm., *E. landsburgii* Harv. and *Pilayella varia* Kjellm., also illustrate the similarity of response to one particular habitat factor that may be shown by different taxa. A better example may be seen in the Laminariales where the formation of a cucullate lamina under conditions of shelter from wave action has been experimentally induced in *Laminaria digitata* (Sundene, 1964), *L. hyperborea* (Svendsen and Kain, 1971) and *Saccorhiza polyschides* (Norton, 1969). It seems very probable that the morphogenetic and anatomical processes associated with cucullation and described in detail by Norton (1969) are common to all three species.

2. Phenotypic Variation in the Fucales

It is hardly surprising that habitat-linked characteristics have been described more frequently in perennial algae than in annuals since they are usually larger and, by virtue of their relatively long lives, experience more prolonged environmental effects. Members of the Fucales provide interesting examples of habitat-linked morphological features which have caused and continue to cause a large measure of taxonomic confusion. This problem may be illustrated by means of a few chosen species.

Halidrys siliquosa has a short stipe from which numerous pinnately branched axes arise. Although quite a common plant, it has a somewhat patchy distribution, locally abundant in some areas, absent from others. In the sublittoral zone, the plants possess long fronds with broad axes and the lateral branchlets arise in obvious succession. Intertidal plants are usually restricted to rock pools where they develop relatively short fronds with narrow axes and a branching pattern in which the pinnate succession is obscured by elongation of secondary laterals. These morphological differences between intertidal and sublittoral populations are illustrated in Fig. 4. When the two populations are compared for length of pinnae and width of axes then the differences are highly significant in both respects ($P<0{\cdot}001$), Fig. 5. The scatter of individual plants shows a weak positive correlation between these parameters and a transition from one form to the other (Fig. 5).

Pelvetia canaliculata has been generally accepted as a morphologically more uniform species than even *Halidrys*. Harvey (1846) certainly regarded it as a homogeneous taxon and Kjellman (1890), who was not known for his hostility to the notion of infraspecific taxa, did not describe any variation within this species. They were correct; *Pelvetia* is indeed one of the more uniform fucoid species, but it is not altogether without variation. Populations from exposed shores consist of very significantly smaller plants than those from sheltered shores ($P<0{\cdot}001$) and possess slightly, but not significantly, greater numbers of dichotomies per unit fresh weight (Fig. 6).

Pelvetia offers fewer morphological characters and is a less interesting object for study than *Fucus vesiculosus*, which is one of the most variable of British fucoids. If one excludes for the moment the estuarine and salt marsh forms of

F. vesiculosus and considers only those found on marine rocky shores, then the range of variation obtained is popularly linked, in a causative way, with differences in wave action. It is generally accepted that the dimensions of plants

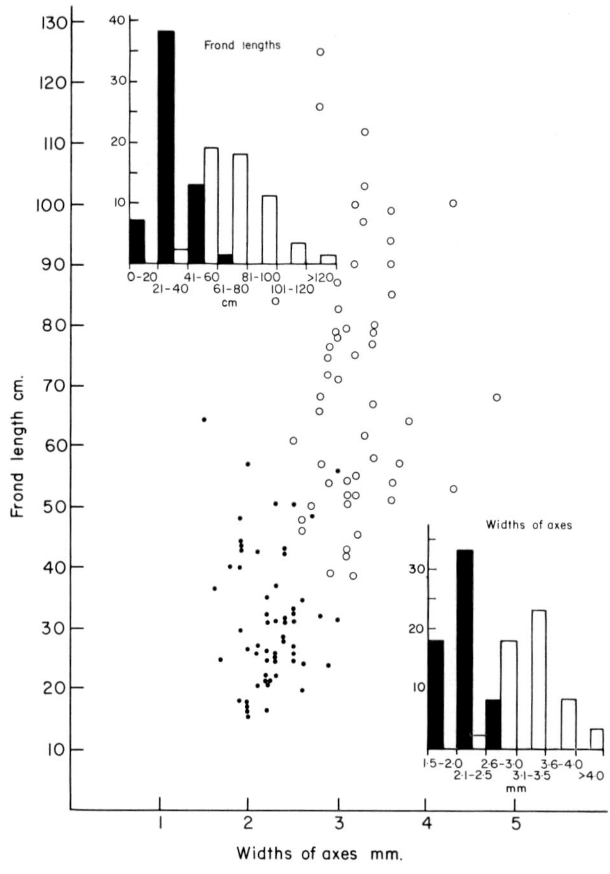

FIG. 5. Scatter diagram of frond lengths against widths of axes in fresh *Halidrys siliquosa* from sublittoral habitats (open circles) and from intertidal rock pools (dots). Inset top left is frequency distribution of lengths and, bottom right, of widths of axes. In histograms, open bars denote sublittoral samples.

and the numbers of vesicles are relatively great in sheltered conditions and that these decline with increasing exposure to an extreme condition in which plants are very small, narrow and evesiculate. This view, which was expressed by Powell (1963), is represented diagramatically in Fig. 7. A similar series of morphological responses to wave action has been reported for *F. distichus* by

Powell (1957). If the simple model in Fig. 7 holds true then it should be possible to establish a number of correlations between particular morphological features and between these and a scale of exposure. In practice, however, this seems to be only partly true.

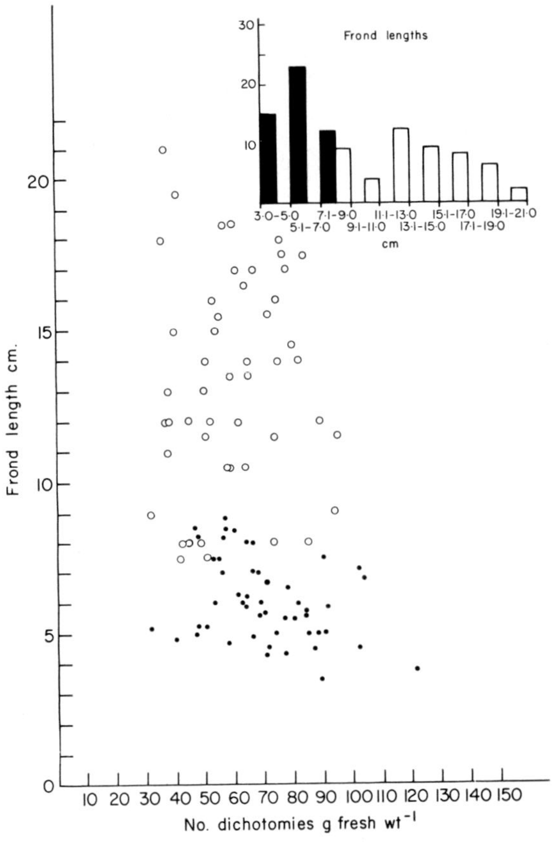

FIG. 6. Scatter diagram of frond lengths against numbers of dichotomies per g. fresh weight in *Pelvetia canaliculata*. Sample from Derbyhaven, Isle of Man (very sheltered site), indicated by open circles; that from Scarlett Point, I.O.M. (exposed site), by dots. Inset top right is frequency distribution of plant lengths.

In randomly sampled individuals from a number of sites of differing exposure on the Isle of Man, there is an obviously strong correlation between frond area and fresh weight of plant (Fig. 8). This correlation reflects some of the growth characteristics of the species and it is perhaps not too surprising that the confidence limits are narrow. The ratio of oven-dry weight to

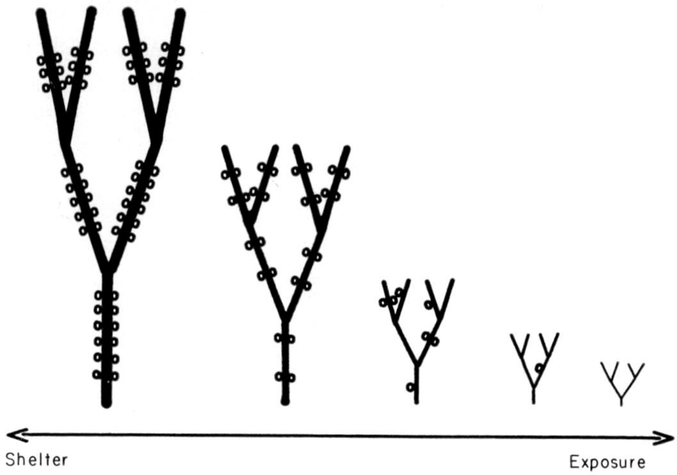

FIG. 7. Diagram of thallus variation in *Fucus vesiculosus* illustrating features associated with different degrees of wave action.

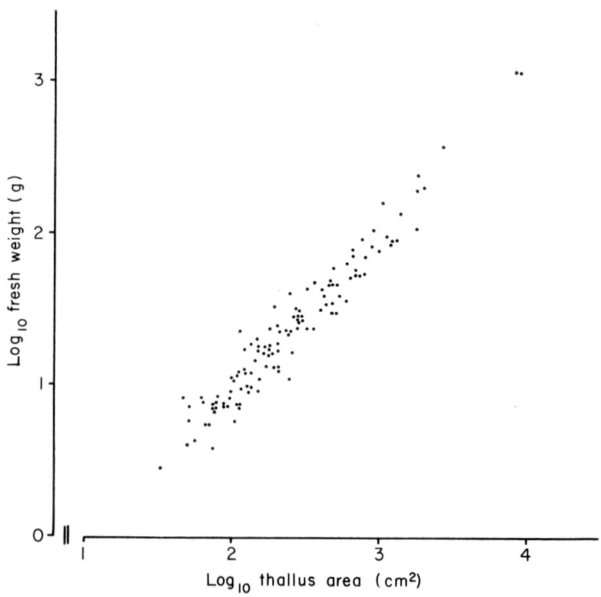

FIG. 8. Scatter diagram of thallus fresh weight (g) against area (cm^2) in randomly sampled *Fucus vesiculosus* from sites on the Isle of Man with different degrees of wave action. Data are $\log_{10}$–transformed.

fresh weight, incidentally, is not a constant but differs, sometimes significantly, from population to population, an observation which differs from that of Kristensen (1965). The correlation between vesicle number and mean frond width, measured at approximately the mid-point of the thallus, is however,

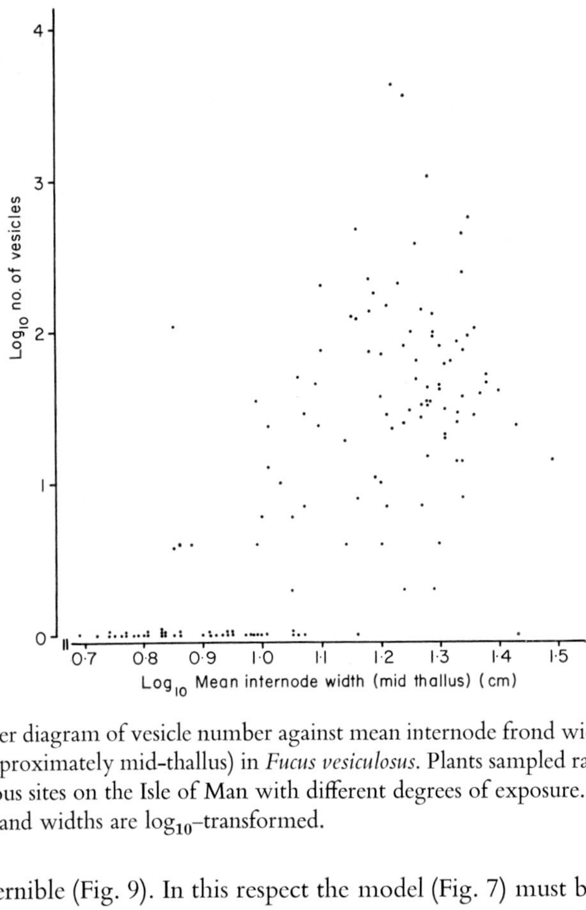

FIG. 9. Scatter diagram of vesicle number against mean internode frond width (measured at approximately mid-thallus) in *Fucus vesiculosus*. Plants sampled randomly from various sites on the Isle of Man with different degrees of exposure. Vesicle numbers and widths are $\log_{10}$-transformed.

barely discernible (Fig. 9). In this respect the model (Fig. 7) must be considered rather unsatisfactory.

Short of making direct measurements of wave action by one or other of the gadgets available for this purpose, there are only two methods of quantifying exposure. These are biological scales such as that of Ballantine (1961) or that of Lewis (1964), and the cartographic method devised by Baardseth (1970). The former approach takes into account morphological features of *Fucus vesiculosus*, which invalidates it for the purposes of this exercise. The Baardseth technique adopted here, involves simply counting the number of 9° compass sectors open to the sea at the required locality on a radius of 7·5 km. An attempt has been made to increase its accuracy by weighting the compass sectors according to prevailing wind pattern. Samples of *Fucus vesiculosus* from several sites on the

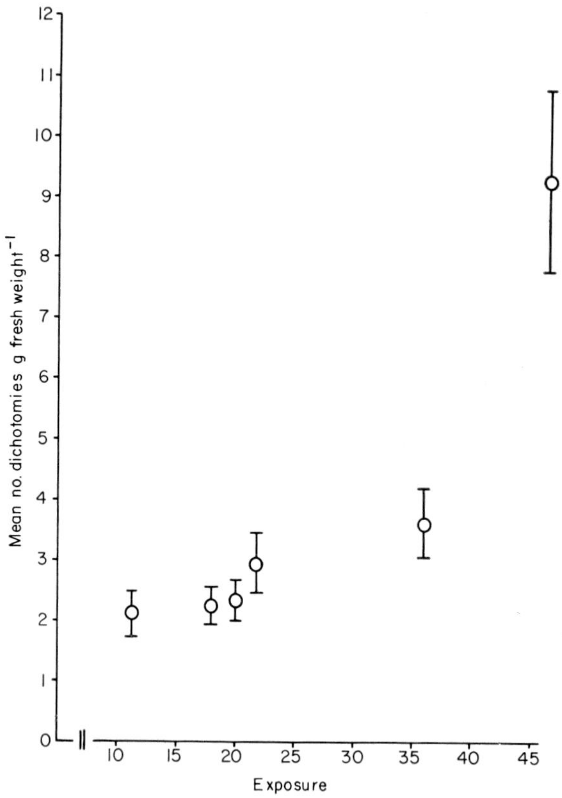

FIG. 10. Mean number of dichotomies per g. fresh weight, and 95% confidence limits, in random samples of *Fucus vesiculosus* from several sites on the Isle of Man. The exposure values are calculated using the Baardseth (1970) method, weighted according to prevailing wind pattern for the Isle of Man.

Isle of Man show a logarithmic correlation of numbers of dichotomies with increasing exposure (Fig. 10), and in this respect *F. vesiculosus* seems to demonstrate, but with greater sensitivity, the response noted in *Pelvetia canaliculata* (Fig. 6).

Other characteristics of *Fucus vesiculosus* correlate less well with wave action. Samples were collected from three sites on the Isle of Man, i.e. Castletown Bay, a sheltered shore with almost complete fucoid cover, Scarlett Point, an exposed shore with *F. vesiculosus* as the only common fucoid, and from The Stack, an intermediate site in terms of wave action but only a few metres in distance from the Scarlett Point population. When the mean frond width of these plants was plotted against plant length, there was an obviously greater number of large and wide-fronded plants at the Castletown site than at Scarlett Point. Of the characters measured, frond width seemed more strongly affected (negatively) by wave action than plant length.

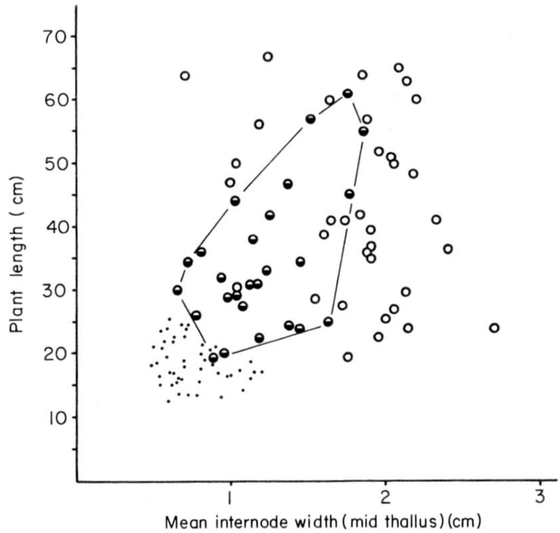

FIG. 11. Scatter diagram of plant lengths against mean internode frond widths (measured at approximately mid-thallus) in *Fucus vesiculosus*. Plants sampled randomly from three sites on the Isle of Man with different degrees of wave action; open circles = Castletown Bay (sheltered), dots = Scarlett Point (exposed), semi-open circles = The Stack (fairly sheltered, but adjacent to Scarlett Point). Data untransformed to illustrate homogeneity of Scarlett Pt. sample.

There was no obvious correlation between frond width and plant length within the populations from Castletown and from Scarlett Point, but there was a weak positive correlation in the sample from The Stack (Fig. 11). The scatter of these points suggests, however, that the correlation may be a false one and that the population is not genuinely intermediate in character, but rather composed of a mixture of plants, some of which have morphological affinities with those of Castletown and others with those of Scarlett Point.

Additional information on the variability of the above mentioned samples may be obtained from vesicle counts. In a uniform population of *F. vesiculosus* the number of vesicles should increase as a function of plant growth and the extent to which a population departs from a linear correlation of these two parameters should serve as an index of its diversity. The scatter of vesicle numbers against fresh weight in the Castletown population shows a rather poor positive correlation (Fig. 12). However, the Stack population gives a barely detectable correlation and is correspondingly more diverse (Fig. 13). In the Castletown sample, evesiculate plants are few and usually small, but at The Stack they are numerous and often quite large. The Scarlett Point plants being almost entirely evesiculate are highly uniform. Therefore the Stack population

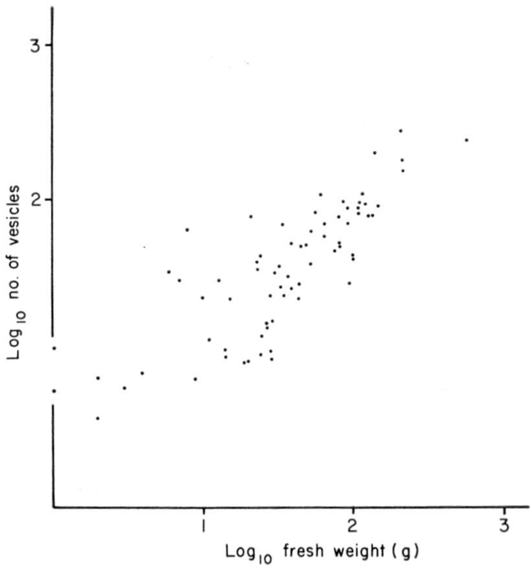

FIG. 12. Scatter diagram of vesicle number against fresh weight in *Fucus vesiculosus* plants sampled randomly at Castletown Bay, Isle of Man. Data $\log_{10}$–transformed.

is more diverse than either of the populations which flank it. This observation runs somewhat contrary to the model in Fig. 7, although it agrees with that of Lewis (1964) who described similar populations from other regions of rapid transition from exposure to shelter. It also suggests that the morphology of *F. vesiculosus* is not altogether the direct product of environmental effects and that other factors of an intrinsic nature are involved.

The statement by McLachlan *et al.* (1971) that phenotypic plasticity is a less important factor in *Fucus* morphology than has hitherto been suggested is almost certainly correct, but only in a general way. Phenotypic variation in *F. vesiculosus* is likely to be, in part, the result of natural selection, through wave action and other factors, operating on the genotypes which arrive at a habitat.

The nature and variety of these genotypes will also depend on the pattern of gene flow from adjacent populations. Conversely, it is inherently improbable that phenotypic plasticity is not involved in the morphology of *F. vesiculosus* and some features, such as relative number of dichotomies and width of frond, are possibly more susceptible to habitat conditions than others. In flowering plants it has long been known that plasticity is a property specific to individual characters in relation to specific environmental influences, and that the degree of plasticity possessed by a character may be itself genetically-determined

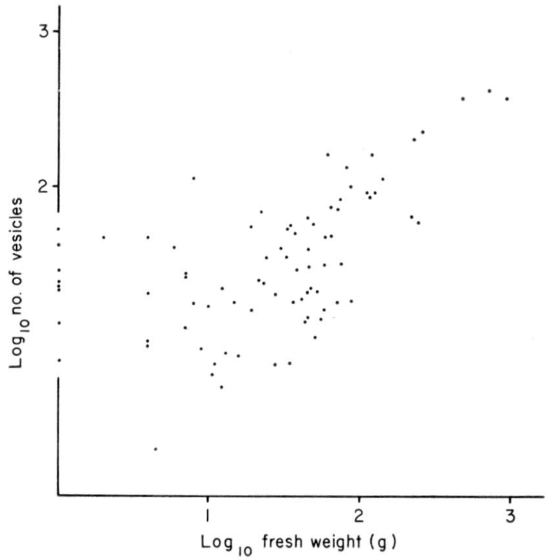

FIG. 13. Scatter diagram of vesicle number against fresh weight in *Fucus vesiculosus* plants sampled randomly at The Stack, Isle of Man. Data log_{10}–transformed.

(Bradshaw, 1965). The field transplant experiments on *F. distichus* by Pollock (1969) suggest that the amount of phenotypic plasticity is of a fairly low order in this species, although all these observations simply argue for more thorough genetic investigations of the genus *Fucus* and a more critical re-examination of phenotypic plasticity in brown algae as a whole.

3. *Taxonomic Status of Intraspecific Variants*

Most attempts to bring some order to intraspecific variation in the brown algae have been made within the framework of classical taxonomy. The problem raises two questions: firstly, should variants be recognized as taxa? Secondly, if so, to what rank should they be assigned? In both respects, however, there has been a history of inconsistency and uncertainty. The three experimental studies of cucullate laminarians, already referred to, illustrate this problem. The results obtained by Sundene (1962) on *Laminaria digitata* caused him to believe that the forma *stenophylla* is merely a growth form conditioned by extremely strong water movements and alterable on transplantation to sheltered habitats (Sundene, 1964). Sundene implies that such forms do not merit separate taxonomic status. Norton (1969) in his experimental work on *Saccorhiza* elected not to assign the different thallus forms described to any formal taxonomic category, but Svendsen and Kain (1971) have used the same sort of results from comparable experiments on *Laminaria hyperborea* to justify the description of a new taxon, forma *cucullata*.

The differences between exposed- and sheltered-shore populations of *Pelvetia canaliculata* (Fig. 6) do not seem to vary either in quality or quantity from those developed under identical conditions by *Fucus spiralis*. Nevertheless, forma *nanus* has been recognized by a number of authors (Powell, 1963; Lewis, 1964) as the appropriate taxon for the reduced, exposed-shore populations of *F. spiralis* while those of *Pelvetia* have been ignored taxonomically.

In the case of *Fucus distichus*, the subspecies *distichus* is a small, narrow-fronded plant of rock pools in the littoral fringe, whereas the larger subspecies *edentatus* is found at a lower level, in the eulittoral zone, and in rather more sheltered conditions (Powell, 1957; Russell, 1974). The differences between these taxa, although quite distinct ecologically as well as morphologically (Fig. 14), are not of a much greater magnitude than those between intertidal and sublittoral *Halidrys* (Fig. 4). The intertidal form of the latter has been referred to as forma or varietas *siliculosus* Stackh. (= forma *minor* Turn.) by a number of earlier authors, but Newton (1931) seems to have been the last in Britain to do so.

The question of the appropriate taxonomic rank has also remained unresolved. Davis and Heywood (1963) have proposed that the rank of subspecies should be reserved for a geographical facies of a species, the varietas for a local facies of a species and the forma for sporadic variants without distinctive spatial patterns of distributions. On this basis it would seem most appropriate to assign all infraspecific taxa of *Fucus* species, including those of *F. distichus*, to the rank of varietas because of their characteristic ecologies, although in practice most have been regarded as formae or subspecies (Powell, 1963). On the other hand, the rank of varietas has been used quite frequently for the infraspecific taxa of the filamentous members of the Ectocarpaceae (Hamel, 1931; Kornmann in Kuckuck, 1962; Cardinal, 1964). These are much less distinct ecologically than their fucoid counterparts and therefore would seem better described as formae.

Experimental taxonomy offers an alternative hierarchy of terms, of which the basic unit, the ecotype, may be regarded as a phenotypic expression determined to some extent by genetic differences. The term ecad may be used for those morphological differences which are due to direct environmental effects. In practice both terms have been abused by adoption without the benefit of

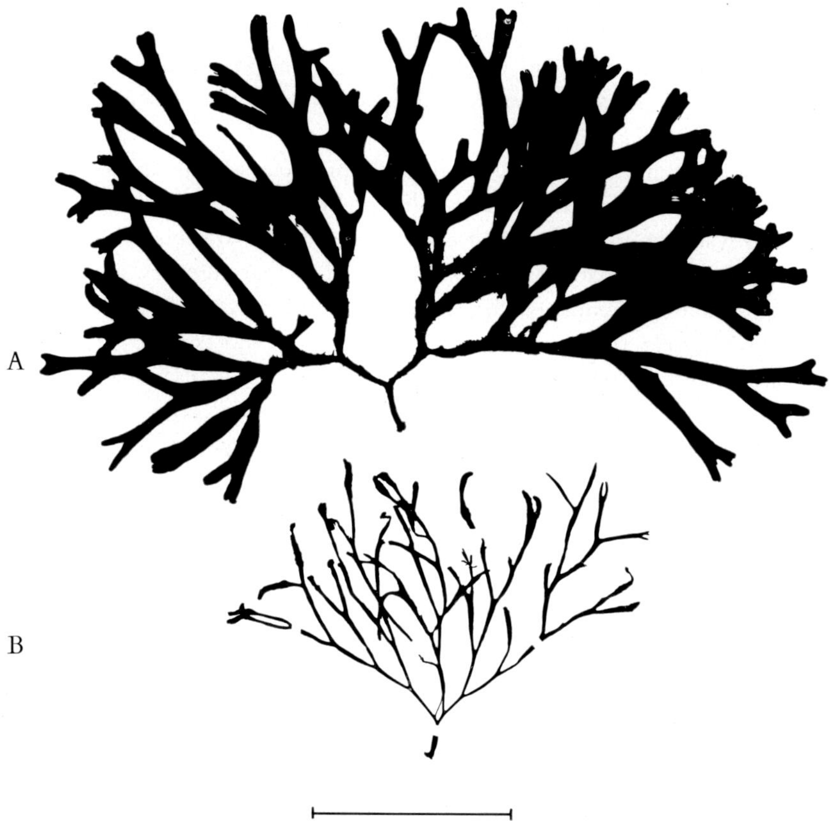

FIG. 14. A. *Fucus distichus* subspecies *edentatus*, *B.F. distichus* subspecies *distichus*; both plants collected from Shetland, and photographed from herbarium sheets. Scale = 10 cm.

supporting experimental evidence. For example, the saltmarsh form of *Fucus vesiculosus* described as the ecad *volubilis* by Baker and Bohling (1916) was elevated to the status of ecotype by Davy de Virville (1944) and accepted as such by Powell (1963), although in no case was any experimentation done. Other workers who have obtained evidence which would allow them to adopt the experimental categories have chosen not to do so, e.g. Sundene (1964), Norton (1969).

The aims of classical and experimental taxonomy are different and consequently there is no reason to suppose that the infraspecific categories should correspond exactly. In practice they do not correspond at all. Thus the subspecies of *Fucus distichus* are all regarded as ecotypes by Powell (1963) as is the forma *nanus* of *F. spiralis*. However, Svendsen and Kain (1971) imply the need for a more consistent attitude to the categories of classical and experimental taxonomy in their statement that because cucullate populations of *Laminaria hyperborea* and *L. digitata* are ecads they should be given the rank of forma and not varietas.

Powell (1963) has also pointed to one of the problems commonly encountered in experimental investigations; the Baltic populations of *F. vesiculosus*, which occur in water of greatly reduced salinity, seem likely to show genetic differences resulting from natural selection of euryhaline characteristics. However, in terms of thallus morphology these plants are not greatly different from others found in the fully saline conditions of W. European shores. There is a similar absence of morphological distinction in the euryhaline ecotypes of *Ectocarpus siliculosus* (Russell and Bolton, 1975) and *Pilayella littoralis*, (Russell 1963). Thus ecotypes may be physiologically rather than morphologically distinct entities, a fact which may have proved a deterrent against the more widespread adoption of the experimental terminology. In any event, there seems little likelihood that the necessary experimental information will be obtained quickly enough for the terminology of experimental taxonomy to make any impact on the classification of intraspecific variation in the brown algae.

CONCLUSIONS

The appeal for a conservative approach to species names was concerned with those simple forms which have life-histories involving morphologically different phases in facultative rather than obligate alternation and where the relationships of the phases are still imperfectly understood. However, the Ectocarpales furnishes other examples in which the transition from one developmental state to another is a much more gradual business, and in these cases the names of the states which precede the "perfect" condition, or depart somewhat from its typical expression, have been generally, probably correctly, discarded. Nevertheless, many of these have a characteristic morphology and may form a recurrent seasonal or ecological facies of a species. A good example of this is *Ectocarpus luteolus* Sauv., a developmental state of *Spongonema tomentosum*; (Kornmann in Kuckuck, 1961), but which forms a characteristic yellow felt on the stipes of *Fucus* species during winter and early spring on rocky shores throughout the British Isles. The abandonment of taxa such as this has parallelled the discarding of those for intraspecific variants of perennial algae such as *Fucus*. This policy has usually been justified on the grounds that the taxa are insufficiently distinct to warrant retention, and put into effect without guidance on the magnitude of difference that qualifies for retention.

As always, the problems of taxonomic recognition depend heavily upon the method of data acquisition. For the plant collector in search of a choice and typical specimen the taxonomic difficulties are probably less intimidating than to the population taxonomist sampling on a more objective basis. Much also depends on the manner of data presentation; the scatter diagram in Fig. 5 makes one less inclined to accept taxonomic distinction than do the same data assembled in histograms and compared statistically.

Few would deny that throughout the brown algae there are populations of plants with characteristic seasonal or habitat features and with characteristic phenotypes. These phenotypes are recurrent and predictable. To dismiss them as of no interest or importance is in effect to discourage investigation in an already much neglected field of study. However, it must be admitted that the categories of classical taxonomy have not proved an acceptable means of bringing intraspecific variation in the brown algae to any semblance of order. This is partly because of a reluctance to draw arbitrary lines of taxonomic distinction between one intraspecific morphology and another, although, as we have seen, the discrimination of species and even genera has been based at times on fairly arbitrary decisions. There has also been a widespread suspicion (e.g. Russell, 1966) that the number of different phenotypes within certain species is, for all practical purposes, limitless and that there is consequently little to be gained from taxonomic discrimination. However, it is important to establish this point on a quantitative consideration of the plants in relation to habitats, seasonality and environmental factors, rather than from selected individuals; over-dependence on the latter approach is as questionable as the use of selected specimens to justify retention of a taxon.

If classical taxonomy has not been a total success, it is equally true that the terminology of experimental taxonomy offers no better prospects for dealing with intraspecific variation. Taking the brown algae as a whole, very little experimental taxonomy has been practised and some of that is fairly inconclusive. Moreover, the adoption of the terms ecad and ecotype without any experimental evidence has probably caused a great deal of lasting damage to their value.

The immediate need is for a system which will permit the orderly treatment of intraspecific variation, in an informal way, unrestricted by the nomenclatural and procedural technicalities of classical taxonomy, and which does not demand a programme of experimentation. The only system which fulfils these requirements is the deme terminology proposed originally by Gilmour and Gregor (1939), developed by Gilmour and Heslop-Harrison (1954) and summarized by Gilmour (1960). Of the numerous categories of deme which have burgeoned over the years only a few are applicable to the brown algal problems raised in this paper. For example, the primary derivatives ecodeme and phenodeme (denoting ecological and phenotypic differences) may be applied respectively to estuarine populations of *Pilayella littoralis* and to the six to eight phenotypes of the *Ectocarpus siliculosus* complex in Britain. For those plants among which the morphological variation appears to be linked with habitat conditions, the secondary derivative phenoecodeme should be used; the reduced, exposed-shore populations of *Fucus vesiculosus* and the free-living plants of *Ectocarpus* and *Pilayella* may be assigned to this category.

It is probable that the deme terminology was devised with only the flowering plants in mind but the system is singularly appropriate to the problem of variation in the brown algae. Equally important, however, are the aims of its proposers. These are to focus attention upon, and stimulate interest in, the processes of variation in plants and their rôle in evolutionary change. This clear purpose has two implications for our present approach to marine algal systematics. It underlines the rather uncritical attitudes to variation which we have adopted and our failure to analyse its causes. Furthermore, it points to the opportunities, scarcely equalled in the plant kingdom, for linking the study of morphological diversity with the various ecological pressures that constitute natural selection.

ACKNOWLEDGEMENTS

I am very grateful to Mr J. S. L. Gilmour for much helpful advice on the deme terminology and for encouraging its application to marine algae. I must also acknowledge the technical assistance of Mrs G. Hayes and the permission of Miss J. Goodwin to make use of the data presented in Fig. 9.

REFERENCES

BAARDSETH, E. (1970). A square-scanning, two-stage sampling method of estimating seaweed quantities. *Norwegian Inst. Seaweed Research Report* **33**, 1–41.

BAKER, J. R. J. and EVANS, L. V. (1971). A myrionemoid variant of *Ectocarpus fasciculatus* Harv. *Br. phycol. J.* **6**, 73–80.

BAKER, S. M. and BOHLING, M. H. (1916). On the brown seaweeds of the salt marsh. Part II. Their systematic relationships, morphology and ecology. *J. Linn. Soc. (Bot.)* **43**, 325–380.

BALLANTINE, W. J. (1961). A biologically-defined exposure scale for the comparative description of rocky shores. *Fld Stud.* **1**(3), 1–19.

BOLWELL, G. P., CALLOW, J. A., CALLOW, M. E. and EVANS, L. V. (1977). Cross-fertilization in fucoid species. *Nature, Lond.* **268**, 626–627.

BRADSHAW, A. D. (1965). Evolutionary significance of phenotypic plasticity in plants. *Adv. Genet.* **13**, 115–155.

BURROWS, E. M. and LODGE, S. (1951). Autecology and the species problems in *Fucus*. *J. mar. biol. Ass. U.K.* **30**, 161–176.

CARDINAL, A. (1964). Étude sur les Ectocarpacées de la Manche. *Beih. nov. Hedwigia* **15**, 1–86.

CHAPMAN, A. R. O. (1975). Inheritance of mucilage canals in *Laminaria* (Section Simplices) in eastern Canada. *Br. phycol. J.* **10**, 219–223.

CLAYTON, M. N. (1972). The occurrence of variant forms in cultures of species of *Ectocarpus* and *Giffordia*. *Br. phycol. J.* **7**, 101–108.

COLE, K. (1967). Chromosome numbers in the Phaeophyceae. *Can. J. Genet. Cytol.* **9**, 519–530.

COLE, K. (1968). Additional chromosome numbers in the Phaeophyceae. *Can. J. Genet. Cytol.* **10**, 670–672.

DANGEARD, P. (1970). Réflexions sur quelques Ectocarpales nées en culture et particulièrement sur les "*Streblonema*." *Botaniste* **53**, 23–61.

DAVIS, P. H. and HEYWOOD, V. H. (1963). "Principles of Angiosperm Taxonomy." Oliver and Boyd, Edinburgh and London.

DAVY DE VIRVILLE, A. (1944). Les *Fucus* des côtes de France. *Rev. sci., Paris,* **3237–3238**, 421–452.

DIXON, P. S. (1963). Variation and speciation in marine Rhodophyta. *In* "Speciation in the Sea" (J. P. Harding and N. Tebble, eds), pp. 51–62. Systematics Association Publ. No. 5.

DRING, M. J. and LÜNING, K. (1975). A photoperiodic response mediated by blue light in the brown alga *Scytosiphon lomentaria*. *Planta* **125**, 25–32.

EDELSTEIN, T., CHEN, L. C-M. and MCLACHLAN, J. (1970). The life cycle of *Ralfsia clavata* and *R. borneti*. *Can. J. Bot.* **48**, 527–531.

EDELSTEIN, T., CHEN, L. C.-M. and MCLACHLAN, J. (1971). On the life histories of some brown algae from eastern Canada. *Can. J. bot.* **49**, 1247–1251.

FJELD, A. and LØVLIE, A. (1976). Genetics of multicellular marine algae. *In* "The Genetics of Algae" (R. A. Lewin, ed.), pp. 219–235. Botanical Monographs, Vol. 12. Blackwell Scientific Publications, Oxford.

FRITSCH, F. E. (1929). Evolutionary sequence and affinities among protophyta. *Biol. Rev.* **4**, 103–151.

GILMOUR, J. S. L. (1960). The deme terminology. *Rep. Scott. Pl. Breed. Stn* **1960**, 99–105.

GILMOUR, J. S. L. and GREGOR, J. W. (1939). Demes: a suggested new terminology. *Nature, Lond.* **144**, 333.

GILMOUR, J. S. L. and HESLOP-HARRISON, J. (1954). The deme terminology and the units of micro-evolutionary change. *Genetica* **27**, 147–161.

GRAN, H. H. (1897). Kristianiafjordens Algeflora. *Skr. Vidensk. Christiania.* I. Mat.-nat. Kl. 1869, **1**, 1–56.

HAMEL, G. (1931). "Phéophycées de France." Fasc. I. Paris.

HARVEY, W. H. (1846). "Phycologia Britannica", Vol. I. Longmans, Reeve and Co., London.

HOEK, C. VAN DEN and FLINTERMAN, A. (1968). The life history of *Sphacelaria furcigera* Kütz. (Phaeophyceae). *Blumea* **16**, 193–242.

HSIAO, S. (1969). Life history and iodine nutrition of the marine brown alga, *Petalonia fascia* (O. F. Müll.) O. Kuntze. *Can. J. Bot.* **47**, 1611–1616.

IRVINE, D. E. G., GUIRY, M. D., TITTLEY, I. and RUSSELL, G. (1975). New and interesting marine algae from the Shetland Isles. *Br. phycol. J.* **10**, 57–71.

KJELLMAN, F. R. (1890). "Handbok i Skandinaviens Havsalgflora." Lamms, Stockholm.

KUCKUCK, P. (1961). Ectocarpaceen-Studies VI. *Spongonema. Helgoländer wiss. Meeresunters.* **7**, 93–113.

KUCKUCK, P. (1962). Ectocarpaceen-Studies VII. *Giffordia. Helgoländer wiss. Meeresunters.* **8**, 120–152.

KRISTENSEN, I. (1965). Surf influence on the thallus of fucoids and the rate of desiccation. *Sarsia* **34**, 69–82.

LEVRING, T. (1945). Zur Kenntnis zweier endophytischen Phaeophyceen. *Acta Horti gothoburg.* **16**, 185–190.

LEWIS, J. R. (1964). "The Ecology of Rocky Shores." English Universities Press, London.

LOISEAUX, S. (1969). Sur une espèce de *Myriotrichia* obtenue en culture à partir de zöides d'*Hecatonema maculans* Sauv. *Phycologia* **8**, 11–15.

LOISEAUX, S. (1970). *Streblonema anomalum* S. et G. and *Compsonema sporangiiferum* S. et G. stages in the life history of a minute *Scytosiphon. Phycologia* **9**, 185–191.

LUND, S. (1966). On a sporangia-bearing microthallus of *Scytosiphon lomentaria* from nature. *Phycologia* **6**, 67–78.

LÜNING, K. (1975). Kreuzungs experiment an *Laminaria saccharina* von Helgoland und von der Isle of Man. *Helgoländer wiss. Meeresunters.* **27**, 108–114.

LÜNING, K. and DRING, M. J. (1973). The influence of light quality on the development of the brown algae *Petalonia* and *Scytosiphon. Br. phycol. J.* **8**, 333–338.

McLACHLAN, J., CHEN, L. C.-M. and EDELSTEIN, T. (1971). The culture of four species of *Fucus* under laboratory conditions. *Can. J. Bot.* **49**, 1463–1469.

MÜLLER, D. G. (1963). Die Temperaturabhängigkeit der Sporangienbildung bei *Ectocarpus siliculosus* von verschiedessen Standorten. *Pubbl. Staz. zool. Napoli* **33**, 310–314.

MÜLLER, D. G. (1972). Life cycle of the brown alga *Ectocarpus fasciculatus* var. *refractus* (Kütz.) Ardis. (Phaeophyceae, Ectocarpales) in culture. *Phycologia* **11**, 11–13.

NEWTON, L. (1931). "A Handbook of the British Seaweeds." British Museum (Natural History), London.

NORTON, T. A. (1969). Growth form and environment in *Saccorhiza polyschides. J. mar. biol. Ass. U.K.* **39**, 1025–1045.

PARKE, M. and DIXON, P. S. (1976). Check-list of British marine algae—third revision. *J. mar. biol. Ass. U.K.* **56**, 527–594.

POLLOCK, E. G. (1969). Interzonal transplantation of embryo and mature plants of *Fucus. Proc. int. Seaweed Symp.* **6**, 345–356.

POWELL, H. T. (1957). Studies in the genus *Fucus* L. I. *Fucus distichus* L. emend Powell. *J. mar. biol. Ass. U.K.* **36**, 407–432.

POWELL, H. T. (1963). Speciation in the genus *Fucus* L., and related genera. *In* "Speciation in the Sea" (J. P. Harding and N. Tebble, eds), pp. 63–77. Systematics Association Publication No. 5.

ROBERTS, M. and RING, F. M. (1972). Preliminary investigations into conditions affecting the growth of the microscopic phase of *Scytosiphon lomentarius* (Lyngbye) Link. *Mém. Soc. bot. Fr.* **1972**, 117–128.

ROELEVELD, J. G., DUISTERHOF, M. and VROMAN, M. (1974). On the year cycle of *Petalonia fascia* in the Netherlands. *J. Sea Res.* **8**, 410–426.

ROSENVINGE, L. K. (1941). The marine algae of Denmark. Contributions to their natural history. Vol. II. Phaeophyceae. *K. danske Vidensk. Selsk. Skr.* **4**(5), 1–99.

RUSSELL, G. (1963). A study in populations of *Pilayella littoralis*. *J. mar. biol. Ass. U.K.* **43**, 469–483.

RUSSELL, G. (1964). *Laminariocolax tomentosoides* on the Isle of Man. *J. mar. biol. Ass. U.K.* **44**, 601–612.

RUSSELL, G. (1966). The genus *Ectocarpus* in Britain. I. The attached forms. *J. mar. biol. Ass. U.K.* **46**, 267–294.

RUSSELL, G. (1967). The ecology of some free-living Ectocarpaceae. *Helgoländer wiss. Meeresunters.* **15**, 155–162.

RUSSELL, G. (1974). *Fucus distichus* communities in Shetland. *J. appl. Ecol.* **11**, 679–684.

RUSSELL, G. and BOLTON, J. J. (1975). Euryhaline ecotypes of *Ectocarpus siliculosus*. *Estuar. cstl. mar. Sci.* **3**, 91–94.

RUSSELL, G. and FLETCHER, R. L. (1975). A numerical taxonomic study of the British Phaeophyta. *J. mar. biol. Ass. U.K.* **55**, 763–783.

RUSSELL, G. and MORRIS, O. P. (1973). Ship-fouling as an evolutionary process. *Proc. int. Congr. mar. Corr. Foul.* **3**, 719–730.

SNEATH, P. H. A. and SOKAL, R. R. (1973). "Numerical Taxonomy." W. H. Freeman, San Francisco.

SUNDENE, O. (1962). Growth in the sea of *Laminaria digitata* sporophytes from culture. *Nytt Mag. Bot.* **9**, 5–24.

SUNDENE, O. (1964). The ecology of *Laminaria digitata* in Norway in view of transplant experiments. *Nytt Mag. Bot.* **11**, 83–107.

SVENDSEN, P. and KAIN, J. M. (1971). The taxonomic status, distribution and morphology of *Laminaria cucullata sensu* Jorde and Klavestad. *Sarsia* **46**, 1–22.

WYNNE, M. J. (1969). Life history and systematic studies of some Pacific north American Phaeophyceae (Brown Algae). *Univ. Calif. Publs Bot.* **50**, 1–62.

16 | Studies on the family Ralfsiaceae (Phaeophyta) around the British Isles

R. L. FLETCHER*

Department of Biology and Geology, The Polytechnic of North London, England

Abstract: A taxonomic study of the brown crustose algae belonging to the family Ralfsiaceae is currently being carried out. All the members of the Ralfsiaceae are characterized by a discoid base which gives rise to erect laterally coherent filaments. Comparative studies, in particular those using culture techniques, undertaken on North Atlantic genera, have revealed marked differences in their morphology, anatomical construction, sporeling development and life-history. Examples of these "differences" are presented and their taxonomic implications discussed.

INTRODUCTION

In the Phaeophyta, the term crustose is generally applied to those small, often microscopic algae which have prostrate growth forms, either represented by discs, gelatinous cushions or (more frequently) by crusts. These elaborate heterotrichous thalli are all basically composed of a horizontally spreading system of branched filaments which give rise to erect filaments that remain fairly short, undifferentiated and usually fairly well packed together. The resultant thallus is, therefore, a pseudoparenchyma, with true parenchymatous divisions playing only a minor rôle in its construction. In this respect the crustose algae are readily distinguishable from other prostrate algae such as *Zanardinia prototypus* (Nardo) Nardo and the *Aglaozonia parvula* phase of *Cutleria multifida* (Sm.) Grev.; the latter in section clearly are not filamentous in derivation but are the result of parenchymatous divisions in the cells immediately behind the outer meristematic apical cells. The undifferentiated nature of the erect filaments in the crusts further distinguishes them from the more elaborate pustules and cushions of algae belonging to the two families Elachistaceae and Coryno-phlaeaceae.

Systematics Association Special Volume No. 10, "Modern Approaches to the Taxonomy of Red and Brown Algae", edited by D. E. G. Irvine and J. H. Price, 1978, pp. 371–398. Academic Press, London and New York.

*Present address: Department of Biological Sciences, Portsmouth Polytechnic, The Marine Laboratory, Hayling Island, England.

The crustose algae are placed in two families, the Ralfsiaceae and the Myrionemataceae. The myrionematoid algae are predominantly epiphytic in habit and generally possess a loosely coalescent discoid base, from the majority of cells of which arise short, vertical files of loosely associated filaments. The ralfsioid algae, on the other hand, are predominantly lithophytes and generally possess a much more compact discoid base, from the cells of which (at least in the central parts), arise short, vertical or curved filaments which are moderately to strongly coalescent to form a polystromatic crust. The validity of distinguishing these two families is, however, questionable and some workers (e.g. Hamel, 1939; Loiseaux, 1968) recognize only a single family, the Myrionemataceae.

The present paper is concerned with some preliminary studies carried out on the crustose species of the family Ralfsiaceae in the North Atlantic region. Excluding the warm-water genus *Nemoderma* Schousboe and the problematical genus *Lithoderma* Areschoug, all the remaining five crustose genera of the North Atlantic are represented in the British flora. Observations have centred mainly on the British species, and the salient features of their ecology, construction and life-history are presented and discussed from a systematic viewpoint.

OBSERVATIONS

Table I lists the British representatives of the family Ralfsiaceae as currently reported in the Parke and Dixon (1976) check-list. It is probable that more species await detection, examples of which might include *Ralfsia lucida* Lund, presently restricted to the Danish coast, *Pseudolithoderma rosenvingii* (Waern)

TABLE I. Representatives of the family Ralfsiaceae recorded for the British Isles (after Parke and Dixon, 1976)

Ralfsiaceae
Petroderma maculiforme (Wollny)Kuck.
Pseudolithoderma extensum (Crouan frat.) S. Lund
P. roscoffensis Lois.
Ralfsia clavata (Harv. in Hook) Crouan frat.
R. disciformis Crouan frat.
R. pusilla (Stromf.) Batt.
R. spongiocarpa Batt.
R. verrucosa (Aresch.) J. Ag.
Sorapion simulans Kuck.
Symphyocarpus strangulans Rosenv.

Lund and *P. subextensum* (Waern) Lund, which have been reported in Sweden (Waern, 1949), Greenland (Lund, 1959) and even on the north-east American coast (Sears and Wilce, 1973—*P. rosenvingii*). Included in the present study are observations on the non-British cold-water crust *R. fungiformis* (Gunn.) Setchell and Gardner, based on material collected in the Bay of Fundy, north-east America (P. R. Newroth) and Iceland (I. M. Munda). This circumpolar species, reported from the north-east and north-west coasts of America, northern Japan, Greenland and Iceland, is known on the European coasts only in northern Norway. Also included is *Microspongium gelatinosum* Reinke, a crustose species recorded for various localities in the north Atlantic and currently retained in the family Myrionemataceae.

1. *Ecology*

With the exception of the two very small and rare crusts (*Ralfsia pusilla* and *Symphyocarpus strangulans*), which are primarily epiphytic in habit, all the remaining crusts are predominantly lithophytes. The principal intertidal species, certainly the most common all around the British Isles, are *R. clavata* and *R. verrucosa* (Figs 1–3). These are both essentially tide-pool inhabitants although frequently they are to be found on open rock particularly in moist conditions, for example under the cover of fucoids. *R. clavata* tends to occupy the lower intertidal regions, often in runnels, and appears to have a preference for soft substrates such as chalk and shale. *R. verrucosa*, on the other hand, extends over the whole intertidal range and is not infrequently recorded from high level exposed shallow pools, usually with very few associated species present. The other two *Ralfsia* species, *R. disciformis* and *R. spongiocarpa*, are much more limited in their occurrence; *R. disciformis* is only known from a single subtidal habitat on the south coast of England (Batters, 1896) whilst *R. spongiocarpa* has been only rarely reported from low level pools at a few widely distributed localities (Fletcher, 1978).

Commonly associated with *R. clavata* and *R. verrucosa* on the shore and equally widely distributed around the British Isles is the soft crust *Petroderma maculiforme*. This species is also essentially a tide-pool inhabitant and is particularly common on hard flint pebbles. However, it can frequently be found colonizing open rock situations, often forming extensive patches quite high up on the shore in some exposed localities such as Thanet, Kent, and the Tilly Whim caves, Dorset. The only other intertidal brown crust which can occasionally be found at various localities all around the British Isles is *Microspongium gelatinosum*. This species is also an intertidal pool inhabitant and is usually closely associated with *R. clavata* on stones at and below mid-tide level.

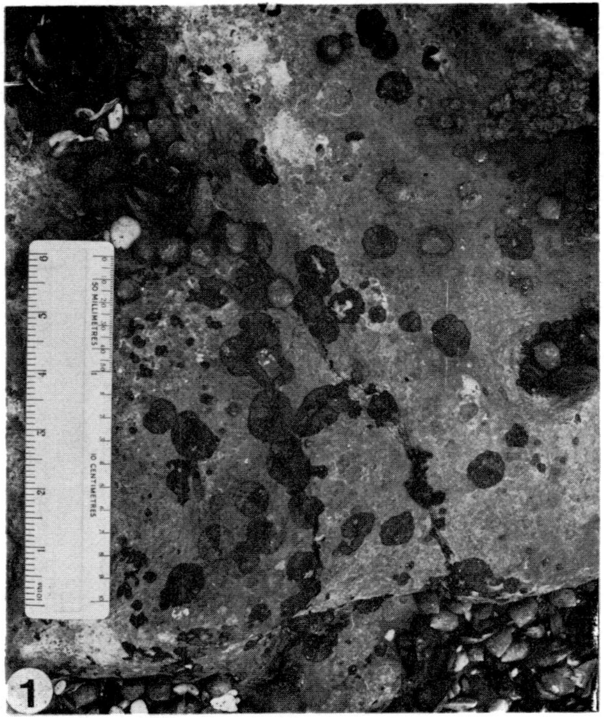

FIG. 1. Chalk boulder with *Ralfsia verrucosa* crusts; Thanet, Kent.

FIG. 2. Surface view of *R. verrucosa* crust. × 3·20.

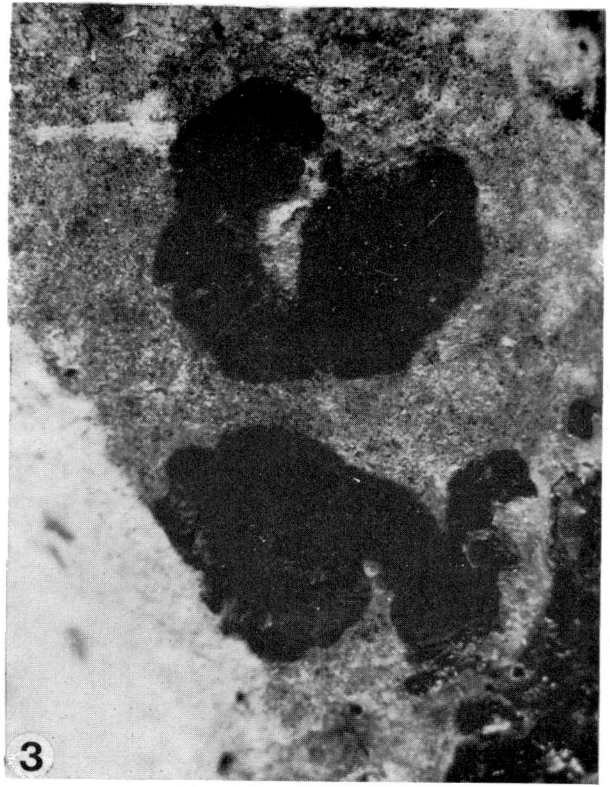

FIG. 3. Surface view of *Ralfsia clavata* crusts growing on chalk. $\times 3{\cdot}2$.

FIG. 4. Surface view of detached crust of *Ralfsia fungiformis*, showing overlapping lobes; collected from Iceland (I. M. Munda). $\times 2{\cdot}4$.

Whilst *Ralfsia* and *Petroderma* are the dominant intertidal crustose brown genera, the subtidal regions are almost exclusively inhabited by the two genera of the so-called *Lithoderma* complex, *Pseudolithoderma* and *Sorapion*. Of these two genera, *Pseudolithoderma* is by far the most common and widely distributed, *P. extensum* abundant in the sublittoral regions all around the British Isles and *P. roscoffensis* is common at low tide level and in the shallow sublittoral in the Solent region of the south coast.

2. *Morphology*

The crustose species display considerable variation in colour, ranging from light brown in young crusts of *P. maculiforme* to dark brown or black in crusts of *Pseudolithoderma* spp. In general all crusts are rounded in shape when young, forming fairly discrete thalli which are usually firmly attached to the substrate. As they enlarge, adjacent crusts frequently overlap to form confluent, irregular expansions up to many mm in diameter. Throughout their life a great majority of the crusts remain closely adherent to the substrate, particularly around the growing margin, and cannot be removed intact. In *R. verrucosa* and *R. fungiformis*, however, the central crust regions usually become bullate in appearance and very loosely attached. This usually results in only the outer, marginal crust regions remaining in many *R. verrucosa* thalli. The crust edge in these two species is also considerably less adherent than in the other crusts examined and usually can be raised intact from the surface.

There are also pronounced differences in the surface texture of the crusts. *Petroderma* and *M. gelatinosum*, for instance, are fairly soft and gelatinous in texture and when fully grown are almost sponge-like in appearance. This is in contrast with the majority of the crusts which tend to be much firmer and even coriaceous in texture. However, marked changes can occur in the nature of the surface during the development of the reproductive sori. In the genus *Ralfsia*, the unilocular sporangia are produced on the surface in fairly gelatinous sori which are sponge-like in appearance, this being particularly evident in the species *R. spongiocarpa*.

These differences in the nature of the crust surface are particularly well demonstrated using scanning electron microscope (S.E.M.) techniques (see Fletcher, 1976, for preparation details). Smooth, coriaceous thalli of species such as *R. verrucosa* and *R. clavata* reveal quite densely packed, vertical filaments with the terminal apical cells closely adherent to give a cobble-stone effect (Figs 5–7). This is in contrast to the sponge-like thalli of soft crusts such as *Petroderma* in which the vertical filaments are more freely projecting from the surface and not noticeably bound together in a cuticular material (Fig. 8).

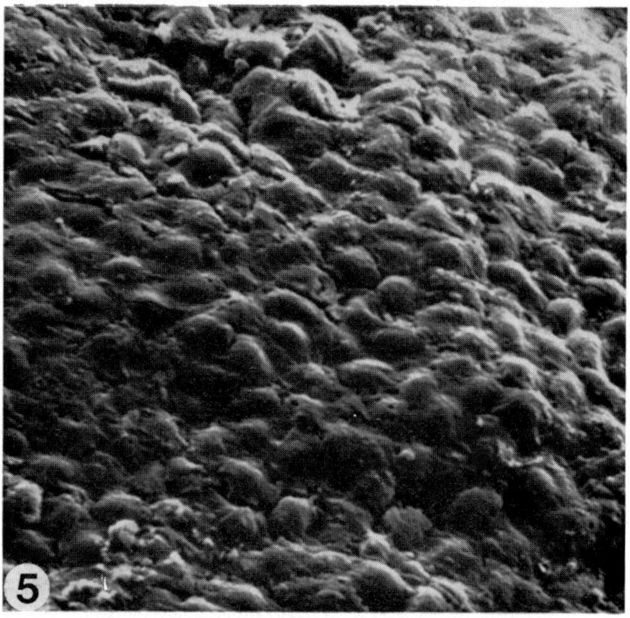

FIG. 5. Scanning electron micrograph (S.E.M.) of surface view of *Ralfsia verrucosa* crust showing terminal cells enclosed in cuticular material. × 1520.

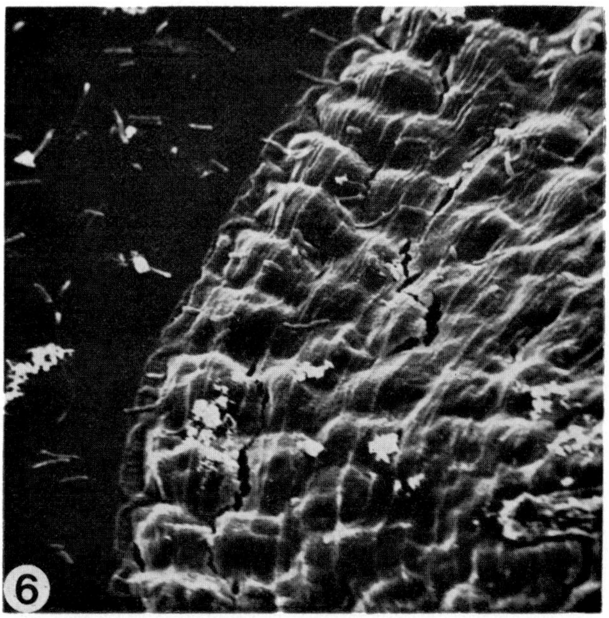

FIG. 6. S.E.M. of margin area of young cultured disc of *R. verrucosa* showing peripheral apical cells secreting outer cuticular material. × 800.

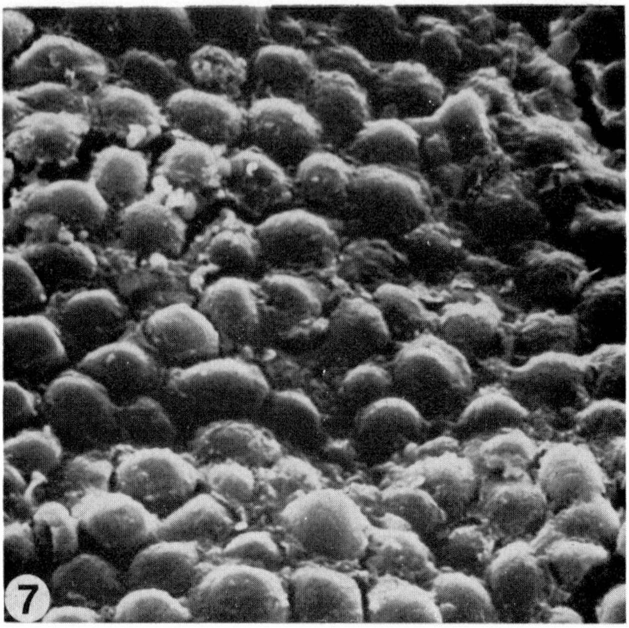

FIG. 7. S.E.M. of crust surface of *Ralfsia clavata* showing terminal vegetative cells enclosed in cuticular material. ×1520.

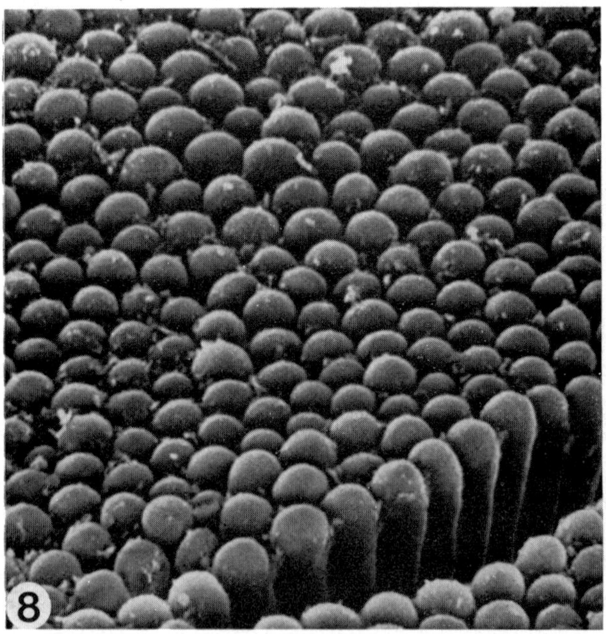

FIG. 8. S.E.M. of crust surface of *Petroderma maculiforme*. Note absence of cuticular material. ×1520.

3. *Thallus Structure*

The surface characteristics of the thalli to a great extent reflect the internal mode of construction of the vegetative filaments. The sponge-like thalli of crusts such as *Petroderma*, *M. gelatinosum* and *Symphyocarpus*, with freely projecting apical regions, are constructed of erect filaments which are only loosely bound together (Figs 9, 10). Under slight pressure, such as the weight of a

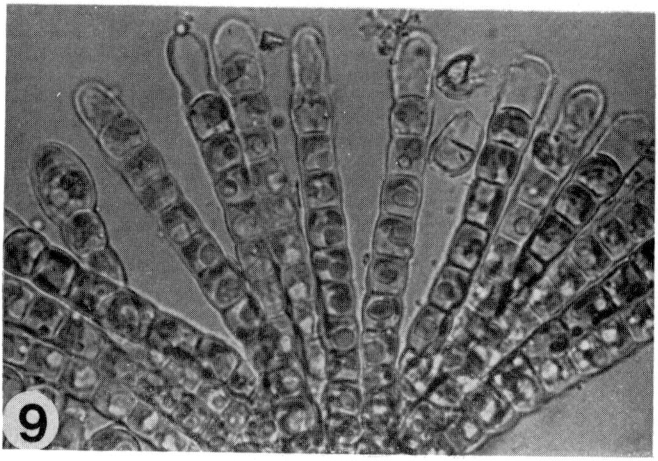

FIG. 9. "Squash" preparation of *Petroderma maculiforme* crust showing free erect filaments. × 700.

FIG. 10. Squash preparation of *Microspongium gelatinosum* crust showing loosely coalescent vertical filaments. × 380.

coverglass, the constituent filaments very easily glide apart, this probably being facilitated by the gelatinous nature of the cell walls. However, crusts such as those of *Ralfsia*, *Pseudolithoderma* and *Sorapion* with smooth cuticular surfaces, have erect filaments which are much more firmly bound together, making squash preparations generally more difficult to prepare (Figs 11–16). Particularly coherent are the erect filaments of *R. verrucosa*, *R. fungiformis* and the

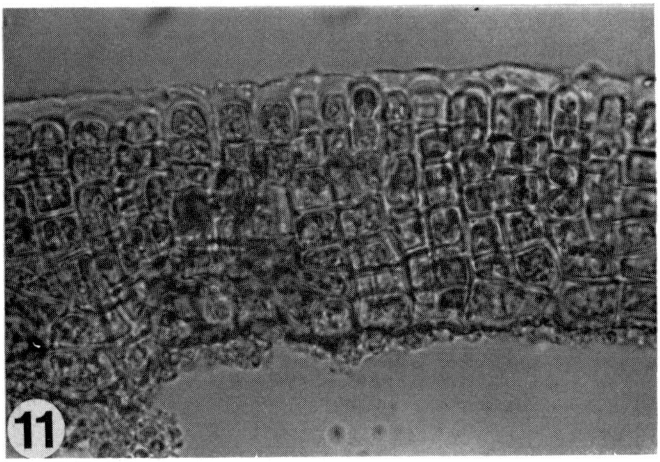

FIG. 11. Part of vertical section through crust of *Ralfsia clavata* showing erect laterally coherent vegetative filaments arising from a monostromatic base. ×635.

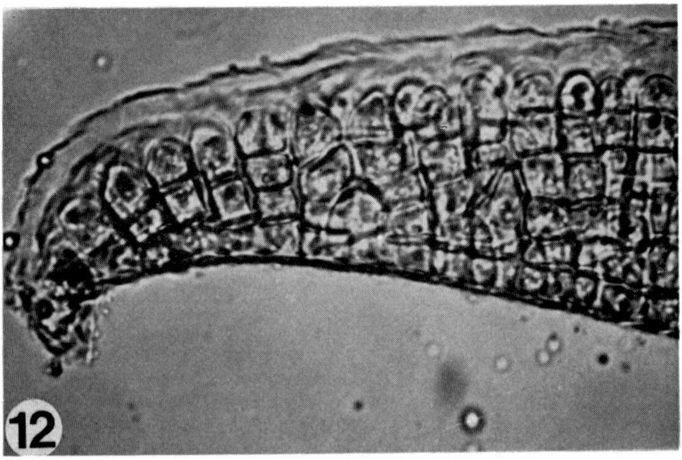

FIG. 12. Vertical section through margin of *Ralfsia clavata* crust showing terminal apical cell and vertically directed vegetative filaments. ×720.

Pseudolithoderma spp., whilst those of *R. clavata* and *R. spongiocarpa* are less tightly bound.

In most of the Ralfsiaceae members, the erect or slightly curved filaments are obviously derived from a single-layered discoid base (the hypothallus). This is particularly well exemplified in vertical sections and squashes of the young

FIG. 13. Part of vertical section through *Ralfsia verrucosa* crust showing multi-layered base giving rise to curved ascending vegetative filaments. × 164.

FIG. 14. Vertical section through margin of *Ralfsia verrucosa* crust showing multi-axial growing region. × 183.

sponge-like crusts of *Petroderma* and *M. gelatinosum*, which show separation of the filaments right down to the basal cell. In older material of these crusts, however, particularly in *M. gelatinosum*, this monostromatic base is less obvious, due to lateral cohesion of some (usually 1–3) of the basal cells of the erect

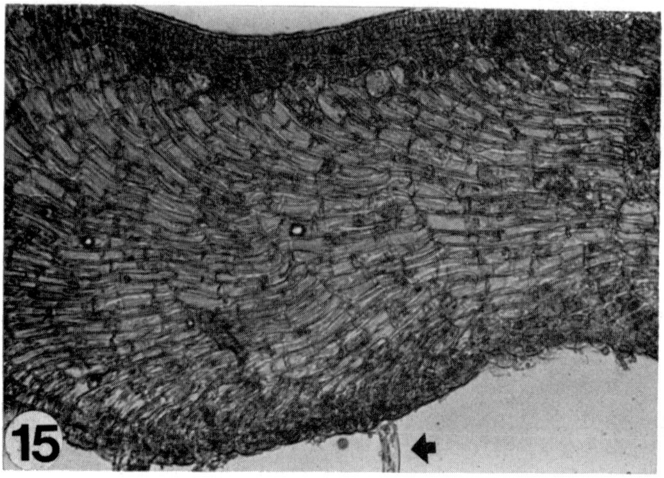

FIG. 15. Part of vertical section through *Ralfsia fungiformis* crust showing both upward and downward growth of cell filaments. Arrow denotes rhizoidal filament. ×168.

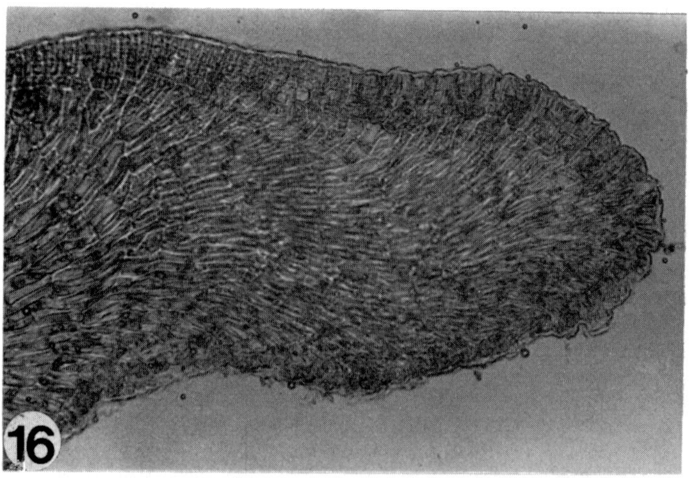

FIG. 16. Vertical section through lobe margin of *Ralfsia fungiformis* showing multi-axial growing region with both assurgent and downwardly curving cell rows. ×187.

filaments. A single-layered base is also present in the majority of *Ralfsia* spp., *Pseudolithoderma* and *Sorapion* spp., although in all these algae the erect filaments are laterally coalescent over their whole length and this is not often very apparent (Figs 11, 12).

A derivation from a single-layered hypothallus is not, however, distinguishable in the two species *R. verrucosa* and *R. fungiformis* (Figs 13–16). Vertical sections through crusts of *R. verrucosa* reveal a horizontally spreading hypothallus composed of several superimposed filaments which gradually arch upwards to produce the erect, strongly coalescent, vegetative filaments of the perithallus (Fig. 13). Cavities and depressions on the substrate surface are usually occupied by additional downward-curving extensions of these hypothallial filaments. The general aspect of this divergence of the hypothallial tissues is characteristic of the more anatomically differentiated species *R. fungiformis*. In vertical section the crusts of this species reveal both assurgent and downwardly curving cell rows, only the upper filaments giving rise to the relatively short perithallus zone (Fig. 15).

This progression of increased anatomical differentiation from the single-layered base of the majority of Ralfsiaceae members to the structurally more complex hypothallial tissues of *R. verrucosa* and *R. fungiformis* is particularly well exemplified in vertical sections through the marginal areas of these algae. Crusts with a monostromatic base such as *R. clavata*, *R. spongiocarpa*, *Pseudolithoderma* spp., *Petroderma* and *M. gelatinosum* usually have a terminally situated, discrete apical cell, the immediate derivatives of which produce the erect or slightly curved perithallial filaments (Fig. 12). In *R. verrucosa* (Fig. 14) horizontal growth does not appear to be carried out by a single terminal apical cell but by several superimposed filaments, whilst in *R. fungiformis* (Fig. 16) these filaments are organized into an almost distinct growth zone.

4. Cytology

A character commonly used to distinguish between Ralfsiaceae members is the number of plastids contained within the cells. All the *Ralfsia* spp. examined appear to have a single, plate-like plastid which usually occupies the upper regions of the cell (Fig. 17). This is particularly obvious in vertical sections and surface views of *R. clavata* and *R. spongiocarpa*, although less pronounced in *R. verrucosa* due to the more diffuse nature of the plastid. In the more elongate hypothallial cells of *Ralfsia*, particularly *R. verrucosa*, the plastid often enlarges considerably and becomes lobed in appearance, sometimes segmenting into 2–3 portions. A single plate-like plastid is also frequently distinguishable in *P. maculiforme* and appears to be present in *M. gelatinosum*, *Sorapion* and *Symphyo-*

carpus. As in *Ralfsia*, the plastids of *Petroderma* are frequently lobed in appearance and in surface view these lobes can be mistaken easily for discrete discoid plastids.

In contrast to all other Ralfsiaceae members, the two *Pseudolithoderma* spp., *P. extensum* and *P. roscoffensis*, were observed to have several discrete discoid plastids variously distributed within the cell body (Fig. 18).

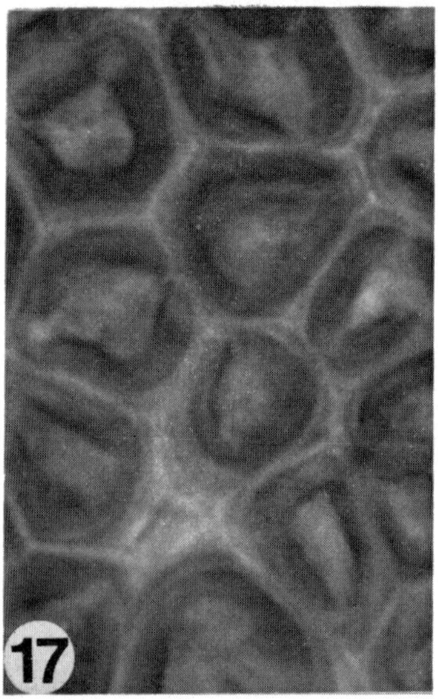
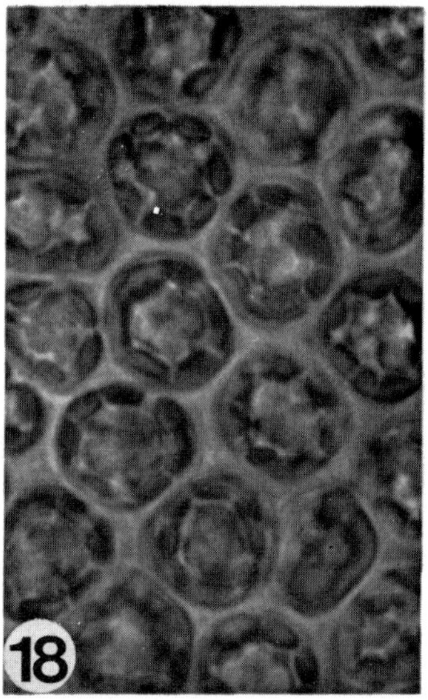

FIG. 17. Surface view of *Ralfsia clavata* crust showing a single parietal plate-like plastid in each cell. × 2200.

FIG. 18. Surface view of *Pseudolithoderma roscoffensis* crust showing several discoid plastids in each cell. × 1530.

5. Reproduction

The Ralfsiaceae members reproduce by spores derived from both unilocular and plurilocular sporangia, which are generally formed in sori on the crust surface. Both types of sporangia are not always present on the crusts of all the species. With the exception of *Ralfsia verrucosa*, only unilocular sporangia have been observed on the *Ralfsia* spp. and also on *M. gelatinosum* and *Sorapion*. On the other hand plurilocular sporangia are the only reproductive organs known in *P. roscoffensis* and *Symphyocarpus* and tend to be much more frequent on crusts of *P. extensum* than are the unilocular sporangia.

Apart from all the *Ralfsia* spp. and *M. gelatinosum*, in which the unilocular sporangia are borne laterally on paraphyses, and *R. verrucosa* which has sub-terminal intercalary plurilocular sporangia, all the remaining Ralfsiaceae members have terminal sporangia. The plurilocular sporangia either project terminally on the erect filaments, forming distinct sori enclosed within a surface cuticle (*P. roscoffensis*—Fig. 19; *P. extensum*) or are produced by direct internal divisions of the upper vegetative cells, usually in a basipetal manner (*Petroderma*). In *Petroderma* this method can result in the formation of long sporangia (up to 135 μm with 32 loculi) (Fig. 20). Unilocular sporangial formation in

FIG. 19. Part of vertical section through *Pseudolithoderma roscoffensis* crust showing terminal plurilocular sporangia. $\times$ 385.

Petroderma is also by direct enlargement and transformation of the terminal vegetative cells (Fig. 21), whereas in other crusts such as *P. extensum* and *Sorapion* sporangia appear to be produced in projected sori on the surface.

In *Ralfsia* and *M. gelatinosum*, the unilocular sporangia, although essentially terminal on the erect vegetative filaments, are positioned lateral to specialized soral filaments termed paraphyses (Figs 22, 23). These paraphyses are initially produced as direct extensions/conversions of the terminal parts of the erect filaments, the basal cell of which usually produces the sporangium as a lateral outgrowth. The gelatinous nature of the paraphysis cell wall suggests that these filaments are protective in function, perhaps a necessary attribute in view of the generally larger dimensions of the sporangia as compared to those of the "*Lithoderma*" complex algae. In *R. clavata*, *R. spongiocarpa* and *M. gelatinosum* the sori are produced in indefinite expansions all over the crust surface, whereas in *R. verrucosa* more discrete, bulbous sori are formed, usually in a zone just behind the growing margin. In *R. verrucosa* the paraphyses also appear to have arisen by upward growth of the vegetative filaments whereas in the other *Ralfsia* spp. and in *M. gelatinosum* they appear to be the result of direct conversion and elongation of erect filament cells. In *R. verrucosa* only the sporangia are produced additionally on 1–3 celled pedicels.

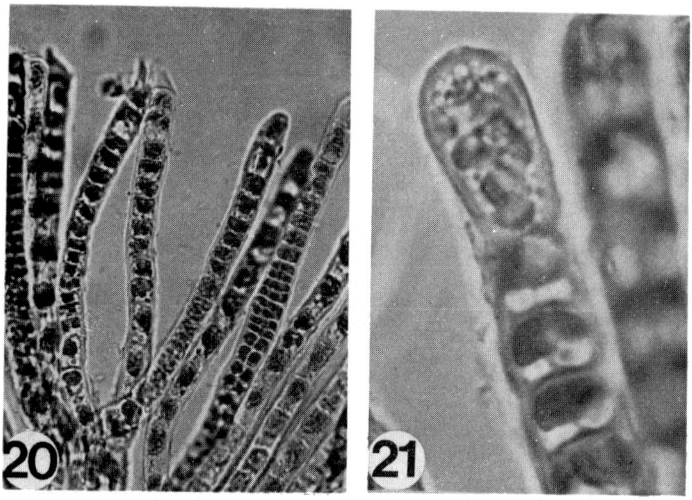

FIG. 20. Terminal plurilocular sporangia of *Petroderma maculiforme*. × 490.
FIG. 21. Terminal unilocular sporangium of *P. maculiforme*. × 1300.

The plurilocular sporangia of *R. verrucosa* have only occasionally been observed on British material and occur as densely packed vertical extensions of the erect vegetative filaments containing up to 26 tiered compartments (135 μm in length) (Fig. 24). The terminal sporangial cell is usually sterile and prominently darker in colour. This columnar arrangement of tiered plurilocular sporangia is also present in *R. fungiformis* but does not appear to occur in all the European *Ralfsia* spp. that have been described.

6. *Culture Studies*

Spore development and life-histories have been investigated in culture for *P. maculiforme* (unispores, plurispores), *M. gelatinosum* (unispores), *P. roscoffensis* (plurispores), *R. clavata* (unispores), *R. spongiocarpa* (unispores) and *R. verrucosa* (unispores, plurispores). All cultures were initiated using the hanging-drop

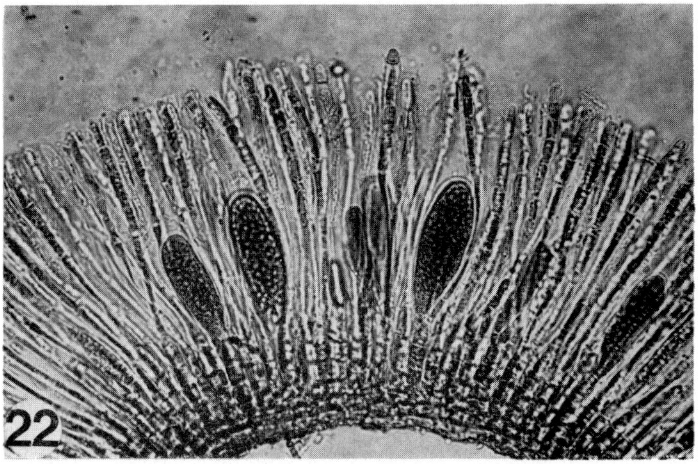

FIG. 22. Part of vertical section through *Microspongium gelatinosum* crust showing unilocular sporangia and associated multicellular paraphyses. ×380.

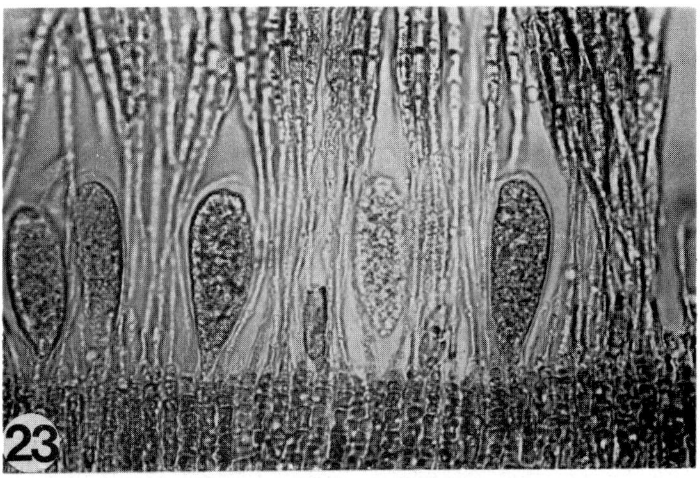

FIG. 23. Part of vertical section through fertile crust of *Ralfsia verrucosa* showing unilocular sporangia and multicellular paraphyses. ×360.

technique of Wynne (1969) and placed into incubators and growth rooms maintained at two distinct regimes:

Régime 1: 10°C, 8–$\overline{16}$ hours daylength conditions.

Régime 2: 15°C, 16–$\overline{8}$ daylength conditions.

Both regimes employed Grolux fluorescent tubes at approximately 1000 Lux

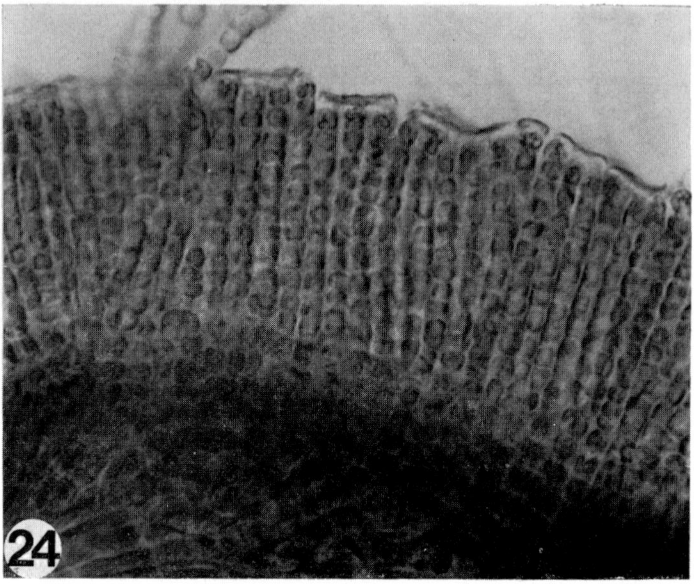

FIG. 24. Part of vertical section through *R. verrucosa* crust showing the terminal production of dense uniseriate tiers of plurilocular sporangia. ×525.

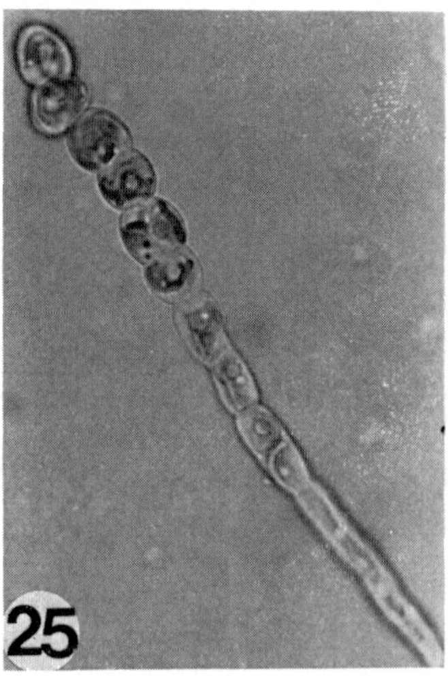

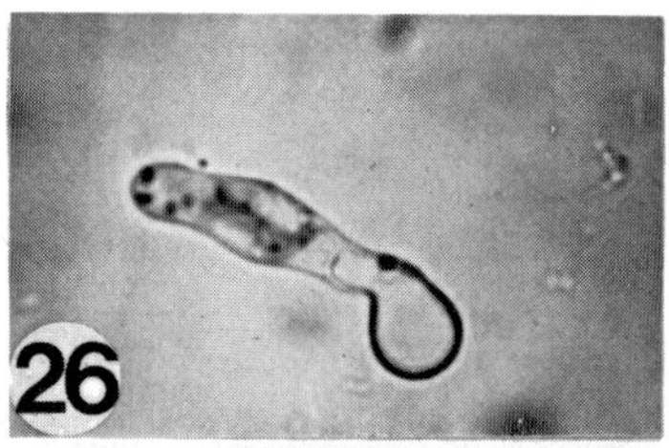

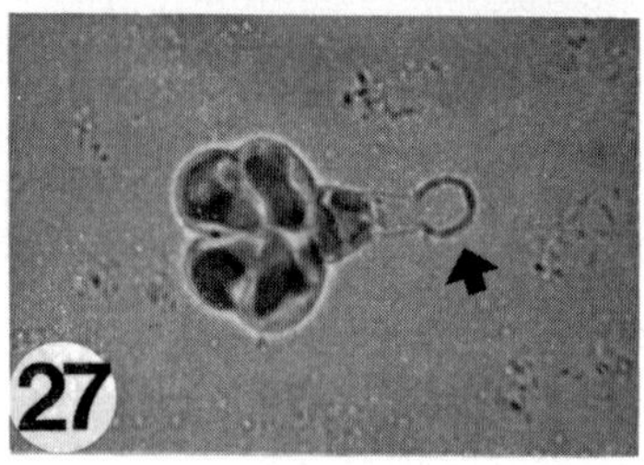

FIG. 25. *Ralfsia clavata*. Seven-day-old germling with hair: Régime 1. ×495.

FIG. 26. *R. verrucosa*. Six-day-old germling showing evacuated spore cell: unispore culture, Rég. 1. ×875.

FIG. 27. *R. verrucosa*. Five-day-old germling: plurispore culture, Rég. 2. Arrow denotes original spore cell. ×648.

and both Erdschreiber (Hammerling, 1931) and von Stosch (1964) culture media.

For both spore types of all algae the early germination sequence was very similar. After settlement, which occurred without sexual fusion, a germ-tube

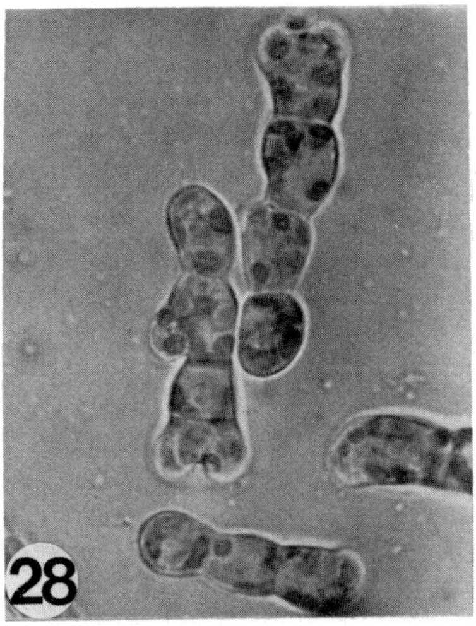

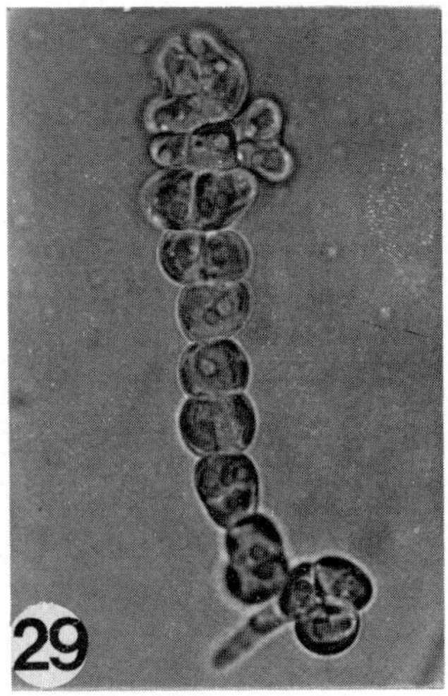

FIG. 28. *Pseudolithoderma roscoffensis.* Five-day-old germlings: Rég. 2. ×770.

FIG. 29. *Ralfsia clavata.* Two-week-old germling at early stage of disc formation: Rég. 2. ×470.

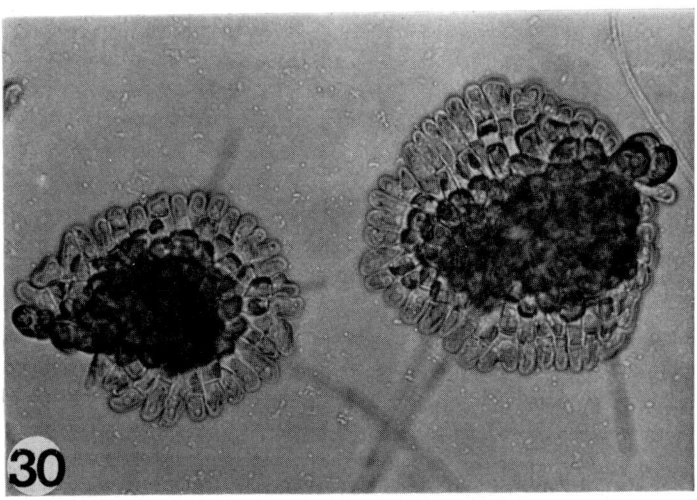

FIG. 30. *R. clavata.* Two-week-old discoid germlings: Rég. 2. ×365.

was produced within 1–2 days. This grew out over the substrate and divided transversely to form a short uniseriate filament (Fig. 25). In the majority of *R. verrucosa* germlings, however, after production of the germ-tube (which enlarged distally), the contents of the spore cell migrated to the distal end of the tube and a cross wall was laid down to seal off the now empty spore cell (Figs 26, 27). From this stage, all algal germlings followed one of two main pathways of development.

In many cultured germlings, the filament remained fairly short and then produced numerous lateral branches which cohered to form a circular disc with

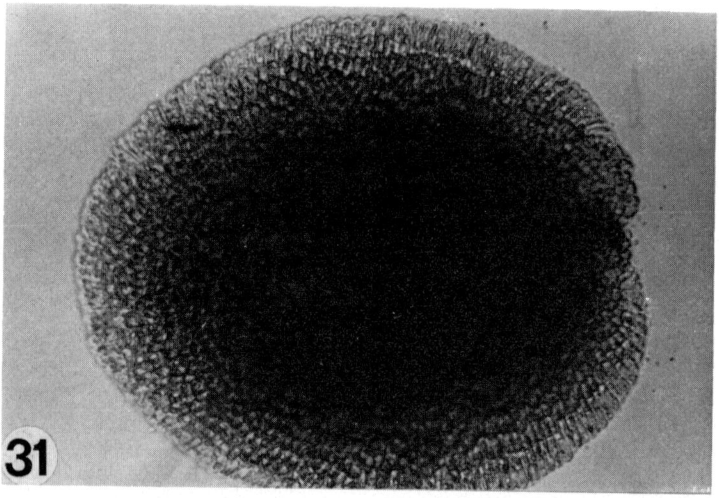

FIG. 31. *Microspongium gelatinosum*. Four-week-old disc. Rég. 1. × 178.

an outer row of apical cells (Figs 28–30). In *R. verrucosa* an initial rosette of four cells was often formed by direct cleavage of the terminally enlarged germ-tube (Fig. 27). The disc then expanded by transverse divisions of the outer apical cells, while oblique divisions formed branches which allowed increase in its circumference (Fig. 31). The discoidal habit of the germlings was maintained by the synchronous divisions of the marginal apical cells and the cohesion of the constituent filaments by the encompassing mucilaginous secretions (see Fig. 6).

Coupled with the horizontal spread of the disc, vertical elongation of the basal cells followed by transverse divisions resulted in the production of the polystromatic crustose thallus. In some species, for example *R. clavata* and *P. roscoffensis*, quite extensive monostromatic discs were initially produced before crust development commenced whilst in others, for example *R. verrucosa*, crust formation occurred immediately behind the growing margin at an early stage of germling development.

Instead of dividing periclinally to form young disc filaments, the germlings very often increased in length and by a series of transverse divisions developed

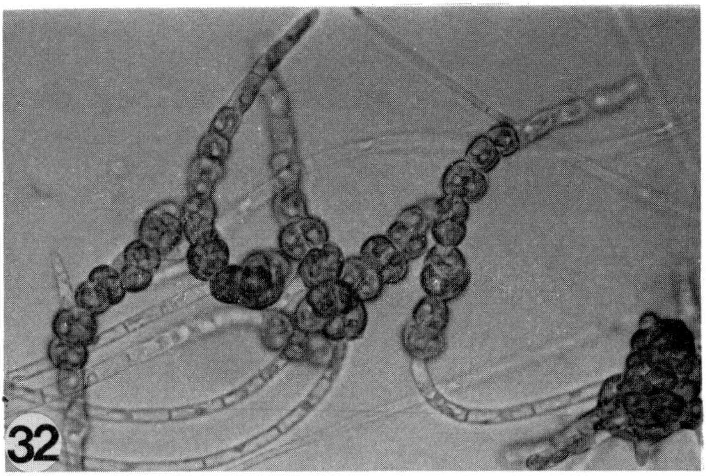

FIG. 32. *M. gelatinosum*. Ten-day-old germlings at early stage of knot-filament formation: Rég. 2. ×380.

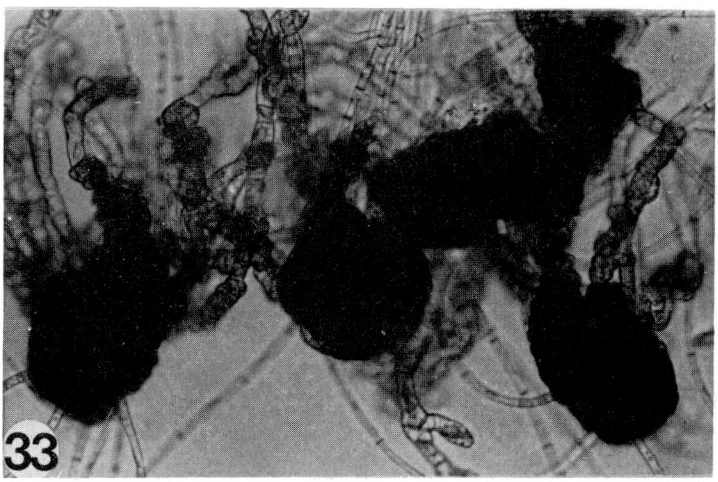

FIG. 33. *M. gelatinosum*. Four-week-old germlings forming knot-filaments: Rég. 1. ×183.

into freely branched, filamentous thalli. Sometimes the thalli remained predominantly filamentous throughout their life, although usually at an early stage the majority divided to form discontinuous parenchymatous knots of tissue along their length (Figs 32, 33). These "knot-filaments" were usually initiated in the centre part of the thallus by a series of periclinal divisions in the main axial cells and then gradually extended outwards along the length of the branch filaments.

Upon squashing, the knot-filaments were seen to consist essentially of outwardly radiating threads of cells which were laterally coalesced. The threads usually bifurcated at the base at least once, to allow expansion of the thallus diameter. This basically pseudoparenchymatous mode of construction would appear to be a three-dimensional representation of the multi-stratose crusts.

The ability of the two morphological expressions, crusts and knot-filaments, to interconvert in culture underlines their essential homology. Individual filaments were often observed to arise from both surface and edge crust cells and to develop into branched filamentous thalli which, subsequently, could form knot-filaments, whereas terminal cells of filaments and knot-filaments in firm contact with the substratum often developed into small discs which later became multi-stratose. Probably the degree of contact between filament cell and substrate determined the subsequent pathway of development.

Close examination of the cultured crusts and even the knot-filaments revealed that they generally displayed morphological, anatomical and cytological features which were similar to those of the parental crust material. The soft crusts of cultured *Petroderma* and *M. gelatinosum* possessed weakly coalesced and gelatinous erect filaments, whilst in cultured *R. clavata*, *R. spongiocarpa* and especially *R. verrucosa* and *P. roscoffensis*, the erect filaments were more tightly bound together. Marginal sections of young *R. verrucosa* crusts also revealed curved filaments arising from the basal layers, not unlike that of field collected material, whilst cultured crusts of all the other Ralfsiaceae members revealed a discrete apical cell and vertical vegetative filaments.

A number of germlings became fertile in culture. Unilocular sporangia, for instance, were formed on knot-filaments and crusts in cultures of *P. maculiforme*, *M. gelatinosum*, *R. clavata* and *R. spongiocarpa*. With the exception of *P. maculiforme*, these sporangia developed on germlings which had themselves been derived from unispores released from field-collected material. The manner of formation of the sporangia was also characteristic of the parental material, with the sporangia lateral at the base of multicellular paraphyses. Only in *R. spongiocarpa* were produced additional germlings which possessed abundant unilocular sporangia terminal on the erect filaments. In *P. maculiforme* only the plurispore-derived cultures became fertile, with the production of terminal unilocular sporangia.

Probably the most significant aspect of the development of the spores, however, was the production, in some cultures, of erect, parenchymatous

FIG. 34. *Ralfsia clavata*. *Petalonia*-like blades arising from 4-week-old cultured disc: Rég. 2. × 180.

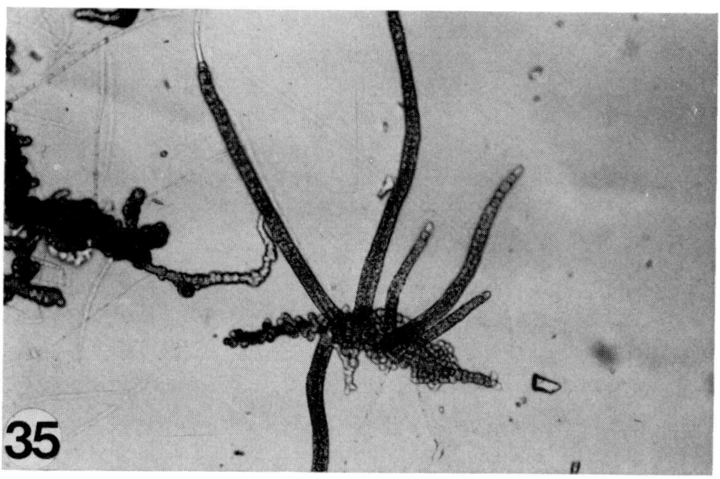

FIG. 35. *M. gelatinosum*. Young *Scytosiphon*-like blades arising from a small 4-week-old cultured disc. Rég. 1. × 48.

blades characteristic of the family Scytosiphonaceae. Germlings of *R. clavata* produced broad, flattened blades identical to those of the species *Petalonia fascia* (Fig. 34), whilst those of *M. gelatinosum* produced narrow, terete blades identical to those of the species *Scytosiphon lomentaria* (Fig. 35). The relationship between these erect blades and the "basal" growths (i.e. branched filaments, knot-filaments, discs and crusts) from which they were derived appeared to be entirely facultative and therefore not governed by any sexual phenomena. Initiation of both blade types appeared to be enhanced under Régime 1 conditions (i.e. winter-like) whilst basal growth, particularly crustose development, was enhanced under Régime 2 conditions (i.e. summer-like).

DISCUSSION

The present paper is concerned with a preliminary investigation carried out by the author on the brown crustose algae of the family Ralfsiaceae in the north Atlantic. Although the crusts appear to form a fairly obvious grouping with their basically similar habit and mode of construction, i.e. prostrate and pseudoparenchymatous respectively, they show considerable variation in the organization of the vegetative filaments, in the structure of the reproductive system and in the spore germination and life-history process. Pronounced differences were observed, for instance, in the strength of cohesion of the erect vegetative filaments, this being relatively weak in crusts of *Petroderma* and *Microspongium* but quite strong in those of *Ralfsia* and *Pseudolithoderma*. Also variable is the manner of initiation of the erect filaments, these arising vertically from a monostromatic base in the majority of species but curving upwards from a multi-layered base in *R. verrucosa* and *R. fungiformis*. Other notable differences include the number of plastids per cell (single in the majority of these algae, but several in *Pseudolithoderma*), the morphology and position of the sporangia (lateral unilocular sporangia in *Ralfsia*; terminal in the other crusts) and the type of life-history operating (monophasic, crustose only, in the majority of algae; diphasic, crustose and erect blade, in *R. clavata* and *M. gelatinosum*). This range of cytological, anatomical and developmental features would certainly appear to undermine consideration of the family Ralfsiaceae as a natural phylogenetic group.

However, despite the range in thallus structure and life-history development within the family Ralfsiaceae there would appear to be little scope for the subdivision of the family into natural groupings. Although classificatory systems have been proposed which involve subdivision of the crusts into two and even three families (usually Ralfsiaceae, Lithodermataceae and Nemodermataceae), they possess little merit and have not generally received widespread acceptance.

Certainly a re-adoption of the two families Lithodermataceae and Ralfsiaceae to encompass the north Atlantic crusts would contradict the findings of the present paper. The Ralfsiaceae, containing the genus *Ralfsia* only, are characterized generally by the presence of a single plastid per cell, laterally positioned unilocular sporangia and intercalary plurilocular sporangia. The Lithodermataceae on the other hand, which would contain all the remaining genera, are characterized by the occurrence of several plastids per cell with terminal unilocular and plurilocular sporangia. The present study has revealed a number of unclassifiable crusts which show inconsistencies in the combination of characters required for entry to either one of these two families. For example, a number of Lithodermataceae crusts with terminal sporangia, such as *Petroderma*, *Sorapion* and *Symphyocarpus*, all possess only a single plastid in each cell. Further evidence for the unification of the families is provided by the production of terminal unilocular sporangia, which are unaccompanied by paraphyses, in cultures of *R. spongiocarpa*.

From a phylogenetic viewpoint the crusts probably represent a number of divergent lines all emanating from a myrionematoid type of ancestor. The more rigid crusts probably evolved from the soft crusts by changes in the cell wall chemistry of the erect filaments resulting in increased lateral cohesion. For instance *Pseudolithoderma* and *Sorapion* probably evolved directly from a *Symphyocarpus*/*Petroderma*-like ancestor. The occurrence of relatively soft *Pseudolithoderma* crusts such as *P. paradoxum* (Sears and Wilce, 1973) and *P. rosenvingii* (Waern, 1949) supports such an evolutionary line. Incorporated into this phylogenetic scheme would be the development of increased numbers of plastids per cell and the formation of distinct sori containing the reproductive organs. The Ralfsiaceae members, on the other hand, could have evolved along a *Myrionema*/*Compsonema*/*Microspongium*/*Ralfsia* line, involving increased development and differentiation of the vegetative thallus, increased lateral cohesion of the erect filaments and the formation of distinct sori containing the reproductive sporangia. Alternatively an all-encompassing *Petroderma*/*Ralfsia*/*Pseudolithoderma* line may well have occurred; support for this comes from the development of *Sorapion*-like crusts with terminal unilocular sporangia in cultures of *R. spongiocarpa*. All the remaining soft crust algae, the majority of which occur in warmer waters, might well represent an independent evolutionary line, also starting from a *Petroderma*-like ancestor. A possible sequence could be *Petroderma*/*Hapalospongidium*/*Mesospora*/*Basispora*/*Nemoderma*, involving development of increased numbers of plastids, terminal to lateral/intercalary positioned unilocular sporangia, and isogamy to anisogamy.

The range in thallus structure and life-history development within the genus

Ralfsia merits particular consideration. With the exception of *R. verrucosa* and *R. fungiformis*, all the north Atlantic species investigated possess moderately coalesced vegetative filaments which arise vertically or slightly curved from a single-celled discoid base, whilst the sporangia are produced in indefinitely-extending sori. *R. verrucosa* and *R. fungiformis*, however, possess strongly coalesced filaments which characteristically curve up from a multi-stratified basal tissue with the sporangia being produced in more delimited, somewhat bulbous sori. These interspecific variations were noted for the British representatives as early as 1889 by Batters (1890) who placed *R. verrucosa* on its own within the subgenus *Euralfsia* and all the other species within a separate subgenus *Stragularia* (Strömfelt, 1886). Newton (1931) and Loiseaux (1968) disagreed and included *R. clavata* (and *R. spongiocarpa* in Newton, 1931) within their concept of *Euralfsia*. These fundamental differences in vegetative construction between the *Ralfsia* species would seem to be sufficient for generic separation and therefore for the acceptance of *Stragularia* as an autonomous genus (see Hamel, 1939). A further subdivision might also be justified within the *Euralfsia* group itself in view of the greater complexity of vegetative construction in *R. fungiformis* than in *R. verrucosa*.

The heterogeneous composition of the genus *Ralfsia* is further substantiated by the results of the present culture studies. In *R. verrucosa*, for example, a different early germination pattern was observed when compared with that of the other *Ralfsia* spp.; germ-tube production was followed by evacuation of the spore cell contents, whereas in *R. clavata* and *R. spongiocarpa* the spore cell retained its protoplasm and appeared indistinguishable from the other subsequently formed basal cells. More significant, however, was the occurrence of a monophasic life-history in *R. verrucosa*, whereas a diphasic one, incorporating both a crustose and scytosiphonoid thallus, was present in the *Stragularia*-type species *R. clavata* and the closely related *M. gelatinosum*. These findings contribute to the increasing number of reports (see especially Nakamura, 1965; Tatewaki, 1966; Wynne, 1969; Edelstein *et al.*, 1970; Roeleveld *et al.*, 1974) which vindicate Wynne's (1969) hypothesis that *Stragularia* has nothing to do with *Ralfsia sensu stricto* (i.e. *Euralfsia* spp.) but comprises an assemblage of crustose entities that are involved in the life-cycles of members of the family Scytosiphonaceae. Present studies of *R. spongiocarpa*, however, failed to detect a scytosiphonaceous phase in its life-history; since that species falls within *Stragularia*, the latter may still be heterogeneous and require further subdivision (see Loiseaux, 1968, and Hollenberg's, 1969, cultural study of strains of *R. clavata* and *R. confusa* respectively).

The evidence from the culture studies suggests that the life-histories of the other crusts studied, *P. maculiforme*, *P. roscoffensis* and *R. verrucosa*, are probably similar to that of *R. spongiocarpa* and are of the "direct" type. There was no indication of a sexual process during the settlement and development of the plurispores, even in *P. maculiforme* and *R. verrucosa* which have separate plurilocular sporangiate thalli, suggesting that they were not gametophytic in origin. Further, in those crusts with unilocular sporangia it seems likely that the spores develop apomeiotically. These results are in agreement with previous studies on these species (see Wynne, 1969 *re P. maculiforme*, Loiseaux, 1968 and Edelstein *et al.*, 1971 *re R. verrucosa* and Loiseaux, 1968 *re P. roscoffensis*).

The present study would appear to offer very little support for Nakamura's (1972) proposed new order Ralfsiales, which incorporates the three families Lithodermataceae, Nemodermataceae and Ralfsiaceae. This order is based upon three principal characters—the possession of an alleged *Ectocarpus* (i.e. isomorphic) life-history, a discoid type of germling development and the presence of a single parietal plate-like plastid which lacks a pyrenoid. Known facts regarding the structure and life-history of the north Atlantic species, including the possession of a "direct" apomeiotic cycle of development and the occurrence of several discoid plastids in the cells of some genera, e.g. *Pseudolithoderma*, tend to refute these proposals. The presence of pyrenoids has also been determined in some crusts such as *R. verrucosa* (Fletcher, unpublished). Although a discoid type of development was consistently observed in all the crusts, that germination pattern is not confined to this group but can also occur in many members of the family Myrionemataceae and the Scytosiphonaceae.

ACKNOWLEDGEMENTS

Sincere appreciation and gratitude is extended to Dr David Irvine for considerable guidance and help throughout this study. I also gratefully acknowledge the use of the Cambridge 180 Stereoscan-electron microscope in the Department of Biological Sciences, Portsmouth Polytechnic, and in particular for technical assistance by Mrs E. Boulter. I would also like to thank Mr K. Purdy and Mr A. Hawton for photographic assistance. The greater part of this work was carried out during the tenure of a Research Assistantship in the Department of Biology and Geology, Polytechnic of North London, for which appreciation is expressed.

REFERENCES

Batters, E. A. L. (1890) ["1889"]. A list of the marine algae of Berwick-on-Tweed. *Hist. Berwicksh. Nat. Club* **12**, 221–392.

Batters, E. A. L. (1896). New or critical British marine algae. *J. Bot., Lond.* **34**, 384–390.

Edelstein, T., Chen, L.-C. M. and McLachlan, J. (1970). The life cycle of *Ralfsia clavata* and *R. borneti*. *Can. J. Bot.* **48**, 527–531.

Edelstein, T., Chen, L.-C. M. and McLachlan, J. (1971). On the life histories of some brown algae from eastern Canada. *Can. J. Bot.* **49**, 1247–1251.

Fletcher, R. L. (1976). Post-germination attachment mechanisms in marine fouling algae. *In* "Proc. 3rd Int. Biodeg. Symp." (J. M. Sharpley and A. M. Kaplan, eds), pp. 443–464. Applied Science Publishers Ltd, London.

Fletcher, R. L. (1978). Observations on the ecology and life history of *Ralfsia spongiocarpa* Batt. *Proc. int. Seaweed Symp.* **8** (in press).

Hamel, G. (1939). "Phéophycées de France", fasc. V, pp. 337–432+I–XLVII, Paris.

Hammerling, J. (1931). Entwicklung und Formbildungsvermögen von *Acetabularia mediterranea. Biol. Zbl.* **51**, 633–647.

Hollenberg, G. J. (1969). An account of the Ralfsiaceae (Phaeophyta) of California. *J. Phycol.* **5**, 290–301.

Loiseaux, S. (1968). Recherches sur les cycles de développement des Myrionématacées (Phéophycées). III. Tribu des Ralfsiées. IV. Conclusions générales. *Rev. gén. Bot.* **75**, 295–318.

Lund, S. (1959). The marine algae of east Greenland. I. Taxonomical Part. *Meddr Grønland* **156**, 1–247.

Nakamura, Y. (1965). Development of zoospores in *Ralfsia*-like thallus, with special reference to the life cycle of the Scytosiphonales. *Bot. Mag., Tokyo* **78**, 109–110.

Nakamura, Y. (1972). A proposal on the classification of the Phaeophyta. *In* "Contributions to the Systematics of Marine Algae of the North Pacific" (I. A. Abbott and M. Kurogi, eds), pp. 147–156. Japanese Soc. Phycology, Kobe.

Newton, L. (1931). "A Handbook of the British Seaweeds." British Museum (Natural History), London.

Parke, M. and Dixon, P. S. (1976). Check-list of British marine algae—third revision. *J. mar. biol. Ass. U.K.* **56**, 527–584.

Roeleveld, J. G., Duisterhof, M. and Vroman, M. (1974). On the year cycle of *Petalonia fascia* in the Netherlands. *Netherlands J. Sea Res.* **8**, 410–426.

Sears, J. R. and Wilce, R. T. (1973). Sublittoral benthic marine algae of southern Cape Cod and adjacent islands: *Pseudolithoderma paradoxum* sp. nov. (Ralfsiaceae, Ectocarpales). *Phycologia* **12**, 75–82.

Stosch, H. A. von (1964). Wirkungen von Jod und Arsenit auf Meeresalgen in Kultur. *Proc. int. Seaweed Symp.* **4**, 142–150.

Strömfelt, H. F. G. (1886). Einige für die Wissenschaft neue Meeresalgen aus Island. *Bot. Zbl.* **26**, 172–173.

Tatewaki, M. (1966). Formation of a crustaceous sporophyte with unilocular sporangia in *Scytosiphon lomentaria. Phycologia* **6**, 62–66.

Waern, M. (1949). Remarks on some Swedish *Lithoderma. Svensk. bot. Tidskr.* **43**, 633–670.

Wynne, M. J. (1969). Life history and systematic studies of some Pacific North American Phaeophyceae (brown algae). *Univ. Calif. Publs Bot.* **50**, 1–88.

17 | Active Speciation in the Taxonomy of the Genus *Cystoseira* C. Ag.

M. ROBERTS

Department of Botany, University of Durham, England

Abstract: Fifty-five species are presently listed as belonging to the genus *Cystoseira*, several very variable in their external morphology. Speciation appears still to be operating and species fall into a small number of fairly clearly defined groups, each group with Mediterranean and Oceanic components. Three groups containing British representatives are examined here in some detail; those containing *C. foeniculacea* (and *C. humilis*), *C. tamariscifolia* and *C. nodicaulis* respectively. *C. baccata*, alone of British species, shows no similarities to any other species. The 14 species not occurring in the Mediterranean or Atlantic appear, where sufficiently known to be assigned to a group, to be associated with the *C. foeniculacea* group. Differences between species within these groups may be slight, often involving aspects of growth and differentiation. There is considerable parallelism between species groups in the variation of some features. More detailed information concerning structure, reproduction and geographical distribution of individual species is needed before relationships within the genus can be better understood.

INTRODUCTION

Cystoseira C. Ag. is a genus of worldwide distribution with about 80% of the species occurring along the Mediterranean and adjoining Atlantic coasts. Of the remaining species, four occur off the Pacific coast of north America, one of these also occurring in the Sea of Okhotsk; one species is endemic to the Indian subcontinent; two others occur in the Red Sea, Persian Gulf, along the coast of east Africa as far as Mozambique, and in the Caribbean or off Florida; one of the latter also occurs in Indonesia, New Caledonia and southern and eastern Australia. Four or five species occur in the Sea of Japan,* and specimens have also been recorded from the Sargasso Sea. Only five species are found in the

Systematics Association Special Volume No. 10, "Modern Approaches to the Taxonomy of Red and Brown Algae", edited by D. E. G. Irvine and J. H. Price, 1978, pp. 399–422, Academic Press, London and New York.

* Some of Yendo's (1907) species need re-investigation.

British Isles; all of these are at the northern limit of their range and occur mainly on the south coast, although some of them extend up the west coasts of England, Wales and Ireland, with sporadic records of their occurrence on the offshore islands of Scotland as far north as Barra. Most British phycologists are thus familiar with only a small sample of the genus, but this is sufficient to give some insight into the problems of the genus as a whole.

THE CONCEPT OF THE GENUS

The genus *Cystoseira*, as originally created by C. Agardh (1820), contained 37 species.* In the period since that date at least 114 entities have been assigned to the genus; 55 of these remain in the genus as now understood, including 14 of Agardh's original 37 species. Of these only nine have undergone no name change and one of the surviving names is illegitimate.

TABLE I. Genera of the Cystoseiraceae. Apical cell three-sided in T.S.; symmetry radial or bilateral; conceptacles in ramuli of lateral branch systems, either scattered on the thallus or aggregated into receptacles

1. *Acrocarpia*	6. *Caulocystis*	11. *Hormophysa*
2. *Acystis*	7. *Coccophora*	12. *Landsburgia*
3. *Bifurcaria*	8. *Cystophora* (incl. *Platylobium*)	13. *Myriodesma*
4. *Bifurcariopsis*	9. *Cystoseira* (incl. *Cystophyllum* and *Myagropsis*)	14. *Scaberia*
5. *Carpoglossum* (incl. *Platythalia*)	10. *Halidrys*	15. *Stolonophora*

N.B. The author separates *Marginariella, Phyllospora, Scytothalia, Seirococcus, Cystosphaera* and possibly also *Axillariella*, into a separate family for which Schmitz's name Phyllosporaceae, in a slightly amended sense, is the most appropriate.

C. Agardh's genus was soon recognized to comprise several genera, and distinctive features of morphology proved acceptable criteria for the separation of such genera as *Hormosira, Bifurcaria, Carpoglossum, Landsburgia, Cystophora* (Womersley, 1964), *Stolonophora* (Nizamuddin, 1969) and *Hormophysa* (Papenfuss, 1968) (Table I). J. Agardh (1848) removed a number of species into *Cystophyllum* but his diagnosis has not been shown by subsequent workers to contain any criteria which do not also refer to *Cystoseira*. Although Fensholt

* Several new species have now been added by Gerloff and Nizamuddin (1975b, 1976).

(1955) re-examined the situation and suggested new criteria for separating two of Yendo's (1907) species of *Cystophyllum* as *Myagropsis* Kützing (1843), the boundaries between the two remain indistinct and the present author prefers to retain all species of *Cystophyllum* in *Cystoseira*. The number of species remaining in the genus has been further reduced as awareness of the extent of variability within the species has increased.

Surprisingly, perhaps, there has been no recent reappraisal of anatomical, morphological or reproductive features which might be used to characterize the genus as a whole. The classic study by Valiante (1883) of ten species found in the Bay of Naples* suggests a line which might be taken, namely the early development of the sporeling with the characteristic mode of origin of the definitive axis—also reported by Mairh and Krishnamurthy (1968) in *C. indica* (Thivy et Doshi) Mairh—and the ensuing patterns of growth during ontogeny (cf. Womersley, 1964, in *Cystophora*), together with the often remarkable similarity of the juvenile plants during their first few months of growth.

Detailed anatomical accounts are few and where they have been carried out on a comparative basis they usually cover too few species for any general principles to emerge (Roy, 1938; Fensholt, 1955; Guern, 1962; Johnson, 1966; Roberts, 1966, 1967a, b, 1968, 1970, 1977). The genus is therefore still only imperfectly known, and this in turn leads to difficulties in selection of suitable criteria for species separation.

There is a wealth of data concerning recognition in the field, morphological variation and distribution; although these accompany related attempts to sort out species limits (Sauvageau, 1912a, b, 1920; Hamel, 1939; Ercegović, 1952), many uncertainties remain. The difficulties that collectors have experienced in identifying their specimens are well illustrated by such statements as that of Sauvageau (1920) that "On sait que le *C. montagnei* de J. Agardh n'est pas celui que Montagne décrit sous le nom; c'était une espèce composite correspondant à la plante de Montagne, au *C. adriatica* et au *C. spinosa* . . . le *selaginoides* cité par Picconé dans plusieurs Mémoires est le *C. spinosa*". More recently, the very long footnote of Papenfuss (1967), still referring to J. Agardh's concept of *C. montagnei*, suggests four entities which he might have confused. Some of this confusion may well have arisen from inadequacies of the specimens and the lack of appreciation of the range of variability within the species. Sauvageau

* viz. *C. fimbriata* (Desf.) Bory (*sub C. abrotanifolia* J. Ag.); *C. barbata* C. Ag.; *C. discors* C. Ag.; *C. hoppii* C. Ag.; *C. brachycarpa* (*sub C. crinita* Val.); *C. mediterranea* Sauv. (*sub C. amentacea*); *C. sauvageaueana* Hamel (*sub C. selaginoides*); *C. spinosa* Sauv. (*sub C. ericamarina* Nacc. and *C. montagnei* J. Ag.); *C. zosteroides* C. Ag. (*sub C. opuntioides*); *C. fucoides* Erc. (*sub C. dubia* Val.).

Table II. The principal species of *Cystoseira* arranged in groups (tentative) on the basis of morphology

---

Group 1

1. *C. foeniculacea*
2. "*C. discors*" *sensu* Sauvageau
3. *C. fimbriata*
4. *C. canariensis*
5. *C. brachycarpa*
6. *C. pumila*
7. *C. humilis*
8. *C. trinodis*
9. *C. osmundacea*
10. *C. geminata*
11. *C. myrica*
12. *C. indica*
13. *C. crassipes*
14. *C. hakodatense*
15. *C. sisymbrioides*
16. *C. turneri* (Yendo) ***comb. nov.***[a]

Group 2

17. *C. tamariscifolia*
18. *C. mediterranea*
19. *C. balearica*
20. *C. caespitosa*
21. *C. Stricta*
22. *C. spicata*
23. *C. corniculatq*
24. *C. amentacea*
25. *C. sedoides*

Group 3

26. *C. nodicaulis*
27. *C. usneoides*
28. *C. sauvageaueana*
29. *C. gibraltarica* (includes *C. mauritanica*)
30. *C. zosteroides*
31. *C. spinosa*
32. *C. squarrosa*
33. *C. adriatica*
34. *C. platyclada*
35. *C. jabukae*
36. *C. platyramosa*
37. *C. elegans*

Group 4

38. *C. barbata*
39. *C. bosphorica*
40. *C. hoppii*
41. *C. crinita*
42. *C. crinitophylla*

Group 5

43. *C. baccata*

The following species are too poorly known for them to be placed in any group.

44. *C. abies-marina* C. Ag.
45. *C. selaginoides* Nacc.
46. *C. hoppii* Valiante
47. *C. neglecta* Setch. et Gardn.
48. *C. pelagosae* Erc.
49. *C. sedoides* C. Ag.
50. *C. setchellii* Gardn.
51. *C. sonderi* Piccone
52. *C. thunbergii* (Mert.) C. Ag.
53. *C. tingitana* Sauvageau
54. *C. filifolia* C. Ag.
55. "*Cystophyllum caespitosum*" Yendo

---

[a] Basionym: *Cystophyllum turneri* Yendo, **1907**, 40–42.

(1912b) pointed out that it is quite possible to come to entirely different conclusions about the relationship between *C. discors* and *C. foeniculacea* if one examines only fragmentary herbarium specimens on the one hand and entire plants in the field on the other.

It is also possible that some of these uncertainties occur because, as suggested by Ercegović (1952), the genus is still in the process of active speciation and that some of the particularly polymorphic species will eventually evolve into new entities.

The present paper aims to look at several of the areas of uncertainty, especially those which involve, or impinge on, species occurring along British shores, and to deal with them in the context of groups of species (Table II) which show similarities in respect of a number of basic features. In this way it may be possible to assess whether the idea of active speciation is valid. The account is primarily concerned with the degree of similarity between species and not with affinities or phylogenetic considerations. Species are thus regarded in the sense of groups of plants which can be consistently given names and which can be consistently recognized by their behaviour (i.e. response to particular conditions); criteria such as hybridization are not considered as no information is available.

TAXONOMIC PROBLEMS IN SELECTED GROUPS OF SPECIES

Group 1. C. foeniculacea *(L.) Grev.*, C. humilis *Schousboe in Kütz.*

C. foeniculacea (L.) Grev. is the type-species of the genus; therefore it is important to resolve problems concerning this species. The problems are of two kinds, one involving a consideration of the criteria distinguishing it from *C. humilis* Schousboe in Kützing (1860) (incl. *C. myriophylloides* Sauv.), the other a major one involving the concept of the species itself.

C. foeniculacea and *C. humilis*, together with the Mediterranean species *C. fimbriata* (Desf.) Bory,* *C. discors* C. Ag. *sensu* Sauvageau,† *C. brachycarpa* J. Ag. and *C. canariensis* Sauv. are of caespitose habit with principal axes bearing prominent circular scars. When mature, the plants possess lateral branch systems which are subdivided pinnately or irregularly dichotomously to a varying degree; they do not at any time bear laterals of limited growth of the type customarily known as leaves. They may, however, bear lateral spinous outgrowths, which Sauvageau considered to be morphologically different structures. Receptacles form seasonally in the ultimate ramuli. They are rarely

* *C. compressa* (Esper) Gerloff and Nizamuddin (1975a).

† *C. ercegovicii* Giaccone nom. nov. (Giaccone and Bruni, 1973).

extensive and vary considerably in appearance but are always recognizably derived from the slightly swollen ultimate ramuli of the branch system.

Sauvageau (1911b) pointed out that these species possess antherozoids which lack eyespots and that after discharge their oogonia are retained at the surface of the receptacle, until after fertilization, by means of solid mucilaginous stalks formed from the middle layers of the oogonial wall. This retention was seen as related to the lesser capacity to swim and to orientate themselves shown by these antherozoids compared with those possessing eyespots. They are also produced in much smaller numbers and on much simpler antheridial hairs than are pigmented antherozoids. A basal cluster of hairs in the conceptacle may be a further feature of this group, the conceptacle itself being relatively small and containing only a small number of basally broad oogonia (Roberts, 1968). Most of the group have been recorded as possessing the structures referred to by Sauvageau (1911a) as "pédicelles cryptifères", which are interpreted as modified branch apices. Guern (1962) has shown that the initial segmentations of the oospore follow the same sequence throughout this group. The work of Roy (1938) and Roberts (1966) has indicated that group members may possess the same type of nuclear structure, namely finely granular in the meristoderm and cortex but containing chromocentres in the medulla; both forms of *C. discors* from the Adriatic also possess this type of nuclear structure.

C. myriophylloides was created by Sauvageau (1912b) for plants of Atlantic distribution, which he believed had been included by some authors under *C. discors* C. Ag.; at the same time he proposed the re-establishment of *C. foeniculacea* (L.) Grev. for plants (previously placed in *C. discors*) found growing off the coasts of the British Isles and the Atlantic coast of France. He reserved *C. discors* for Mediterranean plants which he believed to be of a distinctive form. Sauvageau also questioned whether *C. humilis* from West Africa might not be a dwarf form of this species (*C. myriophylloides*), a doubt later repeated by Hamel (1939) and Dangeard (1949). Ardré (1970) found both forms off the coast of Portugal, the specimens north of the Tagus being larger than the ones to the south. She suggested the reduction of *C. myriophylloides* to the synonymy of *C. humilis* which Levring effected (1976) and Price and John (in Price *et al.*, 1978) have utilized the name *myriophylloides* at varietal level within *C. humilis*. *C. canariensis* also seems to merit consideration here.

There is sufficient difference in form at the latitudinal extremes of the range of *C. humilis* for it to have been originally recorded as two species, but this variation can be interpreted as clinal and intraspecific. On the Channel coasts, where the range of the larger plants of *C. humilis* overlaps with *C. foeniculacea*, the difference between the two species may be so slight as to make determination difficult. Most of these differences are of degree only, such as extent of branching of laterals, overall shape of the plant, shape of aerocysts and abundance of cryptostomata. The differences, however, appear to be genetic and non-adaptive—since the two species may be found growing together in the same pool—and so are of a different type from the differences between the plants of *C. humilis* at their latitudinal extremes, which may be induced ecologically.

The most certain distinction between the two, according to Sauvageau (1912b), is that *C. humilis* never produces bilateral (i.e. foliaceous) lateral branch systems after the primordial phase. The sporeling of both species produces several delicate, cylindrical primordial laterals followed by several distichously-arising, flattened ones; at this stage the sporelings of the two species are barely distinguishable (Sauvageau, 1912b). After this primordial stage, *C. humilis* produces only radially arranged, filiform lateral branch systems. In *C. foeniculacea*, on the other hand, juvenile plants with an axis about 10 mm long may frequently, although not invariably, produce several foliaceous laterals up to 100 mm long. Foliaceous laterals sometimes occur on older plants and, at maturity, their distal ends grow out into filiform branches which terminate in receptacles (Roberts, 1968). Mature foliaceous plants are rare in England and have not been recorded at all from Ireland. Insufficient is known about the occurrence of foliaceous plants on the shore to decide whether they are genetically distinct, although this is an important point in relation to the current concept of the species, still largely that of Sauvageau (1912b).

C. foeniculacea is usually considered to be Atlantic in distribution and not to extend into the Mediterranean. A similar species, *C. discors*, is widespread within the Mediterranean and occurs sporadically in the Atlantic, having been reported near the Canaries. This species is usually characterized by the possession of foliaceous lateral branch systems throughout its life-history; these form distal, filiform fertile branches during the reproductive phase. The reproductive branches closely resemble those found in *C. foeniculacea* in the British Isles and this resemblance is particularly marked in populations of *C. discors* off the coast of Algeria.

Ercegović (1952) made a detailed study of the form and distribution of 15 species of *Cystoseira*, including *C. discors*, from 151 stations in the Adriatic, mainly from Dubrovnik in the south to Rovinj in the north, with especial interest shown in the series of offshore islands just south of Split (Brac, Hvar, Korčula and Biševo) leading out to the mid-Adriatic isolated rocky islets of Palagruž to the south and Jabuka slightly to the north. Ten stations only were visited in the southern region of mainland Jugoslavia (between Dubrovnik and Ulcinj) and one other at Cape Glossa (Strait of Otranto). *C. discors* is widely distributed in the Adriatic, and Ercegović found a definite correlation between the form of the lateral branch systems and the depths at which the plants were growing. Thus, at Jabuka, the plants at 1 m had primary laterals 150 mm long and 25 mm broad. These were elliptical in transverse section and bore elliptical secondaries; from a depth of 100 m they were 600 mm long, 40 mm broad and foliaceous in form, bearing entirely foliaceous secondaries. He did not give figures for an intermediate depth at this locality, but at the islet of Palagruž the comparable laterals of plants from a depth of 30 m were 350 mm long, 40 mm broad, foliaceous and with foliaceous secondaries. Considerable information was presented showing a consistent relationship between depth and flattening, although the degree of flattening did not increase evenly with depth. Further, the surface-living population showed progressive changes as the open sea was approached, with reduction in the flattened laterals and their replacement by cylindrical ones. Off the island of Vis, Ercegović recorded a surface-living form which approached very close to the normal form of *C. foeniculacea* found off the British Isles. This he designated "Geographic forma *tenuiramosa*".

Ercegović considered the increase in flattening with depth to be a direct individual response to diminishing illumination. He also considered that the production of cylindrical ramuli in the ontogeny of the individual was basically genetic, occurring only after a certain minimum size had been achieved. A reduction in the number of foliaceous laterals could occur, then, only if some genetic factor operated to reduce this minimum size.

Ercegović also found that, although the littoral populations possessed long spines on both axes and primary branches, in offshore populations the spines were shorter and generally restricted to the basal regions of the primary laterals, as in *C. foeniculacea.*

This variation in the Adriatic populations apparently encompasses the full range of both *C. discors* and *C. foeniculacea* and makes it necessary, once again (see Sauvageau, 1912b) to consider the relationship between the two. Both species vary considerably in their aspect at different times of the year. Both are eurythermal in their requirements and do not have a period of complete dormancy, but they both tend to have a period of maximum growth which concludes with the shedding of the gametes, followed by a period of less active growth when the branches produced differ in size, form and aerocyst production from those of the preceding period. It is necessary to make comparisons between the two on the basis of comparable growth periods rather than on their appearance in particular months or seasons.

Sauvageau considered the two to be distinct species, and suggested (1912a) that the Mediterranean *C. discors* had arisen from the oceanic *C. foeniculacea* as the species had spread eastwards from the Atlantic into the Mediterranean and had adapted to the new conditions. Ercegović (1952) saw the variability of *C. discors* in the Adriatic as resulting from a polymorphic species in the process of yielding new segregates—a theory which is just an extension of that of Sauvageau. Both these opinions were based entirely on field observations.

Sauvageau observed that *C. discors* from the Mediterranean coast regularly possessed foliaceous laterals, in contrast with *C. foeniculacea* from the Atlantic coast (Biarritz area) in which such laterals only occurred occasionally. The plants from the south coast of England agree with those of the French Atlantic coast in this respect, but plants with foliaceous laterals have never been reported from the west coast of Ireland (de Valéra, personal communication). The situation on the Channel coasts appears to be intermediate in this respect between that in the Mediterranean and that in Ireland; both extremes are found in the Adriatic.

Sauvageau also distinguished *C. discors* on the basis of the numerous "pédicelles cryptifères" ("crypt"-bearing laterals) which it bore, stating these to be lacking in *C. foeniculacea*, and of the presence of much more numerous and longer spines in *C. discors* than in *C. foeniculacea*. I have observed the presence of small "pédicelles cryptifères" in the filiform British specimens of *C. foeniculacea* (Roberts, 1968), and in the foliaceous plants the cryptostomata and spines entirely agree with Sauvageau's diagnosis for *C. discors*. In England, spines and "pédicelles cryptifères" appear to be produced most abundantly during the phase of less active growth following the reproductive period. Specimens from Ireland are much less spiny, whilst a specimen collected by de Valéra from Bayona on the west coast of Spain completely lacks foliaceous laterals, in this way agreeing with *C. foeniculacea*, but also bears long spines and abundant "pédicelles cryptifères", wherein it resembles *C. discors*. Bayona is the furthest south that *C. foeniculacea* has been recorded. Seoane-Camba (1965) recorded *C. discors* from Coruña and Gijon on the north coast of Spain, further north than the previous record of *C. foeniculacea*, so it is possible that a question of interpretation is involved here. He also recorded *C. discors* from Cabo S. Vicente (Portugal) and Cadíz in the south. Ardré (1970) did not record either species from Portugal. Obviously more information is needed about this area.

In the Adriatic populations of *C. discors*, the variation in length and abundance of spines is as great as that used by Sauvageau to designate the opposite extreme conditions of his two species.

If the respective theories of Sauvageau or of Ercegović—i.e. of speciation in the past and of presently active speciation—are to be acceptable, evidence of genetic separation of the extremes is necessary. Unfortunately, the west coast of the Iberian Peninsula is the only place where there might be an overlap in the range of the two entities. Very little is known about the Cystoseiraceae in this area and there are no records of *C. foeniculacea* and *C. discors* occurring together and maintaining their identity. Further, there is no direct evidence from the variable Channel plants, either from transplantation experiments or from observations of foliaceous and filiform individuals growing together under identical conditions, to indicate whether the capacity to produce foliaceous laterals is possessed only by some individuals and is genetically determined, or whether the capacity is possessed by all but only expressed in response to an environmental demand such as diminished illumination at greater depth. That a genetic factor is involved at some level is indicated by the fact that *C. humilis* and *C. canariensis*, which both resemble *C. foeniculacea* in a number of other details, do not produce any foliaceous laterals at all after the primordial stage. *C. humilis* can only be separated from *C. foeniculacea* with certainty by this consistent lack, and filiform *C. canariensis* occurs in the same pools as foliaceous *C. discors* (Sauvageau, 1912b).

The variation seen in *C. fimbriata* is worth considering in this context. *C. fimbriata* also produces foliaceous laterals but is firmer in texture and darker in colour than *C. discors*. Ercegović found that in this species also, the degree of flattening increased with depth; he illustrated a plant from the island of Markenta, collected from a depth of 30 m, which completely lacked any filiform ramuli and bore receptacles on the margins of the flattened ramuli. These plants, which he called subspecies *plana*, looked similar to the small flattened plants of exposed situations, which he called subspecies *rosetta*, subsequently reported from the Aegean by Huvé (1972). The surface-growing populations of the furthest offshore islands tend, as in *C. discors*, to produce fewer foliaceous laterals before forming filiform receptacular branches than do the mainland populations at similar depth.

This parallel behaviour suggests that within this group of species only *complete* absence (other than in the primordial stage) of the capacity to produce foliaceous laterals, as in *C. canariensis* and *C. humilis*, is a valid criterion for species separation unless other distinctive differences exist, which they do not between *C. foeniculacea* and *C. discors*. The production of foliaceous laterals during ontogeny, however few or small, should then be regarded as an indication of the possession of the capacity, with the degree to which it is expressed depending on interaction with the environment. The populations of *C. discors* and *C. foeniculacea* in the three major regions: Mediterranean, Channel Coast (including northern Spain) and the west coast of Ireland, need further investigation, both experimentally and on the shore, before any firm decision can be made as to whether two species or only one should be recognized, since the occurrence of foliaceous laterals, habitual, occasional or not at all, appears to be consistent within each region. It is not known, for instance, whether the Irish plants possess the capacity to produce foliaceous laterals under different conditions, or whether the non-foliaceous plants of the south coast of England only temporarily lack these laterals, having lost the preceding season's growths by erosion. If all forms are to be regarded as part of a single, very variable, morphological species, then the possibility that several genetically distinct components occur within its limits must be taken into account.

Fragmentary information concerning seven other non-Mediterranean species shows that they possess some similarities with the Mediterranean species just discussed. All seven species lack laterals of limited growth and most (except *C. indica*) produce foliaceous laterals. *C. osmundacea* and *C. geminata* both possess mucilaginous retaining stalks but produce antheridia on very bushy hairs which may be related to their dioecious habit; *C. indica*, also dioecious, possesses stalks and very simple antheridia (Thivy and Doshi, 1966) but there is no information as to whether the antherozoids are pigmented. *C. trinodis* has stalks, and the small conceptacle with basal hairs, illustrated by Papenfuss and Jensen (1967), is indistinguishable from those illustrated by Roberts (1968) and Dawson (1941) for *C. foeniculacea*. *C. crassipes* (Turner) C. Ag. and *C. sisymbroides* (Turner) ***comb. nov.**** also possess stalks (Tahara, 1913; Inoh, 1937), whilst *C. myrica* possesses conspicuous "pédicelles cryptifères", conceptacles similar to *C. trinodis*, and non-pigmented antherozoids.

Group 2. C. tamariscifolia *(Hudson) Papenfuss*

C. tamariscifolia and a group of about six principally Mediterranean species possess lateral branch systems which are pinnately subdivided to varying degrees; these are usually clothed with short, stiff laterals of limited growth, often referred to as "leaves" but probably better called "appendages" in order to remove ambiguity from the descriptive terminology.† The plants are usually iridescent, especially in the young growth. The receptacular regions are generally more extensive than in plants of the previous group and vary in appearance according to whether the conceptacles are in the basal parts of the appendages or immersed in the axis itself, with a greater or lesser part of the appendage

* Basionym: *Fucus sisymbrioides* Turner (1809).

† Sauvageau used "Feuillée", i.e. foliose ("with leaves"), to describe plants bearing appendages of this form. The obviously morphological appendages of *C. baccata* he called "courts ramules". Sauvageau also used "foliacée", i.e. foliaceous ("leaflike") to describe the flattened branch systems of the types present in *C. discors*. Subsequent authors have not always differentiated between the two terms.

present as a spinous outgrowth of the receptacle. Information about the oogonia and antheridia is available only for the three species *C. tamariscifolia*, *C. mediterranea* Sauv. and *C. caespitosa* Sauv. (Guern, 1962; Roberts, 1970); in these species, the conceptacles are larger than those of the preceding group and the oogonia are produced in greater numbers per conceptacle being liberated free into the sea without retaining stalks. The antheridia are also more numerous and are produced on very bushy hairs, with the antherozoids possessing eyespots. These features may be connected with the free liberation of the oogonia, again in contrast to the preceding group. Hairs do not seem to figure prominently in conceptacle development in *C. tamariscifolia*, only a single, non-persistent hair being detected. Nuclear structure is known for only a few species; in *C. tamariscifolia* (Roy, 1938; Roberts, 1970), *C. spicata* Ercegović (Roberts, 1966) and *C. corniculata* (Roberts, unpub.), the nuclei throughout the tissues possess conspicuous chromo-centres.

The species being dealt with here include some of those associated by Sauvageau (1920) with *C. tamariscifolia*, together with *C. spicata* Ercegović (Adriatic). *C. tamariscifolia* presents no problems of identification on the northern Atlantic shores since no other foliose iridescent species occur there; however, on the Atlantic coast of North Africa and in the western Mediterranean it overlaps in range with a very similar species, *C. mediterranea* Sauvageau. *C. tamariscifolia* is mainly Atlantic in distribution, occurring off the coasts of the British Isles, France, Portugal, Spain, the Canaries, Morocco and Mauritania, and only extending a short distance into the western Mediterranean (off Malaga; Sauvageau, 1913, 1920) and the coast of Africa, where it lacks aerocysts. *C. mediterranea* occurs widely in the Mediterranean, off the Balearic Islands, from the Gulf of Lions, the Gulf of Genoa and the Bay of Naples, and has only limited distribution in the Atlantic (off the coast of Morocco in the neighbourhood of Rabat; Gayral, 1958). The two differ only slightly in general form: *C. mediterranea* lacks aerocysts and is more regularly pyramidal than *C. tamariscifolia*, has a less rough and spinous texture and differs in time of defoliation; neither has a period of complete dormancy but a period of less active growth follows the reproductive phase. As in the case of *C. foeniculacea* and *C. humilis*, the difference between the two is slight; it appears to be less than the seasonal variation within *C. tamariscifolia*, which exhibits seasonal production of aerocysts and also of branches and appendages of differing form. Some populations, as in North Africa, may completely lack aerocysts. Sauvageau (1920) did question whether *C. tamariscifolia* and *C. mediterranea* represent only one species and suggested transplant experiments between Malaga and the French Atlantic coast.

C. balearica Sauvageau and *C. caespitosa* Sauvageau are two small species which overlap with *C. mediterranea* in distribution and which differ from each other only in small points of detail. *C. balearica* is recorded from the Balearic Islands, from the Mediterranean (Var) coast of France (Huvé and Pellegrini, 1970) and possibly also from Italy and Libya; it is not well-known. *C. caespitosa* is recorded from the Banyuls area and from Algeria. Both differ from *C. tamariscifolia* and *C. mediterranea* behaviourally in having a period of complete dormancy (*C. caespitosa* in autumn and *C. balearica* in winter) and morphologically in having naked apices in contrast to the spinous apices of the former species. Dangeard (1949) considered that *C. tamariscifolia* from the Atlantic coast of Morocco resembled *C. caespitosa*. Receptacles are axial in *C. balearica* but in the bases of appendages in *C. caespitosa*; *C. balearica* has a turf-forming habit and *C. caespitosa* is caespitose, although it tends to be turf-forming in Algeria. *C. stricta* Sauvageau is another turf-forming species, although it is slightly larger than *C. balearica* and *C. caespitosa* and covers extensive areas; it also has a period of complete dormancy at the end of autumn, but possesses a spinous apex. Its range extends further to the east and includes the Gulf of Genoa, Sardinia, Algeria and Libya. In the Adriatic, a capacity for horizontal extension similar to that exhibited by *C. stricta* is seen in *C. spicata* Ercegović (interpreted by Sauvageau as *C. stricta*). Ercegović considered that *C. spicata* was a separate but very variable species which resembled *C. stricta* in one of its forms but was distinguished from it by the lack of spines at the apex of the axis and the absence of foliaceous first-formed primary laterals.

One aspect of variation which Ercegović investigated was the general form of the lateral branch systems, which he expressed as the ratio of the lengths of the primary laterals to the length of the secondaries—G/g. This gives an overall impression of the shape of the system, a low ratio referring to a relatively broad lateral system and a high one to a relatively slender one, without reference to the actual size. Samples from a given population gave a sufficiently constant figure for the ratio to be applied as a measure of the general aspect of that particular population, yet it might range in value from as little as 2·1 to as high as 7·5, as in *C. spicata*, in populations of the same species from different localities. Variation was also seen in the degree of branching, from tufted and bushy forms to slender and flexuous forms with infrequent laterals; in the form of the branches, whether round, ovoid or flattened in section; in the form of the appendages, whether undivided, bifid or multifid; and in the position of the conceptacles, whether terminal on ramuli, basal on appendages or axial.

Such variation in *C. spicata* seemed to have definite geographical or ecological correlations. The lowest values of G/g were found in the littoral populations, with higher values in the populations of the offshore islands. This was particularly pronounced in the mid-Adriatic region, where progressive increase in the value towards the open sea resulted in so much change that the populations at the two extremes might be classified as two separate subspecies. The change in ratio was accompanied by a decrease in the frequency of branching, illustrated by comparison of the densely-tufted plants from Split and the immediately offshore island of Korčula with those of the most distant island of Palagruž. The degree of subdivision of the appendages also decreased and the form of the receptacles changed, those from the open sea being shorter and more slender with the individual conceptacles visible as distinct swellings. Finally, the furthest offshore populations also tended to have flattened terminal ramuli. The combined result of all these trends is illustrated by comparison between the mainland plants from Split and the island populations from Jabuka, which Ercegović recognized as two distinct subspecies, *crassa* and *elegans* respectively. *C. spicata* subsp. *crassa* is very similar to *C. tamariscifolia* in size and general appearance of the lateral branch systems, although the appendages are perhaps slightly longer; the entities differ principally in the caespitose habit, nude apex and short period of complete dormancy of *C. spicata* subsp. *crassa*. The latter possesses similar conspicuous aerocysts but these are of infrequent occurrence. *C. spicata* subsp. *elegans* has a main axis which is almost completely reduced, rarely exceeding 40 mm and usually much shorter; the entity has two differing periods of vegetative growth followed by a short period of complete dormancy.

All these variations described by Ercegović are the results of differing balances between the growth processes giving rise to series of intergrading forms. Observations alone are insufficient to establish whether there is any genetic basis for the differences. Most of the differences between *C. mediterranea*, *C. balearica* and *C. stricta*—i.e. size, caespitose habit, dormant period—could also be regarded in the same way. The two subspecies of *C. spicata* are of interest as they again illustrate the tendency towards shortening of the main axis already mentioned (p. 408) in plants of *C. fimbriata* from exposed situations. In *C. spicata*, the shortening occurs in the furthest offshore populations. In this context, the great capacity of the main axis of *C. tamariscifolia* to subdivide (Roberts, 1970) is of interest, since drastic shortening of the axis with retention of this ability would result in a caespitose habit; possibly the offshore populations illustrate the process repeated to a higher degree. It is not clear that a period of dormancy is a valid criterion for separating species, since this could be a response to conditions suboptimal for growth and could differ in its occurrence in different parts of the range of the species. Aerocysts are infrequent in this group of species within the Mediterranean, where they only occur in *C. spicata* subsp. *crassa*; foliaceous laterals are also infrequent. None of the species discussed occurs in deep water and flattened ramuli are seen only in the offshore populations of *C. spicata* subsp. *elegans*. Presence or absence of spines at the apex appears to be consistent in each species.

Sauvageau (1920) believed all these species to have been derived either directly or indirectly from *C. tamariscifolia* and Ercegović considered this process of speciation to be still active in the Adriatic, where *C. spicata*—itself derived from *C. stricta*—was considered to be in the process of producing two further new species.

Group 3. C. nodicaulis (*Withering*) *M. Roberts*

Most of the tophulose species are placed in this group, largely because they possess these characteristically swollen bases of the lateral branch systems. The taxon now called *C. sauvageaueana* Hamel, described by Sauvageau (1912b, *sub C. selaginoides* Valiante), did not possess these structures. The thickened bases of its lateral branch systems were called "pseudo-tophuli"; their form suggests a possible mode of origin of tophuli (cf. Valiante, 1883). Little is known about most of this group beyond external morphology, which appears to be very variable in the majority. In all but two species (*C. adriatica* and *C. platyramosa* Ercegović) the morphological spectrum includes forms closely resembling *C. nodicaulis*. In several species—*C. sauvageaueana*, *C. squarrosa* de Notaris, *C. spinosa* Sauvageau and *C. zosteroides* C. Ag. (Valiante, 1883), and *C. nodicaulis* (Roberts, 1977)—the juvenile plants produce foliaceous laterals up to 150 mm long before beginning to form adult lateral branch systems. Guern (1962) has investigated the embryology of *C. nodicaulis*; Roberts (1977) has found that the conceptacles of *C. nodicaulis*, like those of the *C. tamariscifolia* group, are considerably larger than those of the *C. foeniculacea* group. These conceptacles follow a distinctive pattern of early development. They contain about 30 oogonia, neither basally elongated nor embedded, which are liberated into the sea; antheridia are borne on much-branched hairs and yield antherozoids possessing eye spots. The mature conceptacle of *C. platyramosa* illustrated by Ercegović (1952) is of the same type. Roberts (1967) found the nuclei of *C. nodicaulis* to be different from both *C. tamariscifolia* and *C. foeniculacea*, being finely granular throughout the tissues; those of *C. jabukae* Ercegović are also of this type. Some species of this group are slightly iridescent but usually only in new growth.

As in Group 2, a few species of very limited distribution occur in the Atlantic and Eastern Mediterranean; one species is fairly generally distributed within the

Mediterranean, and within the Adriatic external morphology shows very great variation.

C. nodicaulis, as found off the British Isles, is morphologically a very variable species both in relation to the pattern of branching of the lateral branch systems and in the form of the receptacles. The latter may be terminal, relatively restricted and externally indistinguishable from those of *C. foeniculacea*, or they may be extensive and have small spinous outgrowths. The plant is completely dormant for a short period during the summer months and shows seasonal production of tophulose branches. The tophules form in the period immediately preceding dormancy and grow out into lateral branch systems after this period; they are not usually formed during the principal reproductive period. Small tubercles are sometimes present on the apices and bases of lateral branch systems. The distribution of the species is restricted, being limited to the Channel coasts of France and England and to the west coast of England, Scotland and Ireland.

Drift specimens similar to *C. nodicaulis* were collected by Sauvageau (1912) from the Gulf of Gascony and referred by him to a distinct species, *C. usneoides* (*sub C. concatenata* C. Ag.); this was distinguished from *C. nodicaulis* by the possession of foliaceous secondaries on lower lateral branch systems, with receptacles which closely resembled those of *C. nodicaulis*, and by more abundant aerocysts. The plant has subsequently been recorded growing off the coast of Spain–Malaga (Sauvageau, 1920); Cadíz (Seoane-Camba, 1965); further north at Pontevedra (Miranda, 1934) and Arosa (Donze, 1968)—and from Portugal south of the Tagus (Ardré, 1970). I have not seen any of the latter specimens but in the drift specimens collected by Sauvageau, the distinctions seem largely a matter of degree, as is the case with *C. foeniculacea* and *C. humilis*. I have recently seen many examples of small, flattened laterals on reproductive plants of *C. nodicaulis*, which suggests that the possession or lack of foliaceous laterals may prove not to be a valid difference. Growing plants of *C. usneoides* were later described by Sauvageau (1920) as possessing fibrous attaching discs.

Dangeard (1949) reported "*C. granulata* Ag." off the coast of Morocco in the Mogador region; plants exhibited iridescence characteristic of *C. nodicaulis* but possessed abundant aerocysts. Sauvageau, incorrectly, referred to *C. nodicaulis* as "*C. granulata*" which is also a synonym of *C. usneoides*. This record clearly needs further investigation. Ardré (1970) recorded both species from Portugal.

Dangeard (1949) also reported *C. gibraltarica*, another tophulose species which was first recorded by Sauvageau (1933) as a variety of *C. sauvageaueana* (*sub C. selaginoides* var. *gibraltarica*) and said to be common along the entire coast of Morocco. This species could be consistently recognized by its spinous tophules, which contrasted with the smooth ones of *C. nodicaulis*. This difference, though slight, may be genetically determined and tophule detail is a consistent feature in Adriatic species. Dangeard also suggested that *C. mauritanica* was probably only a form of *C. gibraltarica* in which he found unisexual conceptacles amongst the usual hermaphrodite ones; the dioecious *C. mauritanica*, known only from two specimens, did not seem to him sufficiently distinct to warrant specific status.

C. spinosa★ Sauvageau (Ercegović, 1952; Valiante, 1883, *sub C. erica-marina* Nacc.) is a tophulose species, very widely distributed throughout the Mediterranean, having been recorded from the Gulf of Lions and Algeria to the Adriatic, Aegean and the coast of Syria. This is a very variable species and some of the variation is described by Sauvageau and Valiante. In many localities the plant closely resembles *C. tamariscifolia* in its iridescence and (apart from the tophuli) in the form of its lateral branch systems, which may be clothed with abundant spinous appendages. At greater depths the appendages are fewer and smaller and there is a tendency to produce foliaceous laterals (Hamel, 1939, pls 8 and 9). The species has a short period of dormancy.

In the Adriatic Ercegović found even greater variability, particularly in relation to degree of branching and flattening of the ramuli. He included *C. spinosa*, together with *C. adriatica* Sauv. ex Ercegović,† *C. squarrosa* de Notaris *C. platyramosa* Ercegović, and *C. jabukae*, in what he called "super-species *C. spinifera* Erc."‡

C. adriatica is found in the northern half of the Adriatic, on both the mainland coastline and the offshore islands. Its morphological variation includes forms similar in size and appearance to *C. spinosa*, but it is generally more slender. The variation is thus very wide (Ercegović, 1952).

C. squarrosa also includes plants similar to the radial forms of *C. spinosa* and, according to Sauvageau (1912), is separated from the latter only by differences of degree in certain features—it is generally smaller, more compact, more spiny and stiffer, but is most clearly distinguished by its characteristic receptacles; these are often several centimetres long with close-set spines, bifid at their tips. *C. squarrosa* usually occurs in exposed situations.

★ The name needs investigation.

† *C. adriatica* was proposed by Sauvageau (1912) but without a Latin diagnosis. Further information was provided by Ercegović (1952) with an accompanying Latin diagnosis.

‡ This is not a valid taxon: the term is also used in angiosperm taxonomy to refer to species which have arisen by polyploidy

C. jabukae and *C. platyramosa* are two species with generally flattened ramuli, both characteristic of the outer islands.

These five Adriatic species have spinous apices and the tophules of all except *C. jabukae* are similarly spinous; if tophules are seasonally lacking, the lower parts of the secondary laterals are also clothed with spines. These species, except *C. squarrosa* which is sub-caespitose, produce a single long axis; all except *C. jabukae* have a period of complete dormancy. Both *C. spinosa* and *C. adriatica* have forms in which the appendages are represented only by scattered and reduced spinous outgrowths, a habit which regularly occurs in *C. jabukae* and *C. platyramosa*. In all species except *C. platyramosa* there are forms which are hardly distinguishable from *C. nodicaulis*.

These species also show great variation of the same features in which *C. spicata* varies, i.e. in patterns of branching, degree of subdivision of the appendages and the degree of flattening of the lateral branch systems. As in *C. spicata*, the variation also shows consistent ecological or geographical correlations. One behavioural difference between these five species and *C. spicata* is that they show a G/g quotient decrease from mainland to open sea, whilst in the group containing *C. spicata* it increases. The behaviour in this group of five species was further complicated by the fact that, in general, all showed parallel responses but the actual range of one species might overlap that of another so that clines appeared to exist. Further, a single species might exhibit this type of intergrading with more than one other species in respect of different characters (Ercegović, 1952). The form of the tophules, however, appeared to be constant. The Adriatic species thus seem to have complex interrelationships and the various formae and subspecies occupy distinct ecological niches or have definite geographical limits.

This Adriatic group is thus of interest morphologically, since it includes species whose lateral branch systems resemble those of *C. tamariscifolia* (with close-set spinous appendages) as well as some which resemble *C. nodicaulis* (lacking appendages); yet others include a range of forms linking the two. One Adriatic species is foliaceous and four of the other Adriatic species discussed possess the capacity to develop foliaceous laterals under certain environmental conditions, a capacity possessed only to a very limited extent by the *C. tamariscifolia* group.

The five Mediterranean species discussed form a closely interrelated group, with forms lacking appendages increasing in the Adriatic region; *C. spinosa*, the least variable species, is found most commonly in the Aegean in its more foliaceous form (Huvé, 1972). To the west, however, the only representative of this group is *C. spinosa*, where it occurs principally in its appendage-bearing form; in the region of the Strait of Gibraltar, the change to species lacking appendages and therefore typical of the Atlantic is abrupt and interspecific, contrasting with the gradation through intraspecific variation in the Adriatic.

Sauvageau (1920) further associated *C. platyclada* Sauvageau and *C. montagnei* Mont. non Ag. with *C. spinosa* and *C. adriatica*, but these two cannot be discussed here. There are many taxonomic problems to be solved in connection with *C. montagnei* and *C. platyclada* is too little known; Hamel (1939) reported that it had been collected off the Basque Coast by trawling. It is possible that, taxonomically, it lies close to *C. nodicaulis* or *C. usneoides*.

Some growth forms of *C. crinita* Bory are similar to *C. nodicaulis* and are similarly slightly iridescent, but the spinous projections are less frequent and the plants are caespitose. Ercegović groups this species with *C. crinitophylla* Ercegović in another "super-species", *C. crinitoides*, which shows many similarities to the other Adriatic species discussed. The receptacles resemble those of *C. nodicaulis*. Zinova and Kalugina (1974) include *C. bosphorica* Sauvageau in *C. crinita*. Some forms of *C. barbata* bear tophules (Ercegović, 1952; Zinova and Kalugina, 1974) which suggests that, although this species has no capacity to produce foliaceous laterals, it may also be associated with this group.

DISCUSSION AND CONCLUSIONS

Sauvageau amassed an enormous amount of useful information about *Cystoseira* in the field and was able to point the way to an understanding of the genus. He recognized the two conditions of species, those with retained oospheres and those with oospheres released free into the sea, together with the two associated antherozoid types, but lack of further information prevented him pursuing this idea very far. In the 57 years since the publication of his last paper on the subject little progress has been made in understanding the genus. Sauvageau (1912a) interpreted the relationship between the Mediterranean species as being one of descent and suggested that the genus had invaded the Mediterranean from the Atlantic, producing new species in response to the new conditions within the Mediterranean. He also pointed out (1920) that the number of species increases towards the east, that they are less well differentiated from each other than the Oceanic species, the differences separating them being minimal, and that the Eastern Mediterranean is particularly rich in species. Ercegović (1952), investigating the Adriatic species, developed Sauvageau's ideas and concluded that the factors which had caused speciation in the past were still operating in this area and that many new taxa would ultimately emerge.

The information now available, though still very incomplete, suggests that the species may fall into a number of very clearly defined groups. Three such groups are outlined and are represented by species from the Atlantic coasts of Europe and North Africa, and throughout the Mediterranean.

The three groups discussed differ from each other in respect of fairly basic features, whilst the differences between their constituent members—with the exception of the non-Mediterranean species—are often very slight. The delimitation of the species within the groups shows certain aspects in common in all groups. Similar non-adaptive features occur in all and form bases for species separation; the occurrence of small spinous protuberances on the apices and bases of lateral branch systems is one example, and in the *C. nodicaulis* group the shape and ornamentation of the tophules is another. Parallel variations in growth patterns are seen in two of the groups. These points indicate that the three groups resulted from an early burst of speciation in the genus whilst their component species are of more recent origin. Seven of the non-Mediterranean/Atlantic species have some similarities to members of the *C. foeniculacea* group but the species are clearly defined and speciation has occurred along different lines from those seen in the Mediterranean. These species must have become isolated from the rest at an early stage—whilst there was still continuity between the Mediterranean and Red Sea—and there is too little appropriate information available for there to be any value in discussing this point at the present; two species in this group are placed by Fensholt (1955) in *Myagropsis* Kützing (1843). These groups could provide a basis for taxonomic subdivision of the genus.

Within the Adriatic, the representatives of all the groups are morphologically very variable, and many subspecies and formae can be distinguished. The differences separating these taxa are often variations in details of growth patterns and, as such, are similar to the features used to separate some species; such features might be taken as illustrative of the thesis that speciation is still occurring.

Some of the variation, however, seems to be an expression of phenotypic plasticity, as, for example, with the production of foliaceous lateral branch systems in deep water and the presence of appendages in *C. spinosa*. The former response, present to some extent in all three groups, is clearly environmentally induced and, in *C. discors* in the Adriatic, appears to be under genetic control, the capacity varying from individual to individual. The mechanism determining the response needs to be established before the taxonomic significance and utility of this feature can be assessed. It is known that in flowering plants a number of distinct mechanisms can determine parallel plastic responses in even closely related species. In the *C. foeniculacea* group, the relationship between *C. discors sensu* Sauvageau and *C. foeniculacea* is the principal taxonomic problem; its solution requires more complete understanding of the mechanisms determining the formation of foliaceous laterals, developmentally, seasonally and in response to depth.

Ercegović (1952) showed how the proportions of the lateral branch systems vary, often in a clinal fashion, in different populations of a species, though no mechanism was suggested. Such variation is also likely to be reflected in details of the receptacles. This is clearly seen in *C. nodicaulis*, which shows both great variation in the morphology of its lateral branch systems and associated variation in receptacle form. The receptacles may be scarcely distinguishable from those of *C. foeniculacea*, which is not surprising since in both species receptacles form in ultimate ramuli of filiform branch systems. Not too much significance should be attributed to receptacle form in taxonomy. In appendage-bearing species, receptacles also vary considerably, even in a single individual, in relation to appendage production in this region and to the precise position of the conceptacles, whether in the axis or in the appendage.

This type of variation might be considered in conjunction with seasonal variations in the form of lateral branch systems, which are frequently seen, and the periods of dormancy which may occur. If optimal and sub-optimal phases, or perhaps vegetative and reproductive phases, are related to climatic conditions, this could have considerable effect on variations in time of fruiting in a species with a wide distribution, or could result in periods of complete dormancy in parts of the range; it could even account for the absence of aerocysts in some populations. External form needs to be related to the growth phase, rather than to the time of year as is frequently done in the descriptions of Sauvageau and Hamel.

A number of investigations into the chemical composition of several species of *Cystoseira* have recently been carried out (Alfimov, 1960; Pellegrini and Pellegrini, 1972, etc.) and seasonal variations have been linked with the growth phases; this information has not yet been utilized for taxonomic purposes. Isolated observations have been made with the electron microscope on structure of pyrenoids (Evans, 1968) and of antherozoids (Manton, 1964) and it would be valuable if such observations could be extended at least to representatives of each of the three groups discussed.

Ercegović (1952) considered that all the taxa he recognized had marked ecological or geographical preferences determining their distributions and he suggested that these were the factors operating to cause segregation of new taxa. He presented considerable information on this aspect. However, as Fisher (1968) pointed out, whilst geographical distribution patterns are extremely difficult to describe verbally, underlying principles may often be readily grasped from pictorial presentation. Mapping of the subspecies and formae in the Adriatic, as well as of the species of the Mediterranean area as a whole, might well throw some light on the processes of speciation, past and present.

The situation in the Adriatic could possibly represent segregation of new taxa but it could also be due to phenotypic plasticity, populations developing in different but constant fashions in the various habitats described by Ercegović, in response to the particular conditions of the locality. Plasticity of the phenotype seems to be an important feature of the genus and in *C. spinosa* (*C. nodicaulis* group) there is the suggestion that even the apparently basic difference represented by presence or absence of appendages, which separates the *C. tamariscifolia* and *C. foeniculacea* groups, could also be the outcome of phenotypic plasticity.

ADDENDUM

Since completing this paper, the author's attention has been drawn to two papers by Giaccone and Bruni (1971, 1973) in which the genus *Cystoseira* is analysed using the methods of numerical taxonomy, and, as a result, six groupings of species are proposed. Some of the groups come very close to those proposed in this paper.

REFERENCES

Agardh, C. A. (1820). "Species Algarum", Vol. 1. Lund.

Agardh, J. G. (1848). "Species, genera et ordines Algarum", Vol. 1. C. W. K. Gleerup, Lund.

Alfimov, N. N. (1960). Biochemistry of *Cystoseira barbata* from the Black Sea. *Bot. Zh. SSSR* **45**, 1196–1198.

Ardré, F. (1970). Contribution a l'étude des algues marines du Portugal. *Port. Acta biol.* (B) **10**, 137–532.

Dangeard, P. (1949). Les algues marines de la côte occidentale du Maroc. *Botaniste* **34**, 89–189.

Dawson, A. E. E. (1941). Some observations on *Cystoseira foeniculacea* (L.) Grev. emend. Sauv. *New Phytol.* **40**, 316–325.

Donze, M. (1968). The algal vegetation of the Ria de Arosa (N. W. Spain). *Blumea* **16**, 159–192.

Ercegović, A. (1952). Sur les *Cystoseira* adriatiques, leur morphologie, écologie et évolution. *Fauna Flora Adriatica* **2**, 1–112.

Evans, L. V. (1968). Chloroplast morphology and fine structure in British fucoids. *New Phytol.* **67**, 173–178.

Fensholt, D. E. (1955). An emendation of the genus *Cystophyllum* (Fucales). *Am. J. Bot.* **42**, 305–322.

FISHER, F. J. F. (1968). The role of geographical and ecological studies in taxonomy. *In* "Modern Methods in Plant Taxonomy" (V. H. Heywood, ed.), pp. 241–259. Academic Press, London and New York.

GAYRAL, P. (1958). "La Nature au Maroc. II. Algues de la côte atlantique marocaine." Société des sciences naturelles et physiques du Maroc, Rabat.

GERLOFF, J. and NIZAMUDDIN, M. (1975a). Bemerkungen zur Nomenklatur einiger Arten der Gattung *Cystoseira* C. Ag. *Nova Hedwigia* **26**, 341–348.

GERLOFF, J. and NIZAMUDDIN, M. (1975b). Three new species of the genus *Cystoseira* C. Ag. *Willdenowia* **73**, 565–582.

GERLOFF, J. and NIZAMUDDIN, M. (1976). New species of the genus *Cystoseira* C. Ag. *Nova Hedwigia* **27**, 165–182.

GIACCONE, G. and BRUNI, A. (1971). Le Cistoseire delle coste Italiane. I. Contributo. *Ann. Univ. Ferrara* (N.S.) Sez. IV. Bot. **4**(3), 45–70.

GIACCONE, G. and BRUNI, A. (1973). Le Cistoseire e la vegetazione sommersa del Mediterraneo. *Atti. Ist. veneto Sci., Cl. sci. mat. nat.*, **131**, 59–104.

GUERN, M. (1962). Embryologie de quelques espèces du genre *Cystoseira* Agardh. *Vie Milieu* **13**, 649–679.

HAMEL, G. (1939). "Phéophycées de France." Fasc. 5. Paris.

HUVÉ, H. (1972). Aperçu sur la distribution en mer Egée de quelques espèces du genre *Cystoseira* (Phéophycées, Fucales). *Bull. Soc. phycol. Fr.* **17**, 22–37.

HUVÉ, H. and PELLEGRINI, L. (1970). Sur la présence de *Cystoseira balearica* Sauvageau le long du littoral varois. *Bull. Soc. phycol. Fr.* **15**, 37–42.

INOH, S. (1937). An embryological study on *Cystophyllum crassipes*. *Botany Zool. Tokyo* **5**, 1821–1829.

JOHNSON, A. (1966). Studies in the conceptacle development and taxonomic position of *Cystoseira prolifera* J. Ag. *Botanica mar.* **10**, 134–140.

KÜTZING, F. T. (1843). Uber die Eigenthumlichtkeit der vegetation in der chinesischen und japanischen Meeren. *Bot. Ztg.* **1**, 53–57.

LEVRING, T. (1976). The marine algae of the Archipelago of Madeira. *Bolm. Mus. munic. Funchal* **28**(125), 1–111.

MAIRH, O. P. and KRISHNAMURTHY, V. (1968). Observations on the germination of oospores and growth of germlings in a *Cystoseira*. *J. Indian bot. Soc.* **47**, 256–263.

MANTON, I. (1964). A contribution towards understanding of "the primitive fucoid." *New Phytol.* **63**, 244–254.

MIRANDA, F. (1934). Materiales para una flora marina de las rias bajas gallegas. *Boln R. Soc. esp. Hist. nat.* **34**, 165–180.

NIZAMUDDIN, M. (1969). *Stolonophora*, a new genus of Cystoseiraceae (Phaeophyta: Fucales) from Guadalupe Island, Mexico. *Phycologia* **8**, 1–9.

PAPENFUSS, G. F. (1967). Taxonomic and nomenclatural notes on three species of brown algae. *Botaniste* **50**, 319–330.

PAPENFUSS, G. F. (1968). The history, morphology and taxonomy of *Hormophysa* (Fucales, Cystoseiraceae). *Phytomorphology* **17**, 42–47.

PAPENFUSS, G. F. and JENSEN, J. B. (1967). The morphology, taxonomy and nomenclature of *Cystophyllum trinode* (Forsskål) J. Ag. and *Cystoseira myrica* (S. G. Gmelin) C. Agardh (Fucales: Cystoseiraceae). *Blumea* **15**, 17–24.

PELLEGRINI, L. and PELLEGRINI, M. (1972). Contribution à l'étude biochimique des

Cystoseiracées méditerranéennes IX. Variations saisonnières des teneurs en quelques éléments minéraux chez *Cystoseira stricta* (Mont.) Sauv. *Bull. Soc. phycol. Fr.* **17**, 46–61.

Price, J. H., John, D. M. and Lawson, G. W. (1978). Seaweeds of the western coast of tropical Africa and adjacent islands: a critical assessment. II. Phaeophyta. *Bull. BM(NH), Bot.* (in press).

Roberts, M. (1966). The structure of the vegetative nuclei of the British species of *Cystoseira* C. Ag. *Br. phycol. Bull.* **3**, 69–73.

Roberts, M. (1967a). Studies on marine algae of the British Isles. 3. The genus *Cystoseira*. *Br. phycol. Bull.* **3**, 345–366.

Roberts, M. (1967b). Studies on marine algae of the British Isles. 4. *Cystoseira baccata* (Gmelin) Silva. *Br. phycol. Bull.* **3**, 367–378.

Roberts, M. (1968). Studies on marine algae of the British Isles. 6. *Cystoseira foeniculacea* (Linnaeus) Greville. *Br. phycol. Bull.* **3**, 547–564.

Roberts, M. (1970). Studies on marine algae of the British Isles. 8. *Cystoseira tamariscifolia* (Hudson) Papenfuss. *Br. phycol. J.* **5**, 201–210.

Roberts, M. (1977). Studies on marine algae of the British Isles. 9. *Cystoseira nodicaulis* (Withering) M. Roberts. *Br. phycol. J.* **12**, 175–199.

Roy, K. (1938). Recherches sur la structure du noyau quiescent et sur les mitoses somatiques de quelques Fucacées. *Revue algol.* **11**, 101–188.

Sauvageau, C. (1911a). Sur le passage des conceptacles aux cryptes pilifères des Fucacées, et sur les pédicelles cryptifères. *C.r. Séanc. Soc. Biol.* **71**, 468–470.

Sauvageau, C. (1911b). Sur les *Cystoseiras* à anthérozoids sans point rouge. *C.r. Séanc. Soc. Biol.* **71**, 472–473.

Sauvageau, C. (1912a). Sur la possibilité de déterminer l'origine des espèces de *Cystoseira*. *C.r. Séanc. Soc. Biol.* **72**, 479.

Sauvageau, C. (1912b). A propos des *Cystoseira* de Banyuls et de Guéthary. *Bull. Stn biol. Arcachon* **14**, 133–556.

Sauvageau, C. (1913). Sur les Fucacées du détroit de Gibraltar. *C.r. hebd. Séanc. Acad. Sci., Paris* **157**, 1539–1540.

Sauvageau, C. (1920). A propos des *Cystoseira* de Banyuls et de Guéthary. *Bull. Stn biol. Arcachon* **17**, 5–55.

Seoane-Camba, J. (1965). Estudios sobre las algas bentónicas en la costa sur de la Península Ibérica (litoral de Cadiz). *Investigación pesq.* **29**, 3–216.

Tahara, M. (1913). Oogonium liberation and embryogeny of some Fucaceous algae. *J. Coll. Sci. imp. Univ. Tokyo* **32**, 1–13.

Thivy, F. and Doshi, Y. A. (1966). *Stokeyia Indica* ***gen. nov. et spec. nov.*** of Cystoseiraceae. *Botanica mar.* **9**, 64–69.

Turner, D. (1809). "Fuci . . .", Vol. 2. J. and A. Arch, London.

Valiante, A. (1883). Le Cystoseirae del golfo di Napoli. *Fauna Flora Golf. Neapol* **7**, 1–30.

Womersley, H. B. S. (1964). The morphology and taxonomy of *Cystophora* and related genera (Phaeophyta). *Auast. J. Bot.* **12**, 53–110.

Yendo, K. (1907). The Fucaceae of Japan. *J. Coll. Sci. imp. Univ. Tokyo* **21**(12), 1–74.

Zinova, A. D. and Kalugina, A. A. (1974). Systematics of *Cystoseira* Ag. from the Black Sea. *Nov. Sist. Nizsh. Rast.* **11**, 116–125.

18 | Experimental and Numerical Taxonomy of the Laminariales: A Review

A. R. O. CHAPMAN

Department of Biology, Dalhousie University, Halifax, Nova Scotia, Canada

Abstract: Members of the Laminariales have been subjected to several experimental taxonomic approaches. Among these are crossability tests, transplant experiments, cultivation under standard conditions and quantitative genetics. In most cases the crossability tests show relatively easy hybridization between taxa having close orthodox taxonomic affinities. Conversely, distantly related taxa show little or no hybridization. In the same way, transplant experiments frequently demonstrate that taxonomically confused species are simply phenotypic manifestations of plastic responses to diverse habitats. Thus the new approaches often confirm taxonomic decisions or suspicions arising previously by use of classical techniques. Where there is orthodox taxonomic confusion, experimental methods sometimes lead to an understanding of population interrelationships, but do not aid in making taxonomic decisions within a Linnean framework of categories.

There has been only one numerical taxonomic study of laminarians, Widdowson's (1971a) analysis of *Alaria* populations. The results are very valuable in understanding the confused taxonomy of the genus.

INTRODUCTION

In cold-temperate seas, members of the Laminariales usually form the dominant component of the sublittoral vegetation. These are large, abundant plants of considerable economic importance, and yet much of their taxonomy is very confused. This is not for want of study. For instance, Wilce (1965) has revised the genus *Laminaria* in the north Atlantic but the taxonomy is still in dispute (South, 1976). The reason for the taxonomic problems is more fundamental than lack of study; it stems from the incompatibility of the Linnean binomial system with the reality of continuously varying populations in the sea. In the face of these problems, experimental and numerical methods may seem, at first glance,

Systematics Association Special Volume No. 10, "Modern Approaches to the Taxonomy of Red and Brown Algae", edited by D. E. G. Irvine and J. H. Price, 1978, pp. 423–432, Academic Press, London and New York.

very promising. However, although each of these methods can provide important evidence for classical taxonomy, they are not a universal panacea. The new methods produce new taxonomic categories, for example, ecotypes, ecospecies and phenons. They do not produce Linnean categories and, therefore, they are not always helpful in classical taxonomy.

In the Laminariales there has been some experimental taxonomic work carried out, mostly in the last 17 or 18 years. The experiments can be conveniently placed in four categories: crossability tests; transplant experiments; cultivation under standard conditions; quantitative genetics. There are other aspects to experimental taxonomy, but these have not been applied to laminarians.

Numerical taxonomy is the grouping by numerical methods of taxonomic units into taxa on the basis of their character states (Sneath and Sokal, 1973). Within this definition there have been only seven or eight numerical taxonomic studies of algae up to the present time. It is therefore fortunate that there has been a single taxometric study of a member of the Laminariales, that of *Alaria* by Widdowson (1971a).

EXPERIMENTAL TAXONOMY

1. *Crossability Tests*

The first tests were carried out as long ago as 1930 (Schreiber, 1930). In an exemplary piece of work he attempted to cross *laminaria hyperborea*, *L. digitata* and *L. saccharina*. These are morphologically well differentiated taxa and Schreiber obtained no hybrids. Later work by Sundene (1958) showed complete interfertility between two previously recognized taxa of *L. digitata*, f. *digitata* Fosl. (= f. *genuinea* Le Jol.) and f. *stenophylla* Harv., and an unidentified morphological form distinguished from the others by a stiffer stipe and more extensive bullation of the lamina.

Very interesting and intricate experimental analyses of hybridization of kelp species have been carried out in Japan, notably by Yabu (1964) and Saito (1972). Yabu isolated gametophytes of five *laminaria* species (*L. diabolica*, *L. japonica*, *L. religiosa*, *L. angustata* and *L. ochotensis*) and one *Alaria* species, *A. crassifolia*. *L. japonica*, *L. ochotensis* and *L. religiosa* readily produced interspecific hybrids, as did crosses between *L. japonica* and *L. diabolica*. These are not morphologically distinct species and Kain (1971) has placed them all in one group, the *L. japonica* group. *L. angustata* is a morphologically well defined species and failed to produce interspecific hybrids. Intergeneric hybrids were not produced.

Probably the most comprehensive piece of work in this field is that of Saito (1972). In Japan there are three species of *Undaria*, *U. pinnatifida* (with two morphological forms), *U. peterseniana* and *U. undarioides*. According to

Okamura (1915). *U. pinnatifida* and *U. peterseniana* are only distantly related. He felt that there are closer relationships between the other taxa of *Undaria*. Saito (1972) crossed all forms and species. The results are shown in Table I. There are two important points. Firstly, crossability is not an all-or-nothing phenomenon. There is a range from 20% successful crosses to 100% successful. Secondly, crossings are most successful within species and least successful between distantly related species. In other words the experimental findings are in agreement with Okamura's (1915) concept of species relatedness based on morphological criteria.

TABLE I. Percentages of successful crosses between *Undaria pinnatifida* f. *distans* (A), *U. pinnatifida* f. *pinnatifida* (B), *U. undarioides* (C) and *U. peterseniana* (D). After Saito (1972)

		Males			
		A	B	C	D
Females	A	100	100	—	—
	B	100	100	65	20
	C	—	60	100	50
	D	—	45	75	100

Saito (1972) carried his work further and took plants through the F_1 and F_2 generations. Most interestingly the hybrids between *U. peterseniana* and *U. pinnatifida* (distantly related according to Okamura, 1915) did not become reproductively mature. The other F_1 crosses became reproductively mature and an F_2 generation was successfully raised. This is an important point and shows that crossing experiments should be continued to the F_2 because meiotic aberrations preventing introgression may not appear until then.

The most recently published crossability studies on *laminaria* are those of Svendsen and Kain (1971), and Chapman (1974, 1975). Svendsen and Kain worked on *L. cucullata* and *L. hyperborea*. These are closely related species and, as would be expected, complete interfertility was found. Similarly Chapman found that a group of poorly-differentiated species in the north-west Atlantic readily produced F_1 hybrids. These were raised to reproductive maturity in the sea, an F_2 generation was then raised and these plants developed normally.

The outstanding point to emerge from all of these studies is that where species can be distinguished readily on morphological criteria, there is usually reproductive isolation. Morphologically confused species are interfertile. Internal isolation is only one barrier to crossability. There may also be other, external, barriers, but these have not been looked at in the Laminariales.

2. Transplant Experiments

It is a commonly held viewpoint that much of the inter-population phenotypic variation in seaweeds results from plastic responses to habitat (Dixon, 1970). One approach to this problem is that of reciprocal transplantation in which the growth form of plants in a new habitat is compared with that of controls. Laminarians readily lend themselves to this approach.

Sundene (1962, 1964) initiated this type of work when he transplanted exposed-coast sporophytes of *Laminaria digitata* (known as forma *stenophylla*) to sheltered conditions and found that plants soon took on the form of *L. digitata*

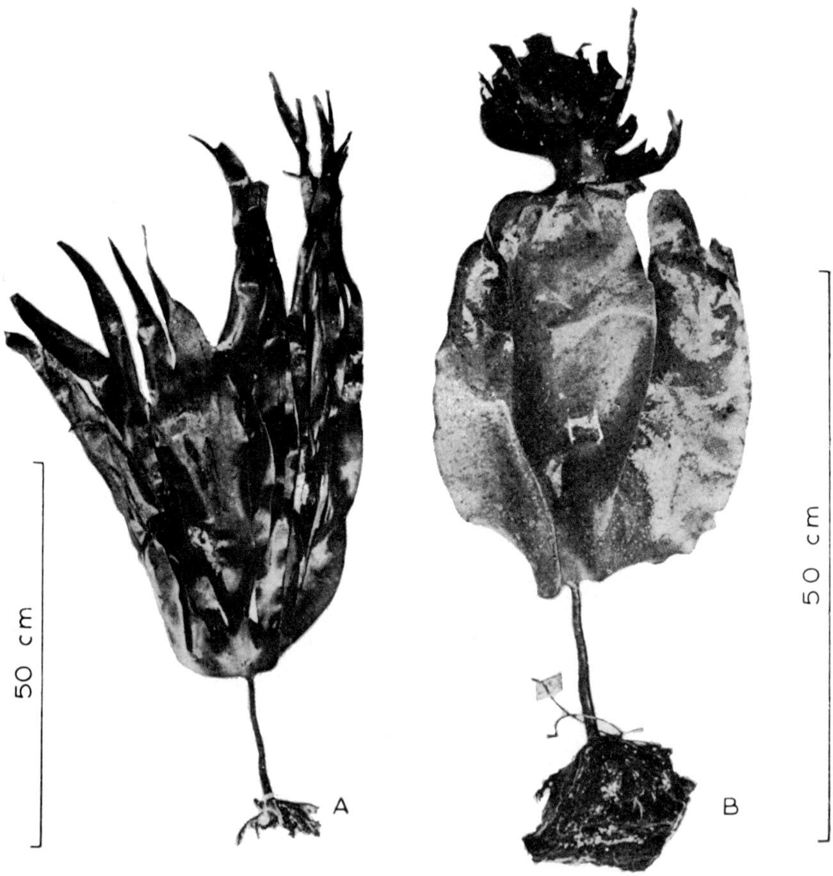

FIG. 1. The effect of transplantation of *Laminaria hyperborea* from exposed to sheltered conditions. Plant A shows a digitate morphology prior to transplantation. Plant B shows a cucullate morphology assumed after transplantation. (A and B are not the same individual.) From Svendsen and Kain, 1971.

f. *digitata*. This latter form occurs naturally in sheltered conditions and is characterized by the possession of a broad lamina.

Svendsen and Kain (1971), also working in Norway, carried out reciprocal transplantations of *L. cucullata* and *L. hyperborea* at a site where the former occurred at 6–20 m depth and the latter around the low tide mark. The deep-water plants quickly disintegrated after transplantation but shallow-water plants moved to deep water grew well and produced a cucullate blade (Fig. 1).

Perhaps the most ambitious transplant programme has been carried out by Markham (1972). Plants of *L. longipes* and *L. sinclairii* were moved between Alaska, British Columbia and Oregon. These two species are best distinguished on the presence or absence of mucilage ducts in the stipe. Transplants did not change with respect to this characteristic when growing in a new habitat.

Similarly, Chapman (1974) found that stipe characteristics of *Laminaria* did not display a plastic response to transplantation. In eastern Canada, populations of the Simplices section of the genus in sheltered conditions develop long, hollow stipes. Plants in exposed conditions have short, solid stipes. The two forms have been assigned to separate species, *L. longicruris* and *L. saccharina* respectively (Wilce, 1965). On transplantation there was no significant change in stipe structure.

Where plants exhibit no response to transplantation the results are ambiguous. They may mean that the phenotypic differentiation observed has a high genetic component. On the other hand it must be remembered that adult plants are used in transplant experiments and there is a distinct possibility that developmental canalization takes place early in development.

3. *Cultivation Under Standard Conditions*

Cultivation of diverse phenotypes under standard garden conditions is a routine genecological technique. It was first used on laminarians by Sundene (1962) who obtained offspring in culture from three morphological forms of *L. digitata*. Hybrids between the forms were also obtained. When large enough the sporophytes were transplanted to the sea and cultivated under standard calm conditions. All of the offspring developed similar broad blades in spite of the differing parental morphologies. However, offspring derived from parents with stiff stipes and bullate blades developed these characteristics in contrast to other offspring. Sundene suggested that some genotypic differentiation had occurred with respect to these characteristics.

A quite different approach has been used by Burrows (1964). She obtained young sporophytes of British and Canadian Laminarias and grew them to large plants under laboratory conditions. She found that offspring of the Canadian parents lacked mucilage ducts. British plants developed mucilage ducts but the extent of their development was temperature dependent. Furthermore, she found that the shape of the base of the blade, blade texture and the extent of peripheral frill development were all easily modified by changes in light and temperature.

In Japan *Undaria pinnatifida* occurs in two forms, f. *distans* and f. *pinnatifida* (Saito, 1972). These are differentiated on morphological characteristics and are found in areas of differing wave exposure. Saito found that young sporophytes retained their distinct characteristics when grown under standard conditions. This contradicts the earlier work of Ii (1964) who obtained similar morphological forms of both taxa when grown under standard conditions. These experiments point once again to the drawbacks of using, as experimental material, plants with a previous history of development in a different environment.

4. *Quantitative Genetics*

The aim of quantitative genetics is to partition phenotypic variance into various components. In the first instance this is usually into genetic and environmental components. This is of course one of the aims of experimental taxonomy. However, the genetic techniques have only rarely been used for taxonomic purposes. A good measure of the genetic component is found through an estimate of heritability, which, in the broad sense, is the ratio of genetic variance/phenotypic variance. There are many ways of estimating heritability but they all involve measuring the degree of resemblance between known genetic relatives.

These methods have been used by Chapman (1974, 1975) to estimate the genetic and environmental components of phenotypic variation in stipe structure and mucilage duct anatomy in non-digitate Laminarias in eastern Canada. These are important taxonomic criteria. The degree of mucilage duct development was found to have a very high environmental component in Newfoundland populations. Heritability is thus low. Conversely, there was found to be a high genetic component in the differentiation of populations in Nova Scotia with respect to stipe length and hollow development.

NUMERICAL TAXONOMY

Widdowson's (1971a) treatment of *Alaria* is highly unusual. He made use of Discriminant Function Analysis and Distance Function Analysis. A linear discriminant function is a linear function z of characters describing Operational Taxonomic Units (OTU's) that weight the characters in such a way that as

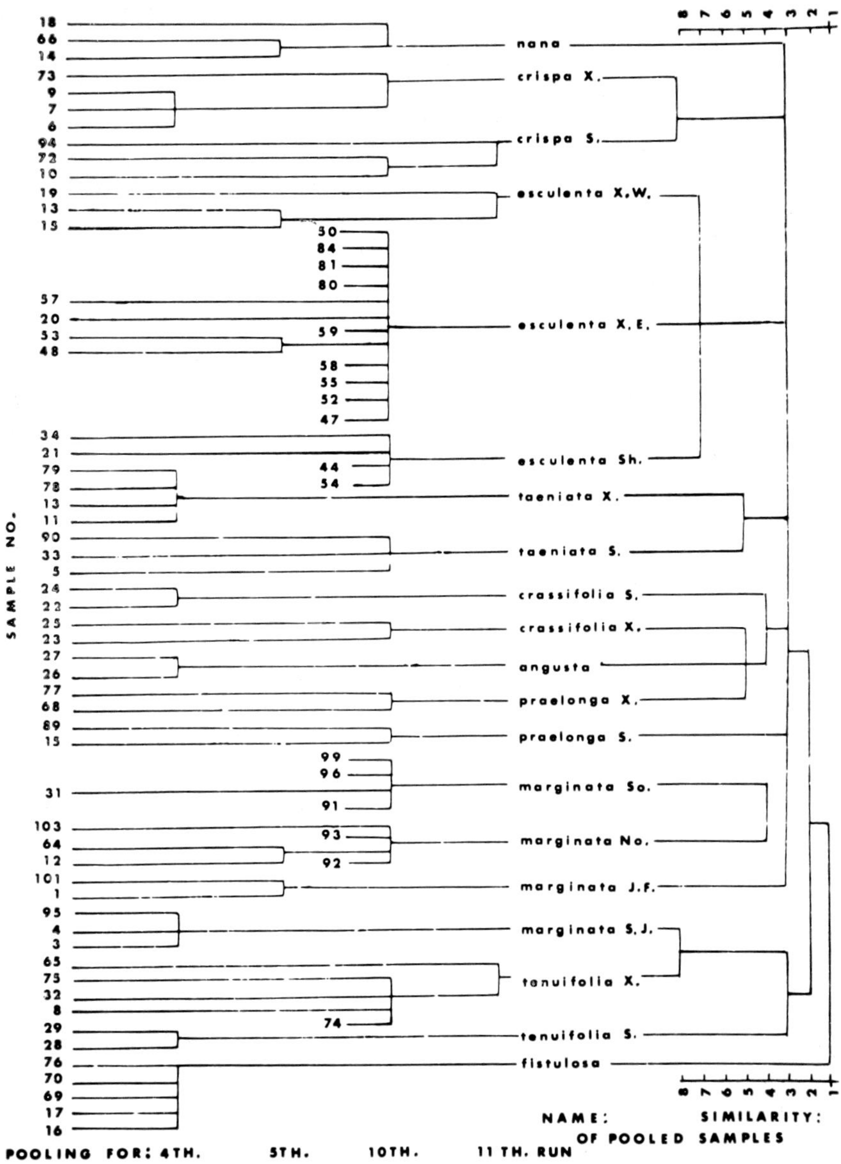

FIG. 2. Hierarchical classification of *Alaria* samples using distance functions. From Widdowson, 1971a.

many as possible of the OTU's in one taxon have high values of z and in another taxon they have low values.

Thus there is a series of weights symbolized as: $z_1, z_2, \ldots z_m$ for characters 1, 2, . . . m. If there are two taxa J and K the discriminant scores DS can be calculated for the centroids by summing the products of the character values and the weights. The absolute difference between DS_J and DS_K is equal to D^2, Mahalanobis' Generalized Distance. The computation of z is explained most clearly by Sneath and Sokal (1973) who also provide a critique of the use of Discriminant Function Analysis in numerical taxonomy.

Widdowson (1971a) had 83 samples of *Alaria* collected throughout the range of the genus in the northern hemisphere. Using conventional methods he identified ten species (Widdowson, 1971b). Twelve metric characters were analysed for each plant. The samples were pooled by an agglomerative technique dependent on the distance functions. The samples were pooled sequentially at maximum distance levels of 4·0 units, 4·9, 5·9, 6·9 and 9·4 to give a dendrogram (Fig. 2).

At this point there were several unpooled samples. Each isolated sample was pooled with the sample nearest to it. At this point Widdowson attached specific epithets to the pooled groups.

Subsequently the degree of similarity between the pooled groups was computed and this is shown on the right of Fig. 2. It is to be noticed that two groups of species cannot be distinguished consistently: *A. angusta*, *A. crassifolia* and *A. praelonga*, the three Japanese species; *A. marginata* and *A. tenuifolia*. Surprisingly, Widdowson feels that in the absence of experimental work, it is best to continue recognizing these as separate species.

CONCLUSIONS

Experimental taxonomy of laminarians has furnished data, firstly, on interfertility of specific and subspecific taxa and, secondly, on genetic versus environmental components of phenotypic differentiation. There are phycologists who advocate that internal reproductive isolation provides the best evidence for the erection of species (Ling and Tyler, 1974). This viewpoint is rejected entirely here because, whereas reproductive isolation often corresponds to morphologically well defined species (e.g. Schreiber, 1930, Yabu, 1964), crossability is not an all-or-nothing criterion (Saito, 1972). Furthermore, meiotic aberrations which might prevent species crossability may not be manifest until the F_2 generation.

Morphological types which arise as a result of environmental modifications are usually called phenecotypes and it is common practice to "sink" them taxonomically (Davis and Heywood, 1963). Thus *Laminaria cucullata* and the various forms of *L. digitata* have been sunk (Svendsen and Kain, 1971; Sundene, 1962, 1964). The concept of phenecotypes may not in fact be tenable, since it is clear from quantitative genetic studies that all metric characteristics are subject to some degree of environmental modification. Heritability varies with environment and the biological characteristics in question. At what level of heritability is a taxon sunk? Clearly, environmental modification, like species crossability, is not an all-or-nothing criterion. The use of the term *plastic* is very vague in this regard and considerable care is needed in its taxonomic application.

Obviously it is not possible to draw any broad conclusions or generalizations from the only numerical taxonomic study of a laminarian (Widdowson, 1971a). It is worth noting, however, that Widdowson did not in fact use numerical methods to modify the orthodox taxonomy of *Alaria*.

It is clear from all of these studies that taxonomic changes that have been published could have been made within the framework of an orthodox taxonomic study. To date the new methods confirm that certain indistinct laminarian taxa should be placed together in larger categories.

REFERENCES

Burrows, E. M. (1964). An experimental assessment of some of the characteristics used for specific delimitation in the genus *Laminaria*. *J. mar. biol. Ass., U.K.* **44**, 137–143.

Chapman, A. R. O. (1974). Genetic basis of morphological differentiation in some *Laminaria* populations. *Mar. Biol.* **24**, 85–91.

Chapman, A. R. O. (1975). Inheritance of mucilage canals in *Laminaria* (section Simplices) in eastern Canada. *Br. phycol. J.* **10**, 219–223.

Davis, P. H. and Heywood, V. H. (1963). "Principles of Angiosperm Taxonomy." Oliver and Boyd, Edinburgh and London.

Dixon, P. S. (1970). A critique of the taxonomy of marine algae. *Ann. N.Y. Acad. Sci.* **175**, 617–622.

Ii, A. (1964). "The Textbook of 'Wakame' (*Undaria pinnatifida*) culture." Hyogo Pref. Federation of Fish Cooperative Assoc. (In Japanese.)

Kain, J. M. (1971). Synopsis of biological data on *Laminaria hyperborea*. *FAO Fisheries Synopsis* **87**, pag. var.

Ling, H. U. and Tyler, P. A. (1974). Interspecific hybridity in the desmid genus *Pleurotaenium*. *J. Phycol.* **10**, 225–230.

Markham, J. W. (1972). Distribution and taxonomy of *Laminaria sinclairii* and *L. longipes* (Phaeophyceae, Laminariales). *Phycologia* **11**, 147–157.

Okamura, K. (1915). *Undaria* and its species. *Bot. Mag., Tokyo* **29**, 269–281.

Saito, Y. (1972). *In* "Contributions to the Systematics of Benthic Marine Algae of the North Pacific" (I. A. Abbott and M. Kurogi, eds), pp. 117–132. Japanese Society of Phycology, Kobe.

SCHREIBER, E. (1930). Untersuchungen über Parthenogenesis, Geschlechtsbestimmung und Bastardierungsvermögen bei Laminarien. *Planta* **12**, 331–353.

SNEATH, P. H. A. and SOKAL, R. R. (1973). "Numerical Taxonomy. The Principles and Practice of Numerical Classification." W. H. Freeman, San Francisco.

SOUTH, G. R. (1976). A check-list of marine algae for eastern Canada—first revision. *J. mar. biol. Ass., U.K.* **56**, 817–843.

SUNDENE, O. (1958). Interfertility between forms of *Laminaria digitata*. *Nytt Mag. Bot.* **6**, 121–128.

SUNDENE, O. (1962) ["1961"]. Growth in the sea of *Laminaria digitata* sporophytes from culture. *Nytt Mag. Bot.* **9**, 5–24.

SUNDENE, O. (1964). The ecology of *Laminaria digitata* in Norway in view of transplant experiments. *Nytt Mag. Bot.* **11**, 83–107.

SVENDSEN, P. and KAIN, J. M. (1971). The taxonomic status, distribution and morphology of *Laminaria cucullata sensu* Jorde and Klavestad. *Sarsia* **46**, 1–22.

WIDDOWSON, T. B. (1971a). A statistical study of variation in the brown alga *Alaria*. *Syesis* **4**, 125–143.

WIDDOWSON, T. B. (1971b). A taxonomic revision of the genus *Alaria* Greville. *Syesis* **4**, 11–49.

WILCE, R. T. (1965). Studies in the genus *Laminaria*. III. A revision of the north Atlantic species of the Simplices section of *Laminaria*. *Proc. Vth Mar. Biol. Symp. Gothoburgensia* **III**, 247–256.

YABU, H. (1964). Early development of several species of Laminariales in Hokkaido. *Mem. Fac. Fish., Hokkaido Univ.* **12**, 1–72.

Author Index

The numbers in *italics* indicate the pages on which names are mentioned in the reference lists

P*

T

U

V

W

Y

Z

Taxonomic Index

H

M

Chemical Index

Metabolites and metabolically active compounds only are included

D

E

F

G

H

I

K

L

M

N

O

P

Subject Index

D

E

N

Q

R

U

V

W

Z

The Systematics Association Publications

2. THE SPECIES CONCEPT IN PALAEONTOLOGY (1956)
   Edited by P. C. Sylvester-Bradley
3. FUNCTION AND TAXONOMIC IMPORTANCE (1959)
   Edited by A. J. Cain
4. TAXONOMY AND GEOGRAPHY (1962)
   Edited by D. Nichols
5. SPECIATION IN THE SEA (1963)
   Edited by J. P. Harding and N. Tebble
6. PHENETIC AND PHYLOGENETIC CLASSIFICATION (1964)
   Edited by V. H. Heywood and J. McNeill
7. ASPECTS OF TETHYAN BIOGEOGRAPHY (1967)
   Edited by C. G. Adams and D. V. Ager
8. THE SOIL ECOSYSTEM (1969)
   Edited by J. Sheals
9. ORGANISMS AND CONTINENTS THROUGH TIME (1973)†
   Edited by N. F. Hughes

LONDON. Published by the Association

Systematics Association Special Volumes

1. THE NEW SYSTEMATICS (1940: reprinted 1971)
   Edited by Julian Huxley
2. CHEMOTAXONOMY AND SEROTAXONOMY (1968)*
   Edited by J. G. Hawkes
3. DATA PROCESSING IN BIOLOGY AND GEOLOGY (1971)*
   Edited by J. L. Cutbill
4. SCANNING ELECTRON MICROSCOPY (1971)*
   Edited by V. H. Heywood
5. TAXONOMY AND ECOLOGY (1973)*
   Edited by V. H. Heywood
6. THE CHANGING FLORA AND FAUNA OF BRITAIN (1974)*
   Edited by D. L. Hawksworth
7. BIOLOGICAL IDENTIFICATION WITH COMPUTERS (1975)*
   Edited by R. J. Pankhurst
8. LICHENOLOGY: PROGRESS AND PROBLEMS (1976)*
   Edited by D. H. Brown, D. L. Hawksworth and R. H. Bailey
9. KEY WORKS (1978)*
   Edited by G. J. Kerrich, D. L. Hawksworth and R. W. Sims
10. MODERN APPROACHES TO THE TAXONOMY OF RED AND BROWN ALGAE (1978)*
    Edited by D. E. G. Irvine and J. H. Price

* Published by Academic Press for the Systematics Association

† Published by the Palaeontological Association in conjunction with the Systematics Association